GW01458531

THE AUTOMOBILE

THE AUTOMOBILE

A textbook for students and those who want to know what an automobile is and how it works

HARBANS SINGH REYAT

M.A., M.I.A.E.

S Chand And Company Limited

(ISO 9001 Certified Company)

S Chand And Company Limited

(ISO 9001 Certified Company)

Head Office: Block B-1, House No. D-1, Ground Floor, Mohan Co-operative Industrial Estate, New Delhi – 110 044 | Phone: 011-66672000

Registered Office: A-27, 2nd Floor, Mohan Co-operative Industrial Estate, New Delhi – 110 044 | Phone: 011-49731800

www.**schandpublishing.com**; e-mail: **info@schandpublishing.com**

Branches

Ahmedabad	:	Ph: 27542369, 27541965; ahmedabad@schandpublishing.com
Bengaluru	:	Ph: 22354008, 22268048; bangalore@schandpublishing.com
Bhopal	:	Ph: 4274723, 4209587; bhopal@schandpublishing.com
Bhubaneshwar	:	Ph: 2951580; bhubaneshwar@schandpublishing.com
Chennai	:	Ph: 23632120; chennai@schandpublishing.com
Guwahati	:	Ph: 2738811, 2735640; guwahati@schandpublishing.com
Hyderabad	:	Ph: 40186018; hyderabad@schandpublishing.com
Jaipur	:	Ph: 2291317, 2291318; jaipur@schandpublishing.com
Jalandhar	:	Ph: 4645630; jalandhar@schandpublishing.com
Kochi	:	Ph: 2576207, 2576208; cochin@schandpublishing.com
Kolkata	:	Ph: 23357458, 23353914; kolkata@schandpublishing.com
Lucknow	:	Ph: 4003633; lucknow@schandpublishing.com
Mumbai	:	Ph: 25000297; mumbai@schandpublishing.com
Nagpur	:	Ph: 2250230; nagpur@schandpublishing.com
Patna	:	Ph: 2260011; patna@schandpublishing.com
Ranchi	:	Ph: 2361178; ranchi@schandpublishing.com
Sahibabad	:	Ph: 2771238; info@schandpublishing.com

© S Chand And Company Limited, 1962

All rights reserved. No part of this publication may be reproduced or copied in any material form (including photocopying or storing it in any medium in form of graphics, electronic or mechanical means and whether or not transient or incidental to some other use of this publication) without written permission of the copyright owner. Any breach of this will entail legal action and prosecution without further notice.

Jurisdiction: *All disputes with respect to this publication shall be subject to the jurisdiction of the Courts, Tribunals and Forums of New Delhi, India only.*

First Edition 1962
Subsequent Editions and Reprints 1995, 97, 98, 2000, 2001, 2003, 2004, 2005, 2006, 2007, 2008, 2009, 2010, 2011, 2013, 2014, 2018
Reprint 2020

ISBN : 978-81-219-0214-4 **Product Code :** H3AUE42AUTO10ENZX0XR

PRINTED IN INDIA

By Vikas Publishing House Private Limited, Plot 20/4, Site-IV, Industrial Area Sahibabad, Ghaziabad – 201 010 and Published by S Chand And Company Limited, A-27, 2nd Floor, Mohan Co-operative Industrial Estate, New Delhi – 110 044.

Dedicated to

Lt. General N. Sen Gupta
D.G., E. M. E. (Retd.)

Who infused in me the spirit of high thinking and service to nation.

PREFACE TO THE SIXTH EDITION

It is heartening to note that this book is becoming more popular day by day amongst the student community and general public who want to know what an automobile is and how it works. This is apparent from the fact that the book has to be printed every or every alternate year to meet the growing demand.

In order to provide uptodate knowledge about the automobile, new information should be added in the book in every edition. In this edition, trouble shooting chapter has been expanded and turbocharger, now being installed in Tata Sumo-DX-Turbo has been discussed. In addition to this some more information about four valves per cylinder, technical terms in air-conditioning servicing has also been added.

Now-a-days the burning point is emission control in automobiles. This point has already been discussed in detail in chapter five under the heading "Air Pollution and Its Control".

I hope that this book shall keep on fulfilling the need of readers.

HARBANS SINGH REYAT

1954, Janta Nagar
St. No. 2, Gill Road,
Ludhiana – 3
PUNJAB

PREFACE TO THE FIFTH EDITION

The present edition includes technical data of new Indian cars and trucks. A chapter 'Air Conditioning of Automobiles' also has been added. Some new topics such as Rotary Distributor Fuel Injection Pump, Glow Plugs, Metric Size Tyres, etc., have been incorporated. The glossary of technical terms has been expanded. Some questions have been modified keeping in view new models of cars, trucks, buses, etc. At the end, a Survey Report has been given to provide information about the modern trends in Indian automobile manufacturing.

I am thankful to my publishers M/s. S. Chand & Company Ltd. for taking interest in publishing this book and promoting its sale.

I hope this book shall go on serving the student community and general public in the years to come.

HARBANS SINGH REYAT

PREFACE TO THE SECOND EDITION

Owing to rapid industrial development in our country, Automobile Engineering is becoming an interesting and useful subject for our young students. Keeping in view the need of the hour, our Government is introducing day by day, Automobile Engineering Courses at Diploma, Post-Diploma Degree and Post-Degree levels in the different institutions of the country. In order to meet the increasing demands of the students the book has been revised to bring it to uptodate level. Efforts have been made to cover up syllabii in Automobile Engineering being followed in various courses. Many new things have been included to provide an uptodate knowledge in Automobile field. It is hoped that the book shall meet the growing needs of Automobile Engineering students.

HARBANS SINGH REYAT

ACKNOWLEDGEMENTS

I am thankfull to the following organizations who supplied me necessary literature which helped a great deal to revise the book and make it up-to-date:

(*i*) Daimler Benz Aktiengesellschaft, West Germany.
(*ii*) Ford Motor Company, U.S.A.
(*iii*) Panhard and Levassor, France.
(*iv*) Fabbrica Italiana Automobile Torino (Fiat), Italy.

I am also thankful to M/s Temple Press Ltd., London, who were kind enough to permit me to reproduce illustrations from their weekly magazine 'Motor'.

AUTHOR

Contents

1

The Automobile

1. **Automobile.** An automobile is a self-propelled vehicle which is used for the transportation of passengers and cargo over the ground.

2. **Vehicle.** Vehicle is a machine which is used for the transportation of passengers and cargo. Vehicle consists of two parts, *i.e.* carriage portion and machine portion. Vehicles used upon the ground contain wheels and axle as the main machine portion.

The development of the vehicle started from Sledge—Wheel Barrow—Cart—Wagon to Modern Vehicle.

3. **Self-propelled Vehicle.** A self-propelled vehicle is that in which power required for propulsion purposes is produced from within. Aeroplane, ship, motor-boat, locomotive, car, bus, truck, motor cycle, scooter etc. are examples of self-propelled vehicles.

4. **Motor Vehicle.** It is a vehicle which contains motor D.C. Motor or Engine) to drive it. Motor Vehicle is another popular name for the automobile.

Vehicle+Motor = Motor Vehicle.

Cars, buses, trucks, motor cycles, scooters etc. are the different types of motor vehicles.

5. **Types of Automobiles.** Automobiles can be classified with different regards which are as under :

(*a*) With regard to the purpose these are built for :

(*i*) Passenger carriers.

(*ii*) Goods carriers.

(*b*) With regard to the fuel used :

(*i*) Steam carriages (Obsolete).

(*ii*) Electric Cabs (Rare).

(*iii*) Petrol or Gasoline Automobiles.

(*iv*) Diesel Vehicles.

(*c*) With regard to the number of wheels :

(*i*) Six-wheeler.

(*ii*) Four wheeler.

(*iii*) Three wheeler.
(*iv*) Two wheeler.

(*d*) With regard to the drive of the vehicle :
(*i*) Six wheel Drive Vehicles.
(*ii*) Four wheel Drive Vehicles.
(*iii*) Two wheel Drive Vehicles.
(*iv*) Single wheel Drive Vehicles.

(*e*) With regard to their construction :
(*i*) Single unit vehicles.
(*ii*) Articulated vehicles and tractors.

(*i*) **Single unit vehicles.** Single unit vehicles range from 2 to six wheels and onward. Usually 4 wheelers and onward are

Fig. 1·1. Panhard CD Front wheel drive car.

considered under this category. Depending upon type of drive, these vehicles are known as two wheel, four wheel and six wheel drive vehicles and are denoted as 4×2, 4×4 and 6×6 respectively, in which the first figure denotes the total number of wheels and the second figure the number of driving wheels. In a 4×2 vehicle, the drive could be at the front wheels or at the rear ones. In majority of two wheel drive vehicles, there are two axles and drive is given to rear wheels. In Ambassador and Premier cars, the drive goes to rear wheels whereas in Panhard CD and D.K.W. cars, the drive is given to front wheels. In order to clarify the drive, the driving axle is denoted by the letter *x* and non-driving axle by the letter *o*. This way Ambassador and Premier cars will belong to *ox* class and Panhard CD and DkW cars to *xo*. Keeping this in view, the single unit vehicles could bec lassified as under :

(*i*) *ox* : 4×2, four wheeler, rear wheel drive vehicle.

(*ii*) *xo* : 4×2, four wheeler, front wheel drive vehicle.

(*iii*) *xx* : 4×4, four wheeler, four wheel drive vehicle.

(*iv*) *oxx* : 6×4, six wheeler, drive to both rear axle wheels.

(*v*) *xxx* : 6×6, six wheeler, six wheel drive vehicle.

(*ii*) **Articulated vehicles and Tractors**. The articulated vehicle is made up of two units, viz. (*a*) tractor unit and (*b*) trailer unit. The tractor unit acts as a horse to pull the trailer unit which is a carrier for load. The trailer is partially superimposed on the tractor which carries an appreciable part of the trailer weight. Thus the articulated vehicle resembles with that of a horse driven carriage or cart. As the horse can be removed away from the carriage, similarly the tractor can be disconnected from the trailer or *vice versa*. The cart is also superimposed partially on the horse back like the trailer on tractor, and carries an appreciable part of

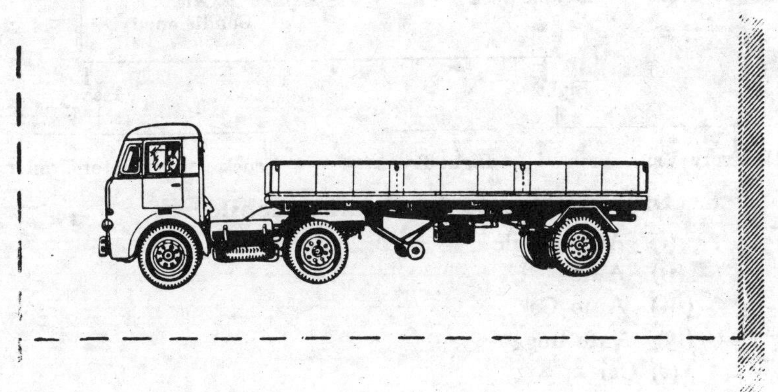

Fig. 1·2. Articulated vehicle

cart load. The tractor takes the place of horse and horse drawn carriage takes the shape of horseless self-propelled carriage known as articulated vehicle.

Usually tractor units of articulated vehicles belong to *ox* category. The coupling arrangement between tractor and semi-trailer is simple. In majority of the cases, all the connections are automatically made by reversing the tractor in position. The tractor is uncoupled through the operation of a lever in driver's cab. The semi-trailer is fitted with a pair of retractable wheels in the front, which are raised and lowered at the time of coupling and uncoupling.

For carrying very heavy loads *independent tractors* are also used. They are of *ox* or *oxx* class. They usually move in pairs, one after the other in tandem form or one acting as *puller* and the other as *pusher* to provide stability where appreciable gradients are to be descended. In such arrangements, there is usually telephonic communication between the two cabs and the braking system could be controlled from the cab in the leading tractor.

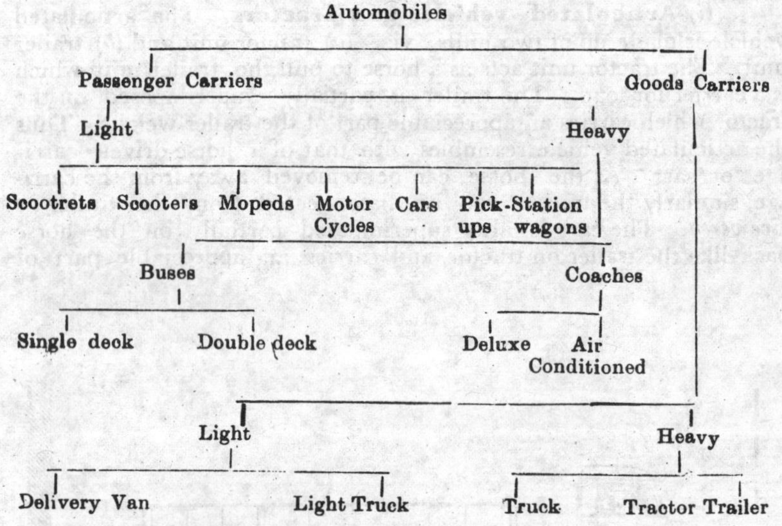

6. Different names for the automobile :

(*i*) Automobile
(*ii*) Auto
(*iii*) Auto Car
(*iv*) Auto Buggy
(*v*) Car
(*vi*) Motor
(*vii*) Motor Coach
(*viii*) Motor Vehicle
(*ix*) Motor Wagon
(*x*) Horseless Carriage

7. Parts of an automobile. Every automobile consists of two main parts *i.e.*,

(*a*) Machine portion, *i.e.* Chassis

(*b*) Carriage portion, *i.e.*, Body

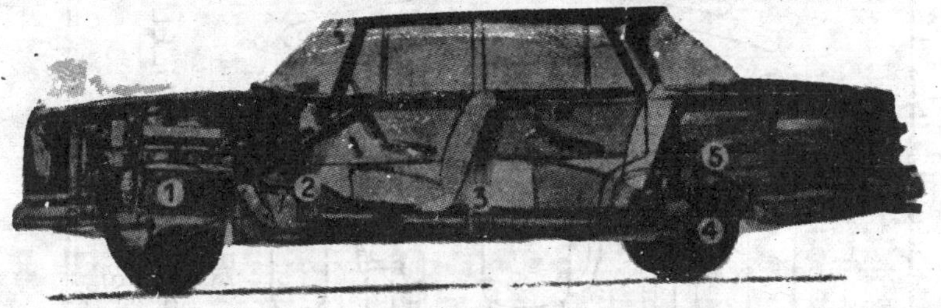

Fig. 1·3. ***An Automobile*** (in cut section).

1. **Engine** 2. **Gearbox** 3. **Propeller shaft**
4. **Rear Axle** 5. **Shock Absorber**

8. Chassis. It is machine portion of the automobile which carries the carriage portion. It is the automobile less body.

Chassis = Automobile – Body.

The chassis contains almost all those parts of an automobile which are necessary to drive the vehicle. It consists of the following main parts :

(*i*) **Frame**. It is the main structure around which all the other parts are connected or suspended to form chassis.

(*ii*) **Springs, shock-absorbers, axles and wheels.** These are the main parts of the suspension system of an automobile with the help of which chassis frame is put on legs and is able to roll on smoothly on the ground.

(*iii*) **Power Unit or Engine.** Power plant to develop the requisite power for the propulsion of an automobile.

(*iv*) **Clutch, Gearbox (Transmission), Propeller Shaft Differential and half-shafts or axle shafts.** These are the main constituents of the transmission line through which power developed by the engine is transmitted to the wheels of an automobile.

(*v*) **Steering, Brakes, Accelerator.** These are the main controls by means of which the vehicle is turned to right or left, stopped and engine speed which ultimately affects the speed of the vehicle, is controlled respectively.

(*vi*) **Fuel Tank.** A tank or reservoir for carrying fuel with the vehicle.

(*vii*) **Battery.** An electro-chemical apparatus to provide electric current for various electrical appliances in an automobile

(viii) **Lamps, Gauges, Switches, Controls, etc.** Lamps ovide eyes to the vehicle whereas gauges serve as indicators and h the help of switches and controls, the vehicle is operated.

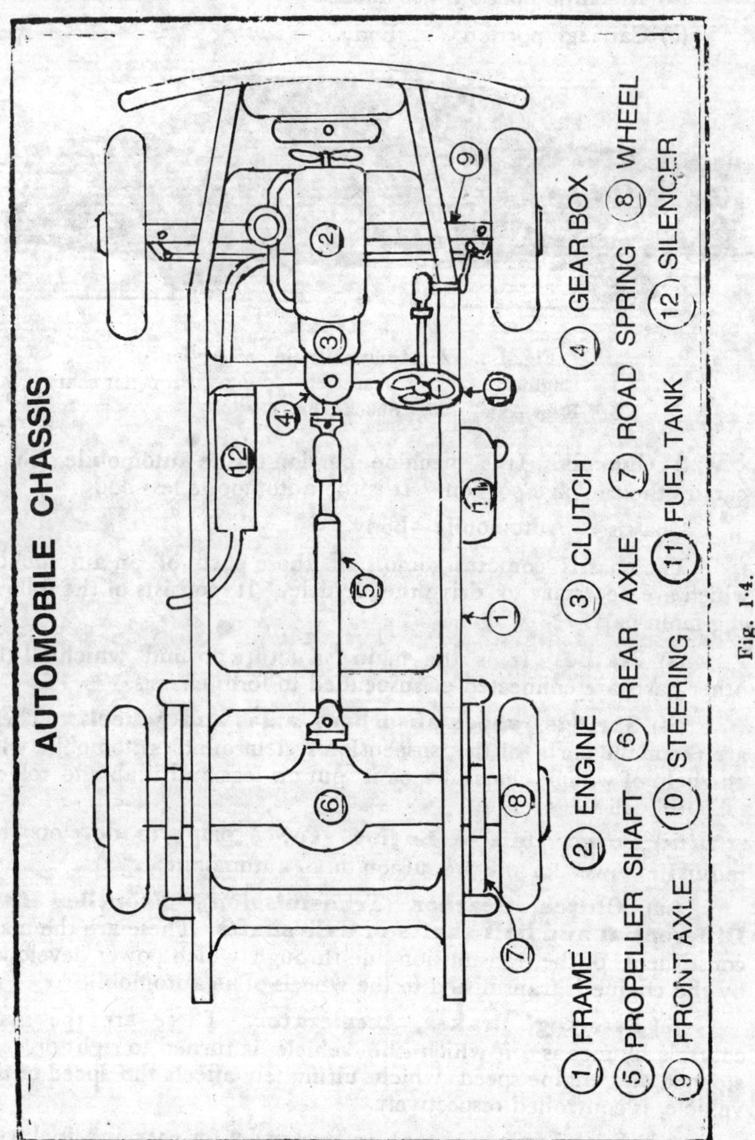

Fig. 1·4.

9. **Body or Carriage Portion.** It is that portion of an automobile where the passengers have their seats or where the cargo to be carried is placed.

The body is designed according to the nature of cargo to be carried. The body of a passenger carrier is much different from

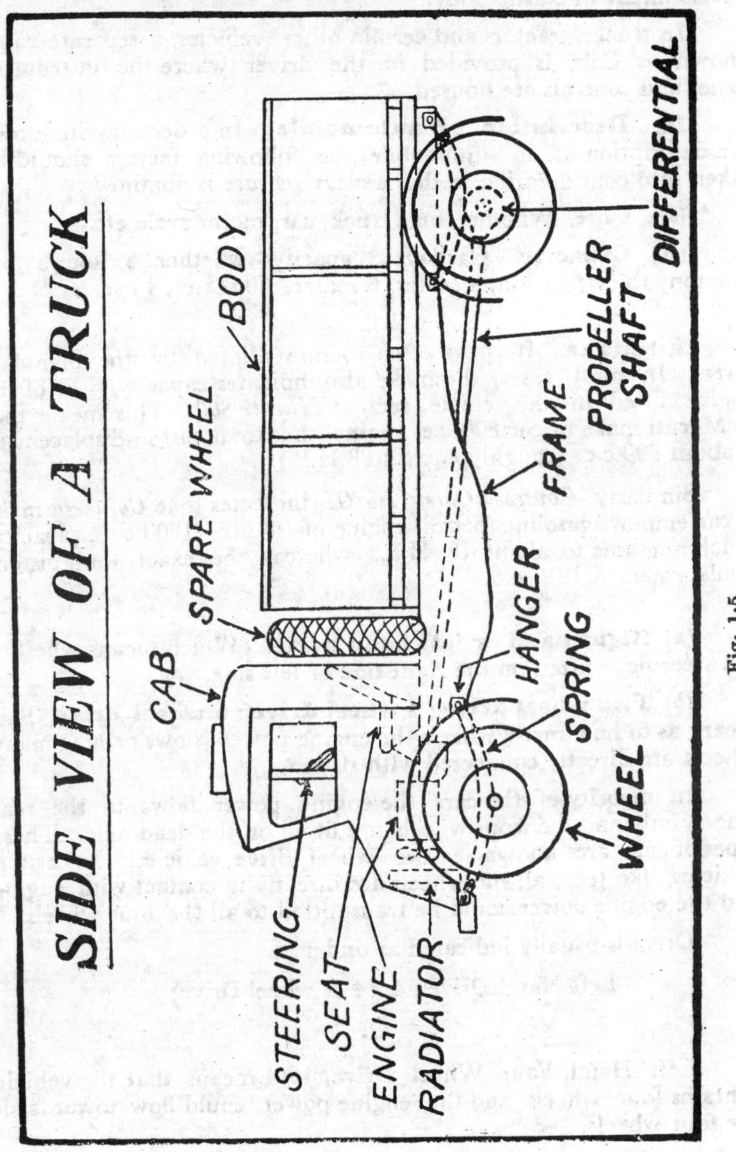

Fig. 1·5

the body of a load carrier, *i.e.*, truck. Its design depends upon the utility for which the vehicle is meant for. It is made of either wood and steel or steel alone. Modern research has led to the development of plastic body.

In trucks, tractors and certain other vehicles, a separate cabin known as Cab, is provided for the driver where the instrument panel and controls are housed.

10. Description of automobile. In order to write down the description of an automobile, the following factors should be taken into consideration so that a clear picture is obtained :

(*i*) **Type.** Whether bus, truck, car, motor cycle etc.

(*ii*) **Capacity.** Carriage capacity—whether 5 ton, 3 ton, one ton, 15 cwt., ½ ton, 4 seater, 6 seater, 30 seater, 45 seater etc.

(*iii*) **Make.** It is the actual name alloted by the manufacturer. In most cases, the make also indicates capacity/H.P. of the engine fitted in the vehicle, such as *Maruti 800*. This means that in Maruti make of car 800 c.c. engine, the total piston-displacement is about 800 c.c. (roughly equal to 8 H.P.).

Similarly, *Contessa Clasic 1·8 GL* indicates that *Contessa* make of car employs gasoline (petrol) engine of 1·8 litre (1800 c.c.) capacity. which amounts to about 18 H.P., whereas the exact total piston-displacement is 1817 c.c.

(*a*) **Right hand or left hand drive.** Which means whether the steering is fitted on the right side or left side.

(*b*) **Two wheel drive ; 4 wheel drive ; 6 wheel drive.** This means as to how many wheels the engine power flows or how many wheels are directly connected with the engine.

In majority of the cars the engine power flows to the rear wheels only and the front wheels are fitted on the dead axle. These types of cars are known as two wheel drive vehicles. In certain vehicles, like Jeep, all the wheels are directly in contact with engine and the engine power could be transmitted to all the four wheels.

Drive is usually indicated as under :

Left Hand Drive ; 4×4 (4 wheel Drive)

Or

Left Hand, Four Wheel Drive, 4×4 means that the vehicle contains four wheels and the engine power could flow towards all the four wheels.

6×4 means that there are six wheels but the engine power could flow towards four wheels only.

(*v*) **Model.** Year of manufacture or special Code Number allotted by the manufacturer.

Hence in order to mention the description of an automobile, the following information shall be required :

(*i*) Type

(*ii*) Capacity

(*iii*) Drive

(*iv*) Make

(*v*) Model

Example, In order to give description of a Jeep, it shall have to mention like this :

Car 6-seater, 4 WD (4×4), Jeep, Mahindra MM-540, DP (Diesel), wherein—

(*i*) *Type*—Car ; (*ii*) *Capacity*—6 seater ; (*iii*) *Drive*—4 Wheel Drive (4×4) ; (*iv*) *Make*—Jeep Mahindra ; (*v*) *Model*—MM-540 DP (employing diesel engine).

1.3 History of Automobile. The invention of the automobile is not the product of a single mind, of a single country or of a single generation in any country. Many men of different generations in different countries are responsible for its evolution and development.

In 15th century, Leonardo Da Vinci felt the possibility of self-propelled vehicles. Sir Issac Newton suggested a steam carriage to be powered by a rearwardly directed jet of steam in 1680, but his suggestion was not seriously considered up to some last years of 19th century.

Christian Huyghens, a Dutch experimenter, described the first Internal Combustion Engine few years before the proposal of Mr. Newton. He showed a single cylinder and single piston engine in which combustion took place by means of gun powder.

These were just the different rising ideas to convert horse drawn carriage into horseless carriage in the different corners of the universe. The dawn of the actual history of automobile is generally agreed to be in 1769 A.D. when Captain Nicholas Cugnot of France, built the actual self-propelled road vehicle. It was a three wheeled carriage propelled by steam engine. Its engine was driving the front single wheel. The boiler of the engine was projecting towards front of the vehicle. This carriage had to stop after about every hundred feet to develop steam.

In 1770 A.D. this carriage was put on trial run in the presence of General Gribeauval, French War Minister and other prominent

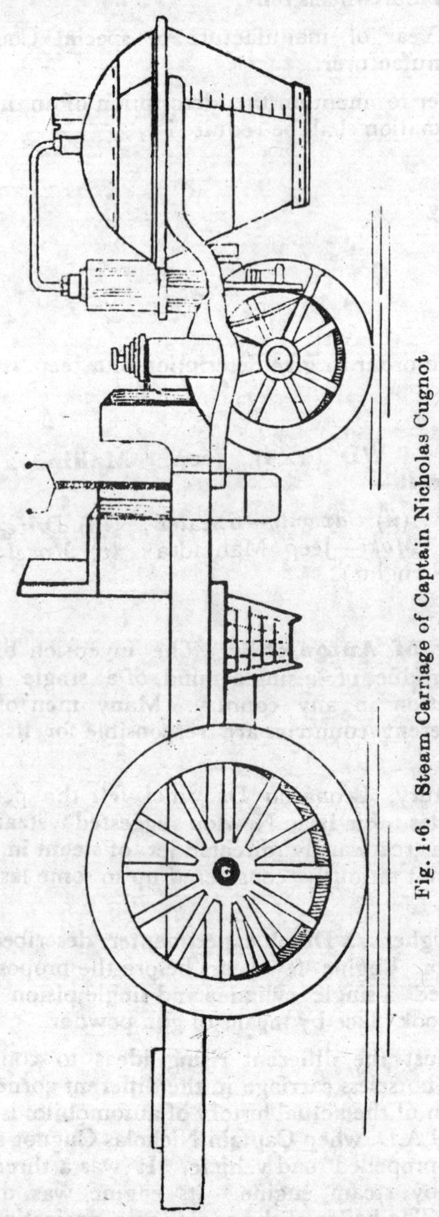

Fig. 1·6. Steam Carriage of Captain Nicholas Cugnot

officers. Four persons were riding in this vehicle and it ran upto the speed of 6 miles per hour. People were very much impressed upon the performance of this vehicle although it struck to a wall. The military officers asked Captain Cugnot to build another vehicle which could run at 8 m.p.h. carrying four to five tons of weight. Their main object was to transport guns and ammunition.

Some historians are of the view that Captain Cugnot's second self-propelled carriage was produced in 1771 and was preserved in Conservatoire Does Arts of Meters, Paris.

The birth of Captain Cugnot's steam carriage gave much inspiration to other persons and as a result efforts were started to convert horse drawn carriage into horseless carriage in the other countries too. This gave rise to the production of steam carriages.

Richard Trevithick built England's first full sized steam carriage in 1801. Other steam carriages in England were built by W.H. James who introduced variable ratio transmission, Walter Hancock, who used the first brake, and Goldsworthy Gurney. All of them

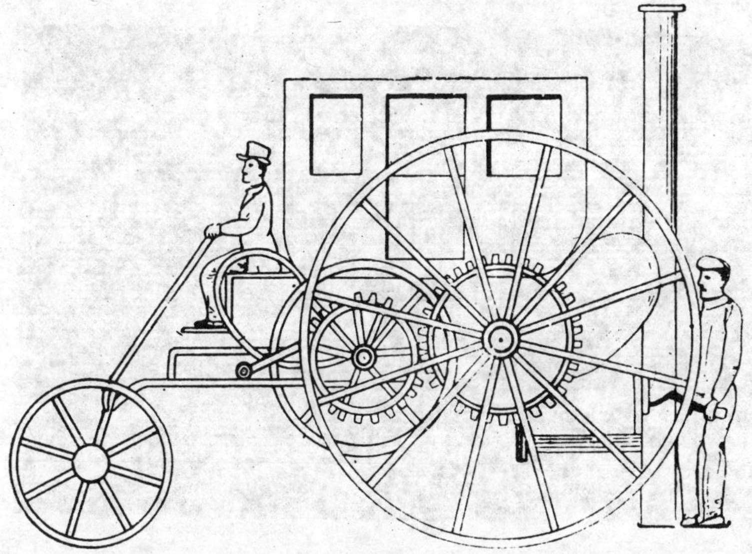

Fig. 1·7. Richard Trevithick's famous steam carriage of 1803.

were working on the development of steam carriages in the period between 1825 and 1836.

In France, Onesiphore Pacquaur took out a patent on the differential for road vehicles in 1827.

Oliver Evans received the first American patent for self-propelled road vehicle in 1789. Sixteen years later, Oliver Evans demonstrated the first steam vehicle in America. In 1829 William James too contributed in the development of steam carriage in America.

Steam vehicles had the disadvantages of a long starting period and limited fuel supply. So electric cabs were thought a better substitute. The electric cabs were in use in Brighton (England), in 1887. Electric cabs were extremely heavy and had a relatively short range of travel between battery chargings.

In 1860, Etienne Lenoir of Paris constructed an Internal Combustion Engine which ran on city gas and was ignited by an electric spark. Two years later, he fitted one of his engines in a vehicle.

In 1866, Nikolaus A. Otto and Engen Langen of Germany, invented four stroke engine which was the predecessor of modern gasoline engine. This engine was run on city gas. It was further improved by Gottlieb Daimler who substituted oil for gas.

The development of automobile was accelerated by the introduction of Internal Combustion Engine.

Fig. 1·8. Karl Benz gasoline engine tricycle of 1885-86

In 1885-86 Gottlieb Daimler patented his high speed I.C. engine which revolutionized automotive transportation. Karl Benz of Germany produced a tricycle with an I.C. engine in 1885-86 and in the same year Gottlieb Daimler built the first motor cycle. In 1386, Daimler produced a four wheeled carriage which impressed the world.

Some historians say that in 1875, Siegfried Marcus of Austria, built a four wheeled vehicle which was powered by an I.C. engine.

Credit is commonly given to Krebs for the first gasoline automobile which incorporated many of the essential features of the

Fig. 1·9. Daimler 4-wheeled gasoline carriage of 1889.

Fig. 1·10. The Marcus of 1875.

modern car. In 1894, he designed the Panhard Car which contained a vertical engine under a hood at the front and a modern type of chassis. The car also had the common type of sliding gear transmission operated by the right hand, clutch and brake pedals and a foot operated accelerator.

In America, gasoline automobiles of horseless type were manufactured in 1890 by Charles Duryea and J. Frank Duryea, Elwood Haynes, Henry Ford, Ransom E. Olds and Elexander Winton.

Fig. 1·11. Duryea Car of 1893 (America's first gasoline automobile).

About 1896 or 1897 considerable work was carried on in Germany, France, England and America on the development of gascline automobiles. Until 1910, the automobile was in the experimental stage. The improvement of roads, rapidity and economy of motor transportation and First World War were the main factors which led to the further development of automobile.

The development of automobile gave birth to a full-fledged automobile industry and production and sale of automobiles became an important business. In 1898, an American Company imported into Bombay three "Oldsmobiles" cars which were sold to Mr. Jamshedji Tata, an industrialist, Mr. Rustom Cama, a solicitor, and Mr. Kawasji Wadia, a merchant.

The early models were just the light wagons having an engine to drive them. Those were just the carriages without horse. Most of the engines were crude, having one cylinder only, mounted sidewise so that power could be transmitted to the wheels by means of a chain and sprockets. No two cars were alike, not even those built by the same manufacturer.

The early automobiles consisted of metal and wood bodies, kerosene or acetylene lamps and solid tyres which have been replaced by modern streamlined cars fitted with all metal bodies, electric lamps and low pressure balloon tyres etc.

Then

(Ford, Model—T—1909)

Now

(Ford Coupe looking like Ambassador Car)

Fig. 1·12. Automobile : Then and now.

The modern automobile has become a moving house furnished with every kind of necessaries and with the every new day new developments in the history of automobile are noticed.

12. A modern car. Mercedes Benz Model 600 is a prestige car from Daimler Benz, West Germany. It is one of the modern cars having the following main features :—

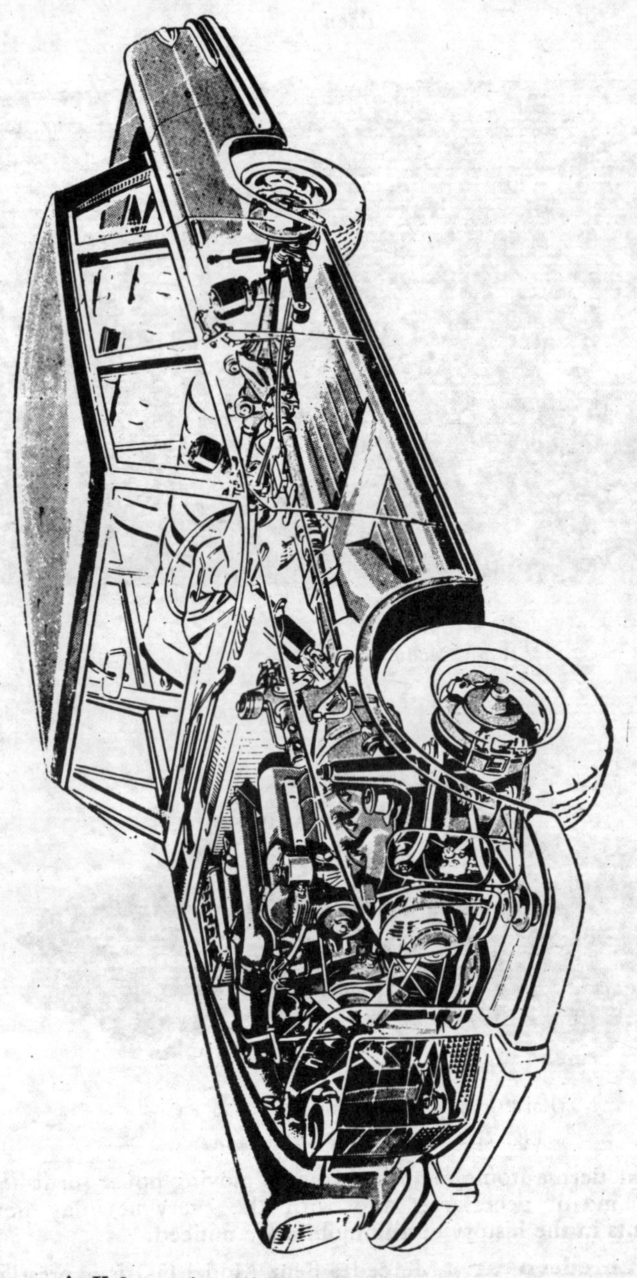

Fig. 1·13. Mercedes Benz 600—A Modern Car.

(*i*) *V*-8 engine with overhead camshaft and induction pipe intermittent fuel injection.

(*ii*) Disc brakes on all four wheels operated by twin hydraulic lines and servo assisted.

(*iii*) Automatic four speed gearbox.

(*iv*) Power steering with adjustable steering wheel.

(*v*) Differential lock to ensure equal power transmission to rear wheels.

(*vi*) Self-levelling and adjustable air suspension system.

(*vii*) Shock-absorbers capable of adjustment by remote control while travelling, through steering-column-mounted lever.

(*viii*) Highly sophisticated central hydraulic system for operating seat adjustment, opening and closing windows, boot-lid, partition, etc.

(*ix*) Electronically controlled heating and ventilating system.

(*x*) Central locking system for all doors, boot and petrol filler cap.

Technical features of the Mercedes-Benz model 600

Engine. 8-cylinder injection engine with a piston displacement of 6·3 litres. Two banks of cylinders form a V of 90°. Every cylinder bank has one overhead camshaft. Induction pipe and injection pump are located inside the V. Eight-plunger injection pump which in order to make full use of the fuel allows timing of the fuel injection with the intake stroke of the individual cylinders. Water cooling with a separate line to each of the cylinder banks. The fan is driven via a thermostatically controlled hydraulic fan coupling, so that it is in operation and produces noise only if actually needed. The powerful engine laid out with a view to supply also the numerous additional units (hydraulic pump, air compressor etc.) has an output of 300 HP/SAE and a maximum permissible engine speed of 4800 r.p.m.

Air suspension. Front wheel suspension by twin wishbones, the lower one also serving as a subframe support. Rear wheel suspension consists of a single-joint swing axle with low pivot point, equipped with suspension units and torsion bar stabilizer. The limited slip differential in the rear axle drive serves to compensate varying road conditions at the rear wheels. Air suspension by means of four bellows supplied with air by the engine driven compressor via three level control valves. Vehicle height and shock absorber adjustment can be regulated from the driver's seat according to road conditions.

Two-circuit booster brake system with four disc brakes. Separate circuits for front and rear wheels. Two callipers for each disc on the front wheels. Compressed air booster brake. The mechanically operated parking brake acts on the disc brakes of the rear wheels. This brake is operated by foot pressure and released either through a lever under the dashboard or automatically when starting.

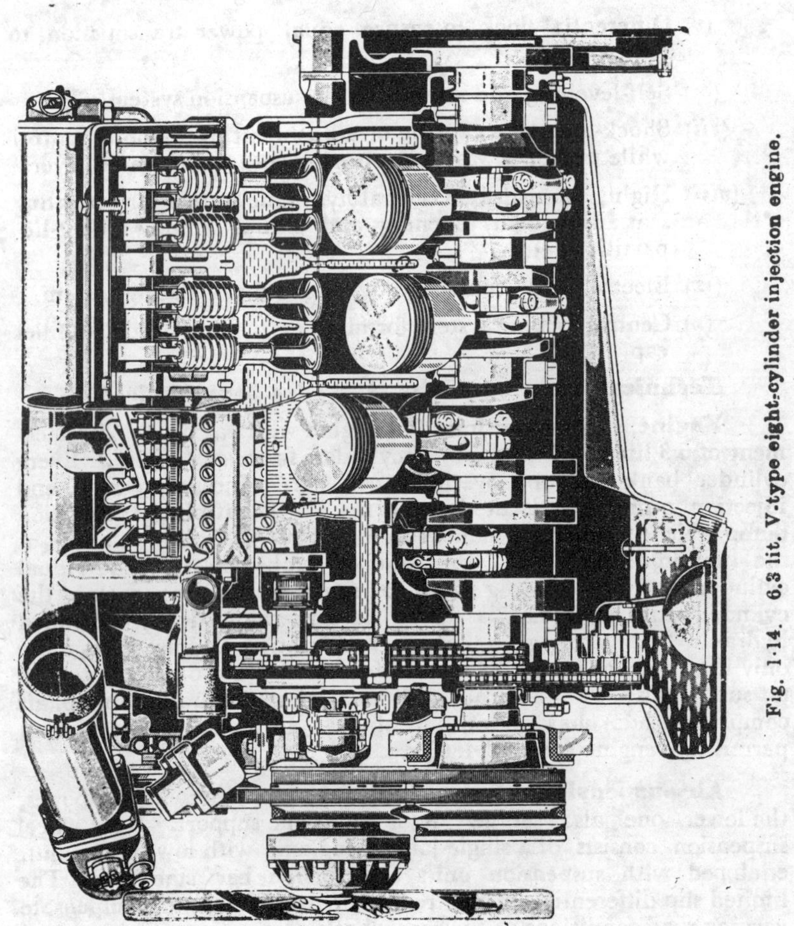

Fig. 1·14. 6.3 lit. V-type eight-cylinder injection engine.

Automatic four-speed transmission and power steering. The automatic four-speed transmission has been adapted to the increased engine torque of the new 8-cylinder V-engine by doubling the number of planetary gears and increasing the size of the multi-plate clutches. New is lay-out of hydraulic clutch with little slip. The automatic transmission offers different selector lever positions and together with the kick down, *i.e.* flooring the accelerator pedal, allows sporty driving. The Daimler-Benz power steering reduces the steering effort but retains at the same time good road-

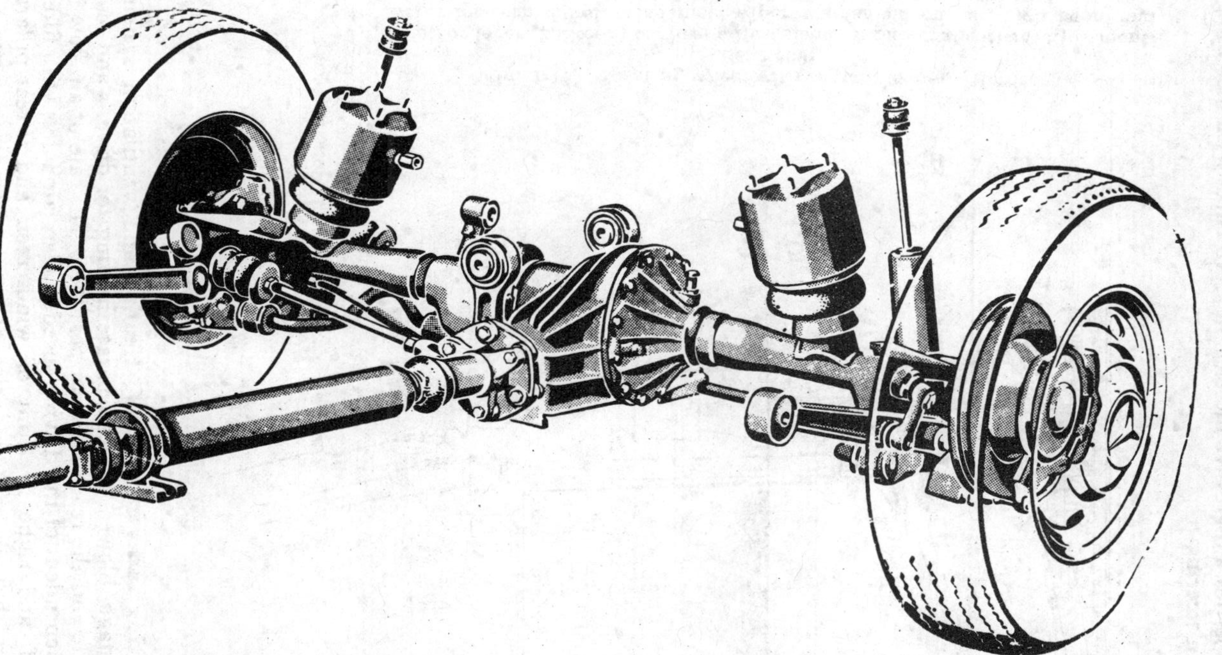

Fig. 1·15. Rear wheel suspension on single joint swing axle with low pivot point, thrust arm with brake torque compression, air suspension, disc brakes.

holding properties. Progressive increase in servo effect with increasing resistance in steering.

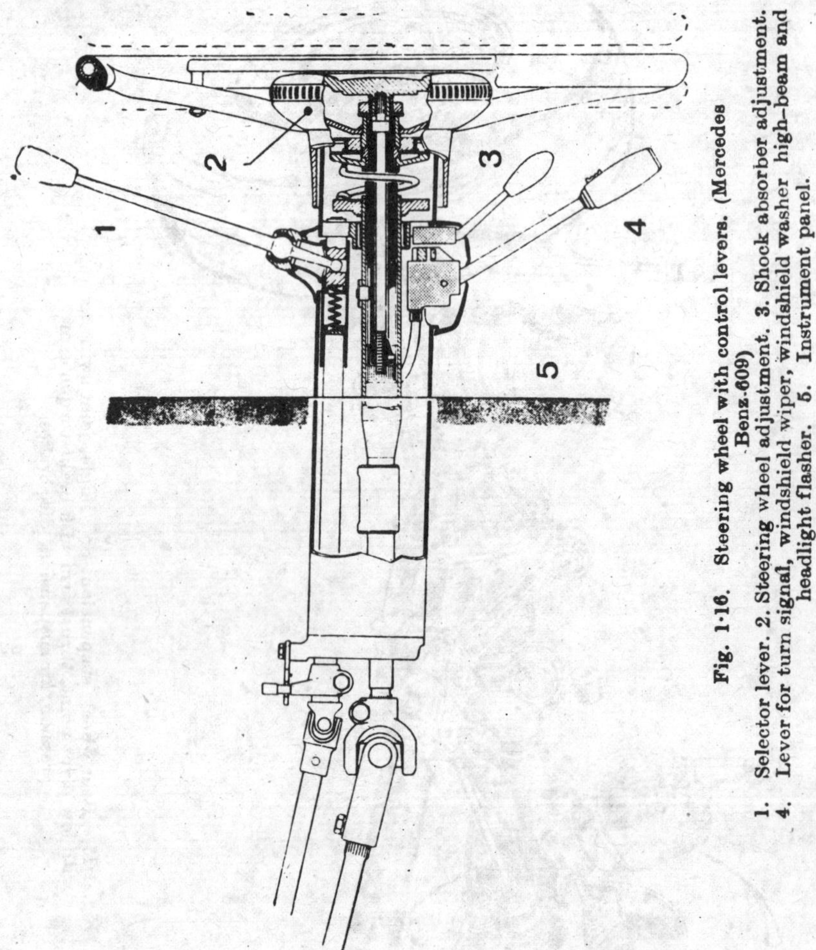

Fig. 1·16. Steering wheel with control levers. (Mercedes Benz-609)

1. Selector lever. 2. Steering wheel adjustment. 3. Shock absorber adjustment. 4. Lever for turn signal, windshield wiper, windshield washer high-beam and headlight flasher. 5. Instrument panel.

Heating and ventilation. Heating and ventilation system with two-stage blower makes separate heating for driver's and co-driver's side and the rear of the car possible. The rate of air flow is almost independent of the driving speed. Two air inlets with filter at the left and right in front of windscreen and at rear pillars Ventilation through roof and slot above rear window. Fresh ai, heating in front via two, in the rear via one heat exchanger. The temperature is electronically kept at a constant level by means of a selector switch Additional dynamic pressure ventilation for nigh outside temperatures in summer with air intake in front of windscreen and air outlet under rear window. Upon special request an air conditioning system can be installed.

Central locking system. Daimler-Benz safety tap lock on all doors which in addition are equipped with a special hydraulic system allowing noiseless closing at slight finger pressure. When the driver's door is locked all 4 doors, the tank filler cap and the luggage compartment lid are also automatically locked by means of a central vacuum-controlled mechanism. When all car doors are centrally locked, the luggage compartment may be opened by means of a separate key. One master key for all locks. With an additional key for garages and workshops, all locks may be opened except that of the luggage and glove compartment.

Interior appointments and additional comfort. A bar is installed in cars with central partition. The partition can be lowered automatically. The version with normal wheelbase is equipped with two folding tables each having glare-free lighting. Remote control for the radio can be installed in the rear of the car. Special compartment for ladies' utensils. Head rests are standard in the rear. in front optional. The floor is covered with velours, the luggage compartment with a Boucle carpet. Rear window can be heated by wires inside the pane. Two outside rear view mirrors adjustable from inside. Reading lamps with adjustable light cones. Axially adjustable steering wheel which together with the seat adjustment operated by means of a push-button guarantees high comfort for the driver.

Technical data type 600

Engine	V-8 engine with intermittent suction pipe gasoline injection
Number of cylinders	8
Valve arrangement	Overhead valves, one overhead camshaft each per cylinder bank.
Piston displacement	386·3 cu. ins.
Engine output	300/4100 HP (SAE)/rpm
Maximum engine speed	4800 rpm
Bore/stroke	4·06/3·74 ins.
Compression ratio	9·0 : 1
Fuel	Commercial premium or gasoline-benzene mixture.
Max. torque	434/3000 ft. lbs./rpm (SAE)
Injection pump	Bosch 8-plunger pump with automatic correction for starting and warming-up

	period (considering gas pedal position, engine speed, air pressure and temperature of cooling water)
Injection nozzles	Bosch
Firing order	1—5—4—8—6—3—7—2
Ignition timing	automatically by centrifugal force and under pressure
Crankshaft bearing	5 multi-layer plain bearings
Connecting rod bearing	Multi-layer plain bearings
Oil filter	Main current and by-pass filter
Oil cooling	Oil-water heat exchanger
Oil filling in crankcase	1·58 US gals./1·05 US gals.
Max./min	1·31 Imp. gals./0·88 Imp. gals.
Cooling	water circulàtion by pump, thermostat with short circuit, ventilator with thermostatically controlled hydraulic clutch
Fuel feed	Electric pump in the rear of car
Electrical system	12 Volt
Battery cypacity	88 Ah
Generator	Two three-phase-current generators

Power transmission

Gearbox	Automatic Daimler-Benz transmission with hydraulic clutch and 2 planetary gears ; selector lever on steering wheel	
Ratio	I.	3·98 : 1
	II.	2·52 : 1
	III.	1·58 : 1
	IV.	1 : 1
	R.	4·15 : 1
Climbing ability	I.	80 %
	II.	42 %
	III.	23 %
	IV.	13·5%
Max. speed in the individual gears	I.	31 mph
	II.	50 mph

*) theoretical value

	III. 80 mph
	IV. 127 mph appr.
Rear axle ratio	3·23 : 1
Differential gear	Limited slip differential
Engine speed at 62 mph	2475 rpm

Chassis :

Frame construction	Frame-floor unit welded to the body, semi-supporting body
Front wheel suspension	Independent wheel suspension with offset pivot pins of wishbones ; above : trapezoidal triangular wishbones ; below : wishbones-support moved further forward ; two rolling bag air springs (manual and self-levelling adjustment of air suspension), rubber helper springs, hydraulic telescopic shock absorbers (adjustable while driving), torsion bar stabilizer, wishbones and engine-gear-box block mounted on to sub-frame, sub-frame rubber mounted on front longitudinal members.
Rear axle	Single joint swing axle with low pivot point and thrust arms, two rolling bag air springs as well as two rubber helper springs, hydraulic shock absorbers (adjustable while driving), torsion bar stabilizer, brake torque compensation, limited slip differential.
Braking system	Two-circuit hydraulic brake with vacuum booster, front and rear disc brakes (in front with two callipers), pedal-operated mechanical parking brake, released either manually or automatieally when starting.
Steering	Daimler-Benz power steering with automatic readjustment and steering shock absorbers ; steering wheel axially adjustable.
Chassis lubrication	Maintenance-free
Size of tyres	9·00—15 Super Sport

Dimension	*Saloon*		*Pullman 600*
Wheel Base	126 ins.		153·5 ins.
Wheel track front		62·53 ins	
rear		62·05 ins.	
Ground clearance		8 ins. appr.	
Overall length	218 ins.		246 ins.
Overall width		76·8 ins.	
Overall height (constant)		59·45 ins.	
Height level control		59·45+2 ins. appr.	
Turning circle diameter	40·7 ft.		47·8 ft.
Tank capacity		29·6 US gals. 24·6 Imp. gals.	
incl. reserve (indicated by warning lamp)		2·6 US gals. appr. 2·2 Imp. gals. appr.	
Weights			
Dry weight (without spare wheel and tools)	5070 lbs.		5489 lbs.
Curb weight (empty weight according to DIN 70020)	5380 lbs.		5800 lbs.
Permissible total weight	6590 lbs.		7230 lbs.
Permissible axle load front	3060 lbs.		3460 lbs.
rear	3530 lbs.		3770 lbs.
Driving performances			
Max. speed		127 mph appr.	
Power weight	18 lbs./HP (SAE)		19·3 lbs./HP (SAE)

13. Famous years in early automobile history :

(*i*) 1769—First self-propelled road vehicle built by Captain Nicholas Cugnot of France.

(*ii*) 1801—First steam carriage built by Richard Trevithick in England.

(*iii*) 1804—First American self-propelled steam vehicle built by Oliver Evans.

(*iv*) 1827—First differential invented by Onesiphare Pacqueur of France.

(*v*) 1832—First 3-speed Transmission patented by W.H. James in England.

(*vi*) 1886—One of the first gasoline engine powered automobiles by Gottlieb Daimler of Germany.

(*vii*) 1893—First American gasoline powered automobile built by Charles Duryea and Frank J. Duryea.

(*viii*) 1894—Panhard and Levassor in France developed a car which incorporated the chief features of the modern automobile.

(*ix*) 1895—First motor car race held.

(*x*) 1897—First car arrived in India.

(*xi*) 1902—First volume of production car "The Curved Dash Oldsmobile" in America.

(*xii*) 1908—Ford 'T' Model car was produced in America by Ford Motor Company.

(*xiii*) 1911—First electric self-starter installed in the automobile.

14. Performance of automobile. The pressures developed by the burning of fuel in the engine are transmitted to the crank-

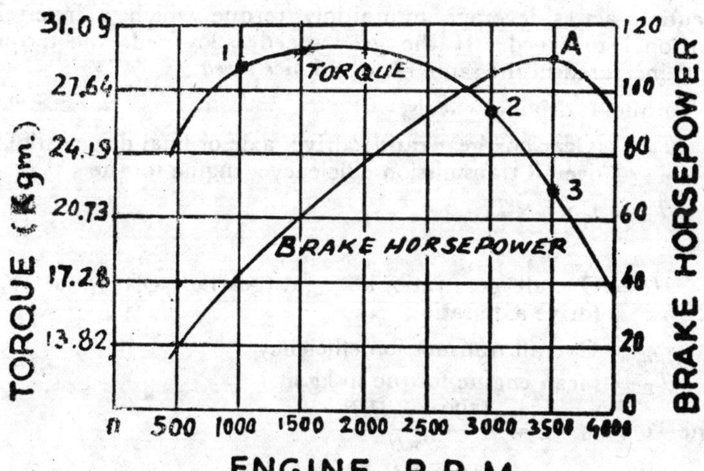

Fig. 1·17. Torque and b.h.p. curves of an engine with relation to speed.

1. 29·02 kgm. torque at 1000 r.p.m.
2. 26·26 kgm. torque at 3000 r.p.m.
3. 22·39 kgm. torque at 3500 r.p.m.

A-Maximum b.h.p. 110 at 35000 r.p.m. after which there is fall in power although engine speed is still increasing.

shaft by the piston and connecting rod and a turning force or effort known as *torque* is produced. The torque sets the crankshaft in motion.

The crankshaft is coupled to the driving road wheels through clutch, gearbox, propeller shaft, differential and axle shafts in an automobile. Thus torque produced by the engine is transmitted through the drive line to the road wheels to propel the vehicle.

The torque depends upon the pressures exerted on the piston and the length of crank arm and is measured in lb. ft. or kg metres. The actual power delivered by the engine is known as *b.h.p.* (brake horse power) and is measured by dynamometer or prony brake. *Torque* is the turning effort whereas *b.h.p.* is the rate at which the work is done.

The torque increases with the increase in engine speed up to a certain point after which it starts to fall down even though the engine speed continues to increase. The number of r.p.m. (revolutions per minute) at which the torque begins to decrease, depends upon engine design. At higher speed, engine vacuum falls down and less fuel enters the cylinders resulting in lesser force available at the piston and hence the fall in torque as shown in figure.

The torque available at the contact between driving wheels and road is known as *tractive effort*. Gearbox and final drive at differential, act as leverage to multiply torque which is inversely proportional to speed. If the gear speed is lowered, the torque shall be increased in the same ratio and *vice versa*.

Torque at driving wheels,

$T_w =$ (Gear box gear ratio × drive axle or final drive ratio) × overall transmission efficiency × engine torque

or $T_w = G \times \eta_t \times T_E$

where,

$G =$ Overall gear ratio, i.e., g.r. (gearbox gear ratio) × a.r. (drive axle ratio)

$\eta_t =$ Overall transmission efficiency.

$T_E =$ mean engine torque in kg.m.

$$\text{Engine Torque, } T_E = \frac{4500 \times B.H.P.}{2\pi N}$$

$$= \frac{716 \times B.H.P.}{N}$$

where, $N =$ r.p.m. of crankshaft.

Since the torque at driving wheels acts through the radius of wheel so the *tractive effort*,

$$F = \frac{T_w}{r} \quad \text{or} \quad \frac{G \times \eta_t \times BHP \times 716}{N \times r}$$

where, r = radius of wheel in metres.

The ratio between engine r.p.m. and vehicle speed depends upon overall gear ratio. A vehicle having four speed gearbox shall have four different speeds and ratio between engine r.p.m. and vehicle speed shall be different.

R.P.M. of driving wheel

$$= \frac{\text{Vehicle speed}}{\text{Wheel circumference}}$$

$$= \frac{V}{2\pi r}$$

where, V = Vehicle speed in metres per minute.

r = Radius of wheel in metres.

$$\text{Vehicle speed} = \text{Wheel circumference} \times \frac{N}{G}$$

or $$V = \frac{2\pi r N}{G}$$

where, N = engine r.p.m.

G = Overall gear ratio.

$$\text{Engine r.p.m., } N = \frac{V \times G}{2\pi r}$$

$$\text{Vehicle speed, } \frac{V \times 1000}{60} = \frac{2\pi r V}{G}$$

where V is in km/hr.

$\therefore$ The ratio between engine r.p.m. (N) and vehicle speed (V)

or

$$\frac{V}{N} = \frac{1000 \times G}{2\pi r \times 60} = 2{\cdot}65\left(\frac{G}{r}\right)$$

$$\therefore \quad V = \frac{2\pi r N}{G} \text{ m/min.}$$

$$\therefore \quad G = \frac{2\pi r N}{V}.$$

The engine torque can be increased by reduction gearing. The torque transmitted by the engine through gearbox and propeller shaft to the final drive is increased in every gear speed except in top (direct) and overdrive. The torque transmitted by propeller shaft is further increased by means of gear reduction of final drive (drive pinion and ring gear at differential). The torque of final drive, provided a differential is fitted, is always equally divided between each axle shaft irrespective of speed of road wheels although this does not apply to limited-slip type of differential.

The speed of propeller shaft is always less than the engine speed except in top gear or when overdrive is engaged. The speeds of axle shafts are always less than the speed of propeller shaft owing to final drive gear reduction. Let us now follow one example to understand torque and speed problem.

Example. The engine of a car develops 50 kg. m. torque at 3500 r p.m. The car is driven in second gear having gear ratio of 4 : 1. If the efficiency of drive is 80% what is the torque and speed of propeller shaft ?

Solution. The torque of propeller shaft is increased by the gear reduction but this torque is slightly reduced owing to mechanical efficiency of the drive as 80%.

Propeller shaft torque = Engine torque × gearbox gear ratio × Drive line efficiency

$$= 50 \times 4 \times \frac{80}{100}$$

$$= \textbf{160 kg. m.} \qquad \textbf{Ans.}$$

The speed of propeller shaft is reduced by gear reduction, hence propeller shaft speed

$$= \frac{\text{Engine speed}}{\text{Gear ratio}}$$

$$= \frac{3500}{4}$$

$$= \textbf{875 r.p.m.} \quad \textbf{Ans.}$$

Power at driving wheels. Some of the power developed in the engine is absorbed or lost due to friction of piston, bearings and gears in the engine. The horsepower available at engine flywheel is about 85%. There is a further loss of power from clutch to drive wheels due to friction in clutch, gearbox, universal joints, final drive, differential and between tyres and ground. Taking into account transmission line losses, the horsepower available at the driving wheels to drive the vehicle ranges from about 60 to 75%. The power lost in the transmission of power from engine to road wheels reflects the transmission efficiency which is taken into account while calculating power available at road wheels.

The thrust known as *tractive effort* provided by the engine at the driving road wheels varies at different engine speeds and gear positions. A moving vehicle is opposed by the various forces known as *resistances.* In order to keep the vehicle moving, a driving force or tractive effort equal to the sum of all the opposing forces has to be applied to it. If the tractive effort exceeds the sum of all the resistances, the excess T.E. (tractive effort) shall accelerate the vehicle. But if the tractive effort is less than the sum of all the resistances, the excess T.E. will decelerate the vehicle.

The following main forces oppose the motion of a vehicle :

(*a*) Rolling resistance.

(*b*) Wind or air resistance.

(*c*) Gradient resistance.

(*a*) **Rolling resistance.** It is mainly due to the friction between wheel tyres and road surface which depends upon the load on each road wheel, type of tyre tread, wheel inflation pressure and nature of road surface.

Rolling resistance is measured in lbs. or kg. and is expressed as lbs. per ton or kg. per tonne of vehicle weight or as a percentage of the vehicle weight. Resistance on average type of road surface is between 1 to 2 per cent of vehicle weight. For good asphalt road surface it is 6·75 kg/tonne ; for best macadam 20·2 kg/tonne; for well rolled gravel 25·7 kg/tonne, and for hard dry clay 45 kg/tonne. For smooth concrete road the average value of rolling resistance is $16.9+0.032\,V$ per 1000 kg. where V is the speed of vehicle in km/hr.

Rolling resistance,

$$R_r = KW$$

where, K = Constant of rolling resistance

W = Total weight of vehicle in kg.

The value of K for best roads and loose sandy roads may be taken as 0·0095 and 0·18 respectively.

(*b*) **Wind or air resistance.** This resistance depends upon the shape and size of vehicle body, air velocity and speed of the vehicle. It increases as the square of vehicle speed owing to which much importance is given to streamlining and frontal area of modern automobiles. In calculating air resistance, air velocity is usually neglected.

Air resistance,

$$R_a = K_a\,AV^2$$

where, K_a = coefficient of air resistance

A = projected frontal area in m^2

V = vehicle speed in km/hr.

For best streamlined cars coefficient of air resistance is 0·0023 5 for average cars 0·0032 and for buses and trucks 0·0046.

(*c*) **Gradient resistance.** This resistance is due to steepness of road gradient. It is subject to vehicle weight and road gradient. It does not depend upon vehicle speed

Gradient resistance,

$$R_g = \frac{W}{G} \text{ or } W \sin\theta \text{ (if gradient is expressed in angular dimensions)}$$

where, W = total weight of vehicle in kg.

G = gradient

θ = inclination.

If the gradient is expressed as 1 in 4, it indicates that for every 4 metres of vehicle travel, it is lifted up by one metre.

When expressed as a percentage, it is per cent gradient.

% gradient $= \tan\theta \times 100$

For small values $\tan\theta = \sin\theta$.

When the vehicle is moving along a level road, it has to face rolling and air resistances but when it moves up a gradient, it has to encounter in addition gradient resistance too. Hence the power required to propel a vehicle is proportional to the *total* or tractive *resistance* to its motion and speed.

Power required to propel a vehicle,

$$P_v = \frac{R \times V}{75^*} \left(\frac{1000}{60 \times 60}\right)$$

$$= \frac{R \times V}{270} \text{ h.p.}$$

Taking into account transmission losses,

$$P_{req.} = \frac{P_v}{\eta_t} = \frac{R \times V}{270 \times \eta_t} \text{ h.p.}$$

where,

P_v = power required by the vehicle.

$P_{req.}$ = engine b.h.p. required

R = total resistance in kg.

V = vehicle speed in km/hr.

η_t = transmission line efficiency.

When the vehicle is moving along a level road,

$$R = R_r + R_a \text{ and}$$

while moving up a gradient,

$$R = R_r + R_a + R_g$$

---

*Metric H. P. is 75 kg force metre per second or 4500 kg force meter per minute.

where,

R_r = rolling resistance in kg.

R_a = air resistance in kg.

R_g = gradient resistance in kg.

If the ***total*** or tractive ***resistance*** on level road is less than the *tractive effort*, the surplus T.E. is used for acceleration, hill climbing and drawbar pull.

Upon acceleration, the rotating parts of a vehicle are also accelerated subject to their moments of inertia and gear ratio. This results in the increase of vehicle weight in use which is known as *effective weight* of the vehicle (W to W_e). When surplus tractive effort is used for acceleration,

$$\text{Surplus h.p.} = \frac{W_e}{g} \times f \times \frac{V}{270}$$

$$\therefore \quad \text{acceleration, } f = \frac{g}{W_e} \times (\text{surplus h.p.}) \times \frac{270}{V}$$

$$= \frac{g}{W_e} \times (P_E - P_{req.}) \times \eta_t \times \frac{270}{V}$$

$$= \frac{g}{W_e} \times (P_E \times \eta_t - P_V) \times \frac{270}{V}$$

$$= \frac{g}{W_e} \times (\text{Tractive effort} - \text{Tractive resistance})$$

$$= \frac{g}{W_e} (F - R)$$

where,

f = acceleration in metres/sec²

g = acceleration due to gravity in metres/sec² (9·81 m/sec²)

W_e = effective weight of vehicle in kg.

P_E = engine b.h.p.

P_{req} = engine b.h.p. required.

P_V = power required by the vehicle.

F = Tractive effort.

R = Tractive resistance.

Let us now follow some examples.

Example 1. The coefficient of rolling resistance for a vehicle weighing 7500 kg is 0·015 and the coefficient of air resistance is 0·00281 in the formula $R = KW + K_a AV^2$ where A is the frontal area in $m^2 + V$ is the speed in km/hr. The transmission efficiency in the top gear 5·5 : 1 is 90% and that in the second gear 11 : 1 is 80%. The frontal area is 5·575 m². If the vehicle has to have a maximum speed of 88 km per hour in the top gear, calculate—

(*a*) The engine b.h.p. required.

(*b*) The engine speed if the driving wheel has an effective diameter of 90 cm.

(*c*) The maximum grade the vehicle can negotiate at the above engine speed in second gear.

(*d*) The maximum drawbar pull available on level at the above engine speed in second gear.

(*A.M.I.E. Sec. B, May, 1975*)

Solution.

In top gear

Resistance $R = KW + Ka.AV^2$

$= 0{\cdot}015 \times 7500 + 0{\cdot}00281 \times 5{\cdot}575 \times (88)^2$

$= 112{\cdot}5 + 121{\cdot}3$

$= 233{\cdot}8$ kg

(*a*) Engine b.h.p.

$$P_e = \frac{R \times V}{75 \times \eta_r}$$

$$= \frac{233{\cdot}8 \times 88 \times 1000}{75 \times 0{\cdot}9 - 60 \times 60}$$

$=$ **84·7 h.p. Ans.**

(*b*) Engine speed,

$$\therefore \quad V = \frac{\pi d N}{G_r}$$

$$\therefore \quad N = \frac{V G_t}{\pi d}$$

$$= \frac{88 \times 1000 \times 5{\cdot}5 \times 100}{3{\cdot}14 \times 90 \times 60}$$

$=$ **2854 r.p.m. Ans.**

In second gear

(*c*) Speed $V = \dfrac{88 \times 5{\cdot}5}{11}$

$= 44$ km/hr.

or $\dfrac{44 \times 1000}{60 \times 60} = 12{\cdot}2$ m/sec.

Resistance $R = 0{\cdot}015 \times 7500 + 0{\cdot}00281 \times 5{\cdot}575 \times (44)^2$

$= 112{\cdot}5 + 30{\cdot}1$

$= 142{\cdot}6$ kg.

Let the maximum grade the vehicle can climb be 1 in x.

$$\therefore \quad R = \left(142{\cdot}6 + \frac{7500}{x}\right) \text{kg}$$

We have, Tractive effort,

$$F = \frac{\text{b.h.p.} \times \eta_t \times 75}{V}$$

$$= \frac{84{\cdot}7 \times {\cdot}8 \times 75}{12{\cdot}2}$$

$$= 416{\cdot}5 \text{ kg}$$

$\therefore$ Tractive resistance = Tractive effort

or $R = F$

$$\therefore \quad 142{\cdot}6 + \frac{7500}{x} = 416{\cdot}5$$

or

$$\frac{7500}{x} = 416{\cdot}5 - 142{\cdot}6$$

$$= 273{\cdot}9$$

$$\therefore \quad x = \frac{7500}{273{\cdot}9}$$

$$= 27{\cdot}4$$

Hence the maximum grade is **1 in 27·4** **Ans.**

(*d*) Maximum drawbar pull on level

= Tractive effort—Tractive resistance

= 416·5 − 142·6 = **273·9 kg.** **Ans.**

Example 2. A fully loaded car weighing 2·032 tonnes has transmission efficiency 88% in top speed, road resistance 23 kg per tonne and air resistance $0{\cdot}00843\, V^2$ where V is speed of the car. Calculate—

(*a*) b.h.p. required for a top speed of 144 km/hr.

(*b*) the acceleration at 48 km/hr. assuming the torque at 48 km/hr. in the top gear 25% more than at 144 km/hr.

(*c*) b.h.p. required to drive the car up a gradient of 1 in 5 at 48 km/hr., transmission efficiency 80% in bottom gear.

Solution.

(*a*) Total resistance at the speed of 144 km/hr.

$R = 23 \times 2{\cdot}032 + 0{\cdot}00843 \times 144 \times 144$

$= 46{\cdot}75 + 175$

$= 221{\cdot}75$ kg

B.H.P. required,

$$P_{req.} = \frac{R \times V}{\eta_t \times 270}$$

$$= \frac{221{\cdot}75 \times 144}{0{\cdot}88 \times 270}$$

$$= \mathbf{134{\cdot}5 \text{ h.p.}} \quad \textbf{Ans.}$$

(*b*) Let engine torque at 144 km/hr be T_{E1}, then engine torque at 48 km/hr. would be,

$$T_{E2} = 1{\cdot}25\, T_{E1} \text{ (25\% more than } T_{E1})$$

Similarly tractive effort $F_2 = 1{\cdot}25\, F_1$

because "*r*", radius of road wheel is the same.

∴ Tractive or total resistance at the speed of 144 km/hr.

$$= 221{\cdot}75 \text{ kg}$$

∴ $$F_2 = 1{\cdot}25 \times 221{\cdot}75 = 277 \text{ kg.}$$

(Tractive resistance being equal to tractive effort)

Total resistance at the speed of 48 km/hr.

$$R = 23 \times 2{\cdot}032 + 0{\cdot}00843 \times 48 \times 48$$

$$= 46{\cdot}75 + 19{\cdot}4 = 66{\cdot}15 \text{ kg}$$

Tractive effort, $F = \frac{W}{g} f + R = \frac{2032}{9{\cdot}81} f + 66{\cdot}15$

By substituting the value of F,

$$277 = \frac{2032}{9{\cdot}81} f + 66{\cdot}15$$

or $$\frac{2032}{9.81} f = 277 - 66{\cdot}15 = 210{\cdot}85$$

∴ $$f = \frac{210{\cdot}85 \times 9{\cdot}81}{2032}$$

$$= \mathbf{1{\cdot}017 \text{ m/sec}^2} \quad \textbf{Ans.}$$

(*c*) For the gradient 1 in 5,

$$\tan\theta = 0{\cdot}2 \quad \text{and} \quad \sin\theta = 0{\cdot}196$$

∴ Total resistance to climb the grade of 1 in 5 at 48 km/hr.

$$= 66{\cdot}15 + 2{\cdot}032 \times 0{\cdot}196 = 66{\cdot}15 + 398$$

$$= 464{\cdot}15 \text{ kg.}$$

B.H.P. required,

$$P_{req.} = \frac{R \times V}{\eta_t \times 270}$$

$$= \frac{464 \cdot 15 \times 48}{0 \cdot 8 \times 270}$$

$$= \mathbf{103 \cdot 1 \text{ h.p.}} \quad \textbf{Ans.}$$

15. **Power-weight ratio.** The performance of an automobile much depends upon its ratio of power to weight. By keeping the weight of the vehicle down to a minimum and installing engines of high b.h.p , the best performance can be achieved. The higher the effective b.h.p. of the engine and the lower the total weight of the vehicle, the better will be its hill climbing abilities, the higher its maximum speed and better its acceleration. A well designed streamlined car having a high power-to-weight ratio registers a low fuel consumption at any given speed.

Improvement in the performance of an automobile has been made due to the progressive increase in power-weight ratio. Owing to the use of lighter materials and improved methods of chassis and body construction, the total weight of the automobile has been reduced considerably. The power output of the engine has been increased by using aluminium and magnesium alloys, improved piston, piston ring, cylinder and combustion chamber design, increased compression ratios due to use of aluminium alloy pistons and cylinder heads, better engine balancing, lubrication and cooling systems, fuels of higher octane values, balanced fuel supply and correct ignition.

Power-to-weight ratio (b.h.p. per ton) in small and medium cars ranges from 30 to 90 ; special high performance cars have the ratios up to 230.

In view of maximum speed, there should be minimum body resistance in addition to high power-weight ratio because the air resistance of an automobile body and chassis increases as the square of speed whereas the power varies as the cube of speed Although performance of automobile is governed by the power-weight ratio and the resistances affecting its movement yet the road performance is usually assessed by its rate of acceleration from rest and the top speed attainable on level roads. The acceleration is usually measured in terms of time the car takes to reach a given speed from rest.

QUESTIONS

1. What is an automobile and how it differs from other self-propelled vehicles ?

2. What is self-propelled vehicle and which are various types of self-propelled vehicles ?

3. What is motor-vehicle ? Which are the other names for it ?

4. Which are the different types of automobiles ? Give examples of each type.

5. Which are the main parts of an automobile ? Why each part is necessary in it ?

6. Explain the factors for the description of an automobile.

7. What is a four-wheel drive vehicle ? How it differs from two-wheel drive vehicle ?

8. Analyse the following nomenclatures of different automobiles :

(*i*) Truck 3 Ton, 4×2, Tata 608.

(*ii*) Car, 4 seater, 4×2, Ambassador Nova—1500

(*iii*) Tempo, 3-wheeler, 3×1, Pickup Van—762 kg.

(*iv*) Bus, 30+1 seater, D.C.M-Toyota, Dyna clipper.

(*v*) Mahindra Jeep, 4 WD, 6 seater—CJ-4 A.

9. Draw the sketch of an automobile showing the various parts.

10. Describe the short history of automobile.

11. Compare early motor vehicle with the modern automobile.

12. Which are the famous years in the evolution of automobile ? Describe the main events.

13. Explain the main features of a modern car.

14. Explain the following :—

(*i*) *ox, xo, xx, oxx, xxx.*

(*ii*) Articulated vehicle.

(*iii*) B.H.P. versus torque.

(*iv*) Tractive effort.

(*v*) Resistances to the movement of vehicle.

(*vi*) Power weight ratio.

(*vii*) Power required at the wheels.

15. The coefficient of rolling resistance for a truck weighing 6350 kg, is 0·018 and the coefficient of air resistance is 0·00281 in the formula $R = KW + K_a AV^2$ kg. where A is m^2 of frontal area and V the speed in km/hr. The transmission efficiency in top gear of 6·2 : 1 is 90% and that in second gear of 15 : 1 is 80%. The frontal area is 5·574 m^2. If the truck has to have a maximum speed of 88 km/hr. in top gear, calculate

(*i*) The engine b.h.p. required.

(*ii*) The engine speed if the driving wheels have an effective diameter of 81·25 cm.

(*iii*) The maximum grade the truck can negotiate at the above engine speed in second gear.

(*iv*) the maximum drawbar pull available on level at the above engine speed in second gear.

[(*i*) 85·25 h.p. (*ii*) 3560 r.p.m. (*iii*) 1 in 17·1 (*iv*) 371 kg. **Ans.**]

2

Construction and Working of Automobile

1. **Construction**. Wheel is considered as symbol of civilization. Wheels and axle is the main machine in a vehicle. The most common type of vehicle used in our country-side is bullock cart. In order to understand the construction of automobile, let us follow the construction of cart.

A cart consists of the following main parts :

(*i*) Frame

(*ii*) Wheels and axle

(*iii*) Yoke

(*iv*) Superstructure and platform.

The cart could be split up into following two main parts :—

(*a*) Machine portion

(*b*) Carriage portion

Machine portion consists of a frame to which wheels are attached through the axle. A yoke is fixed at the front of the frame through which the cart is pulled by the bullocks.

The carriage portion consists of platform and superstructure which is mounted over the frame. The load of the carriage portion and the contents to be transported, is borne by the frame which consists of two long members connected by cross members.

The basic construction of automobile much resembles with that of a cart. An automobile too, consists of machine portion and carriage portion similar to a cart. The difference between a cart and an automobile being that cart is a simple vehicle whereas auto-

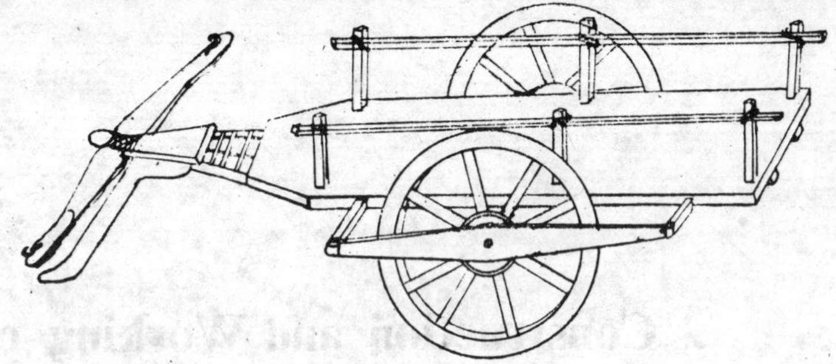

Fig. 2·1. Bullock Cart.

mobile is a motor vehicle. In addition to simple vehicle, an automobile contains a motor or Engine which develops rotary motion to drive the vehicle. For transmitting power to the wheels, clutch, transmission, propeller shaft, universal joints, differential, half shafts etc. are provided in the motor vehicle. To keep the automotive vehicle under control, it is properly bridled. Steering for directional control, accelerator for speed control and brakes for stopping purposes are provided in the automobile.

The cart is a slow coach whereas the automobile is a fast moving machine. Due to fast speed, the automobile is subject to more shocks which put more strains on the frame. In order to overcome this difficulty, the automobile is needed shock proof and its frame should be robust enough to bear all stresses and strains.

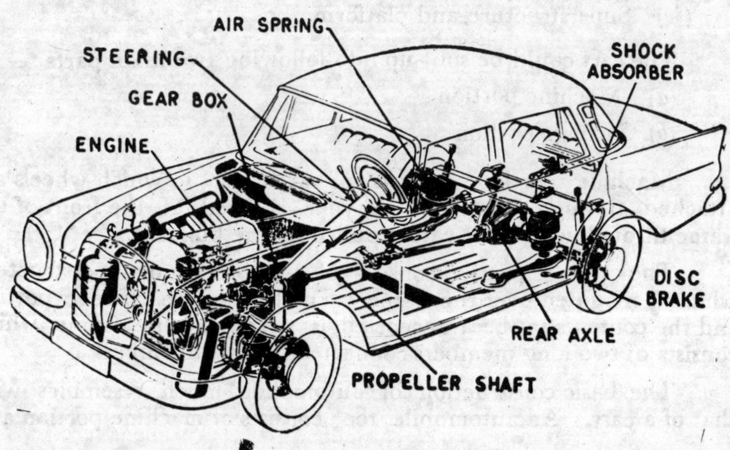

Fig. 2·2. Cut away view of a Car (Mercedes Benz).

In an automobile, axle is not directly fixed with the frame. Axles are suspended with the frame through strong springs. Shock absorbers are further installed to arrest shocks and thus to save the passengers from jerks and jolts due to road irregularities.

Lamps are provided with the automobiles so that these could be driven safely during dark. Horn is installed to provide warning sound to the other road users.

The carriage portion of the automobile is built up to fulfil the complete requirements for which it is built for.

The main additional parts in an automobile, thus, would be as follows :

(*i*) Engine,

(*ii*) Clutch, transmission, propeller shaft, universal joints differential, half shafts etc. (to form drive line)

(*iii*) Controls,

(*iv*) Fuel tank,

(*v*) Battery.

If these parts are properly added to a simple vehicle, it shall become a motor vehicle capable of running with its own power.

2. Chassis Frame. In the construction of an automobile, chassis frame is the basic requirement. It is foundation of the chassis. It serves the following purposes :

(*i*) To form base for mounting engine and transmission units.

(*ii*) To accommodate suspension system.

(*iii*) To take engine and transmission thrust and torque stresses.

(*iv*) To serve as the body, fuel tank and battery mounting unit.

The chassis frame must be strong, light and designed so that it may withstand the shock blows, twists, vibrations and other strains to which it is subjected on road.

During the early years of automobile industry, chassis frames were made of tubular steel, rolled steel sections, wood and armoured wood (wood sills reinforced with steel flitch plates).

Pressed steel frames were introduced on the first Mercedes model in 1900 and their use in passenger cars soon became general,

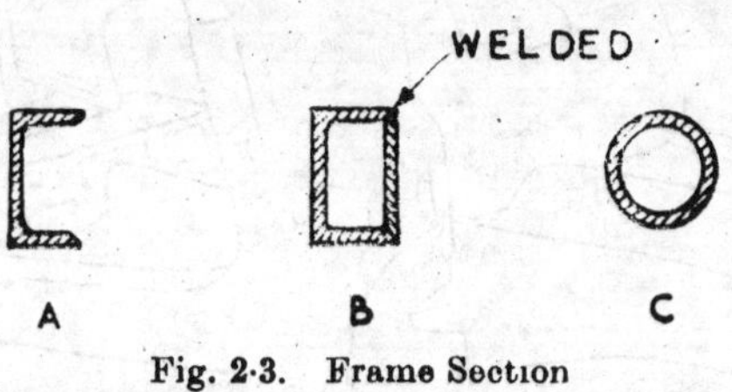

Fig. 2·3. Frame Section
A—Channel B—Box C—Tubular.

but rolled steel frames continued to be used for trucks for many years.

The majority of chassis frames in common use today are of channel or box section and the cross members are the same and are made of pressed steel. In conventional design, the cross members of

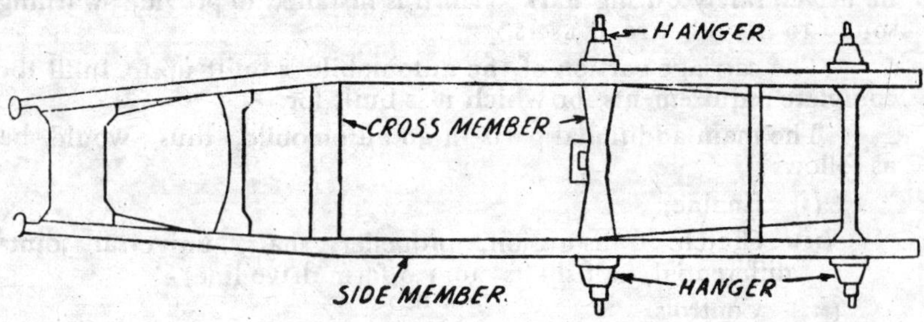

Fig. 2·4. Chassis frame (Conventional).

a chassis frame are at right angles to side members. Several modern chassis frames have cross members that cross in the form of letter '*X*' between the side members. The side members and cross members are rigidly attached to each other by riveting or welding.

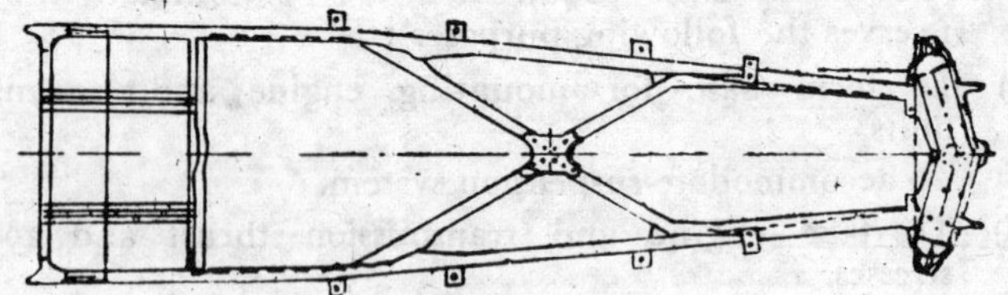

Fig. 2·5. X Type Frame.

Owing to recent advancements in the art of welding and pressing steel in large sheets into complex shapes, chassis cum body

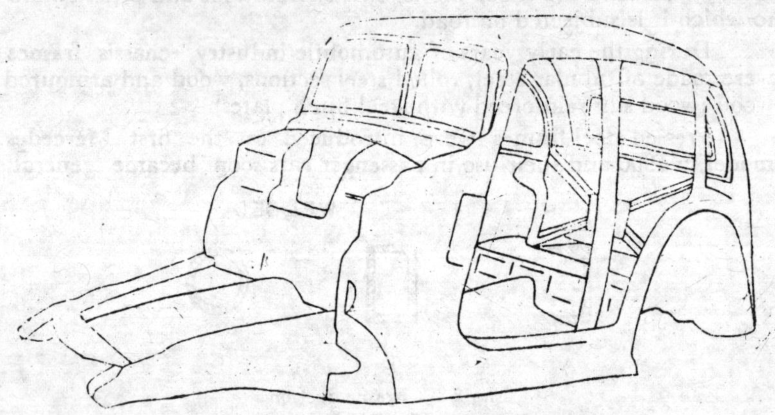

Fig. 2·6. Chassis cum body construction.

construction is coming into favour. In this design, heavy side members are eliminated and cross members are combined with the floor of the body. This has led to much reduction of weight which is main point in design consideration.

Another new thing in chassis frame design is the tubular and backbone type of frame construction. In this design, there is only

Fig. 2·7. Tubular/Backbone type chassis frame ("130 H" MB Car).

one hollow member in the centre running longitudinally, at both ends of which are the end members fitted at right angles for housing suspension system components. The main thing in this design of chassis frame is that it is the best form to resist torsion to save engine and transmission parts from stresses due to road shocks.

3. Loads on chassis frame. A chassis frame is subjected to following loads :

(*i*) Heavy and suddenly applied loads of short duration such as when the vehicle is crossing a broken patch of road.

(*ii*) Combined loads of momentary application at long intervals while negotiating curve, applying brakes and striking a pot hole, all at the same time.

(*iii*) Inertia loads of short duration due to brake application.

(*iv*) Externally applied impact loads, when the vehicle collides with another object.

(*v*) Overloading of the vehicle.

(*vi*) Static load of chassis parts such as engine, transmission, steering, fuel tank, body etc.

4. Loads on axles. The front and rear axles act as beams also to carry vehicle weight. The weight carrying portions of the axles are loaded at spring centres and are subject to the following loads :

(*i*) vertical load at spring centres owing to vehicle weight.

(*ii*) a fore and aft load at the centre of wheel due to driving or braking effort.

(*iii*) a side thrust at the tyre radius owing to centrifugal force while cornering.

(*iv*) torque reactions on driving and braking.

(*v*) shock loading when striking an unexpected obstacle during cross country drive.

5. **Carriage Portion or body**. The body serves as housing for the passengers and cargo to be carried in an automobile. The type of body construted over the chassis gives birth to a particular type of automobile. The body is built according to the utility for which the vehicle is meant. The bodies differ in shape and design. The bodies of all the trucks are not of the same type. Similarly, the bodies of all the cars and buses are not of same nature. There is vast difference between the body of a passenger carrier and goods carrier. Body in other words is just a furnished room providing the necessary amenities. Body is thus a composite of wood and steel work, upholstery work, nickel, paint and decoration work.

The bodies of cars are classified as under :

(*i*) **Sedan**. Enclosed type having two or four doors like Fiat and Ambassador cars in India.

(*ii*) **Coupe**. Having one seat, the rear half of the body being used for luggage spaee, accessible from a separate door.

Fig. 2·8. Coupe (Mercedes Benz. 300 SE).

(*iii*) **Convertible.** In this type of body, the top is of fabric

Fig. 2·9. Convertible (Mercedes Benz, 220 SE).

Fig. 2·10. Limousine (Mercedes Benz, 300 SE).

and could be folded down. The rear is a little narrower to provide room for the folded top.

(*iv*) **Limousine.** This is like a sedan with an intervening window to isolate the driver's compartment. In some cases extra folding seats are provided.

(*v*) **Station wagon.** Originally, station wagons were meant to transport guests and their luggage to and from the railroad station. In most models the centre seats fold down and the rear seat is removed to provide luggage space. Station wagons are built with

Fig. 2·11. Station Wagon (Mercedes Benz).

either two or four doors and all models are provided with a large tail gate.

(*iv*) **Hard Top**. This has the same seating arrangements and the same windows as convertible. When the windows are down, the hard top is very much like the convertible with its top up.

The bodies of the trucks usually come under two categories, *i.e.*, High and half. Full size body trucks are used to carry lighter goods occupying more volume whereas half body trucks are used to carry solid materials such as bricks, pig iron, steel, sand, gravel etc. In trucks, there is a separate cabin for the driver so that the material carried by the vehicle may not roll towards the driver's seat during sudden stoppings, climbing up and down the hill, and at cornering.

Truck bodies contain superstructure to support tarpaulin to protect the cargo from sun, fog, snow and rain etc. A tail board is provided in the body which serves as a gate to load and unload the goods.

Bodies of the buses are enclosed type. Railing is provided at the roof so that the passengers could keep their luggage there. All the doors and windows are provided with safety glasses. The bodies of deluxe and air-conditioned coaches are much improved. Utmost care is taken to provide maximum comfort to the passengers. Adjustable and super cushioned seats is one of the main provisions.

6. Structure and mechanism of a modern mini-car ; Fiat 500 station wagon.

Engine. Two horizontal cylinders in-line flat ; 499·5 c.c., bore 67·4 mm., stroke 70 mm. Maximum output 21·5 h.p. (S.A.E.). Aluminium cylinder block with cast iron liners, Aluminium cylinder head. Two bearing crankshaft, O.H.V. single choke horizontal carburettor. Air intake filter. Mechanical pump fuel feed. Forced lubrication gear pump. Centrifugal oil filter. Air cooled by axial fan. Engine installed in rear of car, beneath loading platform.

Transmission of power to rear wheels through axle shafts connected to the differential by universal joints and to the wheels by flexible couplings.

Single dry disc *clutch* ; 4-speed constant-mesh *gear box* ; Differential and final-drive gears incorporated in gear box. Worm and sector *steering* ; Hydraulic *brakes* ; Parking and emergency brake acting on rear wheel brake shoes.

Front suspension. Independently sprung wheels ; Transverse leaf spring connected to frame at two points. During asymmetric movements of the wheels this spring acts also as a stabiliser. Telescopic shock absorbers.

Rear suspension. Independently sprung wheels with wishbones flexibly attached to body. Coil springs and telescopic shock-absorbers.

Integral construction, sun roof ; *Fuel tank* in front part of car. *Heating* of car interior through warm air taken from the engine cooling circuit. 12 volts *electrical equipment*. *Wheel-base* 1·94 m. *Speed* 60 miles (over 95 km.) per hour. Minimum *maintenance* ; only two grease points.

7. **Working of Automobile.** If a driver is watched carefully while driving an automobile, the following main things shall come into notice :

(*i*) The engine is started.

(*ii*) The gears are changed.

(*iii*) The moving vehicle is controlled by means of various controls ; steering, brakes, accelerator etc.

It is a thing to be studied what actually happens to a stationary vehicle when the driver starts the engine, shifts the gears and uses certain controls in order to operate it.

Wheel and axle is the main machine in a vehicle. If the wheel is rotated, it will push back the earth and the reaction being equal and opposite in direction, the vehicle shall be pushed forward. The pushing force or driving thrust is subject to the turning moment or torque applied at the wheel. Greater the torque at the wheel, greater the driving thrust and more the speed at which the vehicle shall run.

The torque is developed by the engine and is transmitted to the driving wheels by means of transmission system. The torque is not required at the wheels during all the time. When a stop is to be made, the flow of power to the wheels should be discontinued and again continued while moving off. More torque is needed while moving off, climbing a hill and overcoming more resistances of wind, gradient and road. Varied power is required at the driving wheels to drive the vehicle in varied driving conditions of load, speed, road and wind. Hence the following requirements must be fulfilled in an automobile :

(*i*) There should be a means for the development of power.

(*ii*) The rate of power development must be controlled.

(*iii*) An arrangement must exist to transmit the developed power to the driving wheels.

(*iv*) There should be a means to continue and discontinue power flow to the driving wheels.

(*v*) There must be an arrangement to vary the torque.

(*vi*) The driving thrust should be successfully carried in the vehicle.

(*vii*) The vehicle must have a directional control.

(*viii*) There must be a means to stop the vehicle while it is running.

The above requirements are fulfilled as under :

(*i*) The power is produced by the motor or engine.

(*ii*) The rate of power development is controlled by an accelerator.

(*iii*) The power developed by the engine is transmitted to the wheels by means of transmission system.

(*iv*) The flow of power is continued and discontinued to the driving wheels through the clutch.

(*v*) The transmission changes engine torque.

(*vi*) The driving thrust is carried to the chassis frame through the road springs or torque tube.

(*vii*) Directional control is obtained through the steering.

(*viii*) The moving vehicle is stopped and is held up for an indefinite period by means of service and hand brakes.

To sum up, the engine is started to develop the required power to propel the vehicle. The clutch is operated to engage transmission which is done by changing from neutral to low gear speed while moving off. The ratio between the speed of engine and speed of drive line is altered by changing from one gear speed to another. This results in varying leverage or torque. In a four wheeler, power is transmitted to the wheels through the transmission system consisting of clutch, gearbox, propeller shaft and universal joints, final drive, differential and axle shafts. The moving vehicle is kept under perfect control by means of steering, accelerator and brakes.

For moving off, the engine is started, gear is changed through application of clutch and clutch is released by and by. The speed is controlled by means of accelerator and higher gear speed is obtained to run the vehicle faster in accordance with the operation of accelerator. The gears are stepped down while climbing a hill, negotiating a turn or going through a muddy road. The vehicle is turned to the right or left or is held straight through the steering and is held up by means of brakes while it is moving.

In order to understand the detailed operation of an automobile, study of the following systems is necessary :

(*i*) Fuel system
(*ii*) Ignition System
(*iii*) Lubrication System
(*iv*) Cooling System
} Concerned with the operation of engine.

(*v*) Transmission System

(*vi*) Steering System

(*vii*) Suspension System

(*viii*) Brake System

(*ix*) Electrical Systems consisting of starting, charging, lighting, and miscellaneous systems.

These systems have been described in the rest of the text to explain the construction and working of each main part required in the each system of the automobile.

QUESTIONS

1. Describe the difference between the construction of a simple vehicle and a motor vehicle.
2. Discuss the following equation :

   Vehicle+Motor=Motor Vehicle
3. Discuss the construction of automobile.
4. What qualities the chassis frame should possess ? Why ?
5. Describe the different types of chassis frames.
6. Write short notes on the following :

   (*i*) Backbone type frame

   (*ii*) Chassis cum body construction

   (*iii*) '*X*' type frame

7. Describe the different types of car bodies.
8. Describe the difference between the following :
   (*i*) Sedan and Limousine body
   (*ii*) Convertible and Hard Top body
   (*iii*) Coupe and station wagon body.
9. Write short note on the construction of automobile bodies.
10. "The type of body converts the type of vehicle". Discuss.
11. Describe the working of an automobile.
12. Which are the main systems required in an automobile ?
13. Which are the requirements in the operation of an automobile and how they are fulfilled ?
14. Explain the loads on chassis frame and axles.
15. Explain the structure and mechanism of Fiat 500 station wagon.

3

Suspension System

1. **Suspension System.** This system deals with the suspension of wheels and axles with the chassis frame through the road springs. By dint of this system, the frame is put on wheels and becomes capable to roll on It fulfils the following main objects in an automobile :

(*i*) To protect the passengers from road shocks.

(*ii*) To reduce the stresses due to road shocks on the mechanism of the car.

(*iii*) To maintain the body on an even level when travelling over rough ground or when turning so that any rolling, pitching or vertical tendency is minimized.

This system consists of the following main parts :—

(*i*) Springs,

(*ii*) Spring Shackles,

(*iii*) Axles,

(*iv*) Wheels,

(*v*) Shock Absorbers,

(*vi*) Stabilizers.

2. **Springs.** The springs support the chassis frame over which falls the weight of engine, power train components, body, passengers and their luggage etc. They damp the road shocks transmitted to the wheels as they travel over the road, thereby protecting the units supported directly by the frame. In Hotchkiss type of drive the

springs have to bear driving thrust effects in addition to their springing action.

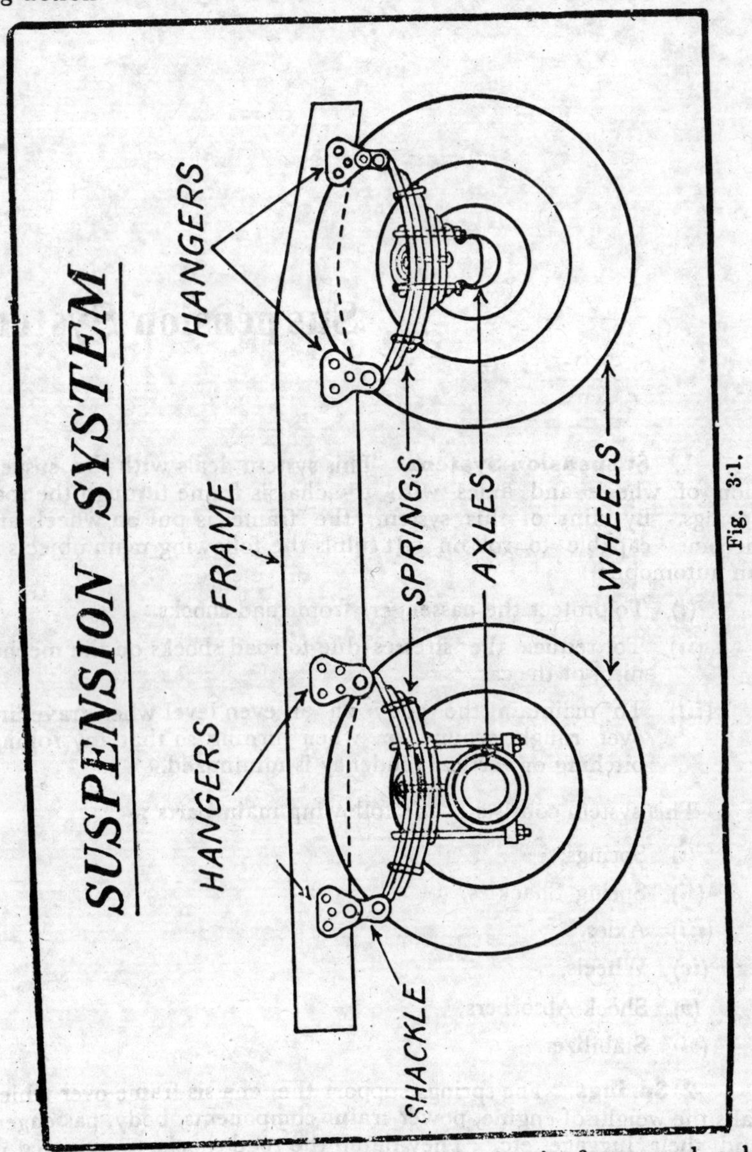

Fig. 3·1.

The springs are placed between the chassis frame and axles. The following types of springs are employed in the automobiles :—

(*i*) Leaf Springs,
(*ii*) Coil Springs,
(*iii*) Torsion Bars/Units.

Each of these types of springs absorbs shocks in a different way. Leaf springs absorb shocks by bending, coil springs by compressing and torsion units by twisting.

3. Leaf Springs. Leaf springs are mad of long, flat strips of spring steel. Several strips are placed one on the other and held together by means of a centre bolt and clamps. Each strip is called a leaf. There is one leaf which extends the full length of spring and usually contains eyes at both ends for making connections with the frame. The other leaves in the spring are assembled with the main leaf by means of centre bolt and clamps. Each succeeding leaf is shorter than the preceding one. The springs which are suspended with the frame through rubber bushings instead of shackles and pins, do not contain loops or es at the ends of the main leaf.

Leaf springs are of elliptical shape. Its camber is a predetermined factor which is set at the time of manufacture or a terwards during recambering. Leaf springs are of the following types :—

(*i*) **Full Elliptic.** Two semi-elliptic springs connected to form the shape of an ellipse.

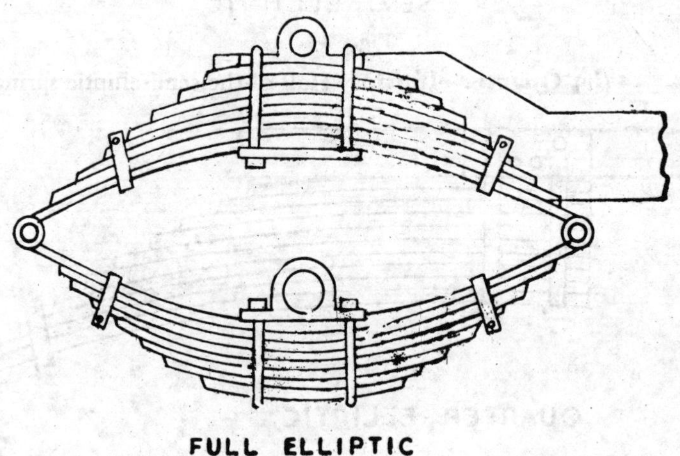

Fig. 3·2.

(*ii*) **Three Quarter Elliptic.** One semi-elliptic spring connected over a quarter elliptic spring.

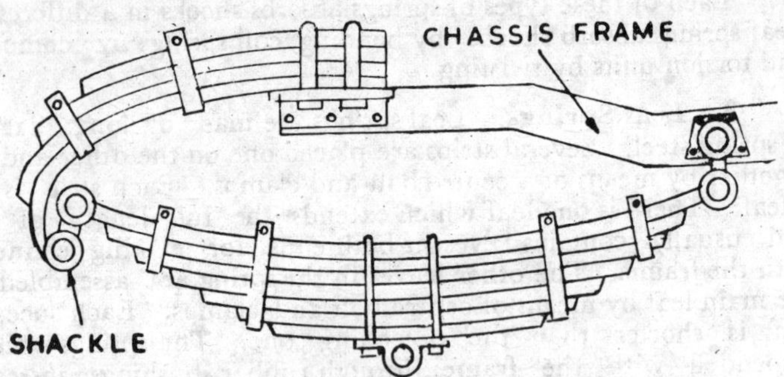

Fig. 3·3.

(*iii*) **Semi·elliptic**. Forming the shape of half ellipse.

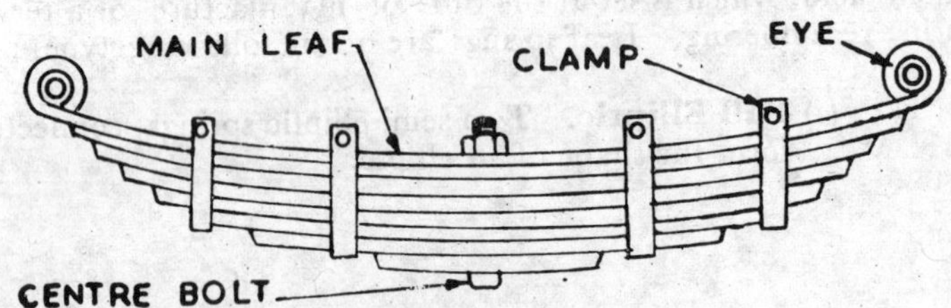

Fig. 3·4.

(*iv*) **Quarter elliptic**. Half of the semi-elliptic spring.

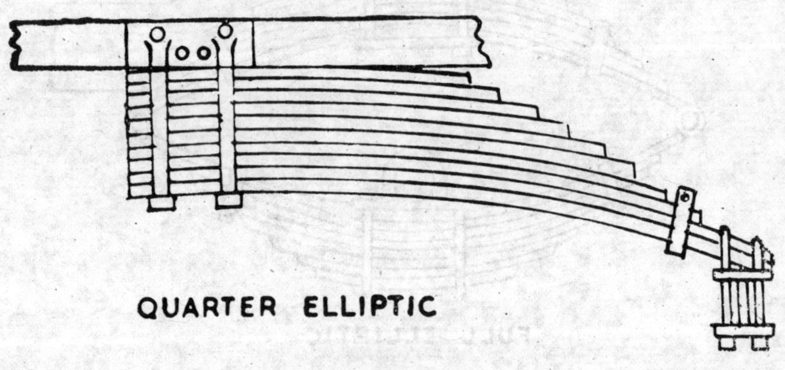

Fig. 3·5.

(*v*) **Transverse.** Semi-elliptic type spring which has the saddle at above forming a bow and is fitted parallel to wheel axle.

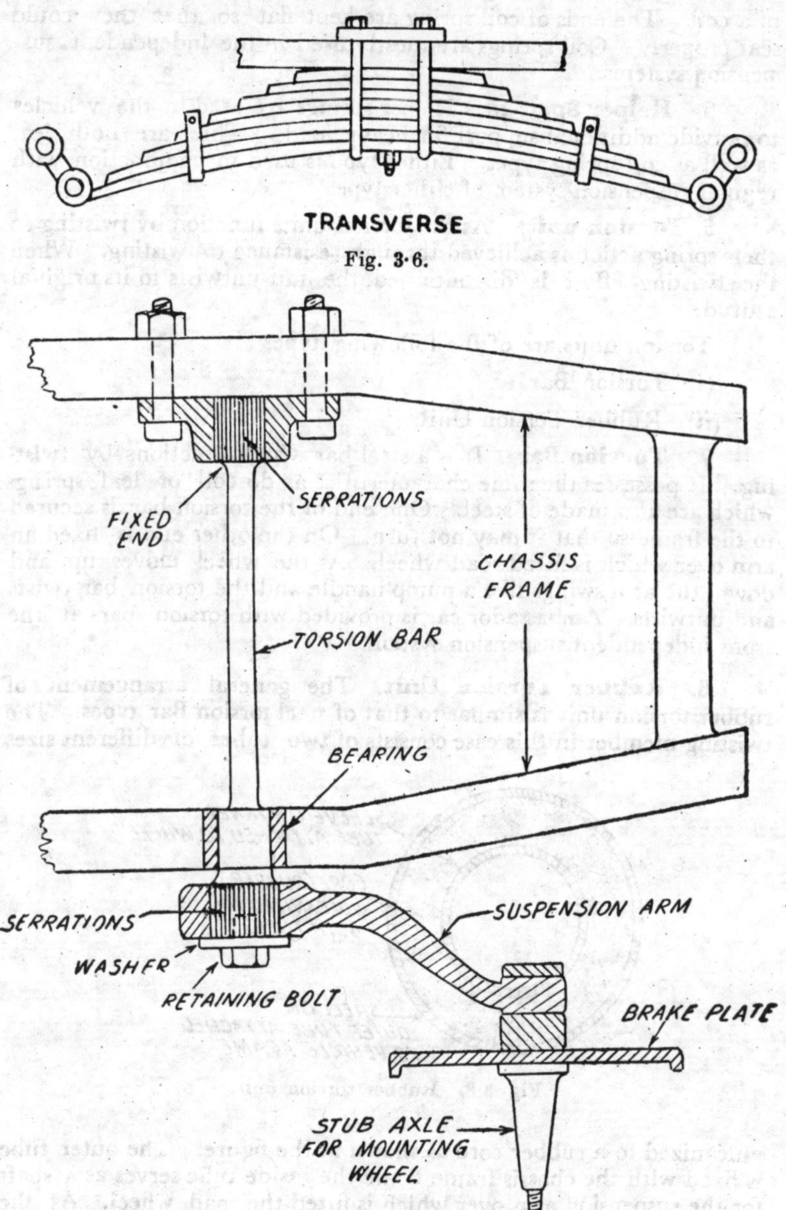

Fig. 3·6.

Fig. 3·7. Torsion bar springing arrangement.

4. Coil Springs. The coil spring is made of a length of special spring steel, usually round in section, which is wound in the shape of a coil. The ends of coil spring are kept flat so that they could seat properly. Coil springs are mostly used in the independent suspension systems.

5. Helper Springs. Helper springs are used in the vehicles to provide additional support for heavy loads. These are both leaf as well as coil spring types. Either type is used in conjunction with regular suspension system of either type.

6. Torsion units. As the torsion units function by twisting so their spring action is achieved through resistance to twisting. When the twisting effort is discontinued, the unit untwists to its original attitude.

Torsion units are of the following types :

(*i*) Torsion Bar

(*ii*) Rubber Torsion Unit.

7. Torsion Bar. It is a steel bar which functions by twisting. It possesses the same characteristics as do coil or leaf springs which are also made of steel. One end of the torsion bar is secured to the frame so that it may not turn. On the other end is fixed an arm over which is fitted road wheel. As the wheel moves up and down, the arm swings like a pump handle and the torsion bar twists and untwists. Ambassador car is provided with torsion bars at the front independent suspension system.

8. Rubber Torsion Unit. The general arrangement of rubber torsion unit is similar to that of steel torsion Bar types. The twisting member in this case consists of two tubes of different sizes

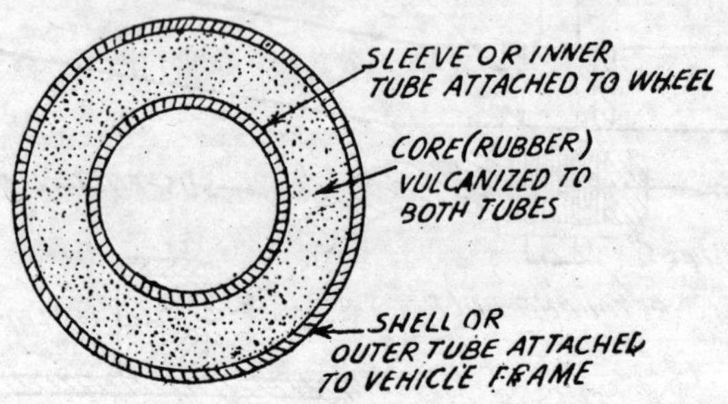

Fig- 3·8. Rubber torsion unit.

vulcanized to a rubber core as shown in the figure. The outer tube is fixed with the chassis frame while the inside tube serves as a shaft for the suspension arm over which is fitted the road wheel. As the wheel moves up and down, the suspension arm swings which twists and untwists the inner tube. Since the inner tube is vulcanized with the outer tube so the media of rubber provides springing action.

9. Spring Shackles. Shackles are a sort of links by means of which leaf springs are connected with the chassis frame. The shackles provide swinging ability to the leaf springs. Due to shock on the road wheel, the spring flattens up and increases in length and during rebound the spring assumes back its shape thereby decreasing in length. The shackles make the springs worthy to swing in and out.

Different types of shackles are used in different vehicles, which are as under :

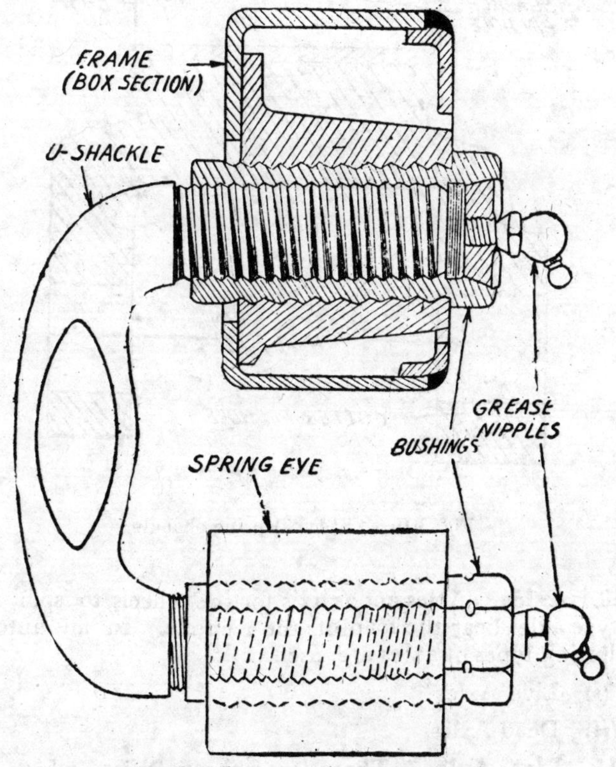

Fig. 3·9. 'U' type shackle arrangement

(*i*) 'U' type

(*ii*) 'Y' type

(*iii*) Link type.

'U' type of shackle is applicable in Jeep whereas link type is provided with Chevrolet vehicles. 'Y' types of shackle is most common in rear suspension.

One end of the shackle is connected with the chassis frame and through the other end connection is made with the spring by means of shackle bolt or pin. The shackle pin contains a hole at which grease nipple is screwed. Lubricant is fed to the shackle or spring eye bushing through the shackle pin hole.

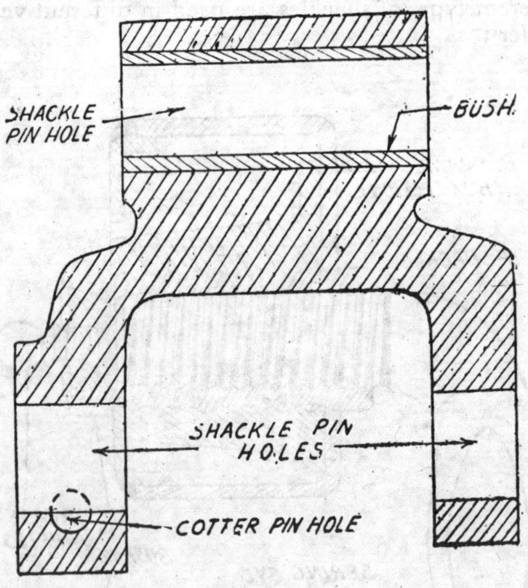

Fig. 3·10. 'Y' type spring shackle.

10. Axles. Axles act as axis for the wheels to spin around. Beam type axles bear the load of the vehicle. In an automobile, the following types of axles are used :

(*i*) Live Axles

(*ii*) Dead Axles.

11. Live Axles. These are those axles which contain differential and through which rotary motion is transmitted to the wheels. In these axles, there is a big housing for enclosing differential. These axles are also known as Drive Axles.

According to construction, the drive axle housing is of two types which is as under :

(*i*) **Banjo Type.** This type of axle housing resembles with banjo instrument. In order to fit the differential in it, the requisite opening is provided. On the back side, a

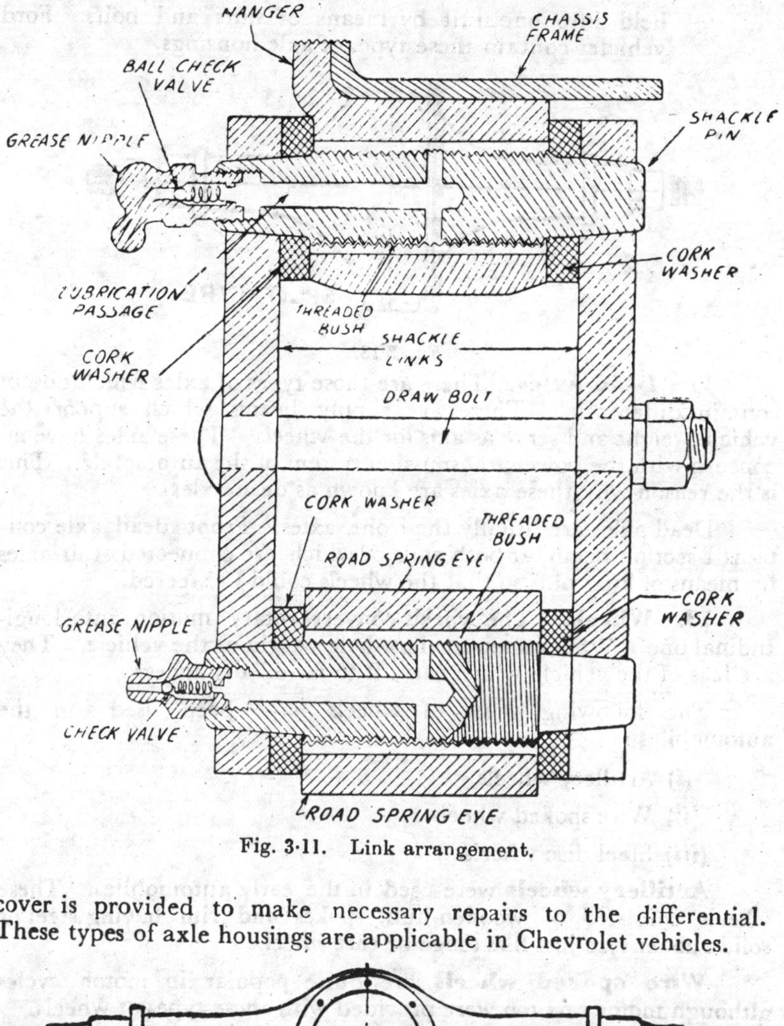

Fig. 3·11. Link arrangement.

cover is provided to make necessary repairs to the differential. These types of axle housings are applicable in Chevrolet vehicles.

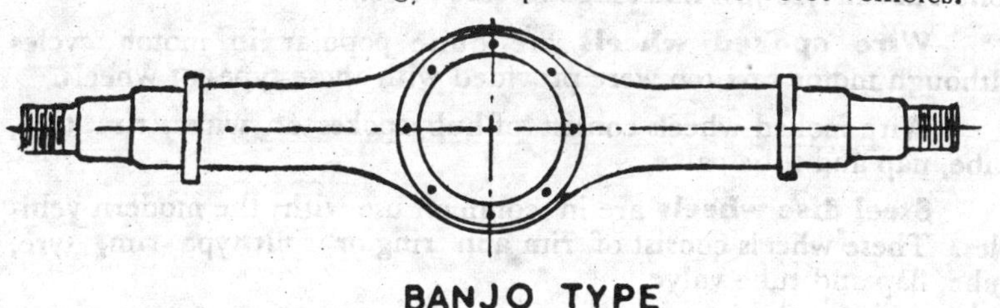

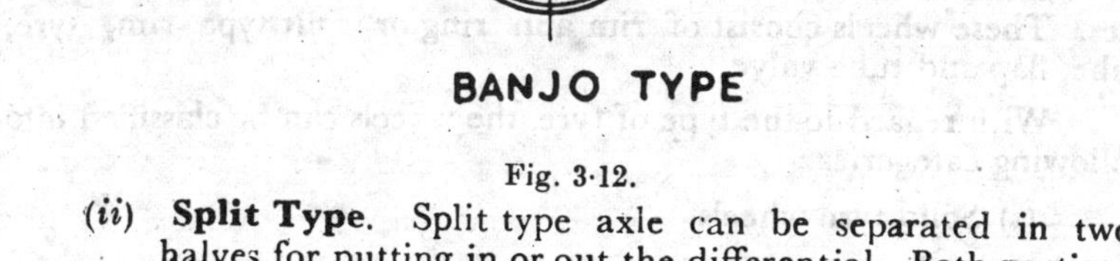

Fig. 3·12.

(*ii*) **Split Type**. Split type axle can be separated in two halves for putting in or out the differential. Both portions held into one unit by means of nuts and bolts. Ford vehicles contain these types of axle housings.

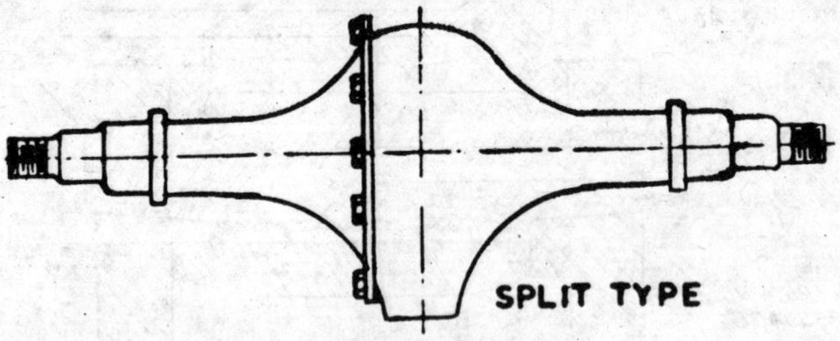

Fig. 3·13.

12. Dead Axles. These are those types of axles which do not contain differential. These are simply beams which support the vehicle weight and serve as axis for the wheels. These axles have no concern with the power transmission system of the automobile. This is the reason why these axles are known as dead axles.

Dead axles are usually the front axles. Front dead axle contains I-section beam, at both ends of which are connected stub axles by means of king pins so that the wheels could be steered.

13. Wheels. The wheels convert rotary motion into longitudinal one Wheels support the whole weight of the vehicle. They are legs of the vehicle which carry it to far-off distances.

The following types of wheels have been used in the automobiles :

(*i*) Artillery wheels,

(*ii*) Wire spoked wheels.

(*iii*) Steel disc wheels.

Artillery wheels were used in the early automobiles. These wheels consisted of wooden hub, spokes and rim having steel or solid rubber tyre just like tonga or cart wheel.

Wire spoked wheels are quite popular in motor cycles although motor cars too were provided with these types of wheels.

Wire spoked wheels consist of hub, spokes or wires, rim, tyre, tube, flap and tube valve.

Steel disc wheels are in common use with the modern vehicles. These wheels consist of rim and ring or split type rim, tyre, tube, flap and tube valve.

With regard to the type of tyre, the wheels can be classified into following categories :

(*i*) Solid tyre wheels,

(*ii*) Pneumatic tyre wheels.

In early vehicles, solid tyre wheels were used. Solid rubber tyre was mounted over the rim of the wheels. These wheels were much heavy and were a hurdle in the matter of speed.

Pneumatic tyre wheels use air as media between tyre, tube and rim to cushion out road shocks. These wheels are much lighter than the solid tyre wheels. All the modern vehicles are equipped with these types of wheels, whether they are wire spoked wheels or steel disc wheels. These wheels contain tyre and tube in place of a solid tyre.

Pneumatic tyre wheels can be further classified as below :

(*i*) **Tube Tyre Wheels** which contain both tube and tyre.

(*ii*) **Tubeless Tyre Wheels** which contain no tube, air being filled into tyre which is sealed with the rim.

14. **Modern Wheels**. The modern wheel is pneumatic tyre type wheel. In motor cycles, wire spoked pneumatic wheels are used whereas in cars steel disc pneumatic wheels are in common use.

Pneumatic tyre is an air bag fitted to the rim of the wheel. It consists of two main components—the outer tyre or cover and the inner tube, both of which are mounted on a rim forming the wheel.

The modern wheel is required to fulfil the following main objects :

(*i*) To absorb road shocks.

(*ii*) To carry load of the vehicle.

(*iii*) To transmit the forces of propulsion and braking.

In order to fulfil these objects, the road wheel should possess the following characteristics :

(*i*) It must be of maximum strength to take the weight, road shocks and driving torque.

(*ii*) It must be strong enough to resist local deformation when the wheel hits against a road kerb or other obstacle.

(*iii*) It must be properly balanced.

(*iv*) It should be easily detachable from the wheel hub with the minimum effort and with the help of a simple tool.

(*v*) It should be easy to clean and of good external appearance.

(*vi*) It should have good tread giving nice gripping in order to avoid slipping.

15. Components of a pneumatic wheel. The components of a pneumatic wheel as shown in the diagram are tyre, tube and valve, flap and rim.

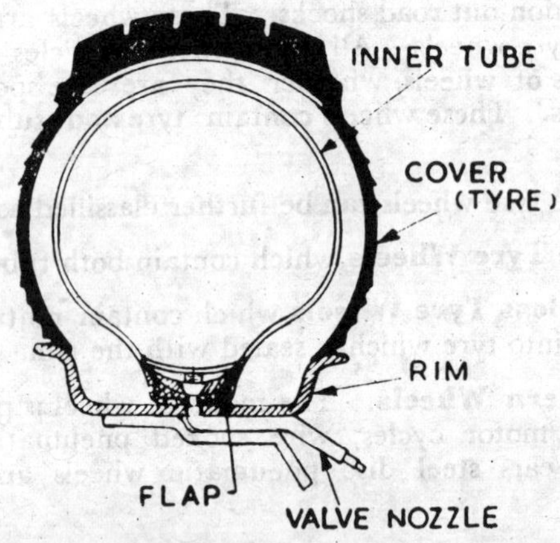

Fig. 3·14. Pneumatic wheel assembly : Parts in section.

16. Tyre. The tyre serves as a cover for the tube and protects it from damage The tyre has to be strong enough to retain the compressed air in the tube and to save it from damage. It has to be pliable enough to withstand the continual flexing produced by the rotation of the wheel under load.

The tyre cover carcase is usually made from rubber impregnated rayon (nylon or cotton are also used) and it is fitted with a rubber tread to provide road grip and to resist the abrasive action of the road surface.

The tyre provides a cushion between the vehicle and the road and thus eliminating road shock and damping down vibrations. It also transmits the power from the engine through the medium of driving wheels to the road. When the brakes are applied, it is the tyre again which transmits the whole of retarding force.

For giving maximum grip on the road surface, a tyre is provided with a good tread pattern which differs from tyre to tyre. The object is to give the best performance under different conditions of road.

The main object of the pneumatic tyre is to float the vehicle on a cushion of compressed air. Air carries the load and therefore the maintenance of correct air inflation pressures is most important.

The tyres are specified to run on recommended inflation pressures by the manufacturers according to the size, quality of the tyre and load to be imposed upon it.

Tyre pressures should be checked when the tyres are cold and not when they have attained normal running temperature. Excessive flexing of the tyre carcase due to underinflation will quickly break or loosen the casing cords and cause premature failure. Underinflation will also cause severe and irregular wear of tyre tread causing a loss in mileage. Overinflation will give an uncomfortable ride, increase vehicle maintenance, and cause rapid wear in the centre of tread.

On the sidewalls of the tyre cover there are several markings indicating tyre size, ply rating, tyre number, tyre make etc. The tyre size is mentioned like this—7·50 × 20.

The first figure (7·50) indicates the nominal cross-sectional width of the tyre in inches. The second figure (20) is the nominal diameter of the tyre from bead to bead in inches and indicates the correct rim diameter.

The ply rating represents an index of tyre strength and does not necessarily represent the number of actual plies of material in the tyre.

On metric size tyres markings such as 155-80-R-13 show 155-*tyre section width* in millimetres ; 80-*aspect ratio* (section height/section widh) ; R-*construction type* (R-radial ; B-Bias Belted ; D Diagonal ; E-Eilliptic) ; B-*rim diametre* in inches.

Tyre when properly maintained gives good service. In order to equalize their wear, it is necessary that they are rotated with one another after every 5,000 miles of run. It is recommended that the tyres may be rotated as shown in the diagram.

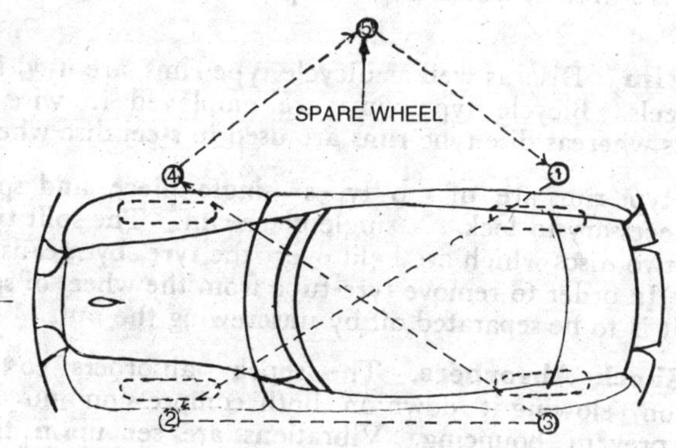

Fig. 3·15. Tyre rotation.

The main causes of the tyre wear are given below :

(*i*) Under-inflation,

(*ii*) Over-inflation,

(*iii*) Excessive road speeds,

(*iv*) Violent acceleration and braking,

(*v*) Presence of oil or grease on tyres,

(*vi*) Overloading,

(*vii*) Misalignment of wheels.

The tyres should be examined periodically for flints, nails and other foreign objects which may be embedded in the tread and also for cuts, penetrations or contamination by oil.

17 Puncture sealing tyres. Some tubeless tyres contain a coating of plastic material in the inner surface. This material fills into the hole left when the nail is removed from the tyre, thus sealing the leak.

18. Tube. The tube is made of rubber and is constructed so as to prevent any loss of compressed air. The tube should be as flexible as possible so that it could be replaced easily.

The tube contains a nozzle in which is fitted a valve. Air is inflated into the tube through the nozzle. The valve is of non-return type and does not allow filled in air to go out.

19. Safety tube. The safety tube is really two tubes in one. One is smaller than the other and is joined at rim edge. When the tube is filled with air, the air flows first into the inner tube. From the inside tube, air passes through an equalizing passage into the space between the two tubes. Thus both tubes are filled with air. Even if any nail embedding into tyre punctures the outer tube, the inner tube is still safe, which keeps on providing a bag of air inside the tyre.

20. Rim. Disc as well as bicycle type rims are used in automobile wheels. Bicycle type rims are employed in wire spoked type wheels whereas disc type rims are used in steel disc wheels.

Disc type rims are of o types—single piece and split type. A ring is necessary to lock t single piece rim. The split type rim consists of two discs which are held over the tyre by means of bolts and nuts. In order to remove tyre-tube from the wheel of split disc type rim, it is to be separated off by unscrewing the nuts.

21. Shock Absorbers. The shock absorbers control the spring action, slowing it down on both compression and rebound and thus prevent bouncing. Vibrations are set up in the road spring while the vehicle is running on shaby road, causing shocks. Similarly when a boulder or stone comes below the wheel or it falls into a pit, a stern shock is experienced which makes vehicle riding an uncomfortable affair. To arrest the shocks, it is necessary that the spring action be controlled. Shock absorber fulfils this requirement.

Shock absorbers can be classified into two categories :

(*i*) **Mechanical Shock Absorbers** in which frictional effect of metallic discs is utilized to control spring action.

(*ii*) **Hydraulic Shock Absorbers** in which fluid is used to resist the spring action.

According to the acting of shock absorbers these can be divided into two classes :

(*i*) **Single Acting Shock Absorbers** which regulate the rebounding action of road springs by controlling the return of the flexed springs to their normal shape.

(*ii*) **Double Acting Shock Absorbers** which control both compression and rebound of the road springs.

22. Mechanical Shock Absorbers. The mechanical shock absorber consists of two links which are connected with each other by means of a pin. Between the link, are placed a number of frictional discs of different metals. One link is connected with the frame whereas the other is fixed with the axle. The frictional discs which are housed between the arms, control the spring action due to their frictional effect and thus help in absorbing road shocks.

23. Hydraulic Shock Absorbers. In hydraulic shock absorber, fluid is tended to pass through a tiny hole which resists the movement of fluid. This resistance creates friction in the fluid which gives rise to heat. Fluid in the shock absorber swallows this heat generated due to spring action and thus energy of motion is absorbed Thus the road shock is absorbed by converting energy of motion into heat which is absorbed in the fluid of shock absorber.

Hydraulic Shock Absorbers can be classified as under :

(*i*) Direct Acting,

(*ii*) Cam Actuated Piston type,

(*iii*) Rotary Vane type.

24. Direct Acting Hydraulic Shock Absorbers. One end of this type of shock absorber is connected to axle whereas the other to chassis frame. It checks both compression and rebound of the road spring.

There are two chambers in this type of Shock Absorber. The outer chamber serves as a reservoir in which the inner chamber is housed which acts as a cylinder. The cylinder contains a combined piston and valve which is connected to a rod. The combined piston and valve divides the cylinder into two chambers—upper and lower. The upper chamber of the cylinder is called rebound chamber whereas the lower one compression chamber as shown in the figure.

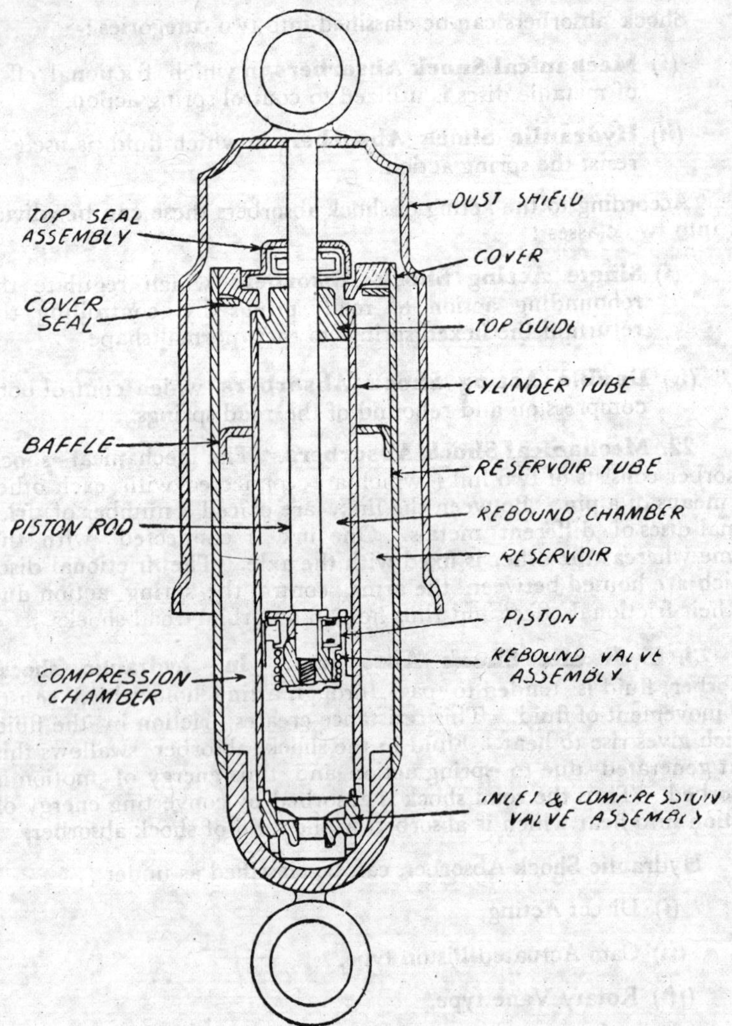

Fig. 3·16. Direct Double acting telescopic shock absorber.

When the road spring is compressed, combined piston and valve move into the lower chamber of the cylinder, compressing fluid in this chamber. Since the fluid is incompressible, so it opens the valve connected with the piston and enters into upper chamber of the cylinder. Some of the fluid runs into the outer chamber by opening the inlet and compression valve located at the bottom of the cylinder.

When the road spring comes at rebound, the fluid resistance in the shock absorber puts its effect on it. As soon as the combined piston and valve is pulled up, valve with the piston is closed but due to increase in pressure, it opens out to other direction as this is double type of check valve. Opening of the valve in the other direction, tends the fluid to flow from upper to lower chamber of the cylinder. So this way this type of shock absorber acts on both ways to check both compression and rebound of the road springs, thereby damping the shocks.

This type of shock absorber is also known as Telescopic shock-absorber as it moves up and down in one line during its working just like the telescope barrel.

25. Cam Actuated Piston Type Hydraulic Shock Absorbers. These types of shock absorbers are both of single acting and double acting type.

(*a*) **Double Acting Shock Absorber.** This type of shock absorber contains two pistons, one of which is rebound whereas the other is compression piston. These pistons are actuated by a cam fitted on a shaft. There is a lever attached to cam shaft, which actuates the cam thereby resulting in the movement of pistons.

Every piston contains one intake and one relief valve each. Fluid is transferred from one chamber to another through these valves.

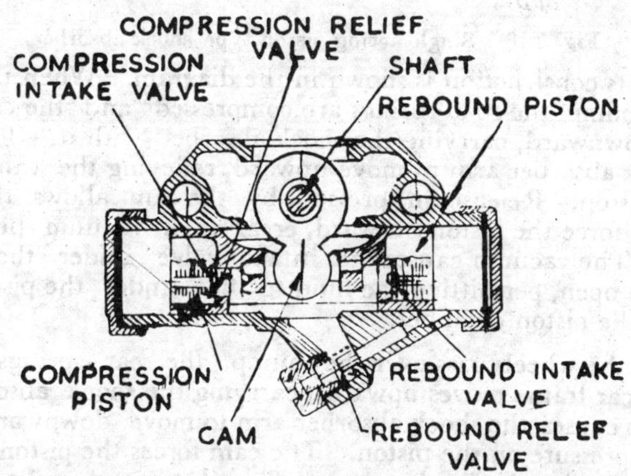

Fig. 3·17. Double acting cam actuatedpiston type shock absorber.

During compression of road spring, the fluid goes out from compression chamber through relief valve and enters into the reservoir. During this process, compression of the road spring is checked up. During the same process, fluid enters into rebound chamber through the rebound valve.

During rebound of road spring, the fluid goes out of rebound chamber through relief valve and enters into the reservoir and rebound of the spring is checked up. In the meantime, fluid enters into compression chamber through the intake valve.

(*b*) **Single Acting Shock Absorber.** In this type of shock absorber, there is only one piston and one intake and one relief

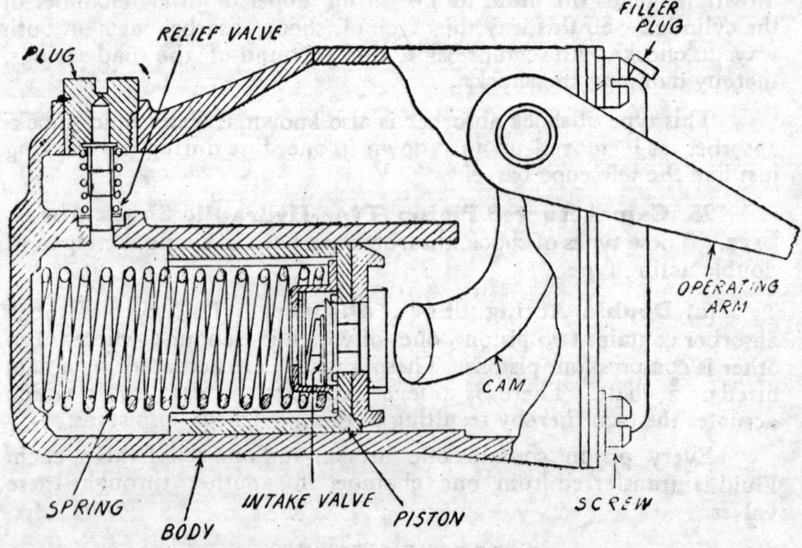

Fig. 3 18. Single acting piston type shock absorber.

valve. Its construction is shown in the diagram. When the wheels strike a bump, the road springs are compressed and the car frame moves downward, carrying the shock absorber with it. This causes the shock absorber arm to move upward, relieving the cam pressure on the piston. Release in pressure by the cam allows the piston spring to force the piston outward, creating a vacuum behind the piston. The vacuum causes the intake valve under the head of piston to open, permitting the fluid to flow under the piston head and fill the piston chamber.

As the wheels pass over the bump, the car springs rebound, and the car frame moves upwards, carrying the shock absorber with it. This causes the shock absorber arm to move downwards, applying cam pressure on the piston. The cam forces the piston into the cylinder, closing the intake valve. The oil trapped in the cylinder, forces the relief valve off its seat and goes out slowly into the reservoir. Rebounding of road spring is thus damped out by this action.

26. Rotary vane type Hydraulic Shock Absorber. This type of shock absorber contains a round chamber in which a two-lobe rotor moves in the viscus fluid. The main chamber contains two-fixed vanes, spring, oil seal, cover and link. When in assembled state, the main chamber is divided into four parts. When the rotor is moved, fluid is under pressure in two chambers because of the construction of valves which allow very little fluid to flow towards chambers having less pressure. When the rotor moves in the reverse direction, high pressure chambers are converted into low pressure ones.

The chambers of the shock absorber are kept filled with fluid automatically so that the effect of its action may not decrease.

The rotor shaft is operated by an actuating arm which is connected with the axle by means of a link. The body of the shock absorber is held with the chassis frame. These types of shock absorber are very popular with Ford vehicles.

27. Stabilizers. Stabilizers or Sway Bars are alloy steel bars which connect shock absorber operating arms or independent suspension control arms. There is one separate bar each to connect rear and front shock absorbers.

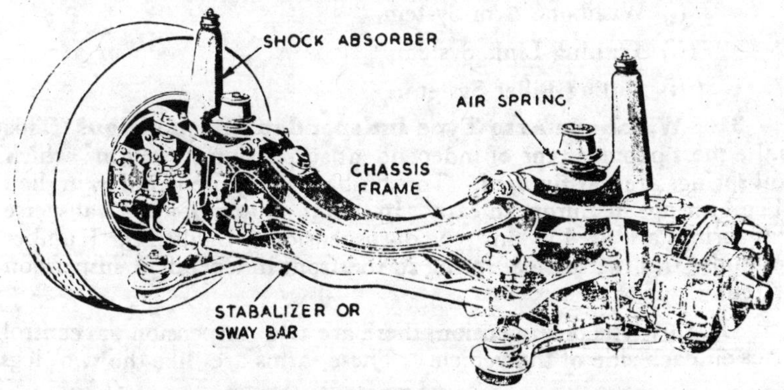

Fig. 3·19. Front independent suspension with air springs showing stabilizer.

The stabilizers are placed parallel to cross members and attached to the front or rear of the frame through rubber bushings or similar material. If one side of the vehicle tries to rise faster than the other, the resultant twist in the bar reacts on the axle or suspension and tends the frame to keep at level. Thus the stabilizers keep the vehicle stable while moving on upheaval road or taking corners.

28. Suspension Arrangements in Different Vehicles. The suspension arrangement differs from vehicle to vehicle. Even the front and rear suspensions are not alike. The following are the different suspension arrangements applicable to different vehicles :—

(*i*) Conventional suspension system at both front and rear.

(*ii*) Conventional suspension at the rear and independent suspension at the front.

(*iii*) Independent suspension at both front and rear.

29. Conventional Suspension System. It is that type of suspension system in which the wheels are fitted on beam type axles which are attached to the chassis frame through road springs. In this type of suspension, the effect on one wheel is directly transmitted to the other side wheel through the axle.

This is the simple type of suspension system and resembles with that of a horse drawn coach.

30. Independent Suspension System. In independent suspension, there is no axle beam running below the frame like the conventional suspension system. Suspension for each wheel is an independent unit and is free from the effect of one another. The linkages used in the construction of this suspension comprise the basic features of the suspension.

The following types of independent suspension systems are applicable to automobiles—

(*i*) Wishbone Arm System,

(*ii*) Trailing Link System,

(*iii*) Sliding Pillar System.

31. Wishbone Arm Type Independent Suspension. This is the most popular type of independent suspension system in which coil springs are mostly used. Torsion bars are quite popular in lieu of coil springs in European cars. In some automobiles, transverse leaf springs are used in this type of suspension. English car Humber contains transverse leaf spring in the front independent suspension system.

In this type of suspension, there are two suspension or control arms on each side of the vehicle. These arms are like the two legs

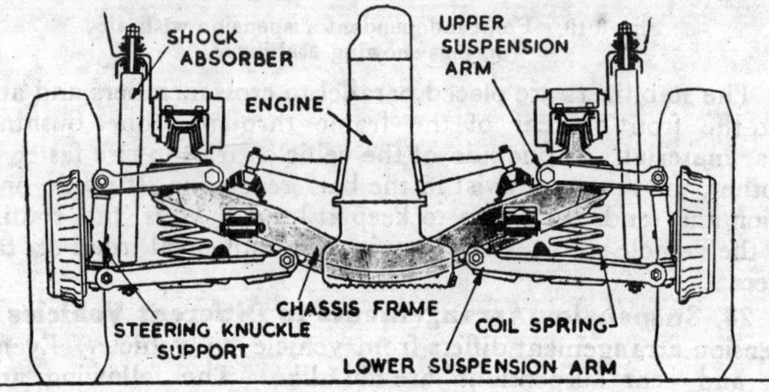

Fig. 3·20. Independent suspension system using wishbone suspension arms.

of chicken wishbone or letter 'V'. These wishbone arms are connected with chassis frame on the open end. The close end spreads out of the chassis frame. One arm is below whereas the other is above the frame. The closed ends of both upper and lower suspension arms are connected with the steering knuckle support to which is attached steering knuckle by means of king pin. A coil spring is placed between the frame and lower wishbone suspension arm. Mostly the open end of upper control arm is connected with the shock absorber shaft which is fitted at the frame.

The upper and lower control arms are connected in position, for the cradle. When there is bump and wheel is tended to go up, the control arms move up and the coil spring is compressed. Since the shock absorber is fitted with the upper control arm, so it damps the vibrations set up in the coil spring due to road irregularities.

The suspension which uses transverse leaf spring, consists of an upper wishbone, steering knuckle support and a transverse leaf

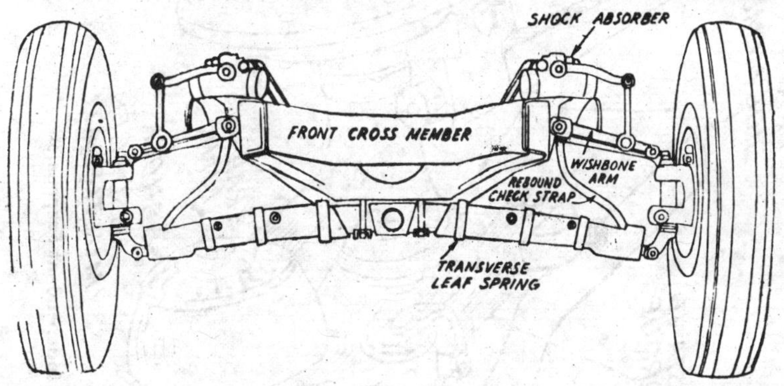

Fig. 3·21. ndependent suspension employing transverse leaf spring.

spring. The transverse spring, serves two purposes of holding the lower end of steering knuckle support and of providing the spring action. The basic construction is similar to the regular wishbone design with the exception that no lower control arm is used.

32. Trailing Link Type Independent Suspension. The trailing link suspension systems use parallelogram linkages lying beside the frame side members. Usually a horizontal coil spring is used in this type of suspension. During compression and rebound, the spring winds and unwinds like the balance spring in an ordinary watch. In some suspensions of this type, the torsion bar is fitted in lieu of horizontal coil spring.

33. Sliding Pillar Type Independent Suspension. The first independent front end suspension ever used was a sliding pillar type. In this type of suspension. the pillar or elongated kingpin is attached to the wheel and slides up and down in the axle type beam affixed rigidly to the vehicle frame.

34. Air Suspension. This is the modern development. In their suspension, the four conventional springs are replaced by four air bags or air spring assemblies. Each air spring assembly is a flexible bag enclosed in a metal dome. The bag is filled with compressed air

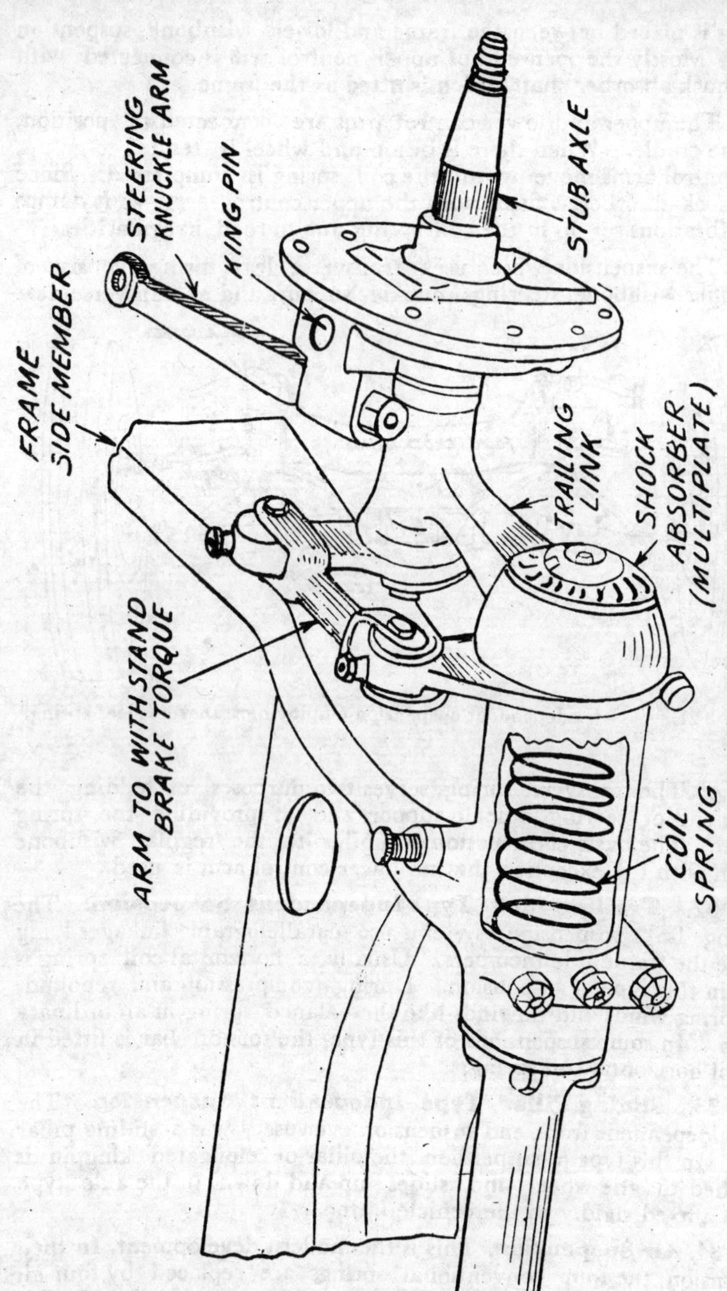

Fig. 3·22. Trailing link independent suspension.

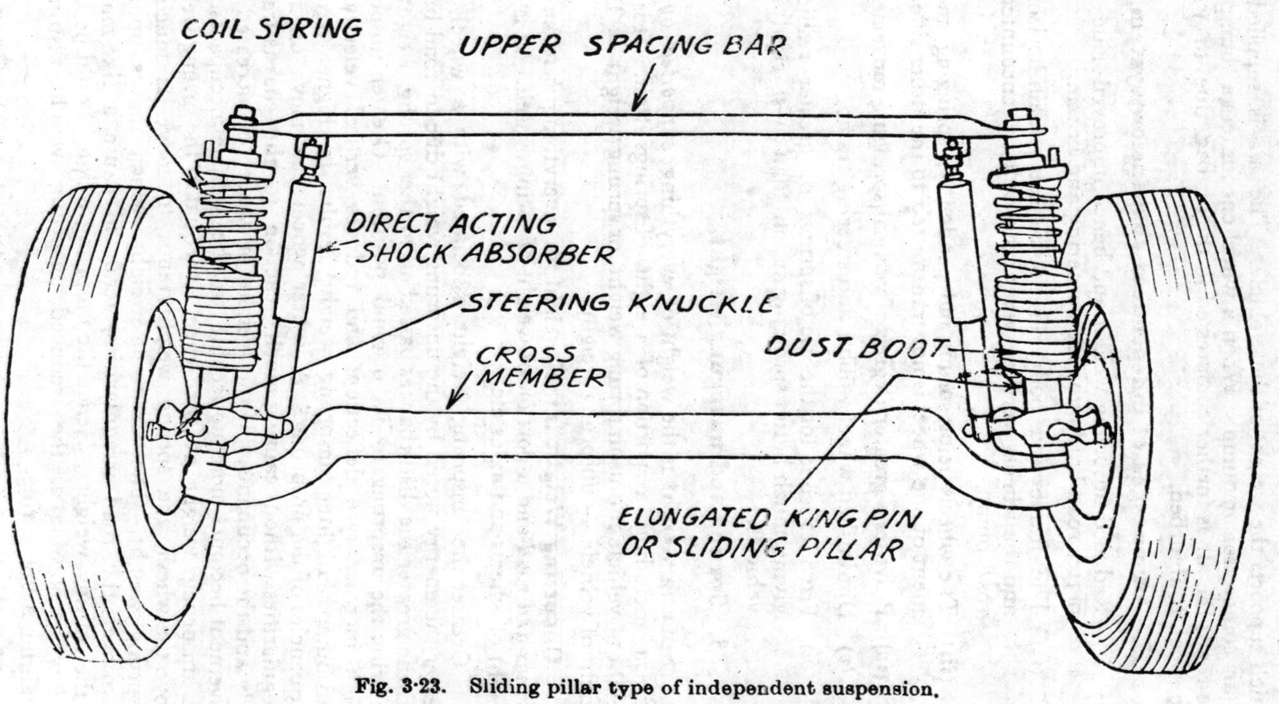

Fig. 3·23. Sliding pillar type of independent suspension.

which supports the weight of the vehicle. The air is supplied by an air compressor or pump. When a wheel encounters a bump on the road, the air is further compressed in the air bag due to which the shock is absorbed.

35. Advantages of independent suspension system.

(*i*) Reduced unsprung weight and hence improved ride and better road holding while cornering and braking.

(*ii*) The frame and body do not tilt but remain horizontal and the wheels vertical when the vehicle encounters a road bump.

(*iii*) The wheels being sprung independently, springing movement of one wheel is not transmitted to the other side.

(*iv*) Provides a greater degree of vertical springing movement.

(*v*) Diminished wheel wobble and steering tramp.

(*vi*) Provides scope for the use of springs of greater resilience giving much better springing action than most rigid axle vehicles.

36. Sprung and unsprung weight.

Sprung weight is the weight of everything *supported by the springs* including a portion of weight of springs itself. It is the weight of vehicle minus unsprung weight, or sprung weight = Total weight of vehicle — unsprung weight.

Unsprung Weight is the weight of everything *between the springs and road* and a portion of weight of springs itself. It is the weight of wheels and axles etc.

Greater the unsprung weight associated with a wheel, the greater the energy stored by the unsprung weight due to road bump and the greater the disturbances passed on to the sprung weight in restoring the unsprung to its original position. Greater unsprung weight increases tyre deflections and reduces vertical velocity on road bump. Lighter unsprung weight results in higher natural frequencies of the unsprung. A lighter wheel can move on road irregularities without causing much reaction to the chassis frame, body and the occupants. If the weight of wheel is increased, its movement becomes more noticeable to the vehicle occupants. If the unsprung weight at the wheel is equal to the sprung weight above the wheel, the sprung weight tends to move as much as unsprung weight. The unsprung weight which moves up and down over the road irregularities, tends to cause a like motion of the sprung weight. Hence the unsprung weight should be kept as low as possible as the reduced unsprung weight results in obtaining a better ride.

37. Basic suspension movements. A vehicle is subjected to move over road irregularities. The road is never cent per cent level. It has pits and falls. The vehicle has to go up and down, cross over bumps and pits in the road and pass through bends, curves and zig-zag path. While in motion, the vehicle suspension is subjected to the following basic movements :—

(a) Bouncing.
(b) Pitching.
(c) Rolling.

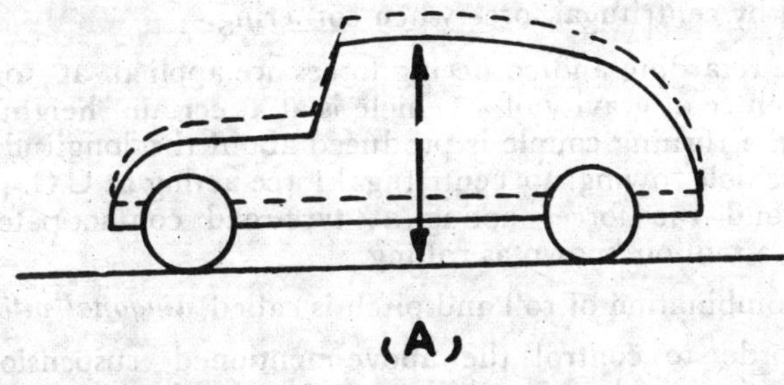

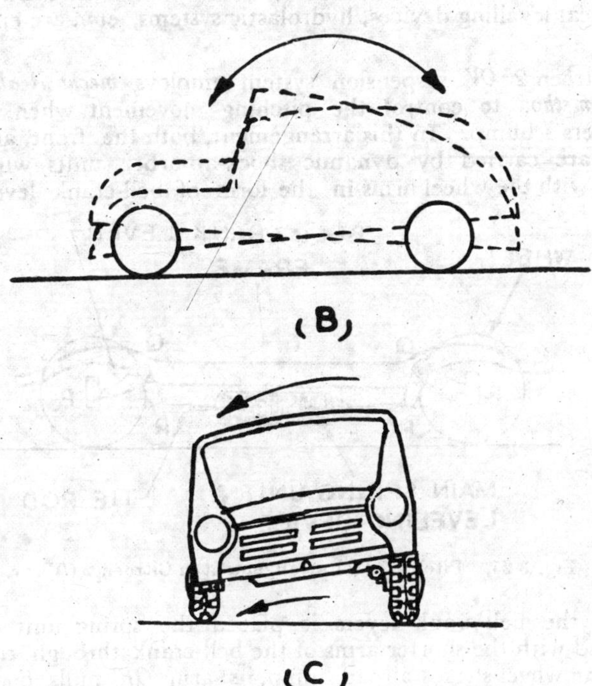

(a) Bounce (b) Pitch (c) Roll
Fig. 3·24. Basic suspension movements.

(*a*) **Bouncing.** It is vertical movement of the complete body. When the complete body of the vehicle rises up and down, it is known as *bounce* or *bouncing*. There may be front end or rear end bounce.

(*b*) **Pitching.** It is rocking chair action or rotating action about a transverse axis through the vehicle parallel to ground. Due to pitching, the front suspension moves out of phase with the rear resulting in rocking effect.

(*c*) **Rolling.** It is movement about a longitudinal axis produced by centrifugal force when cornering.

The retarding and cornering forces are applied at road levels but the centre of gravity of a vehicle is at a certain height. While cornering, a turning couple is produced about the longitudinal axis of the vehicle owing to centrifugal force acting at C.G. (centre of gravity) and the forces acting at tyre-road contact patch. This results in a motion known as rolling.

A combination of roll and pitch is called *diagonal pitch.*

In order to control the above-mentioned suspension movements, anti-sway bars, stabilisers, pitch and roll control bars, mechanical levelling devices, hydrolastic systems, etc., are employed in cars.

Citroen 2 *CV* suspension system employs *mechanical spring control method* to control the pitching movement when the car encounters a bump. In this arrangement, both the front and rear wheels are carried by dynamic shock-absorber units which are integral with the wheel arms in the form of bell-crank levers. In

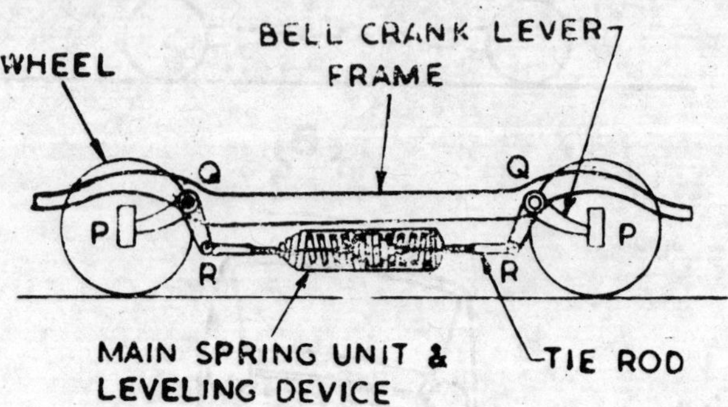

Fig. 3·25. Pitch control arrangement in Citroen 2 *CV* car.

between the bell-crank levers is placed the spring unit which is connected with the shorter arms of the bell-crank through tie rods. When one wheel strikes a road bump, its arm *QR* pulls on its tie rod through the pair of compression springs contained in a cylindrical housing. The pulling effect on tie rod tends to lift the other wheel which results in the reduction of pitching effect.

In ***hydrolastic system*** or ***hydraulic load transfer system***, hydraulic fluid is forced from front wheel unit to rear wheel unit or *vice versa* when the front or rear wheel hits a road bump. Transfer of fluid from one wheel unit to other, tends the other end of car to rise up resulting in reduction or elimination of pitching tendency.

The ***pitch and roll control bars*** are usually placed at the rear end of cars and are mounted transversely. The anti-roll bar is employed to adjust the car roll rates of front and rear. The anti-pitch bars improve pitch stability of the car.

Stabilisers, sway-bars or ***anti-roll bars*** are also fitted transversely at the front or rear or both front and rear to control ***rolling***.

38. **Knee action springing**. The parallelogram or parallel link independent suspension is known as knee action springing system. It contains a pair of radius arms which are hinged at chassis frame and steering knuckle support. In improved designs the radius arms are in the shape of '*V*' and are known as wishbone arms. The upper arm is shorter than the lower one. The springing effect is obtained by the compression of a large coil spring. When the wheel strikes a road projection, it tilts inward at the top due to upper radius arm being short. The lower arm moves the lower part of wheel inwards. The combined action of upper and lower arms maintains the wheel track practically constant and helps in avoiding tyre scuffing.

QUESTIONS

1. What is suspension system and which objects are fulfilled by it ?

2. Which are different types of road springs ? How they absorb shocks ?

3. Describe the construction of leaf springs.

4. Which are the different types of leaf springs ? Explain them.

5. What is helper spring ? Describe its necessity.

6. Which are the different types of torsion units ? Describe their construction and working.

7. Describe the difference between torsion bar and rubber torsion unit.

8. Which are the different types of spring shackles ? Describe them.

9. Which are the different types of axles used in automobiles ? Discuss.

10. Describe the wheels used in different automobiles from time to time.

11. Write a short note on pneumatic tyre wheel.

12. What characteristics a road wheel should possess ?

13. (*i*) Which are the different markings provided on the side-walls of the type ? Discuss.

(*ii*) Analyse the following markings on the side-wall of a tyre :-
195.70 R-15.

Name the cars having such type of markings

14. Which are the main constituents of a modern wheel ? Discuss.

15. Describe the main causes of tyre wear.

16. Write short notes on the following :—

(*i*) Tubeless Tyre

(ii) Puncture Sealing Tyre
(iii) Safety Tube
(iv) Tyre Size Indication
(v) Artillery Wheel
(vi) Wire Spoked wheel

17. What is the object of shock absorber ?

18. Which are the different types of shock absorbers ? Describe them.

19. Classify hydraulic shock absorbers. Explain any two of them.

20. Illustrate the construction and working of direct acting and cam actuated piston type shock absorbers.

21. Illustrate any one type of single acting shock absorber.

22. Describe the construction and working of rotary vane type hydraulic shock absorber.

23. What are stabilizers ? Which objects are fulfilled by them ?

24. Describe the different suspension arrangements. Give examples in which vehicles they are applicable.

25. Classify suspension systems and describe them.

26. What is the difference between conventional and independent suspension system ?

27. Which are the different types of independent suspension systems ? Explain them.

28. Describe independent suspension using transverse leaf spring.

29. Describe air suspension.

30. Write short notes on the following :—
(i) Sliding Pillar Type Suspension
(ii) Trailing Link Suspension
(iii) Double Acting Shock Absorber
(iv) Mechanical Shock Absorber
(v) Telescopic Shock Absorber.

31. Explain the advantages of independent suspension system.

32. Define the following :
(i) Sprung and unsprung weight.
(ii) Bounce, pitch and roll.
(iii) Knee action springing.

33. Explain the various methods to control pitch and roll in automobiles.

4

Engine-I

1. Engine. An engine is a Prime Mover with the help of which heat energy obtained from the fuel, is converted into mechanical energy.

Engine is heart of the automobile. If it fails to work, the vehicle is dead. It is one of the biggest units in the automobile. It is placed mostly in the front part of the vehicle. It is also located in the rear as well as in the middle of chassis frame. In long nose vehicles, the engine is fitted in the nose. In certain vehicles, the cab is built over the engine, known as cab over engine (C.O.E.). The engine in case of box type bodies, is located inside the body. A cover known as bonnet is provided to keep the engine under safe condition.

The engine is fitted over the chassis frame by means of rubber mountings in order to insulate it from shocks.

In certain cases, the engine with clutch, is mounted over the frame as a single unit whereas in some cases engine, clutch and gearbox are assembled into one unit and mounted over the frame. In the latter case, the mountings are at the front end of engine and the gearbox as in the case of Ford vehicles.

2. Types of Engines.

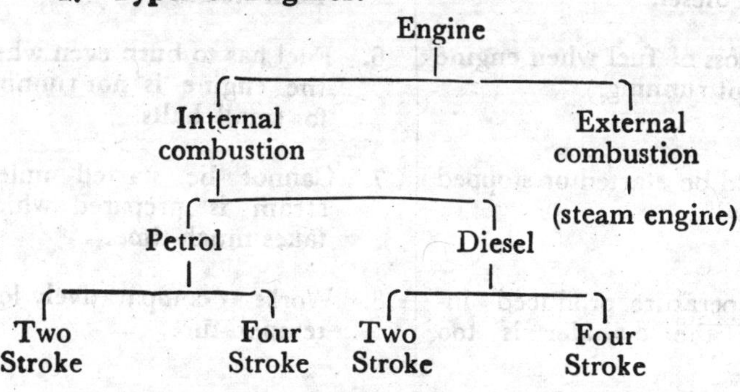

As is apparent from the above chart engines belong to two main classes : Internal Combustion and External Combustion.

External combustion engine is that type of engine in which combustion takes place outside the engine. Steam engine is external combustion engine (E.C.E.).

Internal combustion engine is that engine in which combustion takes place inside the engine. Petrol and diesel engines are internal combustion engines (I.C.E.).

All the different types of engines have been used in the automobiles in different ages. Steam engine was the first to be employed in the first automobile built in 1769 by Captain Cugnot. As the steam engine occupied more space, required more time to prepare steam, needed more than one man to operate, emitted more soot and noise and was very bulky, so it was replaced by the internal combustion gasoline engine. The automobile engines thus stepped into gasoline age after the primary stage of steam.

The comparison of External Combustion and Internal Combustion engines which is given below, shall show the merits and demerits of each category.

Internal Combustion Engine	*External Combustion Engine*
1. Occupies less space.	1. Occupies more space.
2. Lighter in weight.	2. Heavier in weight.
3. High speed engine.	3. Slow speed engine.
4. Combustion of fuel takes place inside the engine.	4. Combustion of fuel takes place outside the engine.
5. Fuels used in being petrol and diesel.	5. Solid or liquid fuels used to form steam.
6. No loss of fuel when engine is not running.	6. Fuel has to burn even when the engine is not running for small halts.
7. Could be started or stopped at will.	7. Cannot be started unless steam is prepared which takes much time.
8. Temperature produced inside the cylinder is too high.	8. Works at comparatively low temperature.

9. Cooling arrangement necessary.	9. No cooling of the cylinders required. Rather it is steam jacketed.
10. Single acting.	10. Mostly double acting.
11. Exhaust gas temperature as high as 300°C.	11. The temperature of exhaust steam is quite low.
12. Thermal efficiency of diesel engine up to 40%.	12. Thermal efficiency up to 24% as that of petrol engine.
13. Needs no boiler, furnace or condenser.	13. Boiler, furnace and condenser are must.
14. Most compact.	14. Not so compact.

Now a days, Internal Combustion Engines are most commonly used in the automobiles. These engines are of many types. These could be classified with regard to the fuel used in them, their cooling arrangement, ignition system, valve arrangement, shape and mode of operation etc.

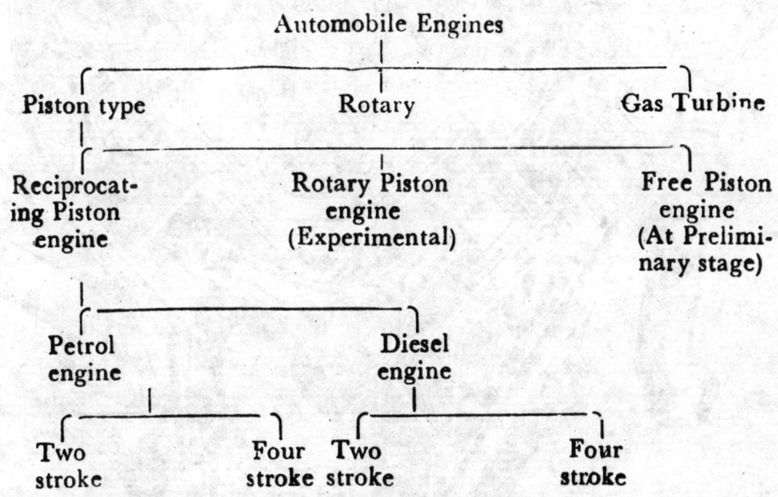

As indicated above, modern automobile engines are of the following types :

With regard to their construction and operation :

(*i*) Piston type.

(*ii*) Rotary.

(*iii*) Turbine.

3. Piston type engines. These engines use piston to transform heat energy taken from the fuel to mechanical energy. Piston is the main acting part which creates different events due to its movement, resulting in the cycle of operation.

Piston type engines are of the following types :

(*i*) Rotary Piston engines.

(*ii*) Free Piston engines.

(*iii*) Reciprocating Piston engines.

The rotary piston engine contains curved pistons which rotate inside the engine. This engine is still in the experimental stage and has not come into use in automobiles.

The free piston engine contains a pair of piston assemblies which oppose each other in a cylinder. The pistons are free to move inside the cylinder. Actually it is not a complete engine for developing power. It is just a device to supply high pressure gas to drive a power turbine.

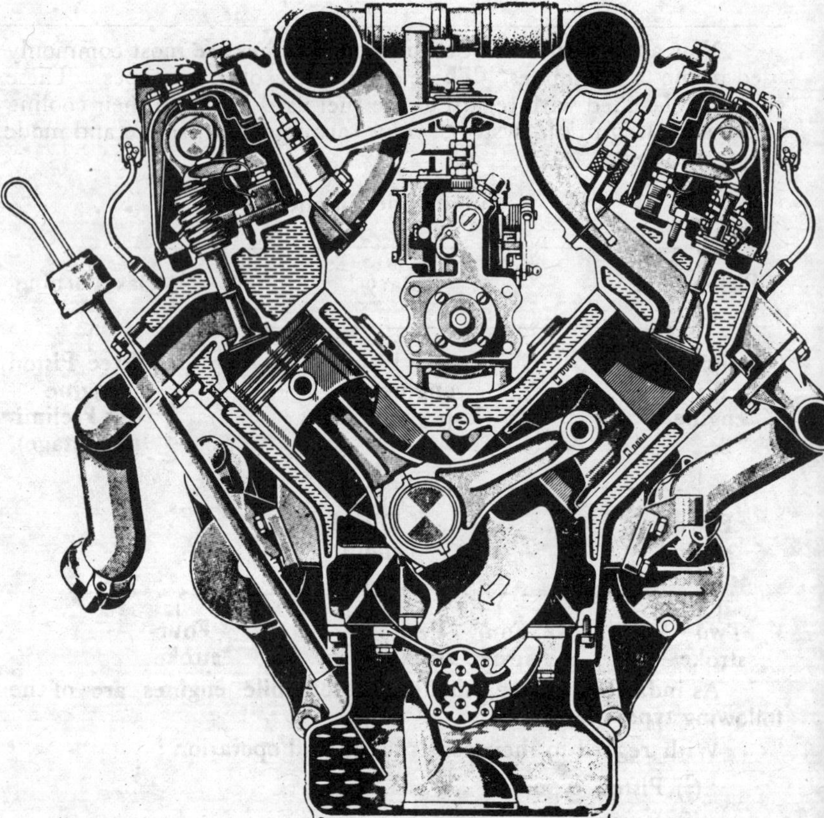

Fig. 4·1. V-8 Engine with overhead camshafts, and gasoline injection (Mercedes Benz).

The reciprocating piston engine is the most popular engine which is commonly used in the modern vehicles. This engine contains pistons which reciprocate inside the cylinders for completing the cycle of operation resulting in the development of power. The pistons are connected with crank-shaft by means of connecting rods. Owing to this attachment, reciprocating movement of the pistons is converted into rotary motion which is required to drive the vehicle.

The reciprocating piston engines can be classified as under :

(1) **With regard to the fuel used in them :**

(*i*) **Petrol or gasoline engines** in which petrol or petrol gas is used.

(*ii*) **Diesel engines** in which diesel is used as fuel.

(2) **With regard to the method of ignition in the engines :**

(*i*) **Spark ignition engines** in which ignition takes place by means of an electric spark. Petrol engines are spark ignition engines.

(*ii*) **Compression ignition engines** in which the injected fuel is ignited due to the temperature of compressed air in the cylinder. Diesel engines are compression ignition engines.

(3) **With regard to their cycle of operation :**

(*a*) **Otto cycle engines or constant volume cycle engines.** The otto cycle comprises of the following events taking place one after the other :

(*i*) Suction of fuel air mixture inside the cylinder

(*ii*) Compression of fuel air mixture.

(*iii*) Ignition.

(*iv*) Power impulse action (working).

(*v*) Exhaust of burnt gases.

The engines which work on this cycle are known as Otto Cycle engines. In Otto Cycle, combustion takes place at constant volume as whole of the fuel is burned instantaneously as an explosion. Hence engines which work on otto cycle are known as constant volume cycle engines. Petrol engines are otto cycle engines.

(*b*) **Diesel Cycle Engines or Constant Pressure Cycle Engines** which work on diesel cycle or constant pressure cycle. In diesel cycle, the combustion takes place at constant pressure because burning takes place gradually without an explosion as the fuel enters. Hence this cycle is known as constant pressure cycle. In diesel cycle, the following events take place one after the other :

(*i*) Suction of only air.

(*ii*) Compression of air.

(*iii*) Injection of fuel.

(*iv*) Action of power impulse (working).

(*v*) Exhaust of burnt gases.

Diesel engines work on this cycle.

(4) **With regard to the number of strokes per cycle :**

(*i*) **Two stroke engines** in which all the events of the cycle are completed in two strokes of the piston.

(*ii*) **Four stroke engines** in which all the events of the cycle are completed in four strokes of the piston.

(5) **With regard to the type of cooling svstem of the engine :**

(*i*) **Air cooled engines** which are cooled by air. Air cooled engines contain fins around the cylinders, cylinder heads and exhaust ports etc. to provide more area for better radiation of heat.

(*ii*) **Liquid or water cooled engines** in which some liquid or water is used to cool them. These engines contain water jackets around the cylinders, combustion chambers and valve ports etc. A radiator is provided with them to cool down hot water.

(6) **With regard to the number of cylinders in the engines :**

(*i*) **Single cylinder engines** which contain only one cylinder.

(*ii*) **Multi cylinder engines** which contain more than one cylinder.

(7) **With regard to the shape of the engines :**

(*i*) **In-line engines** in which the cylinders are in one line or row.

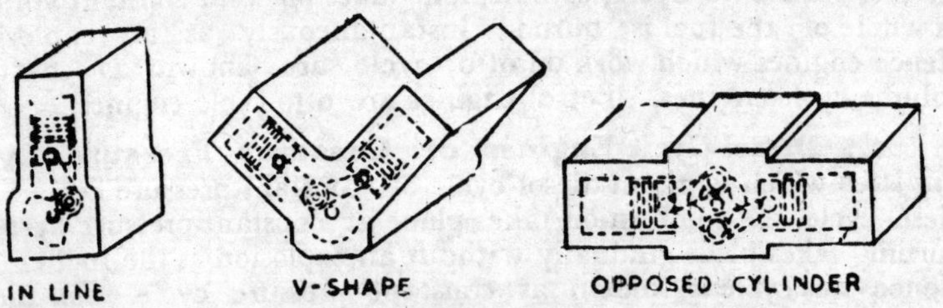

Fig. 4·2

(*ii*) **V shaped engines** in which the cylinders are placed in two rows If centre lines are drawn in both rows of the cylinders, these will meet at the bottom forming the shape of 'V' and hence the 'V' shaped engine.

(*iii*) **Opposed cylinder engines** in which the cylinders are opposite to each other and the crankshaft is placed between them. These are multi cylinder engines and contain even number of cylinders, half the number on opposite direction to the other.

(*iv*) **Radial engines** in which the cylinder radiate from a common centre like the spokes of a wheel. These are not used in automobiles.

(8) **With regard to the arrangement of valves in the engines :**

(*i*) **'L' Head or side valve engines** in which the inlet and exhaust valves are in one row and located by the

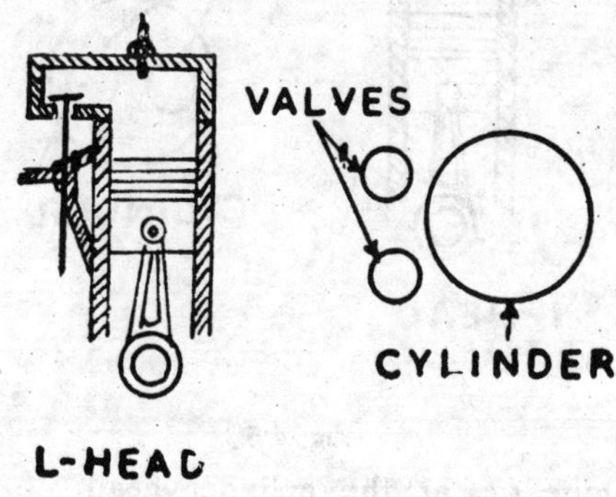

Fig. 4·3

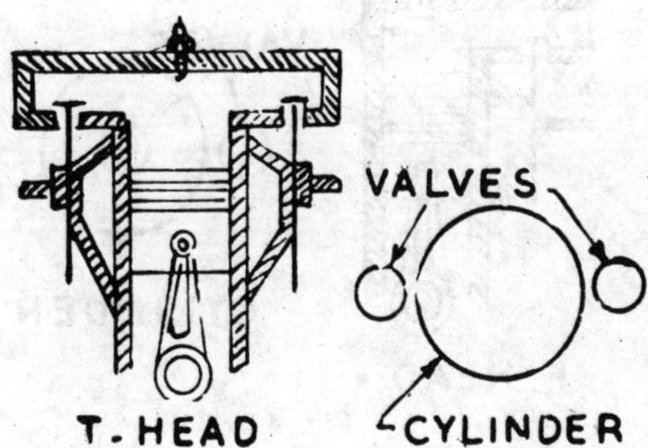

Fig. 4·4

side of cylinders. In these engines, the path of gases travel to and from the engine is of inverted 'L' shape as shown in the diagram. The combustion chamber and cylinder form an inverted 'L' and hence 'L' head engines.

(*ii*) **'T' Head engines** in which intake and exhaust valves are placed in separate rows by the opposite sides of the cylinders. The path of flow of gases to and from the engine is like letter 'T'. The combustion chamber and cylinder form 'T' and hence 'T' head engine.

(*iii*) **'I' Head engines or overhead valve engines** in which the valves are located over the head of the

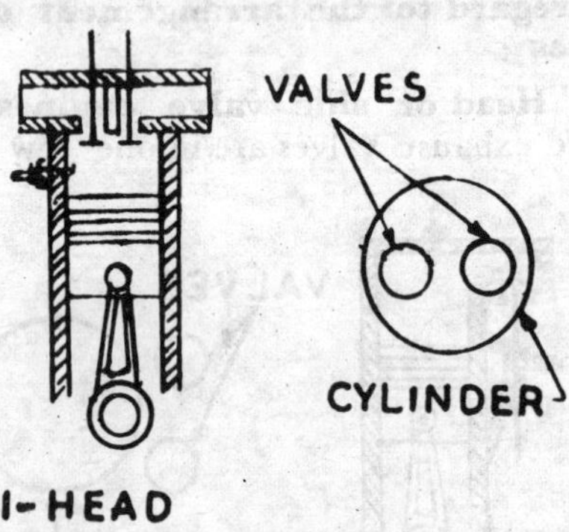

Fig. 4·5

engine, *i.e.*, at the cylinder head. In these engines the fresh charge of gases goes down from the

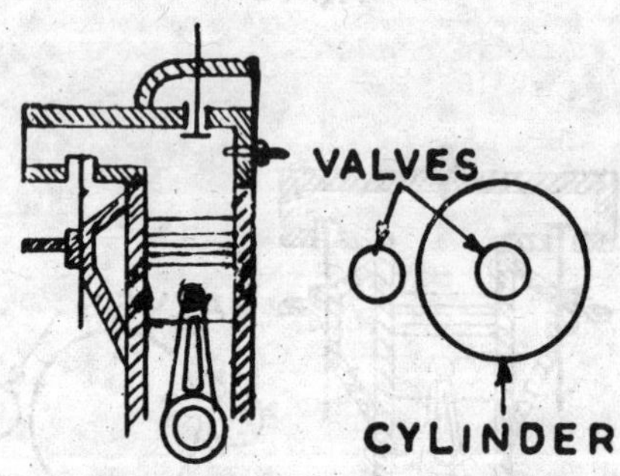

Fig. 4·6

valves into the cylinder in a straight path and similarly the used gases go out of the cylinder to the valve mouth in a straight line. The path of gases in the engine being of 'I' shape, combustion chamber and cylinder forming 'I' and hence 'I' head engine.

(*iv*) **'F' Head engine** in which inlet valves are located in the cylinder head and exhaust valves by the sides of the cylinders. These engines being combinations of L-head and I-head engines are known as F-head engines.

Fig. 4·7. **Working parts of a Wankel Engine.**

1. **Side housing**
2. **Flywheel**
3. **Gas seals**
4. **Rotor**
5. **Centre housing (oval chamber)**
6. **Oil pipe**
7. **Eccentric shaft**
8. **Side housing**
9. **End plate with C.B. point**
10. **Flywheel.**

(9) **With regard to the type of valves used in the engines.**

(*i*) **Sleeve valve engines** containing sleeve valves.

(*ii*) **Rotary valve engines** having rotary valves.

(*iii*) **Poppet valve engines** containing poppet valves.

4. Rotary Combustion Engine. Rotary engine is an invention of the modern age. Dr Felix Wankel of Germany invented this engine in 1954 and Dr. Walter Froede of Germany developed this engine for installation in NSU motor vehicles. This engine was installed in two seater NSU spider sports car for the first time. Several automobile companies in different countries have obtained licences and have started the manufacture of Wankel Rotary Engine.

Wankel Rotary Engine contains a three lobe rotor which rotates eccentrically in an oval chamber. The rotor is attached to the crankshaft through external and internal gears. The rotor lobes seal tightly against the side of oval chamber. The rotor contains oval shaped depressions on its three faces between the lobes. The four cycles of intake, compression, power and exhaust take place simultaneously around the rotor when the engine is working.

The main feature of this engine is that there are three power cycles for each revolution of the rotor and the engine delivers power almost continuously. The other main point in favour of this engine is that all the gases being compressed in three sides of the rotor are ignited by the single spark plug. The main problem in this engine is of sealing between the rotor lobes, the rotor sides and the walls of the oval chamber. This engine works on petrol.

Cycle of operation for this engine is illustrated as under :

Figure I. Inside the oval chamber—

(*i*) Intake starting between A and C.

(*ii*) Compression of fuel and air mixture taking place between A and B.

(*iii*) Power being produced between B and C.

(*iv*) Exhaust being completed between C and A.

Figure II. Inside the oval chamber—

(*i*) Intake continues between A and C.

(*ii*) Compression continues between A and B.

(*iii*) Power development finishing between B and C.

Figure III. Inside the oval chamber—

(*i*) Intake finishing between A and C.

(*ii*) Ignition taking place between A and B.

(*iii*) Exhaust of gases going on between B and C.

Figure IV. Inside the oval chamber—

(*i*) Intake finishing between A and C.

(*ii*) Power being produced between A and B.

(*iii*) Exhaust of gases continues between B and C.

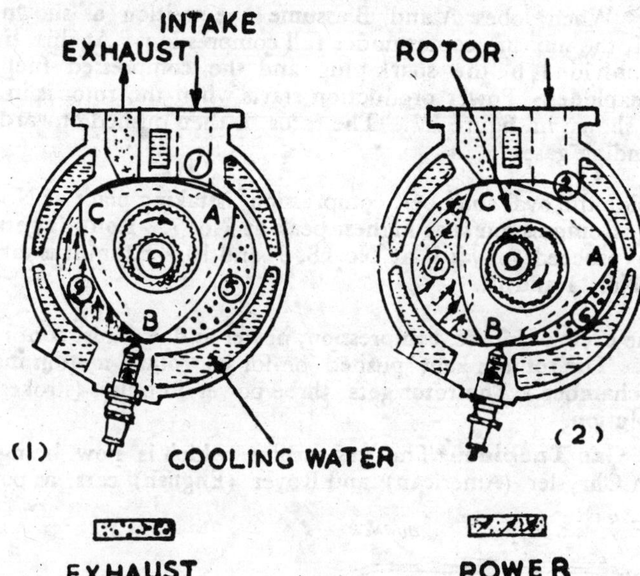

Fig. 4·8. Action in Wankel Rotary Engine : 1st and second stage.

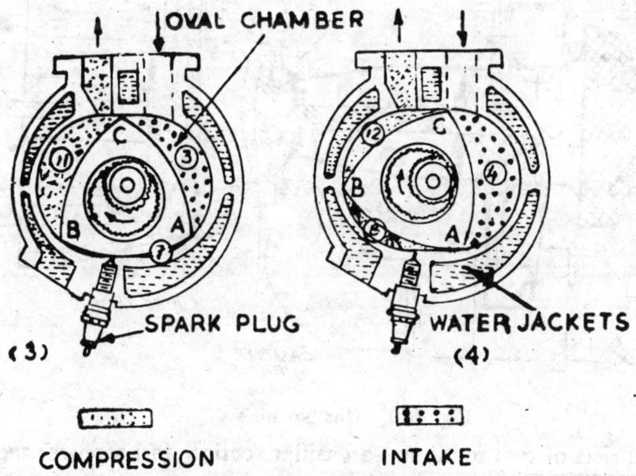

Fig. 4·9. Action in Wankel Rotary Engine : 3rd and 4th stage.

In the first figure intake of fuel air mixture had started between A and C. Intake is at the climax when the rotor reaches the position as shown in figure IV. Incoming fuel air mixture has been shown serialwise in different segments at 1, 2, 3 and 4.

In the first figure, fuel air mixture is being compressed between A and B. When lobes A and B assume the position as shown in figure III, the mixture comes under full compression. At this time, spark is provided by the spark plug and the compressed fuel air mixture explodes. Power production starts when the rotor is in the position shown in figure IV. The rotor is then pushed onward by the expanding gases.

Inside the oval chamber, compression is taking place at No. 5, 6 and 7. Compression is at highest peak at No. 7. Rotor is getting power to proceed onward at No. 8, 9 and 10. Exhaust is taking place at No. 11 and 12.

The cycle of intake, compression, power and exhaust continues this way. The rotor is kept pushed on for its rotary motion inside the oval chamber. The rotor gets three power impulses (strokes) in one revolution.

5. Gas Turbine. The gas turbine which is now being installed in Chrysler (American) and Rover (English) cars, as power

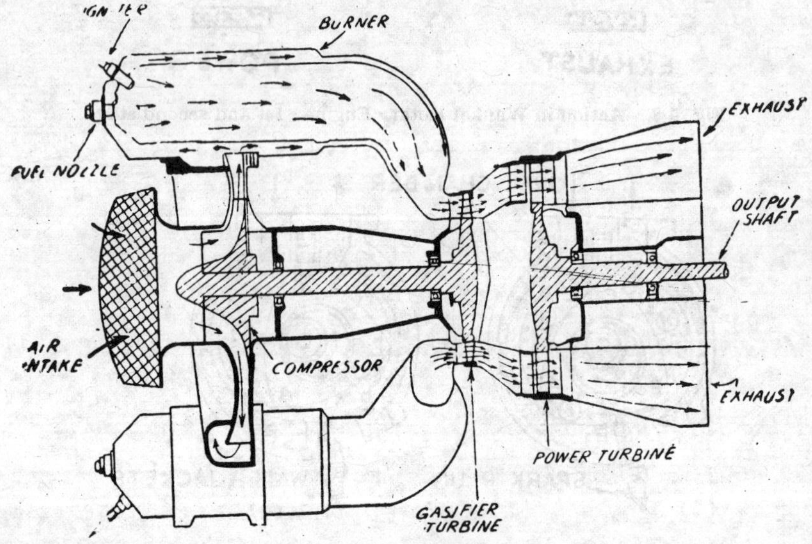

Fig. 4·10. Gas Turbine

plant, consists of two sections—a gasifier section and a power section. In the gasifier section, there is a compressor which contains a rotor having a series of blades around its outer edge. When the rotor rotates, air between the blades is carried around and thrown out by centrifugal force. Air is thus supplied to the burner at relatively high pressure. Fuel, which is gasoline, kerosene or oil, is sprayed into the compressed air. Burning of the fuel increases the pressure further. The gas having high pressure and high temperature then passes through the gasifier nozzle. A series of stationary blades direct the high pressure gas against a series of curved blades on the outer edge of gasifier turbine rotor. The resulting high pressure against the curved blades tends the gasifier turbine rotor to spin at high speed. As the gasifier turbine rotor and the compressor rotor are attached to the same shaft so the compressor rotor is also spun at high speed. As long as fuel is supplied to the burner, sufficient compressed air continues to be supplied.

After the high pressure, high temperature gas leaves the gasifier section and enters into the power turbine where it strikes another series of stationary curved blades. From here the gas is directed against a series of curved blades on the outer edge of the rotor of power turbine. The resulting high pressure against these rotor blades spins the rotor at very high speed up to 50,000 r.p.m. The speed of power turbine rotor is stepped down by a series of transmission gears before the power is transmitted to the road wheels of the vehicle.

6. **Parts of the Conventional Engine (Reciprocating Piston Engine)**. The conventional engine consists of three main parts—cylinder head, cylinder block and oil sump. Multicylinder engine contains an additional main component—induction and exhaust manifold—through which gases flow in and out the engine. All the other engine parts are housed inside the three main portions—head, block and sump.

An engine needs accessories such as fuel pump, carburettor, magneto, distributor, induction coil, oil filter, water pump, fan, self-starter, dynamo, spark plugs, and fuel injection pump and injectors in case of diesel engine. These accessories form the main constituents for the different systems which are required for the working of an engine.

Cylinder Block. It is the main block of the engine which contains cylinders. It provides housing for the crank and cam shafts, and other engine parts. It forms the basic framework of the engine. The other parts of the engine are attached to it or fitted in it. It thus forms the main body for the engine.

The cylinder block is cast in one piece from gray iron alloyed with other metals such as nickel or chromium. Some blocks are made of aluminum in which cast iron or steel cylinder sleeves are employed. Royal Enfield 350 c.c. engine blocks are of this type. In some engines, the cylinder walls are plated with chromium which is very hard metal to reduce wall wear and lengthen their life of service.

Side valve engine blocks contain openings for the valves and valve ports. Watercooled engine blocks are provided with water jackets around the cylinders whereas air-cooled engines contain fins.

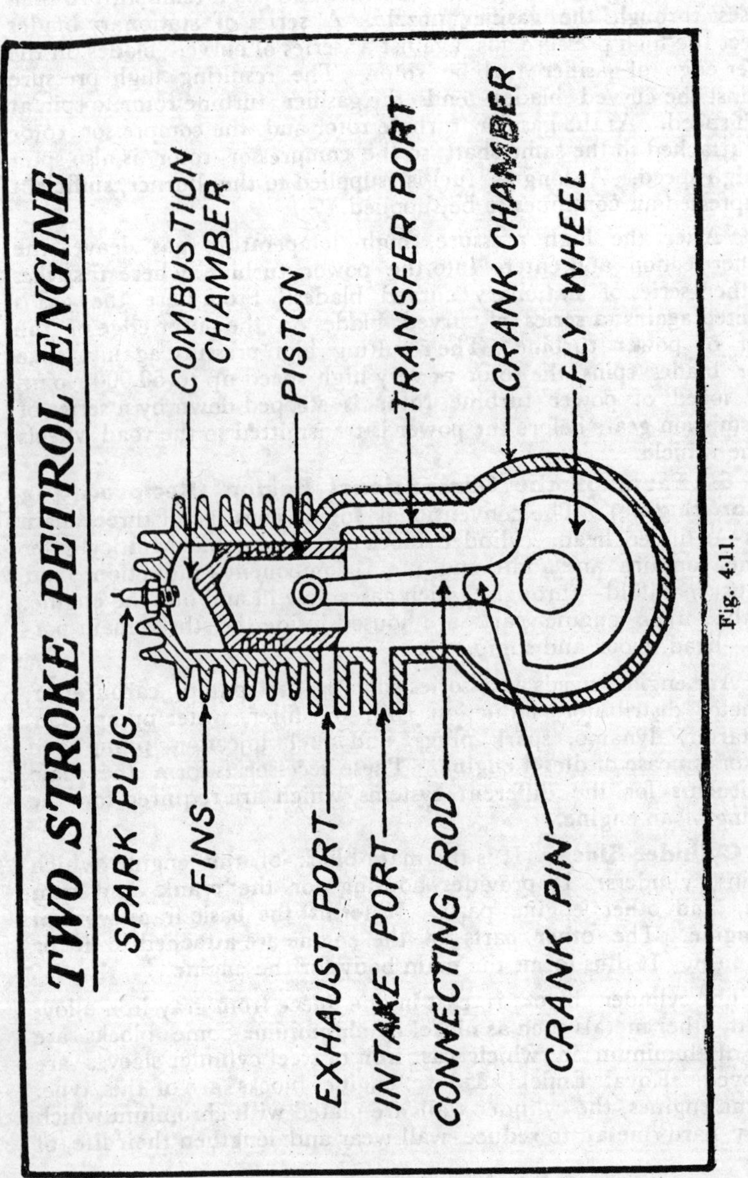

Fig. 4·11.

At the bottom part of the block, crank shaft is held in bearings which are surrounded by the oil sump. At the top of the block, cylinder head is held with it through gasket. The manifolds are further joined to side valve engines.

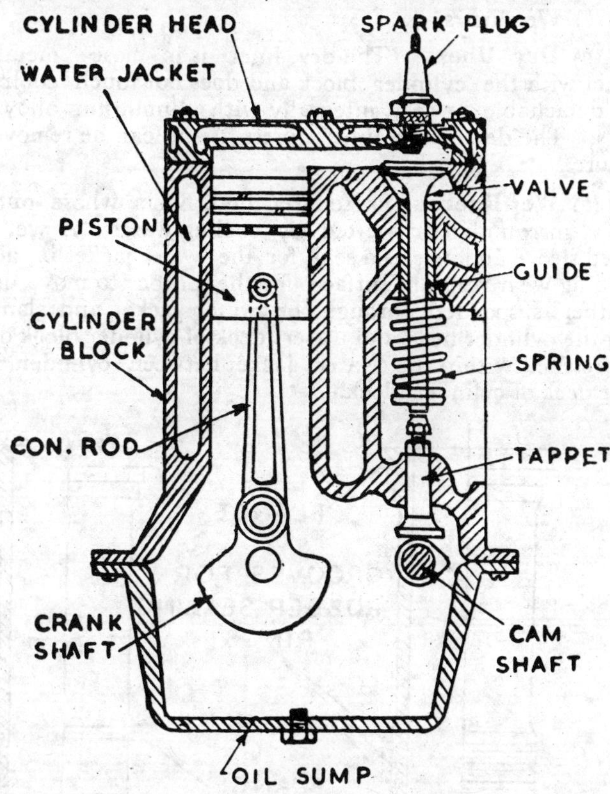

Fig. 4·12. Main parts in an Engine.

The two stroke engine block is little different in construction in some cases where in cylinder aud head is in one piece.

Most of the engine blocks are cast in one piece, whether the block contains cylinder is one row in case of in line engines, two rows in case of V-shape and apposed cylinder engines. In radial or aero engines there are different cylinder barrels which are held with the crank case.

Cylinder liners. A liner is a kind of sleeve which is fitted in the cylinder bore either as original component or as an overhaul feature to obtain standard size bore after using oversize pistons. The liners provide a suitable wear resistant surface for the cylinder bores of aluminium alloy cylinder blocks or simplify the production of cast iron cylinder blocks through open-deck form.

Cylinder liners are of the following two types :—

(*a*) Dry liners.

(*b*) Wet liners.

(*a*) **Dry liners.** The dry liner is in close metal-to-metal contact with the cylinder block and does not touch cooling water. It is detachable or cast integrally with aluminium alloy cylinder blocks. The detachable liner is press fit and can be removed under pressure.

(*b*) **Wet liners.** These are those liners whose outer wall is largely surrounded or wetted by cooling water. A wet liner is a flanged sleeve and acts as a seal for the water jacket in addition to providing wear resistant surface for the piston to move inside. It is either suspended through the water jacket and clamped between the cylinder head and upper deck of cylinder block or held in compression within the water jacket between cylinder head and lower deck of cylinder block.

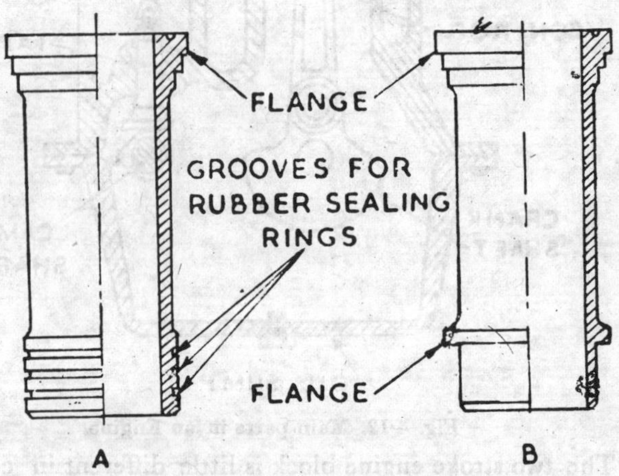

Fig. 4·13. Wet cylinder liners.

(*a*) For closed-deck cylinder blocks.

(*b*) For open-deck cylinder blocks.

Synthetic rubber sealing rings are provided at the lower end of wet liner to seal the leakage of water from the jacket in closed-deck cylinder blocks. A compression sealing gasket is generally used between the flange towards the bottom of liner and its seat in the lower deck of block in case of open-deck cylinder blocks. The cylinder head gasket seals the liners at the top in both closed and open deck cylinder blocks.

Wet liners are commonly used in diesel engines. If the piston and bore clearance exceeds the tolerable limit, liner and piston set is replaced altogether. Dry liners are mostly used in petrol engines. They are thinner than wet liners as gas pressure, piston thrust and impact loadings are resisted by the combined thickness of liner and cylinder. If the cylinder bore exceeds the maximum oversize limit, liner is installed and standard size piston is fitted.

Cylinder liner materials. The materials for cylinder liners vary considerably in view of wear resistance. Cylinder wear is due to two reasons, *viz.*, corrosive wear and abrasive wear. Vehicles moving on unpaved and dusty roads or in deserts are more subjected to abrasive wear than those plying on paved and clean roads. In dusty conditions, the use of cylinder liners having resistance to abrasive wear becomes necessary.

Corrosive wear is due to the leakage of fuel mixture from the combustion chamber to crank chamber which results in acid formation. Combustion products from sulphur are very corrosive. Iron containing about 2% phosphorus has very good corrosion resistance.

Abrasive wear caused by operating vehicles under very dusty conditions, could be reduced by using effective and improved oil and air filters. The other method to reduce wear is to use hard liners or chromium plating of cylinder bores but it is very costly. Comard liners made from chromium plated mild steel tube, have proved exceptionally good in reducing wear rate. Use of chromium plated top piston ring has also been found very effective in wear reduction. Good quality cast iron liners, cast centrifugally, are commonly used in engines due to their low cost and possessing properties to meet ideal requirements.

The cylinder liner materials should possess the following properties :—

(*i*) Should be good resistant to corrosive and abrasive wear.

(*ii*) Should have sufficient strength to withstand the stresses produced by combustion pressures and thermal loads.

(*iii*) The operating surface of linear must be capable to retain adequate quantity of lubricating oil.

(*iv*) Should be tough enough to withstand machining and fitting loads.

Cylinder Head. It is the uppermost part of the engine which covers the cylinders and provides cavity for the ignition of fuel. It is held with the cylinder block by means of a gasket mostly of soft copper and asbestos sheets, to make the joint gas tight so that it may withstand the pressure and heat developed in the combustion chambers.

The cylinder head is usually cast in one piece from iron, iron alloyed with other metals or aluminium. The modern trend is towards aluminium cylinder heads. This is due to the fact that aluminium has the advantage of combining lightness with high heat conductivity due to which it runs cooler.

Cylinder heads for side valve and overhead valves differ much in construction. Cylinder head for side valve engine is comparatively simple as it contains no valves, ports and valve operating mechanism. Side valve engine head contains water jackets for cooling, spark plug holes to which plugs are screwed, combustion chamber cavity, bolt holes through which holding down bolts or studs pass and certain other holes.

Cylinder head for overhead valve engine is more complex as it carries valves and valve operating mechanism. It contains water jackets, valve ports, cavities for combustion chamber, bolt holes and certain other holes. Manifolds are also attached to it.

Cylinder head for air cooled engines is surrounded by fins and contains no water jackets.

Oil Sump. It is the lower bottom part of the engine which carries oil for the lubrication of the engine. It encloses the crank shaft. It is usually made of pressed steel. Cast iron or aluminium sumps are also employed in the engines. They contain fins around them which help in cooling oil contained in the sump.

Manifolds. Manifolds are just complex tubes through which the gases go in and out of the engine. There are two types of manifolds—induction and exhaust. These are used as combined units as well as separate. These are attached over the inlet and exhaust ports in the cylinder head in case of overhead valve engines and cylinder block in case of side valve engines.

The induction manifold provides passage for the flow of fuel air mixture from carburettor to the inlet ports of the engine. The carburettor is installed at the induction manifold.

The exhaust manifold provides passage for the outflow of burnt gases from engine exhaust ports to the exhaust pipe of silencer. In V-8 engines, there are two exhaust manifolds, one for each bank of cylinders. In some V-8 engines, the exhaust manifolds are connected by a crossover exhaust pipe connected to a common muffler and tail pipe as in the case of V-8 Ford engine. In other V-8 engines, each manifold is connected to a separate exhaust pipe and muffler and tail pipe.

In combination unit of induction and exhaust manifold, a *heat control valve* is provided which opens and closes due to heat effect. In such units, the inlet pipe carrying fuel from the carburettor passes through the exhaust manifold so that inlet pipe may be kept warm.

The heat control valve is a flap type valve which is installed in the junction. At low temperatures, it keeps the direct passage in the exhaust manifold closed and the exhaust gases pass out through the bypass which surrounds the inlet pipe. As the temperature rises, the heat control valve begins to open thereby opening the direct passage and closing the bypass.

7. Combustion chamber design. The performance of an engine much depends upon combustion chamber design which includes its geometrical shape, valve arrangement and location of spark plug. The combustion chamber design should ensure smooth combustion, good control over detonation and free flow of gases in and out of the cylinder.

The shape of combustion chamber depends mainly upon the valve arrangement. The various shapes of combustion chambers are *T*-head, *L*-head, *I*-head and *F*-head. The *T*-head design makes it necessary to use lowest compression ratios to prevent knocking with a given fuel. The non-turbulent *L*-head engine is normally of lower compression ratio than *I*-head engine. However, the introduction of high turbulence *L*-head has made these two designs about equal in view of knocking characteristics.

T-head is practically obsolete. *F*-head type has inlet valve in the cylinder head and exhaust valve by the side of cylinder. The primary advantage of *F*-head being that larger inlet valves could be accommodated within a given size of combustion chamber resulting in good volumetric efficiency.

A compact form of combustion chamber is required to obtain smooth combustion and to avoid detonation. The more compact the combustion chamber, the better will be its anti-knock characteristics because the flame travel and combustion time will be shorter. Thus the geometrical shape of the combustion chamber should be such so that the distances over which the flame from spark plug has to travel through unburnt charge are minimum and equal. Owing to lesse· flame travel distances and the combustion time, there is less possibility of delayed burning of remote parts of the charge or *end gases*. If the combustion of end-gases is delayed, high rate of pressure rise along with shock waves, will result in. This shall lead to metallic noise (knocking along) with detonation.

The combustion chamber should be designed as such that the piston head may approach the cylinder head very close at the end of compression stroke to create a highly turbulent motion to the charge during combustion. Thus combustion rate shall be enhanced appreciably and combustion time shall be reduced. The close clearance, if bounded by well cooled walls and piston head, is very effective in retarding the preflame reaction in the end-gases.

The combustion chambers should be fully machined rather than left as cast. Thus there will be accurate control over *clearance volume* of all cylinders and their output shall be equalised for smooth delivery of power.

A certain degree of agitation of fresh charge (fuel air mixture) is helpful in the combustion because it reduces combustion period and any tendency towards detonation. This agitation is promoted by either *compression turbulence* or *induction swirl*.

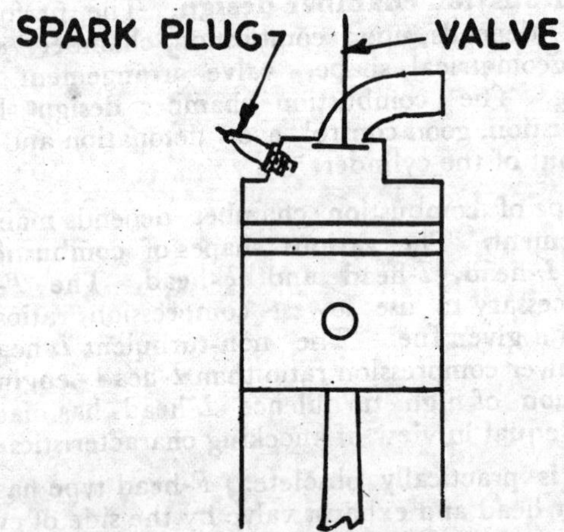

Fig. 4·14. Compression turbulence.

Compression or *"squish" turbulence* is promoted by bringing the piston crown very close to the combustion chamber roof. Thus a *squish* area is formed where the trapped in charge is forcibly ejected across the combustion chamber as the piston reaches at the end of compression stroke. The charge enters the enlarged portion of the chamber resulting in turbulent motion.

Induction swirl is generated by directing the incoming fresh charge to enter the cylinder in a tangential direction. As a result

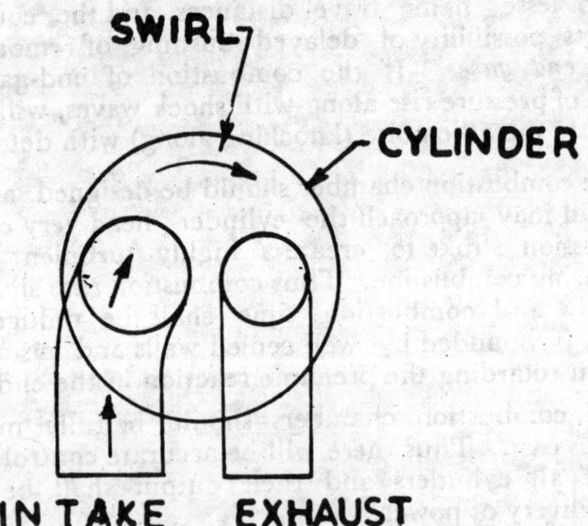

Fig. 4·15. Induction swirl.

of angular momentum imparted to the charge, a helpful vortex action is set up during the intake stroke and continues in the compression stroke.

The knocking characteristics of a combustion chamber can be changed appreciably by changing the location of spark plug. Thus the spark plug should be located at such a place so that the distance of flame travel may be reduced to a minimum. This shall result in the reduction of combustion period and the flame shall pass through the end-gases before detonation.

The spark plug should be located near the hottest spot in the combustion chamber which is the head of exhaust valve. The reason for this location being that during combustion process when end-gases are compressed against the hot exhaust valve, the temperature is raised much higher than compressing it against a cooler surface and the hot mixture burns faster.

A spark plug located in the centre of combustion chamber and affected by high turbulence shall lead to very rapid combustion and result in rough running of the engine. In such a case, the location of spark plug should be changed so that the flame travel is lengthened and combustion time is increased.

Keeping in view combustion chamber design, the power output of an engine can be increased by the following methods :—

(*i*) Decreasing volume of combustion chamber.

(*ii*) Streamlining combustion chamber.

(*iii*) Decreasing combustion time by using more compact combustion chambers.

(*iv*) Using larger valves and ports.

Larger valves and ports will allow more gases to flow in and out the cylinder with less pressure drop resulting in increased volumetric efficiency and hence higher output. By streamlining and using compact combustion chambers, the length of flame travel is reduced which results in combustion time reduction. By decreasing combustion chamber volume, compression ratio is raised which improves engine efficiency and raises power output.

The use of materials of high heat conductivity for combustion chamber walls permit higher compression and greater output without detonation because such material removes heat from the hot spots very rapidly. Owing to this reason aluminium alloy cylinder heads are used in some engines.

In modern engines the following types of combustion chambers are employed :—

(*a*) Hemispherical.

(*b*) Lozenge.

(c) Wedge or ramp.

(d) Piston cavity or bowl in piston.

(a) **Hemispherical**. This type of combustion chamber contains over head inclined valves and provides a high volumetric efficiency. Large and well cooled valves can be accommodated in this

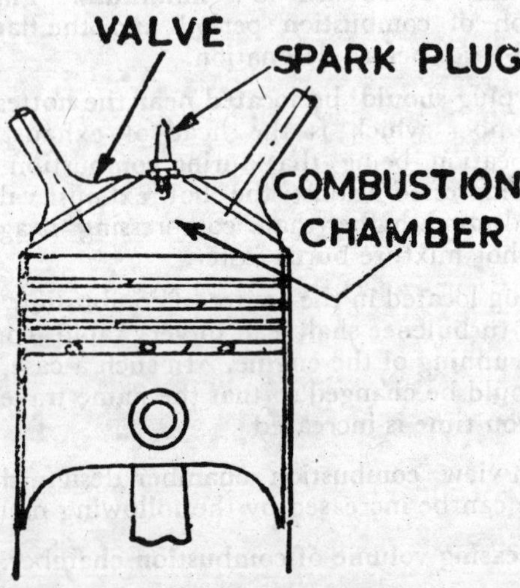

Fig. 4·16. Hemispherical combustion chamber.

design. The valve openings are not masked by the chamber walls due to their radial disposition. The spark plug is located very close to the central axis of the chamber which results in minimizing flame travel distances and good control over detonation.

This design lacks a *squish area* owing to which charge agitation is normally affected by induction swirl. In some designs, the piston dome is offset slightly to promote compression turbulence.

The basic hemispherical combustion chamber has undergone some changes keeping in view the surface area and volume ratio, thermal efficiency and quantity of unburnt fuel.. A domed head piston has been used to reduce clearance volume. A *segmental* form of combustion chamber has been used which assumes a comparatively shallow part-spherical shape with in-line valves rather than inclined ones.

(b) **Lozenge**. This type of combustion chamber resembles with inverted bath tub in which in-line valves are arranged. In order to improve volumetric effciency, the valves are disposed slightly inclined to the cylinder axis, the chamber end walls are recessed to provide wide open areas for valves and the spark plug is located towards the exhaust valve. Thus the combustion process is started in the hottest and completed in the coolest regions of the chamber for good control over detonation.

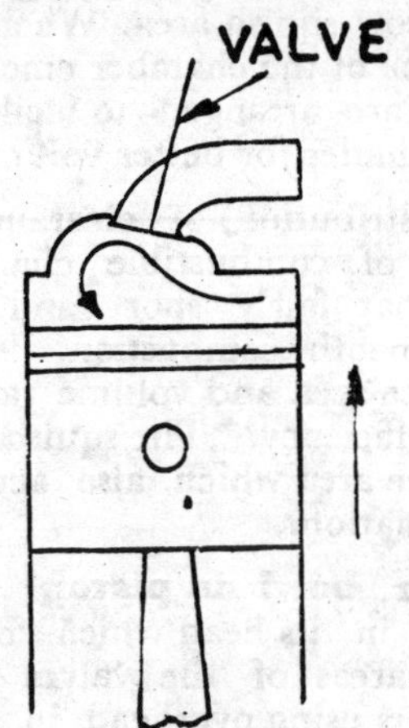

Fig. 4·17. Lozenge combustion chamber.

This design provides a squish area for compression turbulence. The squish area having high ratio of surface area to volume also acts as a *quench area* to cool the remote parts of charge.

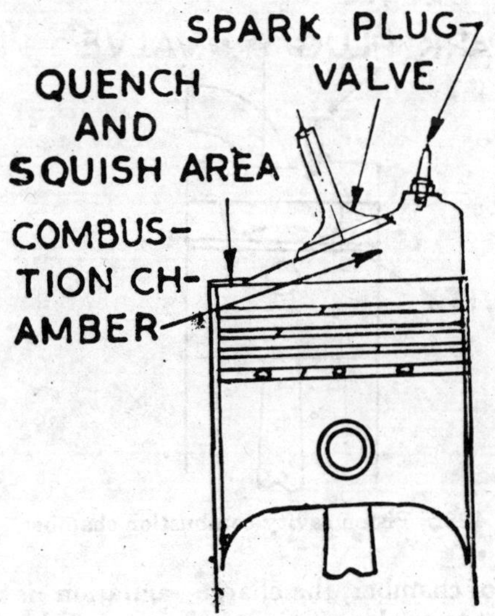

Fig. 4·18. Wedge combustion chamber

(*c*) **Wedge or ramp.** In plan view, this type of chamber is generally of "D" shape, the circular part of which follows the cylinder opening. In cross section, this circular portion tapers uniformly away from the location of spark plug and terminates at the approach to a very narrow squish area. When the piston reaches at top dead centre, 'D' shape of the chamber emerges out. Generally in-line inclined valves are arranged to facilitate the free flow of gases to and from the cylinder for better volumetric efficiency.

In this design, the distribution of clearance volume is such that the greater portion of combustible charge is concentrated around the spark plug so that fairly short and near equal flame travels are achieved for smooth combustion. The sloping roof of the chamber reduces surface area and volume ratio resulting in improvement of thermal efficiency. The squish area covers at least 25% of piston crown surface area which also acts as a quench area for good control over detonation.

(*d*) **Piston cavity or bowl in piston.** In this design, the piston contains a cavity in its head which forms the combustion chamber itself. The open areas of the valves are less masked as compared to other chambers using overhead in line valves. Its flat roof contributes to a low ratio of surface area to volume which results in high thermal efficiency. The spark plug is located close to the central axis of chamber. Thus the geometrical symmetry of this chamber provides minimum and near equal flame travels.

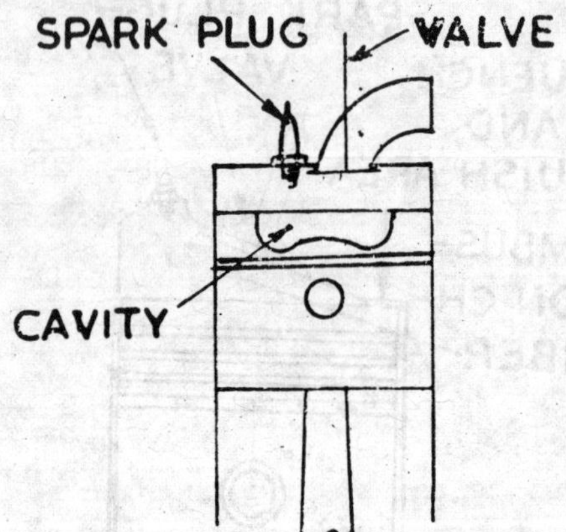

Fig. 4·19. Piston cavity combustion chamber.

In this type of chamber, the charge agitation is initiated by induction swirl due to angled porting and completed by compression turbulence owing to flat topped piston rim coming very closely to the underside of cylinder head and thus providing a squish area. The squish area also acts as a quench area as in other designs.

Combustion chamber design in diesel engines. Diesel engine cycle of operation is different from petrol or spark-ignition. In diesel engine, only air is sucked into the cylinders during intake stroke and compressed to high pressures by upward moving piston during compression stroke. Near compression T. D. C., fuel is injected into the combustion chamber which explodes when it comes in contact with compressed hot air, In spark ignition engine ,a mixture of fuel and air is sucked into the cylinders and compressed into the combustion chamber. The compressed mixture is ignited by spark plug.

In diesel engine, fuel is injected into the combustion chamber against high compression pressures and the combustion depends upon the following factors :—

(*i*) Fine atomization.

(*ii*) High temperature for quick ignition.

(*iii*) High relative velocity between air and fuel particles.

(*iv*) Good mixing of air and fuel particles.

Atomization, penetration and spreading of fuel depends much on injection system. Cylinder bore and stroke, compression ratio and cooling system determine operating temperatures. Mixing depends upon air intake system, injection pattern and combustion chamber design.

The combustion process in a diesel engine is completed in the following four phases :—

(*a*) **First Phase**. During this period some preliminary oxidation takes place but the burning is not sufficient to cause any appreciable rise in temperature or pressure. This is the period between the beginning of injection and beginning of rapid pressure rise and is known as *delay period.* This is a period of *initial* or *ignition lag* and depends upon the following factors :—

(*i*) Type of fuel and its cetane value.

(*ii*) Air inlet pressure.

(*iii*) Compression ratio.

(*iv*) The amount of air in the combustion chamber and its temperature and pressure.

(*v*) Speed and amount of fuel injected.

(*vi*) The amount of atomization taking place.

(*vii*) Type of combustion chamber.

(*viii*) Injection timing.

The higher the temperature and pressure in the combustion chamber, the finer the atomization of fuel particles and higher the turbulence, the shorter will be delay period. Lower the charge temperature, and more advanced the injection timing, the longer will be ignition lag.

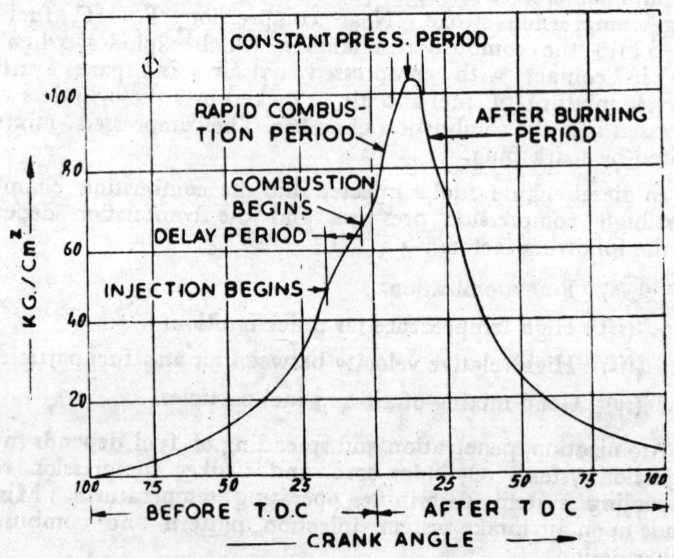

Fig. 4·20. Phases of combustion.

(*b*) **Second phase**. During this period the fuel already injected and the fuel being injected, burns rapidly. There is rapid rise in pressure and temperature. Owing to this fact this period is known as *rapid burning or combustion period.*

The rate of combustion is dependent upon the degree to which the fuel becomes mixed with air through atomization, vaporization, turbulence and the molecular set up of the fuel. The *combustion or diesel knock* in the diesel engine is directly related to the degree of ignition lag and takes place at the end of rapid combustion period *i.e.* second phase.

(*c*) **Third phase.** During this period there is constant pressure rise due to controlled combustion. Injection usually ends just prior to this period. During this phase, the combustion of fuel can be mechanically controlled by controlling the rate of injection.

(*d*) **Fourth phase.** This is the after-burning period during which the piston is rapidly moving downward During this period, the heavier molecules of fuel finally vaporize, mix with oxygen and burn.

The design of combustion chamber plays an important part in the combustion process. In diesel engines, the following types of combustion chambers have been used :—

(*a*) Open combustion chambers.

(*b*) Turbulence chambers.

(*c*) Precombustion or ante-chambers.

(*d*) Air cells

(*e*) Energy cells.

(*a*) **Open combustion chambers.** An open type of chamber is that in which all the air is contained in a single space at the time of injection. It is the simplest form of combustion chamber in which the injection nozzle sprays fuel direct into the combustion chamber. Because of direct injection, the arrangement is known as *open system or direct injection system.*

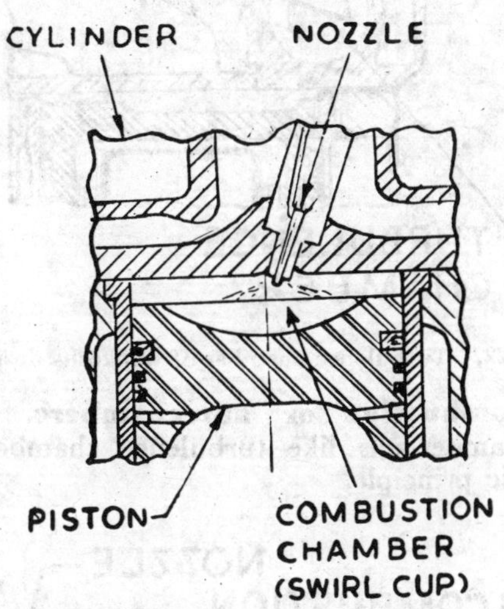

Fig. 4·21. Open combustion chamber with swirl cup.

In this type of chamber, the fuel motion is greater than air upon which the nature of combustion largely depends. In order to bring fuel and air together, the flat head piston has been replaced by concave head piston in modern engines. The deep cut-out swirl cup on the piston crown is being widely used.

Open system combustion chambers are widely used in medium and large-bore engines operating at low and medium speeds.

(*b*) **Turbulence chambers.** In this type of chamber, the fuel is injected into an auxiliary chamber known as turbulence chamber connected with the cylinder by an orifice. The auxiliary chamber houses almost full charge at the end of compression and is nearly spherical in shape. The piston forces air charge into the turbulence chamber and sets up a rapid rotary motion. As the piston rises up, the velocity of air increases through the throat of orifice and reaches at the peak somewhat before T.D.C. Near T.D.C. the injector nozzle injects fuel into the turbulent air currents which results in good mixing during combustion.

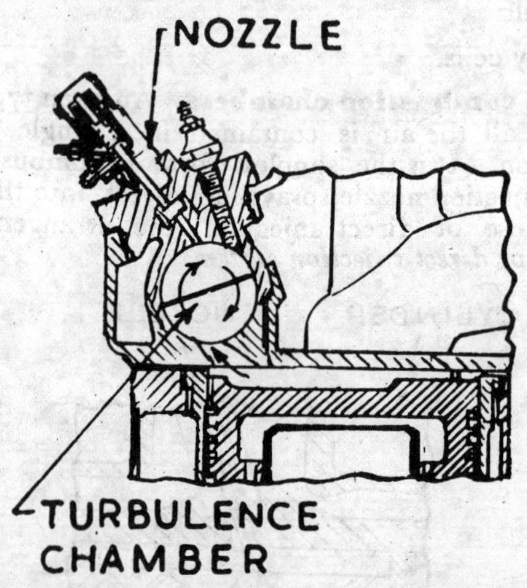

4·22. Turbulence chamber of waukesha design.

(*c*) **Pre-combustion or antechambers.** Although pre-combustion chamber acts like turbulence chambers yet it is quite different in basic principle.

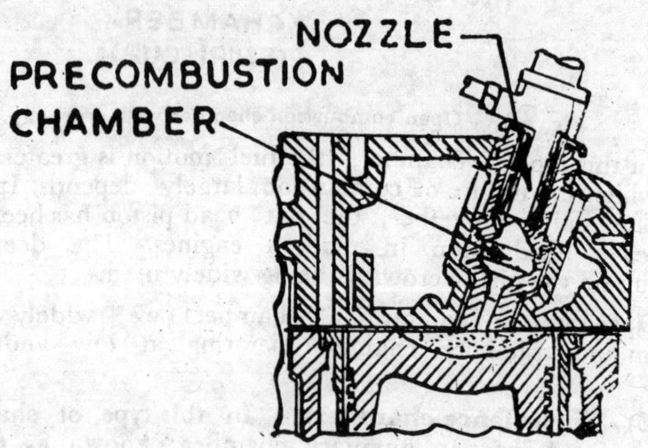

Fig. 4·23. Precombustion chamber.

In pre-combustion chamber system, the auxiliary chamber does not contain the full air charge. Only a part of the charge is contained in the auxiliary chamber and the remainder is kept in the space above piston.

The pre-combustion chamber provides a quiet space in which the fuel starts to burn with an insufficient amount of air. The excessively rich mixture explodes and throws the unburned fuel into the main chamber where it mixes thoroughly with the remaining air.

The pre-combustion chamber shape is more or less cylindrical. At one end of the chamber is located injector nozzle which directs fuel spray at the throat. The chamber is usually water cooled and contains about one-third of air charge after compression.

In Germany, pre-combustion chamber is better known as *Ante-chamber*. In another form precombustion chamber or ante-chamber contains a special cap having holes drilled in it and fitted at the throat. This cap is known as *atomiser*. The chamber is kept at a fairly high temperature and some degree of turbulence is imparted to the air in its passage through connecting holes. Fuel is injected

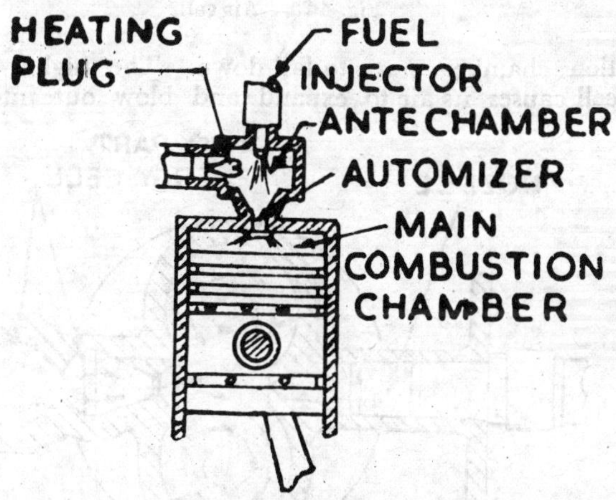

Fig. 4·24. Ante-chamber with atomiser.

into the chamber and ignition of a part of fuel-air mixture results in rapid expansion. About three-quarters of still unburnt fuel particles are driven out through the holes in the atomiser due to expansion in antechamber. Thus a finely divided spray takes place into the cylinder proper where combustion continues.

(*d*) **Air Cells**. An air cell is a space provided in the cylinder head or piston crown in which a large part of air is trapped during compression. In air cell systems, the injector nozzle sprays fuel direct into the main chamber where combustion takes place.

When the piston moves down on its working or power stroke, air pressure is at its maximum in the cell and pressure in the main

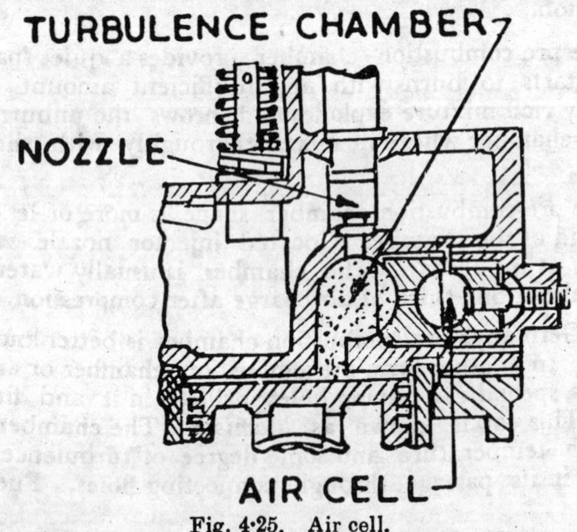

Fig. 4·25. Air cell.

combustion chamber starts to fall down. The higher pressure in the air cell causes its air to expand and blow out into the main

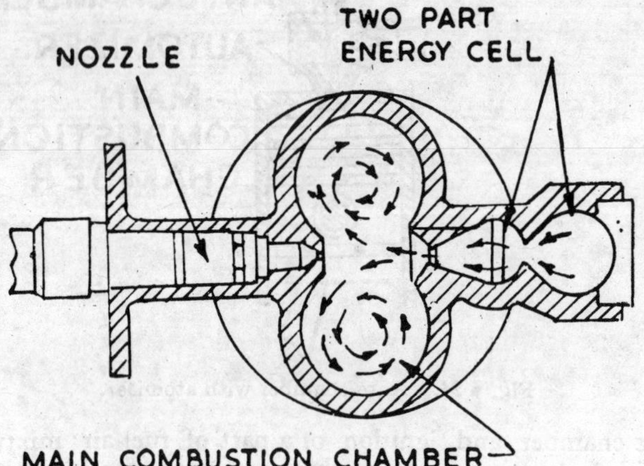

Fig. 4·26. Lanova energy cell.

chamber. Thus an additional turbulence is created and complete combustion of fuel charge is ensured.

As a portion of air remains trapped without combustion in the cell so in improved designs, air cell is used in combination with turbulence or precombustion chamber to obtain better performance.

(*e*) **Enegy cells.** The difference between air cell and energy cell is that fuel is blown into the energy cell where it burns using air in the cell. In air cell system, the cell simply stores and gives up an air charge. The combustion in the energy cell creates a high pressure and greater turbulence and leaves no idle air in the cell.

The popular Lanova energy cell system consists of two rounded spaces cast in the cylinder head. The intake and exhaust valves open into the main combustion chamber. The horizontal injection nozzle points across the narrow section where the lobes join. A two-part energy cell is located opposite to the nozzle. During compression stroke, the piston forces air into the energy cell. Near compression T.D.C., the nozzle sprays fuel across the main chamber in the direction of energy cell mouth. While the fuel charge is passing across the centre of main chamber, nearabout half the fuel mixes with hot air and burns at once. The remaining fuel enters the energy cell and starts to burn there. At this point, the cell pressure rises rapidly, tending the combustion products to flow back into the main combustion chamber at a high velocity. As a result of this, a sharp swirling movement of fuel and air is set up in each lobe of main chamber, promoting final mixing of fuel and air and ensuring complete combustion. The two restricted openings of energy cell control the time and rate of expulsion of blast from energy cell into main combustion chamber.

The energy-cell combustion systems fulfil the requirements of high speed engines and give high power output without high excessive pressures in the main combustion chamber.

8. Working Parts of the Conventional Engine. Working parts are those parts which move in the engine, whether their movement is reciprocating (linear) or rotary (circular). Piston reciprocates whereas the crankshaft rotates in the engine.

The working parts of the conventional engine can be grouped as below according to their attachment relation :—

(*a*) **Group I :**

(*i*) Crankshaft.

(*ii*) Vibration Damper or Harmonic Balancer.

(*iii*) Flywheel.

(*iv*) Connecting Rod.

(*v*) Piston.

(*iv*) Piston Rings.

(*vii*) Piston Pins or Gudgeon Pins.

(*b*) **Group No. II :**

(*i*) Camshaft.

(*ii*) Tappet or Follower.

(*iii*) Valve.

(*iv*) Valve Spring and Retainer.

(*v*) Push Rod.
(*vi*) Rocker.
(*vii*) Rocker Shaft and Spring.
} In case of overhead valves.

(*c*) **Group No. III** :

(*i*) Timing Gears.

(*ii*) Chains and Sprockets.

(*d*) **Group No. IV** :

(*i*) Oil Pump.

(*ii*) Oil Strainer.

(*iii*) Oil Relief Valve.

(*a*) **Group I** :

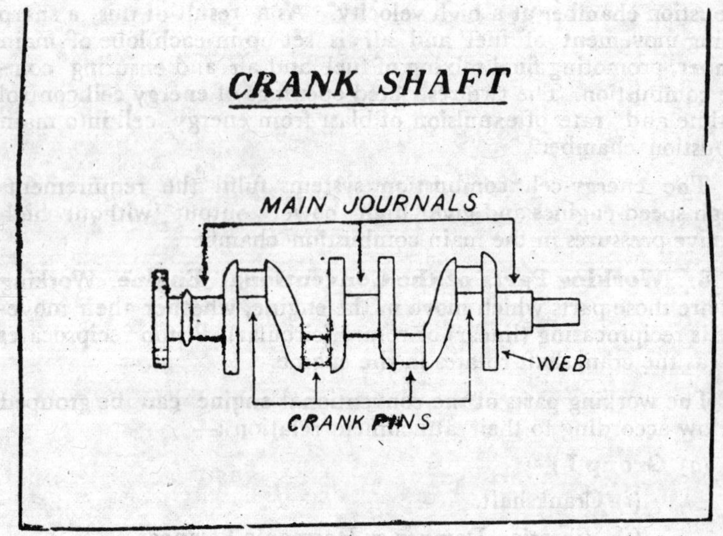

Fig. 4·27

(*i*) **Crankshaft.** It is the main shaft in the engine to which all the other working parts are directly or indirectly related. It is known as backbone of the engine due to its importance. Through the connecting rod, it converts reciprocating motion of the piston into rotary motion.

The crankshaft is a one piece casting or forging of heat treated alloy steel having considerable strength. It should possess the following qualities :

(*i*) It must be strong enough to take the downward thrusts of the pistons during the power strokes without excessive distortion.

(*ii*) It must be well balanced to eliminate undue vibration resulting from the weight of the offset cranks.

The crankshafts are provided with counterweights opposite to the cranks for keeping them in balance. These contain drilled passages through which oil flows from the main bearings to the connecting rod bearings for lubrication purposes.

A flywheel is attached to the rear end of the crankshaft in order to keep it in the regular motion as the flow of power from the engine cylinders is not smooth. The flow of more or less power to the crankshaft at different times tends it to speed up and then slow down. The flywheel absorbs power as the crankshaft tries to speed up and gives back power as the crankshaft tries to slow down. Thus the inertia of flywheel tends to keep the crankshaft turning at the constant speed.

At the front end of the crankshaft, the following parts are carried :

(*i*) The gear or sprocket which drives the camshaft.

(*ii*) The vibration damper.

(*iii*) The pulley which drives dynamo, water pump and fan.

The crankshaft is held with the engine block through split type plain bearings. The bearing back is usually of steel to which is attached the lining. The lining is a combination of several metals such as copper, lead, tin, mercury, antimony, cadmium, silver etc.

(*ii*) **Flywheel.** The *flywheel* stores energy during power stroke and delivers it out during the other strokes to keep the crankshaft moving at the uniform speed.

A ring gear is mounted over the flywheel which meshes with the self-starter drive pinion for cranking the engine. The rear face of the flywheel also serves as driving member of the clutch.

Thus the flywheel plays the following part in an engine.

(*i*) It tends to keep the crankshaft to move at uniform speed.

(*ii*) It acts as a carrier for the ring gear meant for starting motor drive.

(*iii*) It acts as a driving member of the clutch.

(*iii*) **Vibration Damper or Harmonic Balancer.** It damps out torsional vibration in the crankshaft set up by the power impulses. It is usually fitted on the front end of the crankshaft and the fan-belt pulley is incorporated into it.

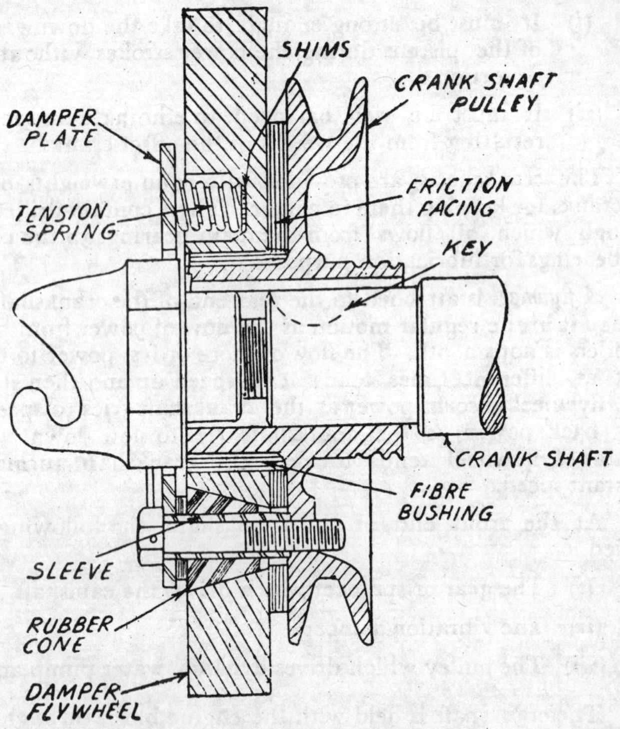

Fig. 4·28. Vibration damper.

A typical damper contains two parts—a small damper flywheel and the pulley which are bonded to each other by a rubber insert. The pulley is fitted at the front end of the crankshaft. As the crankshaft tends to speed up or slow down, the damper flywheel imposes a dragging effect due to its inertia. Due to this effect, rubber insert flexes to a slight degree and the pulley and crankshaft are tended to run at a constant speed. This action tends to check the twist, untwist or torsional vibration of the crankshaft resulting in the relief from stresses.

(*iv*) **Connecting rod**. It connects piston with the crankshaft. Power impulses produced during combustion are transmitted to the crankshaft by this rod. As this rod is to carry power thrusts from the piston to crank-pin so it should possess the following qualities :

(*i*) It should be strong and rigid enough to carry the power impulses.

(*ii*) It should be light in weight to minimize vibration and bearing loads.

The connecting rod contains bearings in its big and small end. The small end usually contains a bush through which gudgeon pin passes for making connection with the piston. It is connected with the crank pin through its big end which is of split type. The cap of the big end is held with the connecting rod by means of bolts and nuts or studs and nuts having locking arrangement.

Some connecting rods contain a direct drilled hole from big end to small end for providing passage for lubricating oil. In some designs there is one small hole each at big end on the junction to form jet or nozzle for the oil to be sprayed out at piston bosses, gudgeon pin and cylinder walls.

The connecting rods are mostly of I-section and are made of forged steel.

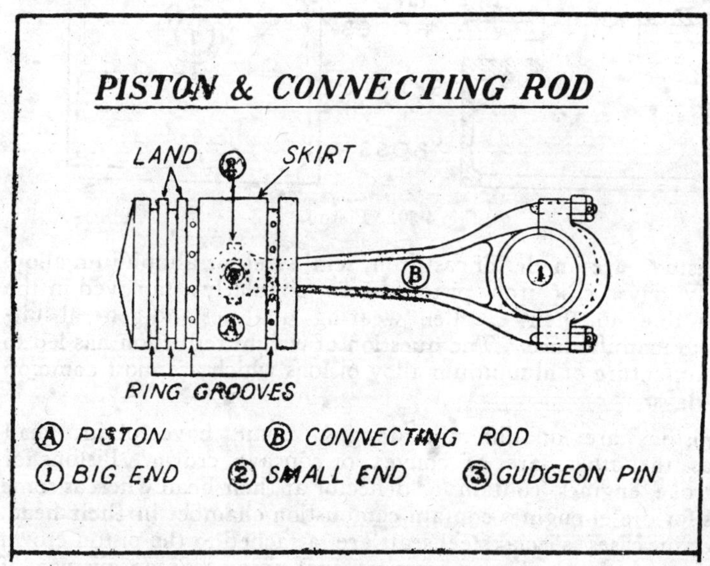

Fig. 4·29.

(*v*) **Piston**. It is a cylindrical plug or bucket which moves up and down in the engine cylinder. It serves the following purposes :

(*i*) It serves as a movable gas tight plug to keep the gases inside the cylinder.

(*ii*) It transmits the force of explosion to the connecting rod.

(*iii*) It acts as a guide and bearing to the small end of connecting rod and bears the side thrust when the rod is oblique.

(*iv*) It acts as a carrier for piston rings.

The piston looks like a bucket, which carries rings at the upper part to provide a good seal between the cylinder wall. Inside the open end, connection of the connecting rod is made with it through the gudgeon pin. The lower part of ring grooves is known as skirt which provides a bearing and guiding surface in contact with the cylinder wall. The top of the piston is called head or crown. The projections inside the open end, through which gudgeon pin passes are known as bosses as shown in the diagram.

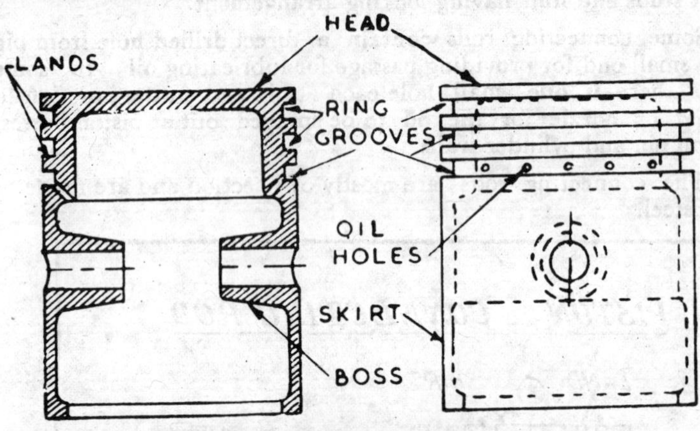

Fig. 4·30. Piston.

Pistons are made of cast iron, semi-steel or aluminium alloy. In early days, cast iron pistons were popularly employed in the engines due to their excellent wearing qualities and general suitability to manufacture. The question of weight reduction has led to the manufacture of aluminium alloy pistons which are most common now-a-days.

Pistons are of different designs. Some have plain head whereas the others are of convex or concave crown. Pistons for two stroke engines contain a deflector at their head whereas some pistons for diesel engines contain combustion chamber in their head. In certain cases special steel seats are attached to the piston crown to withstand the high pressures of fuel spray. Some pistons of overhead valve engines contain depressions in their crown to allow the free movement of the valves downward.

According to the skirt constructions, the pistons can be divided into the following two classes :

(*i*) Solid Skirt Pistons

(*ii*) Split Skirt Pistons.

Solid skirt pistons, as the name indicates, have their skirt without any break whereas the split skirt pistons contain 'T', 'I' or 'C' type slots in their skirts to allow for expansion.

In some heavy duty pistons an oil ring is provided below the gudgeon pin holes.

(*vi*) **Piston Rings**. Ring is a type of bangle which is cut at one point to provide two ends. Rings are fitted in the piston ring grooves to provide a seal between the piston and cylinder wall to avoid the escape of gases from the combustion chamber to crank chamber.

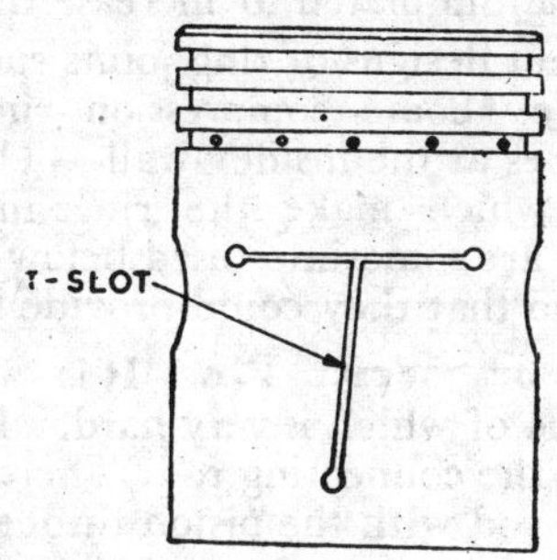

Fig. 4·31. Split Skirt Piston.

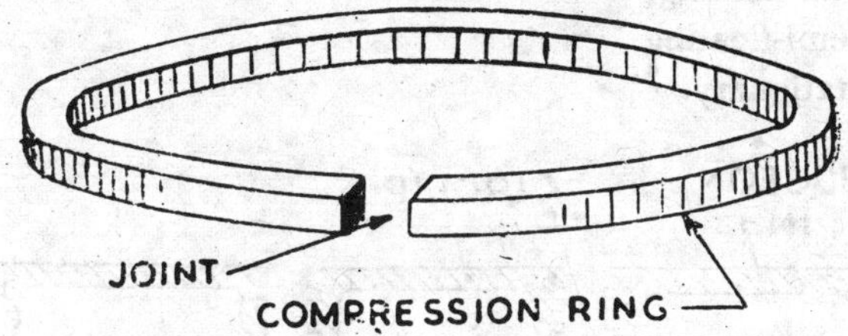

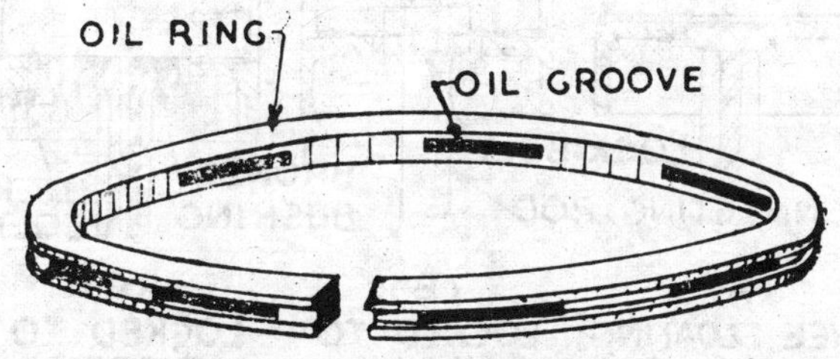

Fig. 4·32. Piston rings.

The piston rings are of the following types :

(*i*) Compression rings,

(*ii*) Oil rings

The compression rings seal the gases in the combustion chamber whereas the oil rings scrape off excessive oil from the cylinder wall which is returned to the oil sump. The oil rings control the flow of oil towards the combustion chamber. The oil control rings as well as the grooves in the pistons, contain holes or slots through which the scraped off oil goes back to the crank chamber.

The rings are usually made of cast iron. Compression rings of some engines are chromium plated to increase their life.

There are different designs of ring joints such as butt, angled, lapped or a sealed type. Some compression rings contain vertical lines at shorter distances at the inside wall. These depressions are known as *hammerings* which make the rings more elastic. Spring steel strip helper rings are sometimes used below the main rings to keep them expanded so that they could provide better seal.

(*vii*) **Gudgeon or Wrist Pin.** It is a hollow pin made of steel, the upper surface of which is very hard. It holds the piston with the small end of the connecting rod. There are different ways of holding connecting rod with the piston through the gudgeon pin. Owing to this holding down arrangement, the piston pins are classified as below :—

(*a*) Full floating,
(*b*) Semi-floating,
(*c*) Stationary.

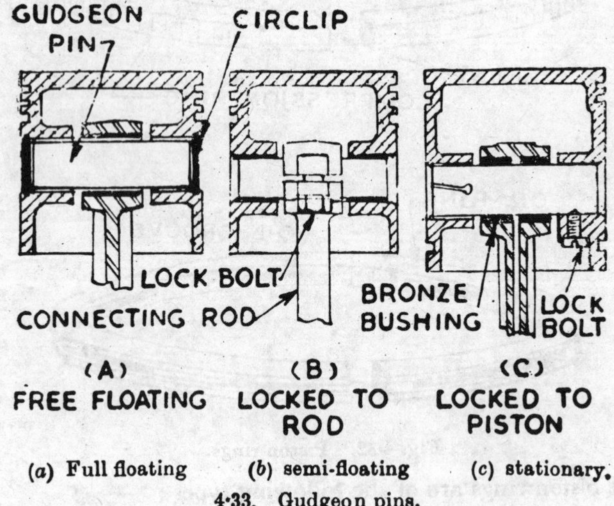

(*a*) Full floating (*b*) semi-floating (*c*) stationary.

4·33. Gudgeon pins.

(*a*) **Full floating.** It is that type of gudgeon pin which is free to rotate inside the small end of connecting rod as well as in the piston bosses. It is held in the piston by means of circlips which keep the pin inside the bosses.

(*b*) **Semi-floating.** It is that type of pin which is fixed in the small end of connecting rod and is held tight by means of a cotter bolt. It is free to rotate in the piston bosses.

(*c*) **Stationary.** This type of gudgeon pin is bolted to the piston and cannot rotate. The connecting rod moves around it.

(*b*) **Group II**

(*i*) **Camshaft.** It is that type of shaft which contains cams to change rotary motion into straight line or linear motion. The camshaft contains so many cams as the number of valves in the engine. An additional eccentric is provided to drive the fuel pump. The camshaft causes the intake and exhaust valves to open. A gear is also provided at the camshaft to give drive to the distributor or oil pump as the case may be.

The shaft rotates inside the plain bearings and is driven by the crankshaft through gears or sprockets and chain. In four stroke engine the camshaft gear or sprocket contains double the teeth than the gears or sprockets on the crankshaft as the camshaft is required to rotate at half the crankshaft speed.

The camshaft is usually located in the crank chamber. In case of V-shape engine, it is installed between the two banks of cylinders. In case of overheat camshaft engines, it is located at the cylinder head.

(*ii*) **Tappet or Follower.** It is a circular steel barrel closed at both ends which acts as a lifter for the valve. It is located between the cam and valve or push rod. It rises up due to cam pressure and lifts the valve off its seat.

The tappets can be divided into the following classes :

(*a*) Non-adjustable,

(*b*) Adjustable,

(*c*) Self-adjustable,

(*d*) Roller type.

(*a*) **Non-adjustable.** Tappets whose lengths are fixed and cannot be adjusted to reduce or increase clearance between valve stem and the tappet are known as non-adjustable tappets. Such tappets are applicable in Ford engines.

(*b*) **Adjustable.** The tappets which could be adjusted to increase or decrease tappet clearance are called adjustable tappets. A screw having lock nut is provided at the head of adjustable tappet, which could be screwed up or down to adjust clearance between the tappet and the valve stem or rocker arm and valve stem as the case may be. After the required adjustment, the screw is locked with the tappet by means of the lock nut. These types of tappets are most commonly used in the engines.

(*c*) **Self-adjustable.** These types of tappets are adjusted automatically to suit the desired needs. Cadillac engine employs these types of tappets.

Self-adjustable tappets are adjusted automatically by the oil pressure from the pressure feed lubrication system of the engine. Following are the main parts of this type of tappet :

(*i*) Body,

(*ii*) Plunger,

(*iii*) Spring,

(*iv*) One way check valve.

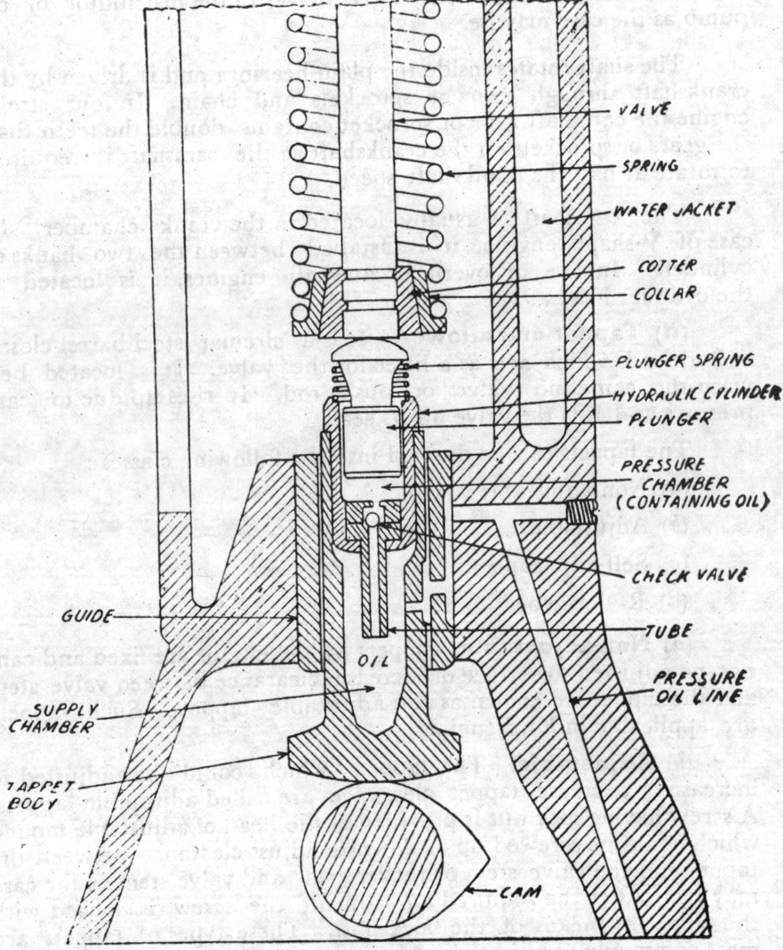

Fig. 4·34. Self-adjustable tappet arrangement.

The upper part of the tappet acts as a cylinder which contains valve and plunger. There is a spring inside the plunger head and tapped body, which keeps the plunger raised up, making its contact with the valve stem end. Below the plunger is housed the one way check valve. There is a hole at the lower part of the cylinder through which oil enters into the cylinder from the reservoir. The oil reservoir is located on one side of the tappet which keeps oil under pressure at all times. The check valve opens due to oil pressure and the oil enters into the upper part of the cylinder. This oil keeps the plunger raised up and acts as a cushion to absorb any noise.

When the pressure of cam is released from the tappet, it falls down due to which the oil hole in the tappet coincides with the oil reservoir. During this period the deficiency of oil between the plunger and the cylinder is made up.

(*d*) **Roller type.** If a roller is employed in the foot of non-adjustable tapper, it shall become roller type tappet. This type of tappet is prevented from rotating by the engagement of the roller in a cross slot in the guide.

(*iii*) **Valves.** The valves act as gates in the engine through which gases go in and out. There is one way traffic for the gases. The fresh gases enter into the combustion chamber through the inlet valves and used gases go out of the combustion chamber through the exhaust valves. Usually there is one intake and one exhaust valve for each cylinder but in high performance engines two intake and two exhaust valves per cylinder are employed. Inlet and exhaust valves differ in size and even in material. The exhaust valve is subject to more heat and hence it should be capable to withstand high temperatures. Several types of valves have been used in the engines. According to their construction, the following are the various types of valves: (*a*) Sleeve Valves, (*b*) Rotary Valves, and (*c*) Poppet or mushroom valves.

(*a*) **Sleeve Valves.** Sleeve valve is a long barrel which contains holes. This valve is located in the engine in such a way that its holes may face towards combustion chamber on one side and induction and exhaust manifold holes on the other side. When the holes of the valve coincide with the induction manifold holes and combustion chamber, fresh charge of fuel enters into the combustion chamber. During this occasion, exhaust manifold holes are not coinciding with the combustion chamber on the other side. When the sleeve is moved up and down, its holes coincide either with induction manifold or with exhaust manifold holes and on the other side with the combustion chamber. When the exhaust manifold holes come into coincidence, the used gases go out of the engine.

Sleeve valves open and close by sliding as explained above.

(*b*) **Rotary Valves.** Rotary valves open and close due to their rotation. These are of the following types:

(*i*) Cylindrical,
(*ii*) Disc type.

The cylindrical rotary valve resembles with that of a sleeve valve. Sleeve valve opens and closes by sliding it up and down whereas the cylindrical rotary valve opens and closes down when it is rotated. This valve rotates in a cylindrical housing, the gases entering and leaving the valve in an axial direction.

Disc type rotary valves are mostly used in two stroke engines. This type of valve contains a disc having holes. In some cases, a part of the disc is removed. The disc is rotated by the operating mechanism to open and close the valve. When the hole in the disc or its cut away part coincides with the hole in the manifold, the gases come in or go out of the engine.

This type of valve is applicable in two stroke Pearl Yamaha Scooter engine.

(*c*) **Poppet or mushroom valves**. These types of valves are most commonly used in the conventional engines. When this valve opens and closes down, it creates a popping noise. That is why

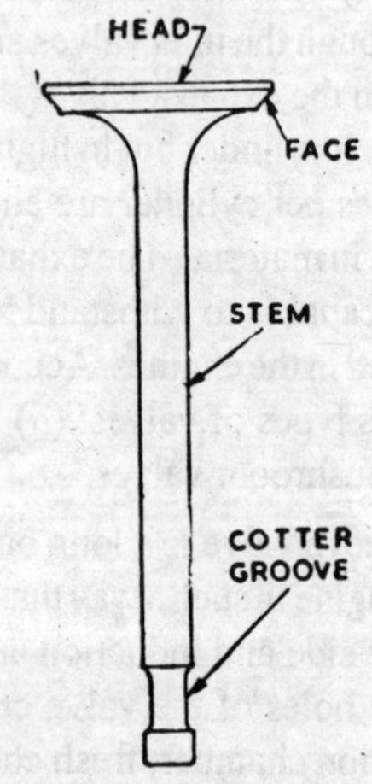

Fig. 4·35. Poppet Valve.

this mushroom type valve is known as poppet valve. The various parts of this valve have been shown in this figure. The valve stem moves up and down in the valve guide which is fixed in the engine. The valve seats at the valve port. In opens when it rises up the seat. For closing down, it falls over the seat.

Holding Down Arrangement. The poppet valve is held with the engine by means of one or two springs, collar cotters or *spring retainer*. Three types of valve spring retainers are used in different engines. These are horse shoe type, pin type and conical type. *Conical valve spring retainers* which are in two halves, are mostly used in the engines. In *pin type* valve spring retainer, a steel pin is pushed into the hole provided in the lower part of valve stem. *Horse shoe type retainer* is pushed into the groove at the lower part of valve stem.

Operating arrangement. The poppet valves are operated by the camshaft through tappets in case of side valve engines. In case of overhead valve engines .these are operated by the cams on the camshaft through tappets, push rods and rocker arms as shown

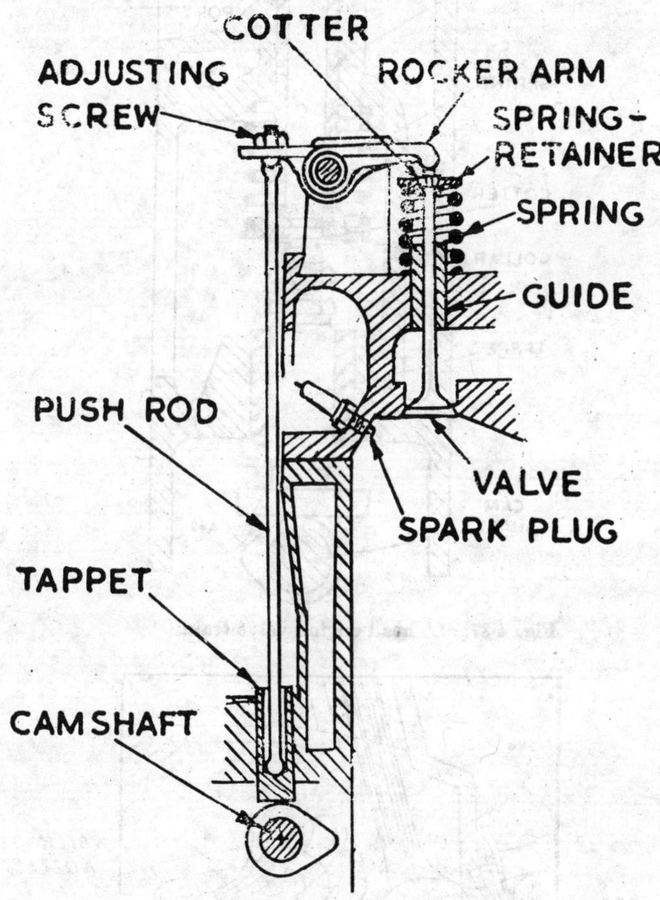

Fig. 4·36. Overhead valve (O.H.V.) operation.

in the diagram. The rocker arms are fitted over the shafts at the cylinder head.

Cooling arrangement. There is a problem to cool down poppet valves as their heads are inside the combustion chamber.

Exhaust valves are under more temperature effects. One way to make exhaust valves worthy to face the heat effects is to use improved

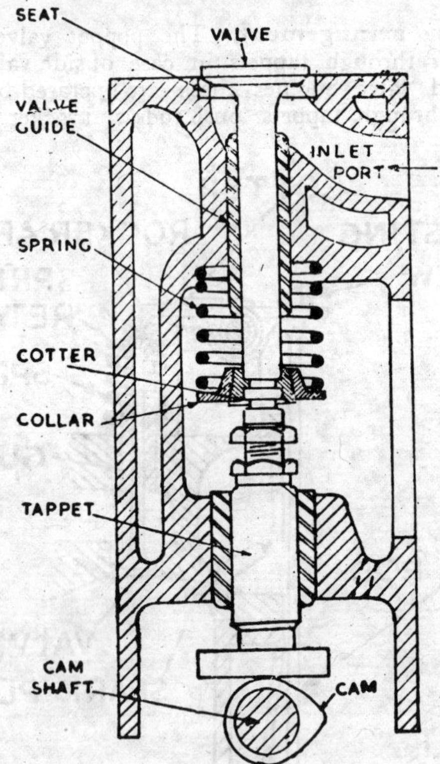

Fig. 4·37. 'L' head engine valve train.

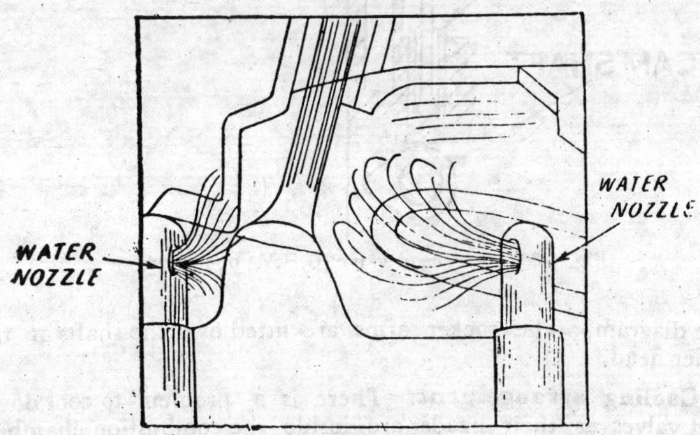

Fig. 4·38. Water nozzles to cool valve seats.

material in their construction. The other ways are to provide effective cooling arrangements for them. In some cases, jets of cooled water are provided inside the water jackets. These jets spray water around the exhaust valve ports thereby accelerating their cooling. In some cases sodium is used for cooling inside the valves which are made hollow. Such valves are known as sodium cooled valves.

Sodium cooled valve keeps a hollow stem which is partly filled with metallic sodium. Since the melting point of sodium is

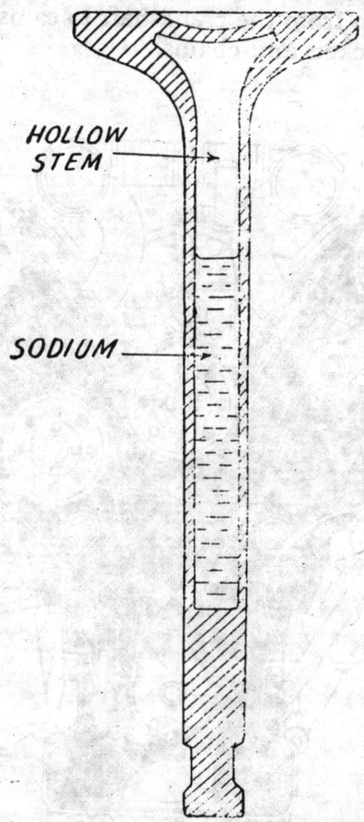

Fig. 4·39. Sodium-cooled valve.

208°F, so at operating temperatures, it is liquid. When the valve moves up and down, the sodium is thrown upward into the hotter part of the valve. Sodium absorbs heat and gives it up to the cooler stem as it falls down into the stem again. This circulation cools the valve head and the valve, and hence it runs cooler resulting in longer life.

(c) Group III

Timing Gears, Sprokets and Chains. These are the gears or sprockets and chains through which drive is given from crankshaft to camshaft. These are known as timing gears or sprockets and chains as the case may be, because these set the timing of the engine. One each gear or sprocket is fitted at the crankshaft and camshaft. Either the drive is through the gears or sprockets and chain or chains. The movement of the piston which is fitted by the rotation of crankshaft, is directly related to the opening and closing of the valves affected by the rotation of camshaft, in the operation of the engine. Their relationship is set up by fixing camshaft gear or sprocket chain with the crankshaft gear or sprocket. This is known as *timing* and the gears or sprockets and chain or chains, through which the drive is given from the crankshaft to camshaft are known as timing gears or sprockets and chains.

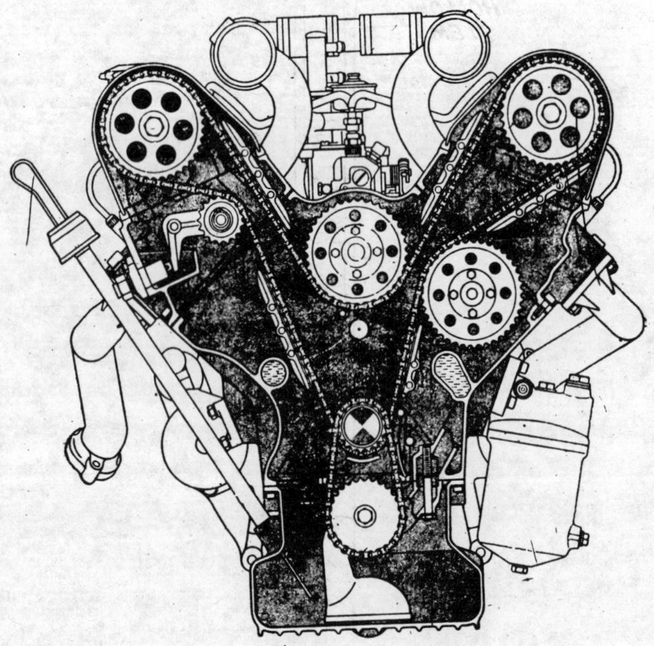

Fig. 4·40. Chain drive of the V 8 engine. (Mercedes Benz 600).

(*d*) Group IV

Oil Pump, oil strainer and oil relief valve. These parts relate to the lubrication system of the engine. Oil pump is fitted in the engine to feed oil to the pressure feed lubricating system. The strainer is attached before the inlet of oil pump so that oil is strained before it goes inside the pump. The oil relief valve is provided in the oil pump or pressure lubricating system to release oil when more pressure is built up in the system.

Engine Mountings. The engine is subject to complex vibration effects. As a result of vibrations, the following free motions are produced :

(*a*) Bounce and yaw about a vertical axis.

(*b*) Fore and aft movement and roll about a horizontal longitudinal axis.

(*c*) Sideways shake and pitch about a horizontal lateral axis.

The three inertia axes of the above six motions—(*i*) bounce, (*ii*) yaw, (*iii*) fore and aft movement, (*iv*) roll, (*v*) sideways shake (*vi*) pitch, intersect at the centre of gravity of the engine as shown in the diagram.

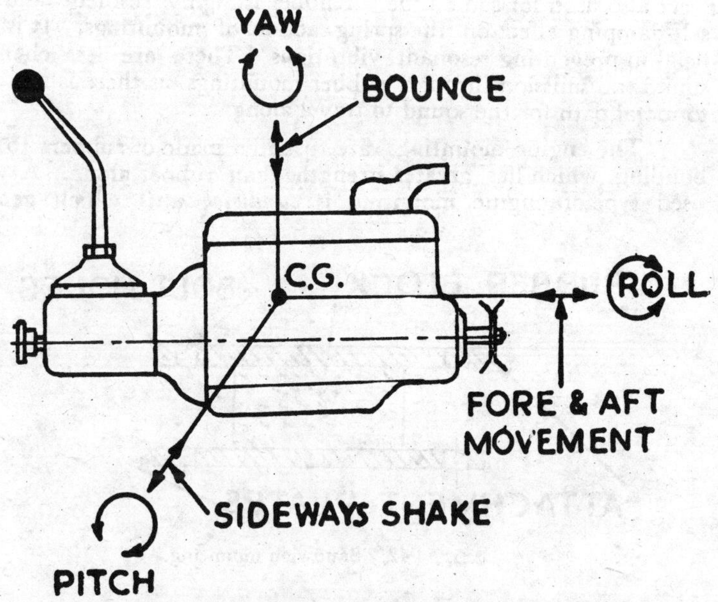

Fig. 4·41. Vibratory motions of engine.

This warrants a ***resilient anti-vibration system*** for mounting the engine so that the vibratory forces are reduced to the relatively small spring forces transmitted by the mountings themselves.

The mountings support the static load of the engine unit and isolate the chassis frame from engine vibrations. They also insulate the engine against vibrations of vehicle mechanism and structure while moving on road. They minimise undesirable movements of the engine and under all working conditions and reduce vehicle shake.

The engine mountings must be capable—

(*i*) to carry static or dead weight of the engine unit,

(*ii*) to withstand torque reactions,

(*iii*) to bear fore and aft thrust when the propeller shaft is pushed back and forth due to road irregularities,

(*iv*) to lower the natural frequency of the engine and the mountings to well below 0·7 times the engine frequency in whatever direction it occurs, and

(*v*) to provide sufficient flexibility to save the engine from any kind of stress when the chassis frame is distorted in an accident.

By virtue of high ratio of deflection to load, rubber is used as spring medium in modern engine mountings although coil springs were also used for some time. Rubber is highly resilient and has a self-damping effect on the spring action of mountings. It is beneficial in preventing resonant vibrations. There are less chances of sound transmission through rubber mountings as there is no metal-to-metal path for the sound to travel along.

The engine mountings are usually made of rubber to metal bonding, which has greater strength than rubber alone. A widely used type of engine mounting is *sandwich unit* which generally

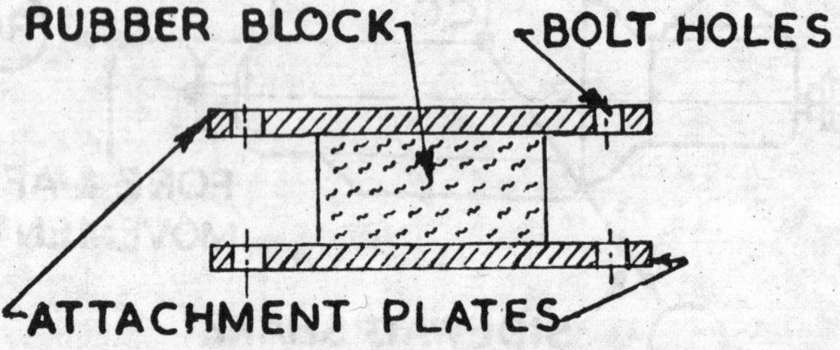

Fig. 4·42. Sandwich mounting.

consists of either a rectangular or a circular block of rubber bonded to metal attachment plates. In order to provide a combination of shear and compression loading, the limbs of U-shaped carriers are inclined in some designs.

Another type of engine mounting is cylindrical unit. The earlier form consisted of upper and lower flanged rubber elements which were subject to only compression loading. The improved type consisted of a conical rubber bushing bonded both to a central sleeve and an outer attachment casing. This mounting may be loaded partly in shear and partly in compression.

In early vehicles, the crankcase was usually cast with four integral supporting arms which were rigidly bolted to the chassis frame. The object being that the engine may act as a sturdy cross member to resist torsion at the front end of chassis frame. Later on flexible engine mountings were introduced to prevent the trans-

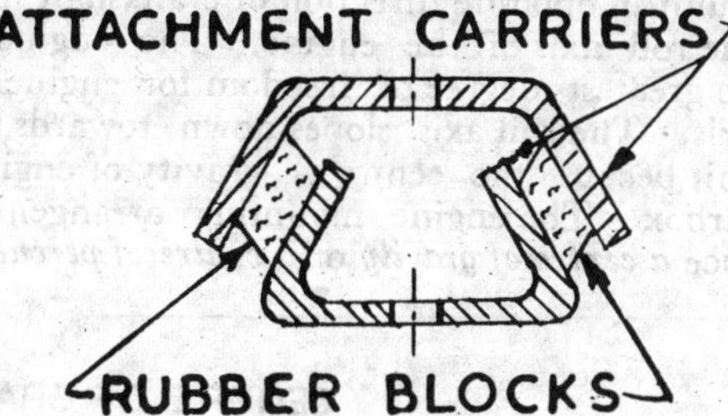

Fig. 4·43. Double sandwich mounting (inclined).

mission of engine vibrations and noises to the body as well as to lessen the effect of road shocks upon engine. Since the engine is

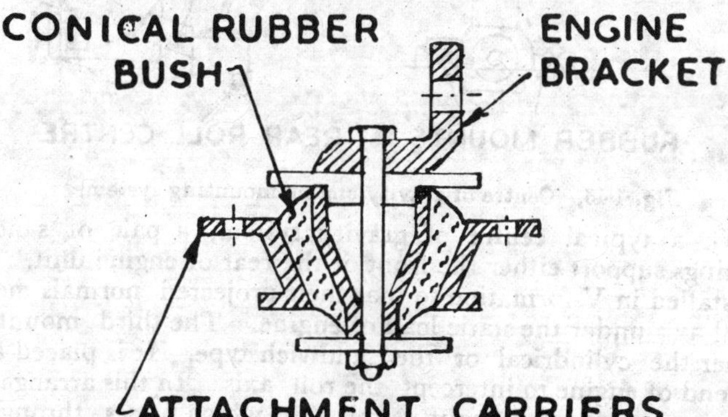

Fig. 4·44. Conical compression and shear mounting.

freed from the effects of chassis stress by the use of flexible mountings, the front portion of chassis frame is made of sufficient strength and rigidity to withstand any normal road shocks without distortion. All the externally connected parts such as exhaust pipe, cables, fuel pipe, etc., are attached flexibly to relieve them from any kind of stresses. The various alternative methods of mounting the engines are as under :

(*i*) At two front and two rear places as in Tata bus or truck.

(*ii*) At one central position on the front cross-member and at two rear places as in Chevrolet and Oldsmobile 6 cylinder in-line engine cars.

(*iii*) At two forward places, one on either side of engine, and at one rear place corresponding to gearbox end as in Jeep.

It is the usual practice to employ *three-point mounting system* for the engine together with gearbox. Owing to greater mounting flexibility, this system is being preferred over *four-point mounting system.* The torque fluctuations induced by the power impulses from each cylinder are a major source of engine vibration. They tend to rock the engine unit in opposite direction of crankshaft rotation and thus establish the roll axis of the engine. The engine mountings must permit the greatest degree of freedom for engine movements about the roll axis. The roll axis slopes down towards the gearbox end of engine unit because the centre of gravity of engine is higher than that of gearbox. The engine mounting arrangement may be devised to produce *a centre of gravity* or *a centre of percussion* system of suspension.

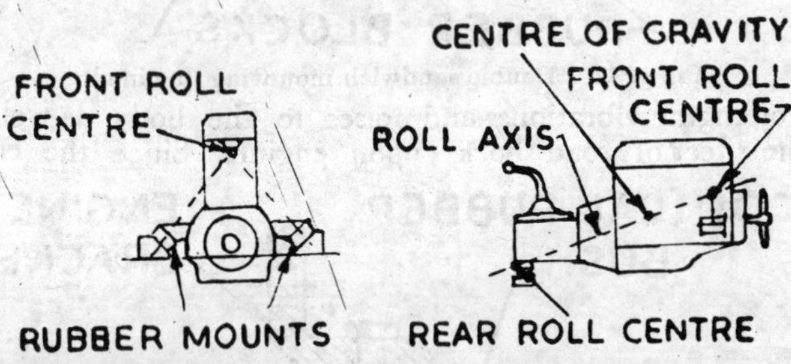

Fig. 4·45. Centre of gravity engine mounting system.

In a typical centre of gravity system, a pair of sandwich mountings support either the front or the rear of engine unit. They are installed in V-formation so that their projected normals meet on the roll axis under the static load of engine. The third mounting is of either the cylindrical or the sandwich type. It is placed at the other end of engine to intercept the roll axis. In this arrangement, the engine is supported about its roll axis which passes through the centres of gravity of both engine and gearbox as in Jeep. As the engine unit rocks on its mountings, the sideways shake is reduced.

In a centre of percussion mounting system too, the engine is supported about its roll axis but the transverse mounting planes are so disposed that no interaction from disturbing forces takes place between the front and rear mountings. In this arrangement, the rear mounting is located at the centre of percussion so that the disturbing forces may act upon the front mountings and *vice versa.* This system is mainly applicable to in line four cylinder engines where the vertically acting unbalanced secondary inertia forces are totally absorbed by the front mountings.

In front wheel drive and rear engined cars, the torque reaction imposed upon the mountings is greater than the front engined rear wheel drive cars because the final drive assembly is integral with the engine unit. The engine mountings must be capable to accommodate the full torque applied to road wheels. In order to minimize the torque reaction loads upon the mountings, the distance between the front and rear mounting planes is kept as large as possible. In case of transversely mounted engines, it is difficult to increase the distance between the mountings due to space limitations in the engine unit compartment. For proper controlling of torque reaction

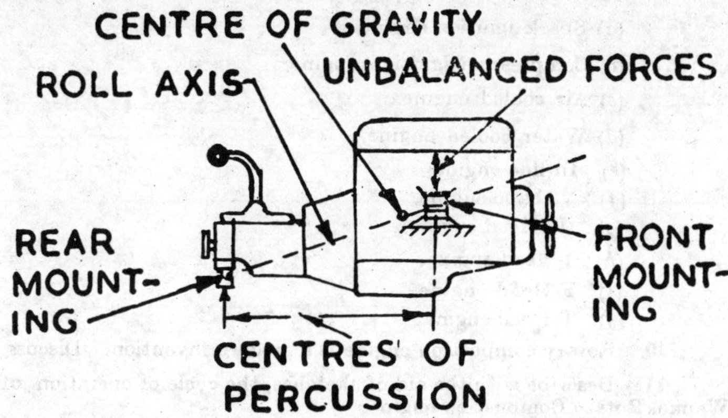

Fig. 4·46. Centre of percussion engine mounting system.

movements, a fore and aft stay rod, in addition, is pivoted between the car body and engine cylinder head as shown in the diagram.

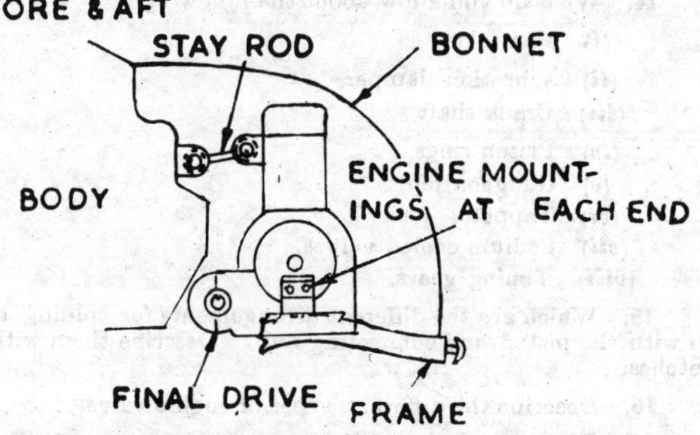

Fig. 4·47. Mounting in transverse engine and transmission unit (Front wheel drive).

QUESTIONS

1. What is an engine ? Describe it.

2. Classify the engines. What is I.C.E. and E.C.E. ?

3. Compare I.C.E. and E.C.E.

4. How would you classify automobile engines ? Show with the help of their family tree.

5. What is a conventional engine ? Describe its main parts.

6. Describe the different types of piston engines.

7. Show the main parts in an engine by means of a suitable sketch.

8. Which are the different types of reciprocating engines with different regards ?

9. Describe the following engines :

(*a*) Spark ignition engine.

(*b*) Compression ignition engine.

(*c*) Air cooled engine.

(*d*) Water cooled engine.

(*e*) In-line engine.

(*f*) V-shape engine.

(*g*) 'L'-Head engine.

(*h*) I-Head engine.

(*i*) F-Head engine.

(*j*) T-Head engine.

10. Rotary combustion engine is a modern invention. Discuss.

11. Describe with the aid of sketches, the cycle of operation of NSU/Wankel Rotary Combustion engine.

12. What do you know about gas turbine ? Describe it in short words.

13. Which are the working parts in a conventional engine ? What is the object of each main part ?

14. What do you know about the following :

(*i*) Manifold.

(*ii*) Vibration damper.

(*iii*) Crank shaft.

(*iv*) Piston rings.

(*v*) Gudgeon pin.

(*vi*) Tappet.

(*vii*) Sodium cooled valves.

(*viii*) Timing gears.

15. Which are the different arrangements for holding down gudgeon pin with the piston and connecting rod ? Describe them with the aid of sketches.

16. Describe the different types of engine valves.

17. Which are the different types of tappets ? Describe the self-adjustable tappet with the help of a sketch.

18. Write short notes on the following :

(*i*) Engine block.

(*ii*) Flywheel.

(*iii*) Semi-floating gudgeon pin.

(*iv*) Roller type tappet.

(*v*) Rotary Valve.

(*vi*) Sleeve Valve.

19. Explain the various types of cylinder liners used in engines.

20. What are the characteristics of a good liner ? Explain the materials used in the manufacture of cylinder liners.

21. Explain the different types of combustion chambers employed in, (*a*) petrol engines (*b*) diesel engines.

22. Explain the phases of combustion in C.I. engine.

23. Explain the following :

(*a*) Swirl

(*b*) Squish

(*c*) Quench

(*d*) Delay period.

24. Engine is subject to complex vibration effects. Which are these effects and how they are neutralised ?

25. Explain the various types of engine mountings applicable in vehicles.

26. Explain the centre of gravity and centre of percussion systems of engine mounting.

27. (*i*) Transverse engine and its advantages.
(*ii*) OHV 5, OHC engines. (*iii*) D.I and I.D.I. engines.

5

Engine-II

1. **Working of conventional engine.** As already discussed, the conventional engine transforms heat energy from the fuel to mechanical energy. To understand how an engine works, is to see how heat energy is converted into mechanical energy. In order to achieve this, the following processes are carried out in the petrol engine :

(*i*) **Suction of fuel air mixture.** Fuel air mixture is sucked into the cylinder. The fuel air mixture is supplied by the carburettor and it reaches the cylinder through induction manifold, inlet port and intake valve. The fresh charge is in the gaseous shape and is readily combustible. When the piston goes downward from top dead centre (T.D.C.) to bottom dead centre (B.D.C.), it creates suction due to which fuel is sucked into the cylinder when intake valve is open.

(*ii*) **Compression of fuel air mixture.** During backward movement of piston from B.D.C. to T.D.C., the drawn in fuel air mixture is compressed into combustion chamber or upper part of the cylinder. During this occasion both the intake and exhaust valves are closed. There is rise in the temperature of fuel air mixture during this process and the fuel becomes very much capable to catch fire and burst at once.

(*iii*) **Ignition of fuel air mixture.** When the piston reaches near about T.D.C. during compression process, ignition of the mixture is affected by means of a spark provided by the spark plug. Ignition results in the combustion of mixture. The burning gases expand and exert pressure at the piston.

(*iv*) **Power development or conversion of heat energy into mechanical energy.** The ignition process starts delivering out heat energy from the fuel air mixture which is readily available at the piston for conversion into mechanical energy.

The burning and expanding gases exert pressure at the piston which is driven down with force. This force is transmitted to the crankpin through the connecting rod, due to which crankshaft is rotated. This is just like the human leg exerting force at the pedal of cycle crank to rotate the wheel.

The power impulse available at the piston crown during ignition process, moves the piston downward in a line. This linear motion is converted into rotary motion by the crankshaft through the connecting rod.

It could be thus concluded that heat energy obtained from the fuel air mixture due to its burning inside the combustion chamber, is converted into mechanical energy through piston in the cylinder, connecting rod and crankshaft. The power available at the piston is thus transmitted to the crankshaft through the connecting rod. During this process power is stored by the flywheel to drive the crankshaft and piston at the uniform speed. From the crankshaft, the power is transmitted further to drive the vehicle.

(*v*) **Exhaust of used gases.** The used gases are expelled out of engine when the piston again rises up from B.D.C. to T.D.C. During this occasion, exhaust valve is open. The cylinder is cleared of the burnt gases for making it ready to admit fresh charge for the next cycle.

The processes are repeated one after the other and the development of power goes on.

In diesel engine, the processes are little different. Instead of fuel air mixture, only air is drawn into the cylinder during suction process. The same air is compressed into the combustion chamber during compression process. The temperature of the air is thus raised high during this process. When the piston reaches near about T.D.C., fuel is injected into the combustion chamber by the injector. The injected fuel at once catches fire when it comes into contact with the compressed hot air. The burning fuel expands and exerts pressure on the piston which is driven down with force. The heat energy thus obtained from the burning of fuel is converted into mechanical energy as explained above. The burnt gases are removed from the cylinder, when the piston travels upward from B.D.C. to T.D.C. and exhaust valve is open.

In four stroke engines, one cycle of these processes is completed in four strokes of the piston or in two revolutions of the crankshaft. In case of two stroke engines, one cycle is completed in two strokes of the piston or in one revolution of the crankshaft. (Stroke is piston travel from T.D.C. to B.D.C. or *vice versa*).

2. **Cycle of operation for Four Stroke Petrol Engine (Otto Cycle), (4 stroke, 5 Event Cycle) :—**

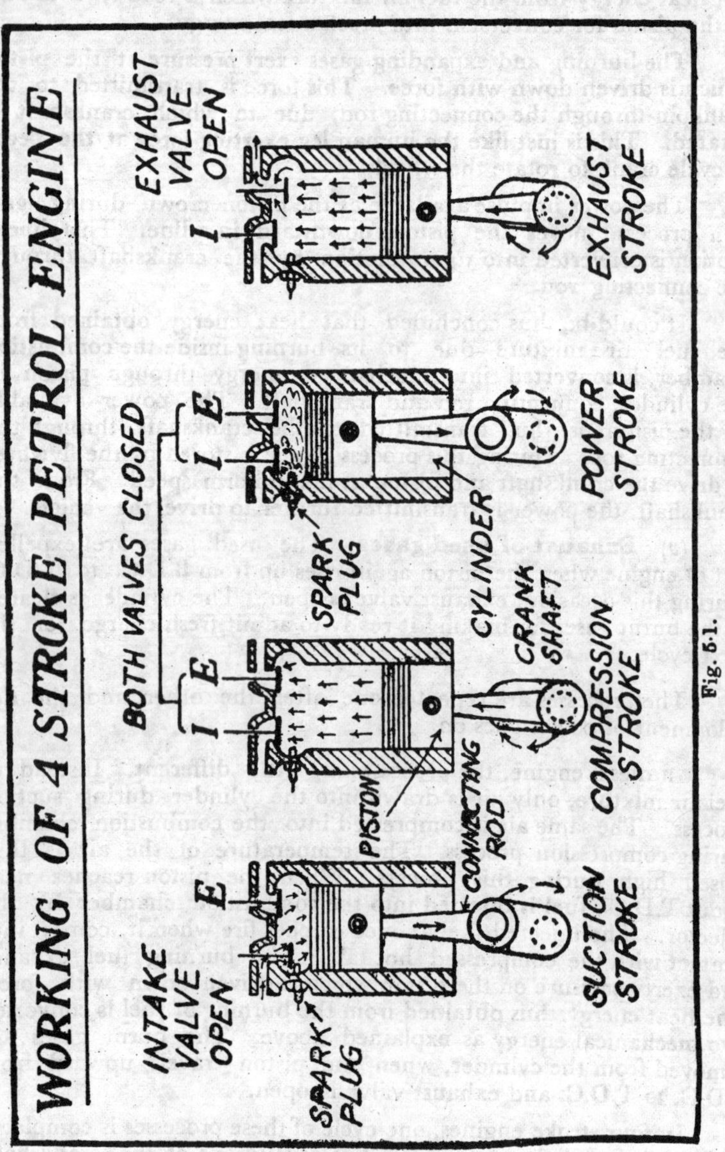

Fig. 5·1.

(*a*) **First Event (suction stroke) :—**

(*i*) Piston moves downwards from T.D.C.

(*ii*) Intake valve is open and the exhaust valve is closed.

(*iii*) Fuel air mixture is sucked into the cylinder.

(*b*) **Second Event (Compression Stroke) :**

(*i*) Piston moves upwards from B.D.C.

(*ii*) Both intake and exhaust valves are closed.

(*iii*) Drawn in fuel air mixture is compressed into the combustion chamber.

(*c*) **Third Event (Ignition).** When the piston reaches near compression top dead centre, a spark is provided by the spark plug, which ignites the compressed fuel air mixture.

(*d*) **Fourth Event (Working or Power Stroke) :**

(*i*) Rapidly expanding burning gases, thrust the piston downward from T.D.C. to B.D.C.

(*ii*) Both intake and exhaust valves are closed.

(*iii*) Power impulse is provided to the piston which is transmitted to the crankshaft through connecting rod.

(*e*) **Fifth Event (Exhaust Stroke) :**

(*i*) Piston moves upwards from B.D.C.

(*ii*) Exhaust valve is open while the intake valve is closed.

(*iii*) Burnt gases are expelled out of the cylinder.

3. Cycle of operation for two stroke valveless petrol engine (Clerk Cycle) (Two stroke, 6 event cycle) :

(*a*) **First stroke** (Induction and Compression)

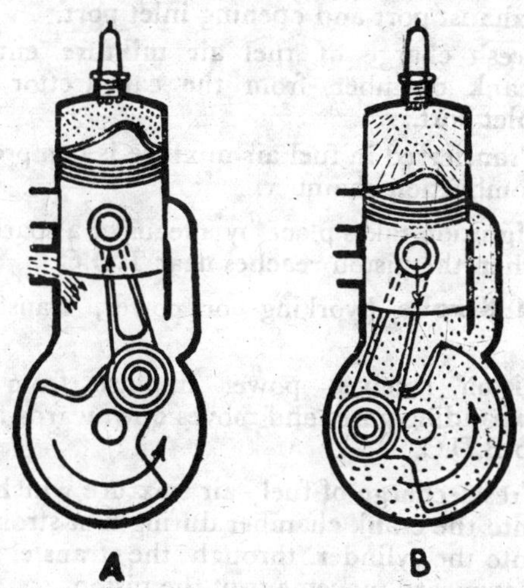

Fig. 5·2. Working of 2 stroke petrol engine.
A—Induction and compression
B—Firing and power
(Piston coming down)
A and D—Upward one stroke

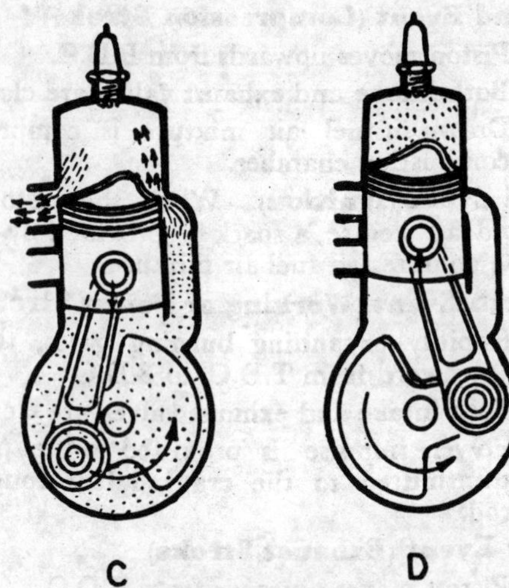

Fig. 5·2. C—Exhaust and transference.
(Piston coming down)
D=Compression
(Piston moving up)
B and C=Downward one stroke

(*i*) Piston travels upwards from B.D C. to T.D.C., closing exhaust port and opening inlet port.

(*ii*) Fresh charge of fuel air mixture enters into the crank chamber from the carburettor through the inlet port.

(*iii*) Transferred in fuel air mixture is compressed into the combustion chamber.

Ignition. Ignition takes place by means of a spark supplied by the spark plug when the piston reaches near T.D.C.

(*b*) **Second Stroke** (working or power, transference, and exhaust) :

(*i*) Piston receives power impulse from the rapidly expanding gases and moves downwards from T.D.C. to B.D.C.

(*ii*) Fresh charge of fuel air mixture which had entered into the crank chamber during first stroke is pumped into the cylinder through the transfer port, by the downward movement of the piston.

(*iii*) Exhaust port is cleared up and the burnt gases go out of the engine.

4. Cycle of operation for Four Stroke Diesel Engine (Diesel Cycle) ; (4 stroke, 5 event cycle) :—

(*a*) **First Event (suction stroke)** :

(*i*) Air from the atmosphere is sucked into the cylinder through the air cleaner, by the downward movement of the piston from T.D.C. to B.D.C.

(*ii*) Inlet valve is open and the exhaust valve is closed.

(*b*) **Second Event (compression stroke)** :

(*i*) Piston moves upwards from B.D.C. to T.D.C.

(*ii*) Both inlet and exhaust valves are closed.

(*iii*) The drawn in air is compressed into the combustion chamber.

(*c*) **Third Event** (Injection) :

When the piston reaches near compression top dead centre, fuel is injected into the combustion chamber by means of Fuel Injection system.

(*d*) **Fourth Event** (Working or Power Stroke) :

As soon as the injected fuel comes in contact with compressed hot air, it catches fire. The gases expand rapidly and provide power impulse to the piston.

(*i*) Piston moves downwards from T.D.C. to B.D.C carrying the power produced due to the combustion of fuel.

(*ii*) Both inlet and exhaust valves are closed.

(*e*) **Fifth Event (Exhaust stroke)** :

(*i*) Piston travels upwards from B.D.C. to T.D.C.

(*ii*) Exhaust valve is open and intake valve is closed.

(*iii*) Used gases are expelled out of the cylinder.

5. **Cycle of operation for two stroke diesel engine** : Cycle of operation for two stroke diesel engine differs slightly from that of two stroke petrol engine. Two stroke diesel engines require a supply of air to blow out exhaust gases and to fill the cylinder with clean air. Air is usually supplied by a blower or air compressor which is driven by the engine.

(*a*) **First stroke (compression)** :

Piston moves upwards from B.D.C. to T.D.C. covering both the intake and exhaust ports.

Injection. When the piston reaches near compression top dead centre, fuel is injected into the combustion chamber by means of Fuel Injection system.

(*b*) **Second stroke (Power, Exhaust and Induction)** :

(*i*) **Power or working.** As soon as the injected fuel comes into contact with the compressed hot air, combustion starts. The burning gases expand rapidly and force the piston downward on its power stroke.

(*ii*) **Exhaust.** As the piston moves downwards towards the end of power stroke, the exhaust port is uncovered and the used gases pass out of the cylinder.

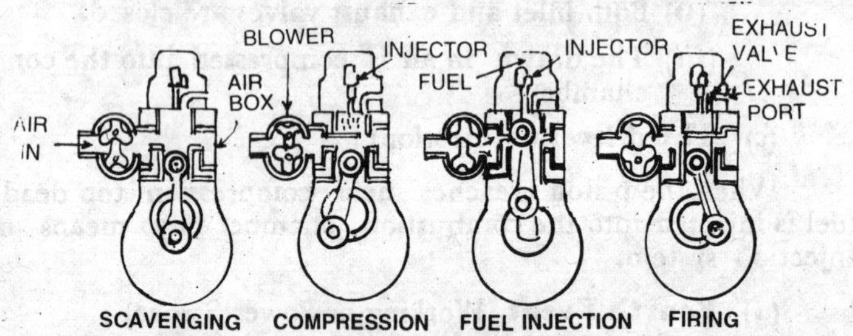

Fig. 5·3. Working of two stroke diesel-engine.

(*iii*) **Induction.** The descending piston uncovers the inlet port soon after the exhaust port. Air under pressure enters the cylinder and helps in expelling out the used gases as well.

6. Scavenging in two stroke engine. The operation of clearing the exhaust gases from the cylinder and filling in it with fresh mixture is known as *scavenging*. In all two stroke engines, air or air fuel mixture is supplied to the cylinder at a higher pressure than the exhaust gases. The transfer or scavenging port begins to open soon after the opening of exhaust port. As soon as the cylinder pressure falls below the scavenging pressure, fresh mixture starts flowing into the cylinder. The flow continues as long as the scavenging port is open and the total scavenging pressure exceeds the pressure in the cylinder. The exhaust gases continue to flow out of the exhaust port due to their flow to this direction at high velocity during *blow-down* period and building up a higher pressure in the cylinder by fresh mixture.

In order to avoid dilution of fresh charge in the crankcase, the exhaust port is uncovered up to about 15° of crankshaft rotation in advance to scavenging or transfer port. The transfer port is covered till the cylinder pressure drops below the existing pressure in crankcase. This is known as *blow-down* period. However, as a result of symmetrical port timing, this *lead* in exhaust port opening conversely becomes a *lag* in closing. As a result, part of scavenge charge transferred from the crankcase tends to run away across the cylinder and is lost through the exhaust port before it closes. Owing to loss of unburnt charge, fuel consumption in two stroke engine rises up.

The early two stroke engines contained a deflector-head piston to promote a *cross-scavenging* effect on the burnt charge going

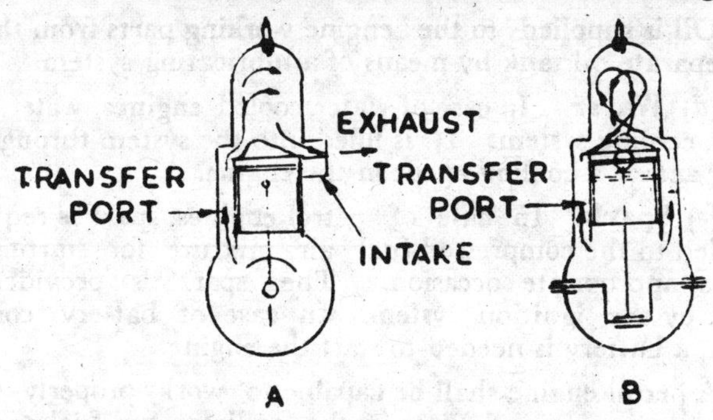

(*A*) Cross scavenging (*B*) Loop scavenging

Fig. 5·4. Scavenging in two stroke engines.

out of the cylinder. Near about 1920, Mr. Schnuerle of Germany devised an improved method of scavenging known as *loop-scavenging*. In this system, the deflector on the piston has been removed and two transfer ports with angled passages are disposed on either side (instead of opposite) of exhaust porta.s in Lambretta and Vijay scooters. The effect of loop-scavenge is such that the two streams of fresh charge converge upon the cylinder wall at a point farthest away from the exhaust port before they intermingle.

7. Requirements for the operation of an engine. In order to put an engine into working order, the following requirements are needed :

(*a*) **Fuel.** It is the main requirement of the engine. This is needed to provide heat energy to the engine. Petrol and diesel are the main fuels used now-a-days. Fuel is contained in the fuel tank and is supplied to the engine by a full-fledged fuel system.

(*b*) **Air.** Air is needed to form combustible mixture of fuel and air for the engine. In case of petrol engine, air and fuel mixture is prepared by the carburettor. Air is well cleaned by the air cleaner before it enters into carburettor or engine. Air is also required for cooling the engine.

(*c*) **Oil.** Oil is needed for the engine—

(*i*) to act as a lubricating agent to decrease wear and tear.

(*ii*) to act as cushioning agent to decrease noise.

(*iii*) to act as a cooling agent to eliminate heat generated by the working parts.

(*iv*) to act as a sealing agent between the piston and cylinder wall.

(*v*) to act as a cleaning agent to move dirt and grit from the passage of working parts.

(*vi*) to decrease the power loss by lowering the friction rate and thus to increase efficiency of the engine.

Oil is supplied to the engine working parts from the oil sump or a separate oil tank by means of a lubricating system.

(*d*) **Water.** In case of water cooled engines, water is required for the cooling system. It is filled into the system through radiator. Water acts as a cooling agent in the engine.

(*e*) **Spark.** In case of petrol engines, spark is required to be provided to the compressed fuel air mixture for starting ignition at the appropriate occasion. The spark is provided in the engine by the ignition system. In case of battery coil ignition system, a battery is needed to start the engine.

A petrol engine shall be capable to work properly if there is oil in the oil sump and water in the radiator and fuel and spark are being supplied regularly. To ensure this, it is essential that the following systems should function properly :

(*i*) Fuel system,

(*ii*) Ignition system,

(*iii*) Lubrication system,

(*iv*) Cooling system.

8. Intake and Exhaust of Gases in the Engine. In petrol engine, the fresh charge of fuel-air mixture is supplied to the engine cylinder by the carburettor through induction manifold, intake port and intake valve. The used gases are expelled out of the engine cylinder into the atmosphere through exhaust valve, exhaust port, exhaust manifold, exhaust pipe, muffler and tail pipe.

In petrol engine, air from atmosphere enters into carburettor where a mixture of fuel (from fuel tank) and air is prepared in correct proportions. The fuel air mixture from the carburettor is sucked into the engine cylinder by the downward going piston when intake valve is open during intake or suction stroke. When the piston moves upwards on its exhaust stroke and exhaust valve is open, the used gases are forced out of the engine cylinder and let out in the atmosphere through the silencer. The flow of fresh charge and used gases in the engine is as under :

Intake of **fuel air mixture** from	**Exhaust** of **used gases** from
Carburettor	The Cylinder & Combustion Chamber
↓	↓
Induction manifold	Exhaust valve
↓	↓
Inlet port	Exhaust Port

↓	↓
Intake Valve	Exhaust Manifold
↓	↓
Combustion Chamber and cylinder	Silencer
	↓
	Atmosphere

In diesel engine, only air is drawn into the cylinder from the atmosphere through air cleaner, induction manifold, intake port and intake valve. The used gases are expelled out of the engine through exhaust valve, exhaust port, exhaust manifold and silencer, into the atmosphere.

The induction manifold provides passage for the flow of fresh gases from the carburettor or fresh air from atmosphere through air cleaner, as the case may be, towards the engine cylinder. The intake port in the cylinder block or cylinder head, as the case may be, leads the fresh charge from induction manifold into combustion chamber and cylinder through intake valve. The intake valve provides entrance for the fresh charge into the combustion chamber and cylinder.

The used gases go out of the cylinder and combustion chamber through exhaust valve which acts as a gate to provide exit for the gases. The gases flow out of the exhaust valve mouth to the connecting passage of exhaust port into the exhaust manifold.

The exhaust manifold in the combined induction and exhaust manifold design contains two passages ; one direct and the other bypass. The flow of gases through the exhaust manifold is controlled by a *heat control valve* which directs the flow to any one passage.

The heat control valve is usually thermostatically operated and opens and closes the direct passage or bypass due to temperature

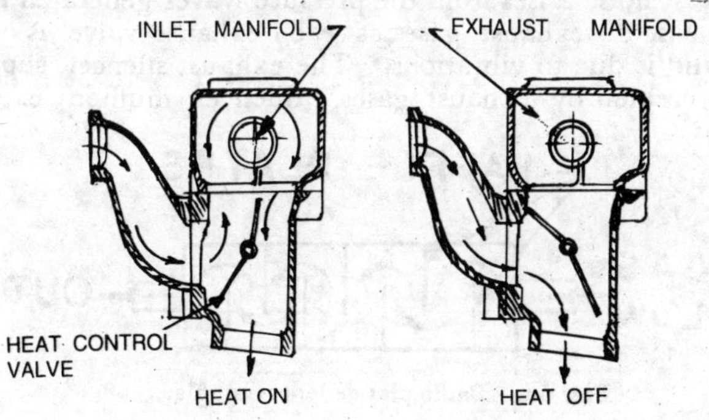

Fig. 5·5. Action of heat control valve.

effect. In combination manifold, a portion of induction manifold known as *hot-spot*, passes inside the passage of exhaust gases in the exhaust manifold. When the engine is cold, the heat control valve is closing the direct passage due to the winding effect of thermostatic coil spring attached to the valve shaft. The exhaust gases pass out of the exhaust manifold through the bypass as shown in the diagram. The exhaust gases thus help in warming the ingoing fuel air mixture through the induction manifold for easy combustion. As the engine reaches its operating temperature, the thermostat coil spring unwinds due to heat effect and opens the direct passage for the flow of exhaust gases out of the exhaust manifold into the exhaust pipe of silencer.

The used gases from the exhaust manifold are let out into the atmosphere through the silencer so that their noise is silenced and do not create nuisance in the public.

The silencer consists of an exhaust pipe, muffler and tail pipe. The muffler is central part and is connected to the exhaust manifold through the exhaust pipe. At the other end of the muffler, an outlet or tail pipe is provided which directs the exhaust gases to the atmosphere.

The *muffler* silences the noise of exhaust gases by reducing the pressure of exhaust gases by slow expansion and cooling. This is attained with the aid of a number of expanding chambers. The gas is allowed to expand from first passage into a much larger second one and then to a still larger third one and so on to the final and largest passage which is connected to the tail or outlet pipe of the muffler. In a simple construction, a section of exhaust pipe having a large number of holes around it, is passed through a chamber which is further split up into smaller chambers through the baffle plates. Before going into atmosphere the exhaust gases pass out into the various chambers where they find chance to expand and cool down resulting in decreased noise.

Exhaust noise arises from the pressure waves generated by the sudden release of exhaust gases as each exhaust valve is opened. All the sound is due to vibrations. The exhaust silencer suppresses the noise created by exhaust gases. Silencer muffler explained

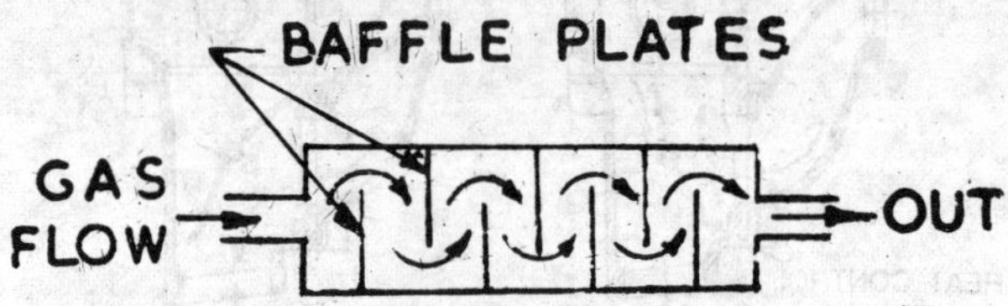

Fig. 5·6. Baffle plate silencer muffler.

above is of *baffle-plate type*. Its main drawback is its low efficiency. Owing to restricted flow of exhaust gases, back-pressure increases causing loss of engine power. In modern exhaust silencers, damping of sound is accomplished by a combination of acoustic filters of resonance and absorption types. In view of this the exhaust silencer can be classified as under :—

(*a*) Resonant

(*b*) Absorbent

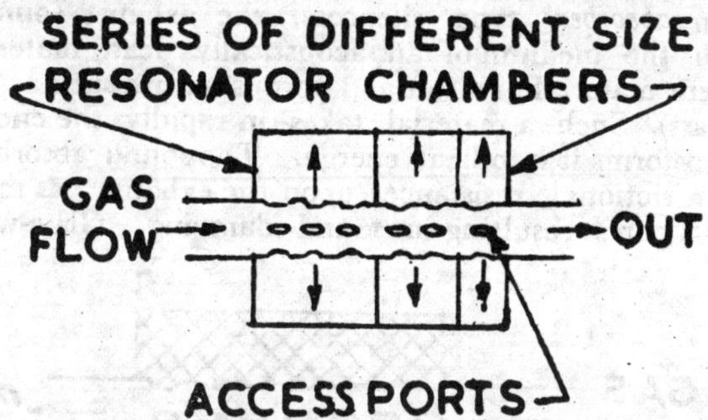

Fig. 5·7. Straight through-flow resonant silencer.

Both the above types of silencers are of *straight through-flow* as well as *reverse flow*. In straight through-flow, the exhaust gases travel straight through the silencer muffler. In reverse-flow silencers, there are separate inlet and outlet tubes, placed one above the other. The exhaust gases have to flow in the reverse direction for entering into the outlet pipe. In some cases there 'is another tube fixed parallel in between the inlet and outlet tubes. Such an arrangement creates more interference in the flow of exhaust gas and helps in damping various frequencies of exhaust noise.

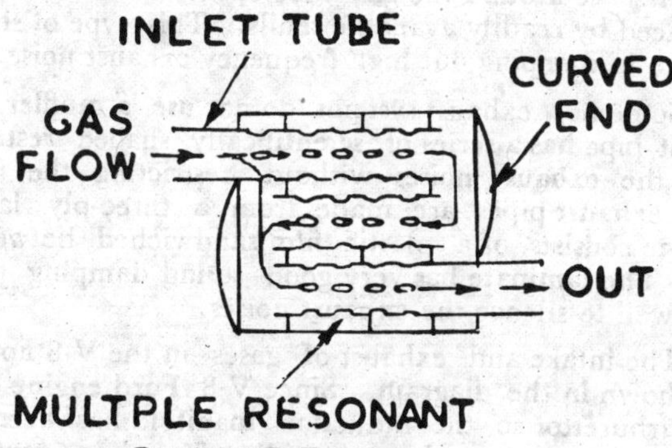

Fig. 5·8. Reverse flow resonant silencer.

In *resonant* type of silencer, there are a series of resonant chambers of different volumes which surround the exhaust gas tube or tubes. (One tube in through-flow and more than one tube in reverse-flow). As the sound waves travel along the pipe/pipes in the resonant chambers, the pulsating gas flows at the entrance to the cavity chamber creates pressure variations and vibrations of gas within it like a coil spring. The resonators absorb energy from the vibrations and the exhaust noise is suppressed over a wide range of firing frequencies through a series of different size chambers.

In *absorbent* type silencers, the exhaust sound is damped through the medium of an acoustically dead material (a sound absorbent material of a soft and porous nature such as glass-wool or fibre-glass). Such a material takes in rapidly the energy of sound and transforms it into heat energy. The sound absorbent material imposes frictional resistance upon the exhaust gas moving in and out of its pores resulting in sound damping. Glass-wool which is

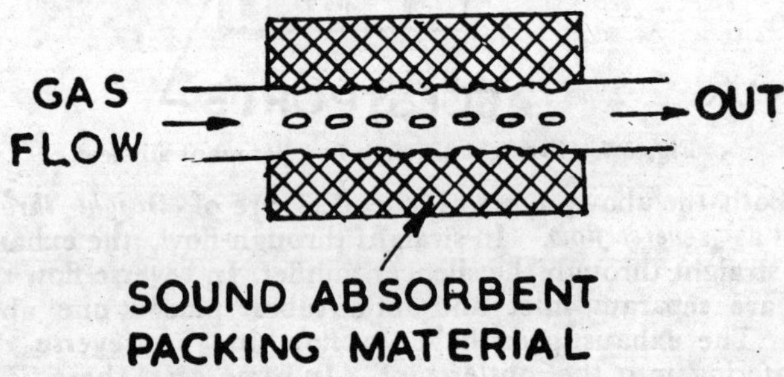

Fig. 5·9. Through-flow absorbent silencer.

heat resistant and sound absorbent material, is packed into the annular space around the gas tube/tubes. The damaged glass wool is replaced by readily available refills. This type of silencer is quite effective in damping out high frequency exhaust noise.

Some new exhaust systems do not use a muffler. Instead, the exhaust pipe has a series of scientifically shaped restrictions which damp the exhaust noises without restricting the flow of gases. Many exhaust pipes are made from a three-ply laminate. The laminate consists of a plastic film sandwiched between two metal skins. The laminate has very good sound damping properties and works well to silence the exhaust noises.

The intake and exhaust of gases in the V-8 Ford engine has been shown in the diagram. Since V-8 Ford engine is fitted with dual carburettor so the induction manifold has been so designed that each carburettor throat supplies fuel air mixture to alternate cylinders in the firing order. R-throat of the carburettor supplies fuel air mixture to the end cylinders on the left bank and two central cylinders on the right bank. L-throat of the carburettor supplies fuel air mixture to the two central cylinders on the left bank and the end cylinders on the right bank. According to the numbering of the cylinders, the L and R throats of the carburettor supply fuel air mixture as under :—

(*i*) L-throat of the carburettor→

Cylinder No. 1-4-6-7

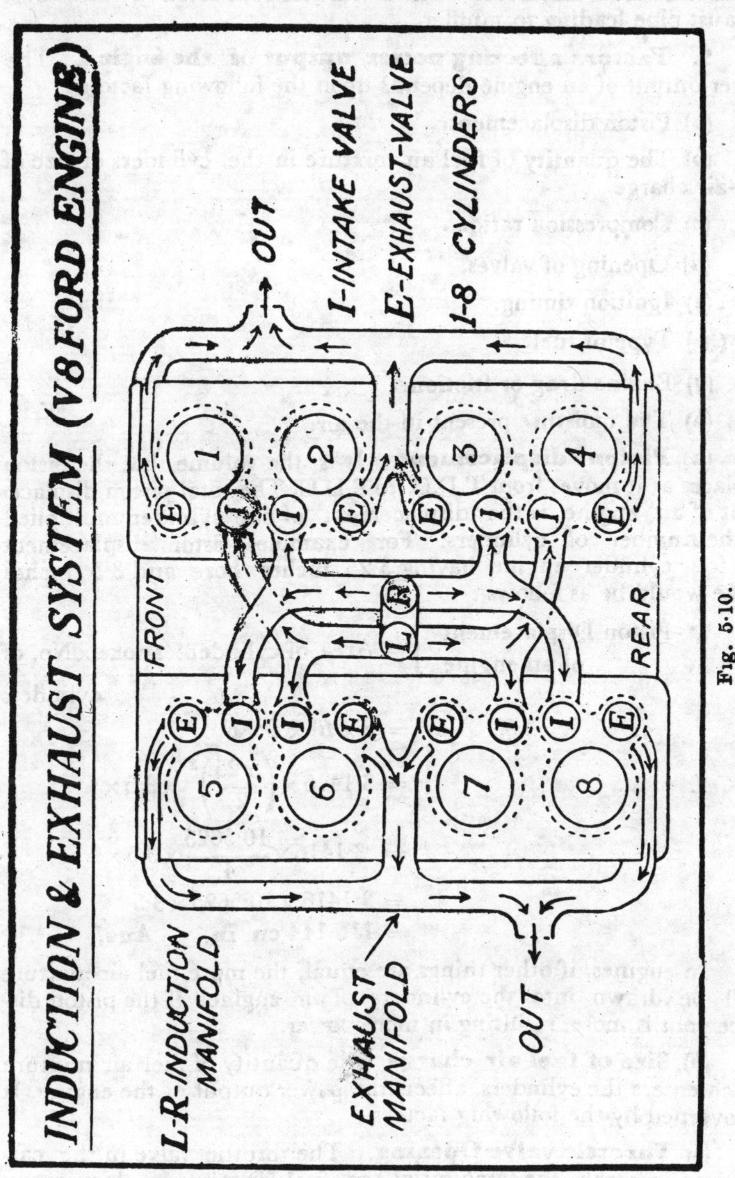

Fig. 5·10.

(*ii*) R-throat of the carburettor→

Cylinder No. 2-3-5-8.

The same procedure is followed in alomst all engines employing dual carburettors. In this case each bank of cylinder contains a separate exhaust manifold which is further connected by cross-over exhaust pipe leading to muffler.

9. Factors affecting power output of the engine. The power output of an engine depends upon the following factors :

(*a*) Piston displacement.

(*b*) The quantity of fuel air mixture in the cylinders or size of fuel-air charge.

(*c*) Compression ratio.

(*d*) Opening of valves.

(*e*) Ignition timing.

(*f*) Type of fuel.

(*g*) Engine drag or friction.

(*h*) The moisture present in the air.

(*a*) **Piston displacement.** It is the volume that the piston displaces as it moves from T.D.C. to B.D.C. The total piston displacement of an engine is the displacement of one cylinder multiplied by the number of cylinders. For example, piston displacement of a four cylinder engine having 3·25 inches bore and 3·5 inches stroke would be as follows :

$$\because \left.\begin{array}{r}\text{Piston Displacement}\\ \text{of an engine}\end{array}\right\} = \text{Area of cylinder} \times \text{stroke} \times \text{No. of cylinders}$$

$$\text{or} \quad ,, \quad ,, \quad = \pi \times R^2 \times L \times N$$

$$\therefore \quad ,, \quad ,, \quad = 3{\cdot}1416 \times \left(\frac{3{\cdot}25}{2}\right)^2 \times 3{\cdot}5 \times 4$$

$$= 3{\cdot}1416 \times \frac{10{\cdot}5625}{4} \times 3{\cdot}5 \times 4$$

$$= 3{\cdot}1416 \times 10{\cdot}5625 \times 3{\cdot}5$$

$$= \mathbf{116{\cdot}144 \text{ cu. in.}} \quad \textbf{Ans.}$$

In engines, if other things are equal, the more fuel air-mixture shall be drawn into the cylinders of an engine, if the piston displacement is more, resulting in more power.

(*b*) **Size of fuel air charge.** The quantity of fuel air mixture which enters the cylinders, affects the power output of the engine. It is governed by the following factors :

(*i*) **Throttle valve Opening.** The throttle valve in the carburettor controls the amount of the fuel air mixture which enters the cylinders. If the throttle valve is wide open, more fuel shall go in and if the throttle is less open, less charge shall be induced.

(*ii*) **Engine speed.** At low engine speed there is more time available for the induction of fuel and comparatively more fuel is fed to the engine than the high speed. As engine speed increases, the time available for fuel induction decreases resulting in the cylinders being only partially filled.

(*iii*) **Atmospheric Pressure.** Due to fall in atmospheric pressure, less air (by weight) is taken into cylinders and hence less fuel supply. In some engines, super-chargers are employed to supply air to the engine at pressures greater than the atmospheric, resulting in a greater fuel air charge.

More the fuel-air charge into the engine, more the power output.

(*c*) **Compression Ratio.** It is the ratio of the total volume of cylinder and combustion chamber when piston is at B.D.C. to the volume of combustion chamber when piston is at T.D C.

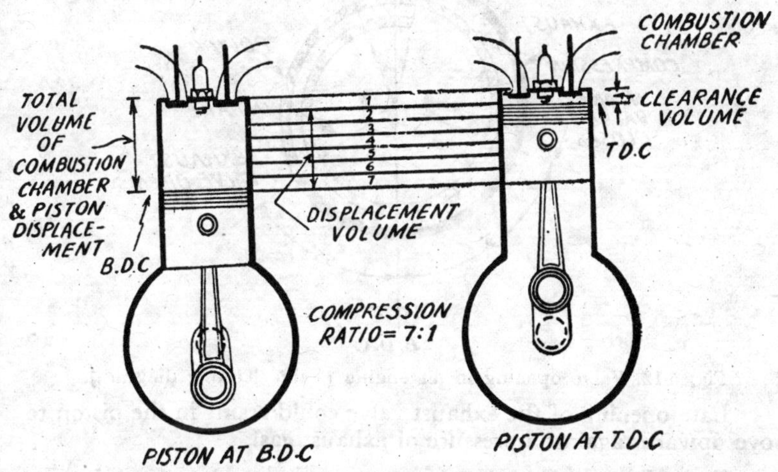

Fig. 5·11. Compression ratio.

In other words,

Compression Ratio = Piston Displacement or swept in Volume + Clearance volume : clearance volume

Higher the compression ratio, higher the degree of compression and higher the rate of expansion due to faster combustion and hence the more output.

(*d*) **Opening of valves.** The inlet valve is generally opened several degrees before the piston reaches at T.D.C., for starting in-take stroke. This is done to keep the valve at least partially open by the time the piston starts downward travel. This prevents the piston from working against a closed inlet valve and helps in the induction of a greater amount of fuel air mixture into the cylinder. Hence more the charge, more the power.

Length of power stroke is determined up to the point at which the exhaust valve opens Opening of the exhaust valve also affects the removal of exhaust gases from the cylinders. The exhaust valve too is opened several degrees before the piston reaches at B.D.C. to start for exhaust stroke. Early opening of the exhaust valve helps the remaining pressure in the cylinder to force the exhaust gases out of the cylinder. By the time, the piston reaches B.D.C., the exhaust gas pressure is reduced to near about atmospheric pressure. When the piston moves upwards, the remaining exhaust gases are forced out of the cylinder.

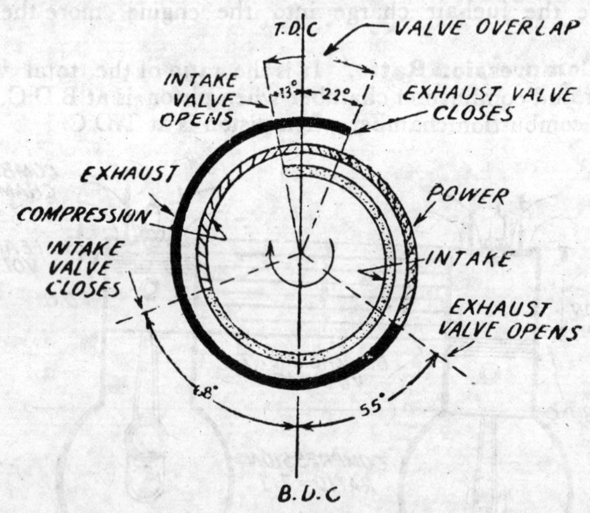

Fig. 5·12. Valve opening in an engine (Valve Timing diagram).

Late opening of the exhaust valve could result in the piston to move upwards against a pressure of exhaust gas.

The exhaust valve is kept opened up to several degrees after the piston starts at its suction stroke. This is to clear the exhaust gases out of the valve opening. If this is not done and the flow of gases is restricted in the exhaust system, a back pressure shall result in which shall reduce the power and speed developed by the engine.

If the exhaust valve is not kept open for the sufficient time, the used gases will not be driven out up to full extent. The fresh charge of fuel air mixture shall mix with the exhaust gases left in the cylinders, resulting in loss in volumetric efficiency and engine power.

(*e*) **Ignition Timing**. For obtaining maximum power output, it is essential that ignition of the fuel starts at the proper time at all engine loads and speeds.

The speed of combustion of fuel air mixture depends upon its compression in the cylinders. The degrees of compression depends upon, among other factors, throttle opening. With a given throttle opening, the amount of fuel-air mixture which could flow into the cylinder, is controlled by the time available for induction. The induction time depends upon the speed of engine. The speed controls the degree of compression which in turn determines the speed of combustion.

When full force of the expanding gases is available while the piston just starts its downward travel from T.D.C., it is considered to be an ideal condition. In order to achieve this condition, it is desirable that the ignition be started few degrees before T.D.C. so that burning of the mixture could be completed by the time piston reaches T.D.C. The starting point of ignition should suit both the load and speed of the engine as both load and speed affect the rate of combustion and the time available for it. The ignition timing is controlled by the distributors which contain automatic retard and advance mechanisms to change the ignition timing according to load and speed of the engine. Setting of correct initial ignition timing is very necessary to develop full power in the engine.

(*f*) **Type of fuel.** The type of fuel used and the speed at which combustion is completed in a cylinder are important factors in the development of power. The travel of the flame across the combustion chamber should progress at a uniform rate. Combustion which progresses too rapidly causes detonation or knocking

Detonation is instantaneous burning of fuel air mixture which creates hammer like striking noise. It is due to fuels having low octane number. Detonation results in loss of power and engine damage. Detonation occurs, if the last of the unburned portion is compressed and heated to such an extent that it explodes before the moving flame reaches it. Detonation generates extremely high temperatures and pressures which result in hammer like blows at the pistons and cylinder walls. Use of antiknock or high octane fuels and adjustment of ignition timing to suit the fuel, is the remedy for detonation.

(*g*) **Engine drag or friction.** A part of the power developed by the engine is consumed in engine friction. It is not possible to finish engine drag completely. However, proper lubrication and establishment of correct clearances could help in minimizing engine drag to a great extent.

(*h*) **Moisture present in the air.** The engine seems to run smoother and with increased power on a rainy day. This could be due to greater weight of air due to moisture present in it while entering the cylinders, increasing volumetric efficiency. It could be due to the fact that moisture present in the air changes into steam during combustion, resulting in greater pressures in the cylinders and hence more power.

10. Power overlap. If one piston is going on power stroke and the other piston starts on its power stroke before the end of

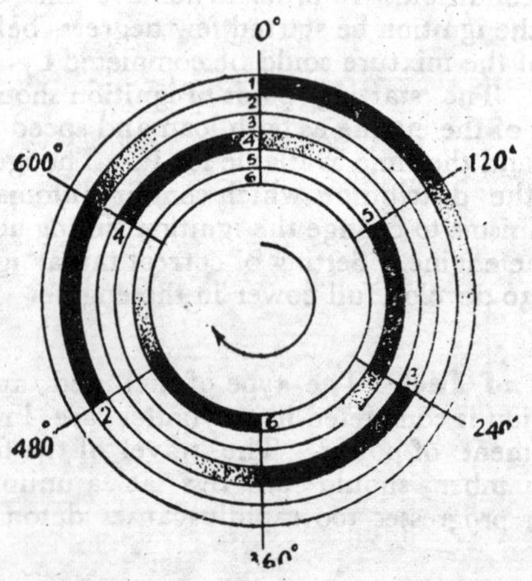

Fig. 5·13-A. Power flow (in shade) in 4-cylinder engine.

power stroke of the previous piston, a period shall be noticed when both the pistons in the engine shall be on power stroke. This is known as power overlap or power being produced during the same time in two cylinders. The more the number of cylinders in an engine, the more continuous shall be the flow of power and smoother shall be its operation due to power flow. The number of power strokes per two revolutions of crankshaft in a four stroke engine are increased by the increase in the number of cylinders. In a two cylinder 4 stroke engine, there is one power stroke in every 360° or one revolution of crankshaft, there being no power overlap. In a four cylinder 4 stroke engine, there is one power stroke in every 180° or half revolution of crankshaft, there being no power overlap as each new power stroke begins exactly the same time when the previous one ends.

In a 6-cylinder, 4-stroke in-line engine, there are six power strokes in two revolutions of the crankshaft or in 1/3 revolution or

ENGINE FIRING ORDER	TWO REVOLUTIONS OR 720° OF CRANK SHAFT ROTATION			
	ONE REVOLUTION OR 360°		ONE REVOLUTION OR 360°	
1	POWER	EXHAUST	INTAKE	COMPRESSION
3	COMPRESSION	POWER	EXHAUST	INTAKE
4	INTAKE	COMPRESSION	POWER	EXHAUST
2	EXHAUST	INTAKE	COMPRESSION	POWER

Fig. 5·13-B. Flow of power in 4-stroke 4-cylinder engine.

120° of crankshaft revolution. This means that a power stroke begins after every 120 degrees of crankshaft rotation and continues

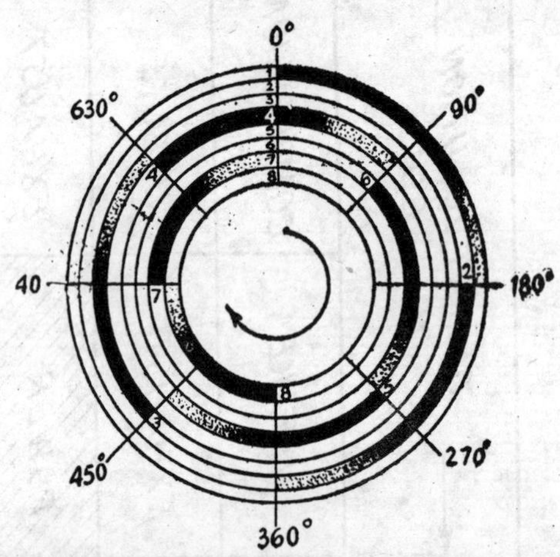

Fig. 5.14 A. Power flow (in shade) in 6-cylinder engine.

for 180 degrees or half the crankshaft revolution. This gives a 60-degree overlap between power strokes that follow each other in the firing order of the engine.

ENGINE FIRING ORDER	TWO REVOLUTIONS OR 720 OF CRANK SHAFT ROTATION											
	ONE REVOLUTION OR 360°						ONE REVOLUTION OR 360°					
	180°			180°			180°			180°		
	120°		120°		120°		120°		120°		120°	
	60°	60°	60°	60°	60°	60°	60	60°	60°	60°	60°	60°
1	Po	P	Po	E	E	E	I	I	I	C		C
5	C	C	Po	P	Po	E	E	E	I	I	I	C
3	I	C	C	C	Po	P	Po	E	E	E	I	I
6	I	I	I	C	C	C	Po	P	Po	E	E	E
2	E	E	I	I	I	C	C	C	Po	P	Po	E
4	Po	E	E	E	I	I	I	C	C	C	Po	P

P = POWER Po = POWER OVERLAP E = EXHAUST I = INTAKE C = COMPRESSION

Fig. 5.14 B. Flow of power in 4-stroke 6-cylinder engine.

In an eight cylinder 4 stroke engine, there are eight power strokes in two revolutions of the crankshaft or one power stroke in

90° or 1/4 revolution of the crankshaft. Thus a power overlap of 90 degrees (180 degrees—90 degrees) is produced.

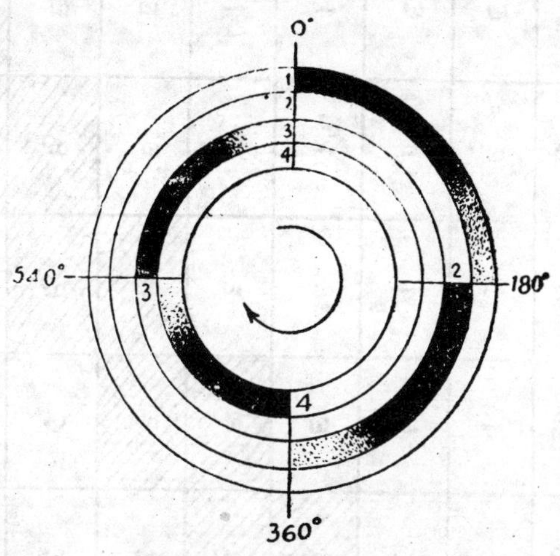

Fig. 5·15 A. Power flow (in shade) in 3-cylinder engine.

11. Engine balancing. For smooth working of the engine, it is necessary that it should be balanced, both power and mechanical. An engine is said to be in *power balance* when the power impulses occur at regular intervals with relation to crankshaft rotation and each power impulse exerts the same force. An engine is known to be in *mechanical balance* when all its working parts are set in such a way so that they counterbalance in operation and thereby minimize vibration.

The rotating parts of the engine are balanced mechanically by bringing to *static* as well as *dynamic* balance. The main parts in the engine which are to be balanced mechanically are crankshaft and flywheel. To obtain static balance, the weight must be equal in all directions from the centre when the crankshaft is stationary.

Dynamic balance relates to balance while the part is rotating, It is attained when the centrifugal forces of rotation are equal in all directions at any point. Special machinery is required in this type of balancing.

12. Multi-cylinder engines. A four-stroke single cylinder engine delivers only one *power impulse* for every two revolutions of crankshaft due to which it produces uneven torque. A multicylinder engine provides one or more power impulses for every revolution of crankshaft owing to which torque irregularities are diminished and a much better standard of *engine balance* is achieved.

A *reciprocating piston engine* contains two types of working parts, *i.e.*, reciprocating and rotating. Complete piston with rings

ENGINE FIRING ORDER	TWO REVOLUTIONS OR 720° OF CRANKSHAFT ROTATION							
	ONE REVOLUTION OR 360°				ONE REVOLUTION OR 360°			
	180°		180°		180°		180°	
	90°	90°	90°	90°	90°	90°	90°	90°
1	P_O	P_O	E	E	I	I	C	C
6	C	P_O	P_O	E	E	I	I	C
2	C	C	P_O	P_O	E	E	I	I
5	I	C	C	P_O	P_O	E	E	I
8	I	I	C	C	P_O	P_O	E	E
3	E	I	I	C	C	P_O	P_O	E
7	E	E	I	I	C	C	P_O	P_O
4	P_O	E	E	I	I	C	C	P_O

P_O = POWER OVERLAP E = EXHAUST I = INTAKE C = COMPRESSION

Fig. 5·15 B. Flow of power in 4-stroke 8-cylinder engine.

and gudgeon pin plus upper part of connecting rod (about $\frac{1}{3}$) is included in the *reciprocating parts*. Big-end assembly, the lower part of connecting rod, crank-pin, the rest of crankshaft and flywheel come under the category of *rotating parts*. The engine must be properly balanced to minimize the effects of vibrations caused by the following :

(*i*) Unbalanced centrifugal forces and couples set up by the rotating parts.

(*ii*) Unbalanced forces and couples set up by inertia effects of reciprocating parts.

For smooth running of the engine all the above forces and couples must be neutralised so that stresses in the engine are reduced. The number and arrangement of cylinders in the engine and arrangement of crank-throws plays an important part in the smooth delivery of power. The choice of a particular arrangement and number of cylinders must take into account the presence of *primary and secondary inertia forces.* The primary inertia forces arise from the force which is applied to accelerate the piston over the first half of its stroke and similarly from the force developed by the piston as it decelerates over the second half of the stroke. When the

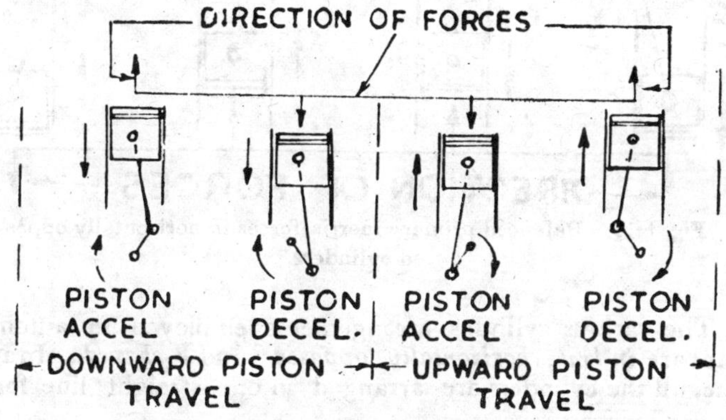

Fig. 5·16. Unbalanced primary inertia forces.

piston is around the mid-stroke position, it is usually moving at the same speed as the crankpin and no inertia force is generated. Thus an unbalanced primary inertia force oscillates once along the axis of the cylinder for every revolution of crankshaft. The arrangement and number of cylinders should be such so that primary inertia forces generated in any particular cylinder are directly opposed by those of another cylinder. An engine is said to be in *primary balance* in which primary inertia forces are neutralised by one another.

The angular variations occurring between the connecting rod and the cylinder axis give rise to secondary inertia forces. As the connecting rod departs from straight line motion, the piston is tended to move more rapidly over the upper half of its stroke than the lower half. Thus the piston travel differs at the two ends of stroke for the same angular movements of crankshaft. The resulting inequality of piston accelerations and decelerations produces corresponding differences in the generated inertia forces. Thus a residual unbalanced secondary inertia force oscillates twice along the cylinder axis for every revolution of crankshaft. The engine is said to be in *secondary balance* in which the differing secondary inertia forces can be matched and opposed in direction between one cylinder and another.

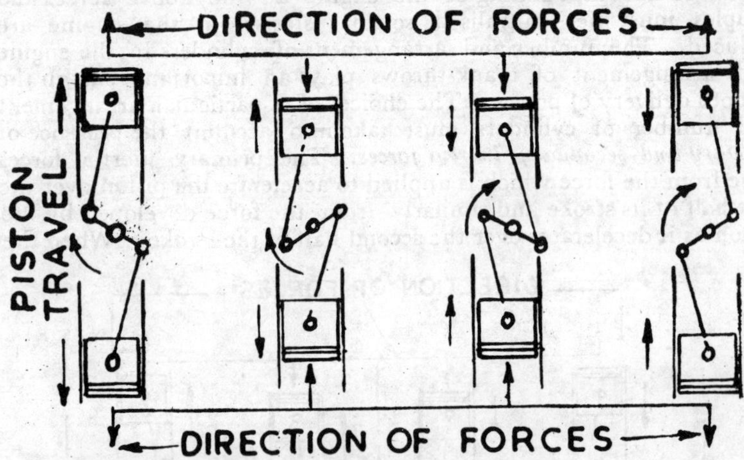

Fig. 5·17. Balanced primary inertia forces in horizontally opposed cylinders.

The various cylinder arrangements employed in automobile engines are *in line*, *horizontally opposed* and *V-shaped*. In in-line engine, all the cylinders are arranged in one straight line forming

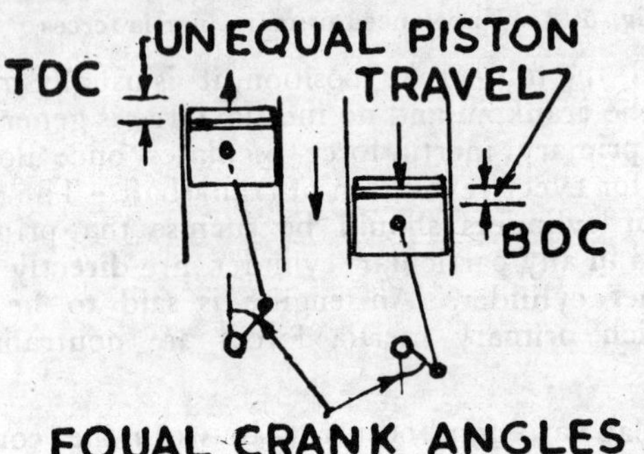

Fig. 5·18. Unbalanced secondary inertia forces.

one bank. In flat engines, there are usually horizontally opposed cylinders. Volkswagen car engine contains horizontally opposed two parts of cylinders. V-shape engines are quite common in modern cars such as Ford, Studebaker, Mercedes-Benz, Cadillac, Pontiac, Chrysler etc.

The in-line cylinder engine is generally confined to two, four and six cylinder versions although 8-cylinder in-line has also been employed in some automobiles. Two cylinder in-line or parallel twin engine is rarely employed in modern automobiles, due to lack of primary balance as both pistons travel in the same direction to obtain equal firing intervals.

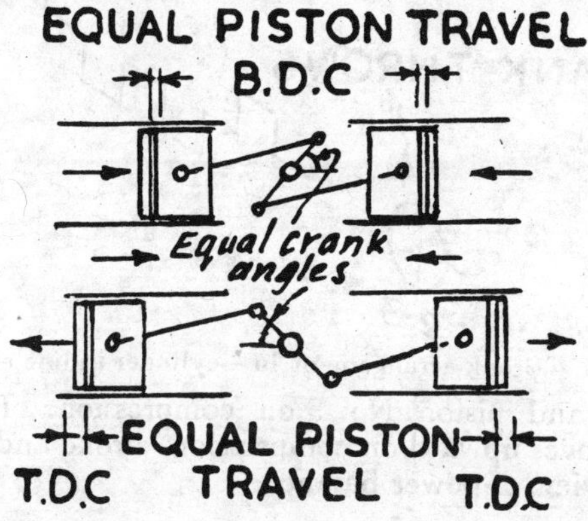

Fig. 5·19. Balanced secondary inertia forces in horizontally opposed cylinders.

In case the cranks are 180° apart in two cylinder engine, power balance shall be as shown in the diagram.

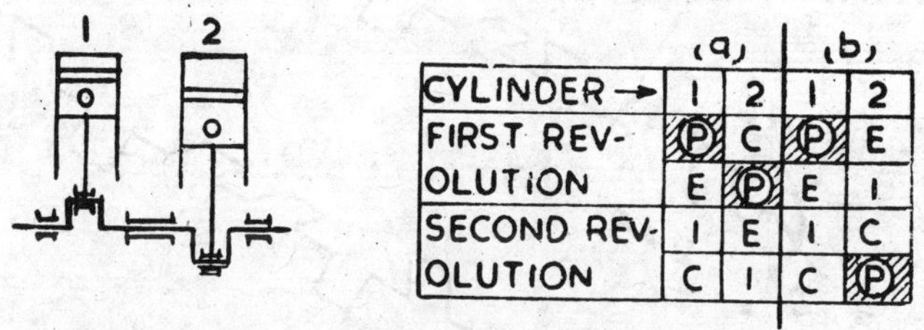

	(a)		(b)	
CYLINDER →	1	2	1	2
FIRST REVOLUTION	P	C	P	E
	E	P	E	I
SECOND REVOLUTION	I	E	I	C
	C	I	C	P

(*a*) No. 1 piston on power stroke and No. 2 piston on compression;
(*b*) No. 1 piston on power and No. 2 piston on exhaust.

Fig. 5·20. Power balance in twin cylinder engine. *a* and *b* two alternate conditions.

In four cylinder in-line engines, the cranks are 180° apart, cranks no. 1 and 4 falling on one side, cranks no. 2 and 3 on another. This arrangement possesses equal firing intervals to provide two power impulses for every revolution of crankshaft. It has the advantage of being in primary balance. Although the inner and outer pairs of pistons (1 and 4 ; 2 and 3) move in opposite directions to give primary balance, their accelerations and decelerations at the two ends of the stroke are not matching due to which there is secondary unbalance. In modern engines secondary vibrational disturbances are isolated with flexible engine mountings. Formerly harmonic balancer was used to counteract secondary unbalance.

In four cylinder in-line engines, the order of firing could be 1-3-4-2 or 1-2-4-3. In former case piston No. 2 moves upward on

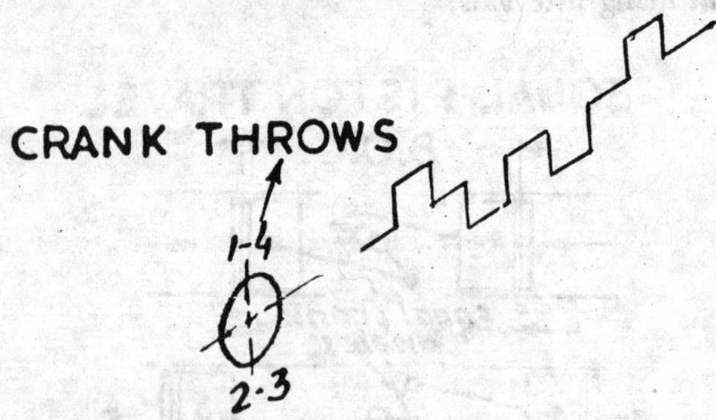

Fig. 5·21. Crank arrangement in 4-cylinder in-line engine.

exhaust stroke and piston No. 3 on compression. In second case, piston No. 2 moves upward on compression stroke and piston No. 3 on exhaust in view of power balance.

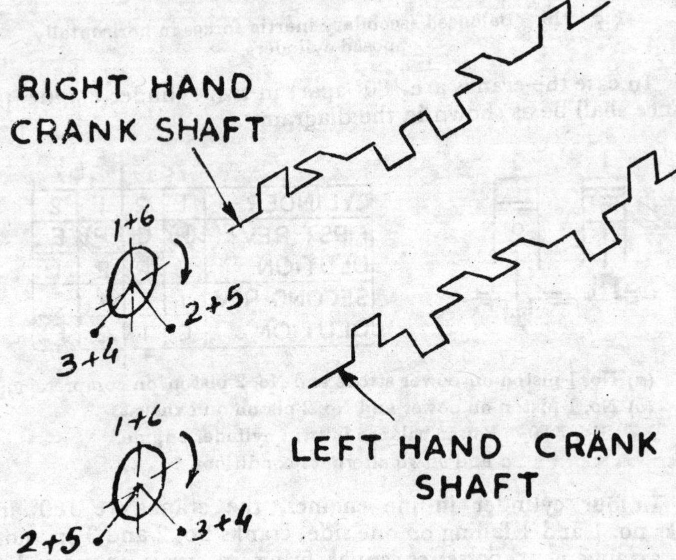

Fig. 5·22. Crank arrangement in 6-cylinder in-line engines.

Six cylinder in-line engines provide three power impulses for every revolution of crankshaft as shown in the power flow chart. In order to obtain even firing intervals, the cranks are paired for the outermost, intermediate and innermost cylinders and spaced equally 120° apart in 3 planes (1 and 6 in one plane ; 2 and 5 in other plane and 3 and 4 in third plane). Owing to this symmetry, the in-line six cylinder engine has both primary and secondary balance and hence smooth delivery of power.

Six cylinder in-line *left-hand* crankshaft has third and fourth throws to the left of No. 1 and 6 throws as viewed from the front side In this arrangement when pistons No. 1 and 6 start to move downwards, No. 2 and 5 are completing the downward stroke and No. 3 and 4 are on their upward stroke. In this arrangement four different firing orders are possible with good engine balance. The standard firing order is 1-4-2-6-3-5.

A *right hand* six-cylinder engine crankshaft has No. 3 and 4 throws to the right of No. 1 and 6 throws as viewed from front. In this design, when No. 3 and 4 pistons are completing their downward stroke, No. 1 and 6 are starting downward movement and No. 2 and 5 are on their upward stroke. In this arrangement too, four different firing orders are possible whereas the standard one is 1-5-3-6-2-4.

Eight cylinder in-line engines provide four power strokes for every revolution of crankshaft as shown in power flow chart. Although these engines provide smooth operation because 2-4-2 or *split four* disposition of crankshaft throws possessed both primary

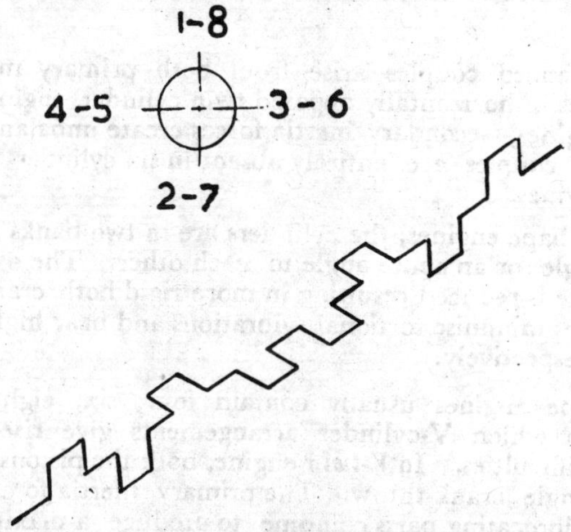

Fig. 5·23. Crank arrangement in eight-cylinder in-line engine.

and secondary balance yet due to very large displacement they are not widely favoured. They are no longer employed in modern cars owing to their great length and consequent demands on body space. Keeping in view the demands of larger body space. tendency to position the engine transversely is gaining momentum Latest models of Austin and Morris cars contain *transverse engines* with front wheel drive.

Due to relatively long crankshafts required in six and eight cylinder in-line engines, tortional vibrations creep in which are suppressed by a *vibration damper* at the front end of crankshaft.

Horizontally opposed cylinder engines have their cylinders in two opposite banks on the crankcase. The opposing cylinders are slightly offset from one another due to constructional reasons. These engines are known as *flat* or *boxer* units and are popular for good balance of reciprocating parts, low centre of gravity for better stability of car and a short engine structure occupying less space. Due to shorter size, this engine is suitable both for front-wheel drive and rear engined cars. It can be fitted either ahead or behind the centre line of driven wheels with minimum overhang. Owing to its low overall height, a sloping bonnet can be designed for the car for better streamlining.

Horizontally opposed cylinder engines may have two, four or six cylinders. Each of them have even firing intervals and possesses primary and secondary balance because the corresponding pistons in each bank move in opposite directions to one another and their accelerations and decelerations at the two ends of the stroke are matched. Owing to offset cylinders, the otherwise balanced reciprocating parts tend to create a rocking movement or couple on the crankshaft which forces the engine to oscillate or *yaw* about a vertical axis.

Unbalanced couples arise from both primary and secondary inertia forces in horizontally opposed twin cylinder engines. In four cylinder engines, secondary inertia forces create unbalanced whereas unbalanced couples are entirely absent in six cylinder horizontally opposed engines.

In V-shape engines, the cylinders are in two banks set at either a right angle or an acute angle to each other. The overall length of the engine is reduced resulting in more rigid both crank-shaft and structure to minimise tortional vibrations and bear higher combustion loads respectively.

V-shape engines usually contain four, six, eight or twelve cylinders in which V-cylinder arrangements give rise to certain balancing difficulties. In *V-twin* engine, both the pistons are connected to a single crank-throw. The primary inertia forces generated by the reciprocating parts combine to produce a maximum inertia force when the crank-throw bisects the V-angle. This force can be balanced out by fixing counter-weights at the opposite end of crank-throw owing to the fact that this force remains constant and rotates at crank speed. However, it is not possible to balance the secondary inertia forces in V-twin engine.

An acute angle of 60° is generally adopted between the two banks in four or six cylinder V-shape engines. For obtaining equal firing intervals in four cylinder engines, the crank-throws are unequally spaced rotationally which lacks in symmetry. This gives rise to rotational unbalance. The relative motions of pistons create both primary and secondary unbalance.

In 60° V-6 engine, the crank-throws are equally spaced rotationally and both primary and secondary balance of reciprocating parts can be achieved. The secondary inertia forces impose a rocking moment upon the crankshaft but the resulting vibration is absorbed by the engine mountings. In view of this, V-6 engine is not so well balanced as six cylinder in-line engine.

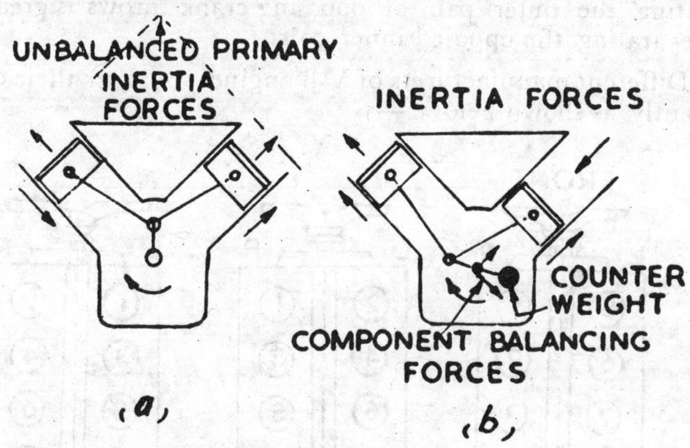

Fig. 5·24. V-twin engine.
(a) Unbalanced primary inertia forces.
(b) Primary balance by attaching counterweight.

V-8 engine has been widely favoured for higher displacement (capacity). An angle of 90° has been adopted between the cylinder banks for obtaining even firing intervals. Like other V-shape engines, the cylinder banks are slightly offset to each other along the axis of crankshaft to facilitate side-by-side fitting of connecting rod pairs.

Formerly, a single-plane crank-shaft having four crank throws as used in four cylinder in-line engines, was employed in V-8 engines. Two connecting rods from opposite pistons were connected at one crank-pin. In this arrangement, the primary inertia forces

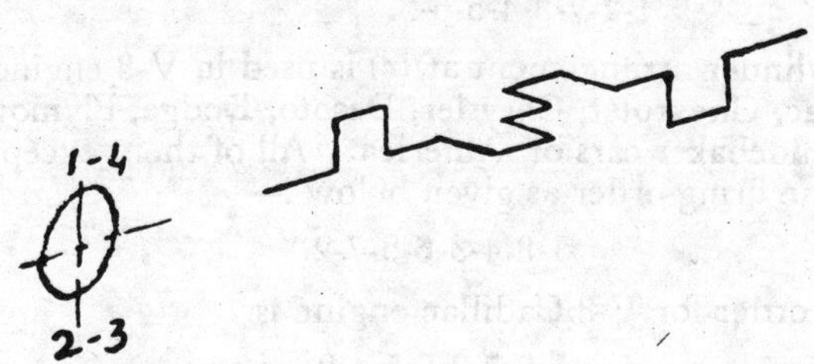

Fig. 5·25. Two plane crank-shaft in V-8 engine.

were balanced because the inner and outer pairs of pistons in each bank travelled in opposite directions. But the secondary inertia forces generated in each cylinder bank could not be balanced. They imposed a rocking movement on the crankshaft in the horizontal plane. In order to provide better balance, two plane crankshaft having two crank throws at right angle to the other two, has been brought in use. With this arrangement of crank-throws, both primary and secondary balance has been achieved. However, the crankshaft is subject to an appreciable rocking moment because the distance separating the outer pair of opposing crank-throws is greater than that separating the opposed inner pair.

Different manufacturers of V-8 engines number their cylinders differently as shown below :—

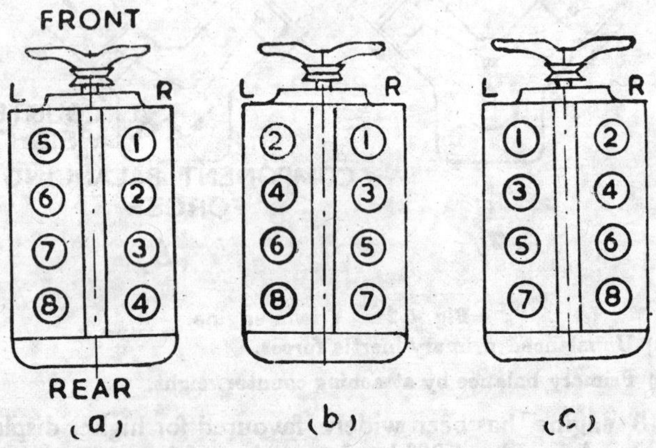

Fig. 5·26. Numbering of cylinders in V-8 engines.

Firing orders differ according to the cylinder numbering system. The cylinder arrangement at (*a*) is used in V-8 Ford, Mercury and Lincoln engines having a firing order of 1-5-4-8-6-3-7-2.

The cylinder arrangement at (*b*) is used in Buick V-8 engine firing in the following order :—

1-2-7-8-4-5-6-3.

The cylinder arrangement at (*c*) is used in V-8 engines employed in Cadillac, Chevrolet, Chrysler, Desoto, Dodge, Plymouth, Pontiac and Studebaker cars of America. All of them except Cadillac have the same firing order as given below :—

1-8-4-3-6-5-7-2.

Firing order for V-8 Cadillac engine is

1-8-7-3-6-5-4-2.

13. Engine Cranking. For starting an engine, it is first cranked after putting on fuel and ignition. Since diesel engines are compression ignition engines, so there is no ignition switch for them. Before cranking, engine is disconnected from the drive line through clutching arrangement. In case the engine is installed on an automobile, it is ensured that the gear is in neutral position before it is cranked. As soon as the engine fires, cranking is stopped.

There are different ways to crank the engines. Rope, cranking handle, cranking motor (self-starter), kick starter, air starter, starting engine etc. are employed in engine cranking. *Rope* is used to crank engines in small generating sets, lawn-mowers, pumping sets, tractors etc. The rope is wound around the cup or pulley at the engine crankshaft and pulled up forcefully. Thus torque applied at the crankshaft sets it in motion. The process is repeated again and again till the engine starts.

Starting handle is a sort of crank which is a common tool with the vehicle to crank the engine. It is fixed in the dog-nut fitted in front of crankshaft and rotated manually. As soon as the engine fires, the cranking handle slips out of the dog-nut due to difference of speed.

Kick starter is quite common with motor cycles and scooters. It is a ratchet type arrangement which works through foot operated pedal. The engine is cranked by kicking the pedal.

Cranking-motor is employed in almost all modern automobiles to crank the engine. It is an electrically operated motor fitted with drive pinion unit on its shaft. By switching on motor, the drive pinion is engaged with the flywheel ring gear and the engine is rotated. As soon as the engine fires, the drive pinion goes out of mesh due to difference of speed.

Air starter works under air pressure supplied from air cylinder or air pressure line. It is usually employed in armoured tanks and aeroplanes. The air under pressure provides thrust at the blades of rotor fitted on crankshaft due to which the engine is rotated. The flow of air is discontinued as soon as engine is started.

Starting engine is used to crank heavy diesel engines employed in earth-movers. Caterpillar tractor D-7 employs twin cylinder petrol engine to crank the main diesel engine for starting purposes. In order to start the main engine, the starting engine is first started and then connected with the main engine through the drive unit. The starting engine transmits its power through a clutch and two speed gearbox to a sliding pinion. The pinion is engaged with the diesel engine flywheel-ring-gear by means of linkage controlled by a hand operated lever. It is automatically released by centrifugal force acting on the pinion latches when the diesel engine starts.

In case of difficult engine starting, the vehicle is pushed or pulled (towed). As soon as it gains momentum, gear is engaged and clutch released for engine cranking. The gear lever is put in neutral position after the engine is started.

14. Engine tuning. An engine should produce rythm while running. It should not misfire, vibrate or run out of tune. Engine tune-up or tuning is not just a carburettor adjustment. In a petrol engine, it covers the following four phases :—

(*i*) Compression.

(*ii*) Ignition system.

(*iii*) Fuel system.

(*iv*) Cooling system.

Excellent performance depends upon how well the engine is tuned. It is therefore necessary that the four phases of an engine tune-up be properly synchronized to ensure and maintain the high performance and economy built into the engine.

Before making any checks or adjustments in the engine, check up the battery first. Examine the battery connections for tightness and cleanliness. If the specific gravity is below 1·200, the battery should be recharged before proceeding further. If the reading is above 1·200, connect voltmeter with battery terminals, press the starter switch button for 15 seconds and note down reading at voltmeter. In case of 6 volts and 12 volts batteries, if the reading is below 5 volts and 10 volts respectively, the battery should be recharged.

If the battery is alright, the next step is the *compression test* to determine whether the engine is in a tuneable condition or not. Proceed as under to check up compression :—

(*i*) Bring the engine to *operating* temperature by running it for some time.

Fig. 5·27. Compression gauge.

(*ii*) While engine is warming up, lubricate valve stems and ensure that the tappets are not riding.

(*iii*) Switch off ignition and remove spark plugs.

(*iv*) Insert compression gauge in spark plug hole of cylinder No. 1, fully open the throttle and crank the engine with self-starter until the gauge attains highest reading. Note down the reading and repeat this operation on all cylinders. Compression pressure in all the cylinders should read alike to within 10 lbs. The average pressure should be near about 100 lbs. per square inch.

If the compression readings are low and vary widely, inject engine oil (mobil-oil SAE 30 or 40) into the cylinders through spark plug holes and crank the engine several times. After this, check up engine compression again. If there is practically no difference or rise or fall in compression, it indicates sticking or poorly seating valves. If there is rise in compression, it indicates compression loss past the piston and rings.

The cause of low or uneven compression must be corrected before proceeding further. If the compression is satisfactory, take up each system one by one and service them.

Servicing ignition system :—

(*i*) Examine the spark plugs. Clean and test them on a *spark plug cleaner and tester.*

Replace the defective spark plugs with those having specified *heat value*. Adjust gap as per specifications with the help of *spark-plug gap gauge.*

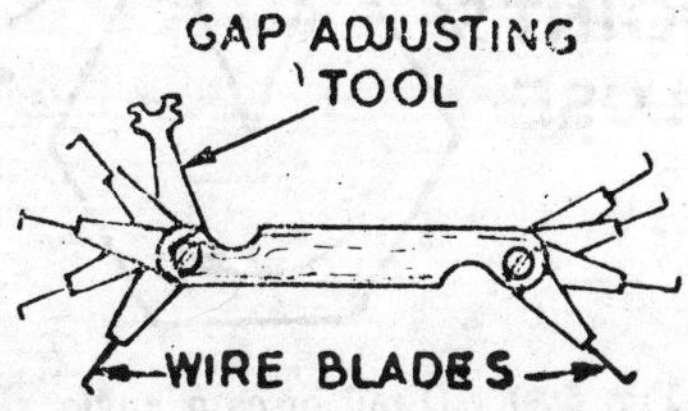

Fig. 5·28. Spark plug gap gauge.

(*ii*) Examine contact breaker points. Replace the badly

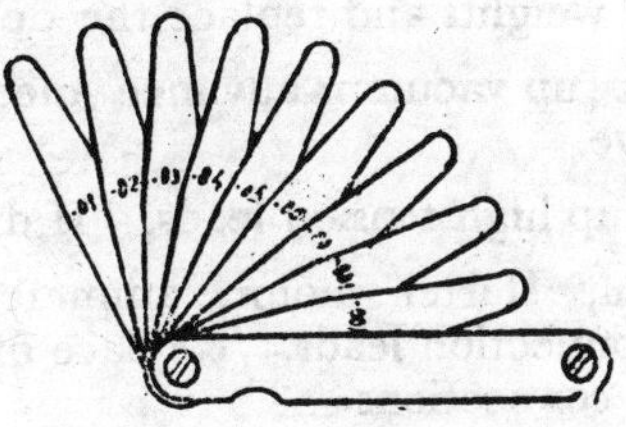

Fig, 5·29. Feeler gauge.

pitted or worn points. Reface or clean if slightly pitted or dirty. Adjust gap as specified with the help of a *feeler-gauge.*

(*iii*) Check insulation of condenser pig tail. If damaged, remove the condenser and check up its serviceability with the help of *coil-candenser tester*.

Replace the defective condenser or repair the leaking cable, as the case may be.

(*iv*) Check up distributor cap and rotor for cracks. Replace the defective parts. Clean the rotor tip and cap segments with zero number emery-cloth.

(*v*) Check up distributor shaft for wear. Replace the bush if the play is excessive.

(*vi*) Check up *dwell angle* which is the number of degrees of cam rotation when the points are closed as shown in the diagram.

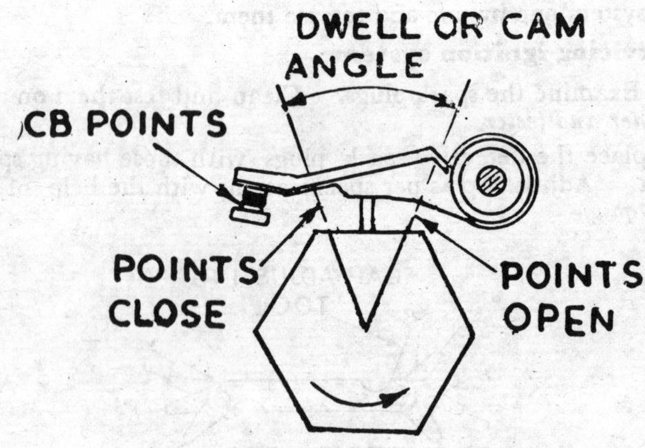

Fig. 5·30. Dwell or cam angle.

Replace the worn out cam.

(*vii*) Check distributor governor. Clean and lubricate the sticking governor weights and replace the defective springs.

(*viii*) Check up vacuum advance mechanism and repair or replace if defective.

(*ix*) Check up high tension leads. If defective, replace them.

(*x*) Check up starter motor, ammeter, ignition switch, coil and distributor connection leads. Replace or repair defective cables and tighten loose connections.

(*xi*) Check generator and regulator connections. Tighten loose connections and repair any short circuit.

(*xii*) Check up and set ignition timing with the help of a *timing-light*.

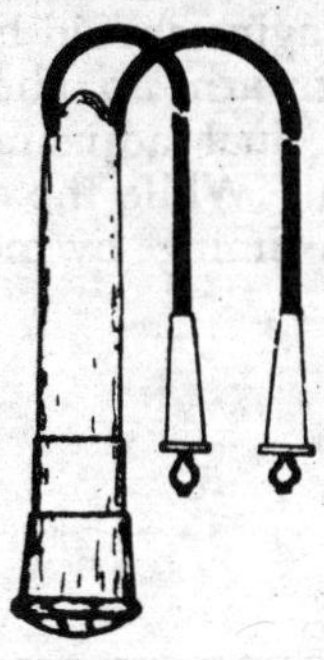

Fig. 5·31A. Neon Timing light (Torch type).

Servicing fuel system :

(*i*) Check up fuel feed lines for any kind of leakage or restriction. Blow out the fuel lines with air under pressure.

(*ii*) Check up fuel pump. Unscrew glass bowl and remove sediment. Clean the filter screen and replace defective gasket of glass bowl.

(*iii*) Remove and clean air-filter or cleaner. Refill oil up to marked level.

(*iv*) Remove carburettor and disassemble it. Clean all the parts in petrol. Clear all the holes in the jets and passages in the carburettor body with air under pressure. ***Never pass wires through the jets or calibrated orifices lest they are damaged.***

Replace the defective parts and gaskets and reassemble the carburettor. Before putting on float-chamber cover, check up and adjust *float-level and height of metering rod* if applicable.

(*v*) Remove and clean positive-crankcase-ventilation valve (PCV valve) if applicable.

Servicing Cooling System :

(*i*) Flush out radiator.

(*ii*) Check up water hoses. Replace the defective rubber hoses or tighten the loose clamps in case of leakage.

(*iii*) Check up fan belt play. Adjust if loose.

(*iv*) Check up water pump. Replace seal and bearing if defective.

After servicing the above systems, start the engine and bring it to working temperature. In order to eliminate all unequal expansion of engine parts, the engine should be run for about half an hour on fast-idle speed. After warm up, check up and adjust *tappet-clearance* as specified. The final adjustments of idle and slow-running should be made then. While the engine is running, check up the correctness of ignition-timing by means of a Neon Timing light.

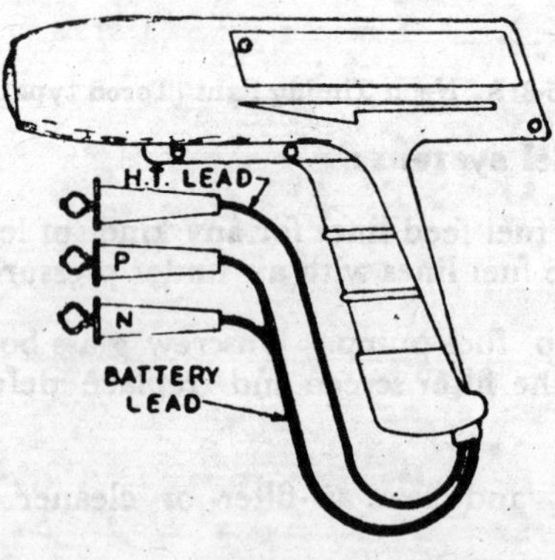

Fig. 5·31 B. Timing light (Pistol type).

After ***final adjustments***, the vehicle should be road tested to check up the performance of engine. The engine should "*ping*" slightly when the vehicle is quickly accelerated from 15 km/h in top gear. This ping should disappear at about 30 km/h speed. Adjust ignition-timing if required.

Conditions differ in single cylinder air cooled engines wherein there is no radiator, water pump, over flow tank, water hoses etc. In two stroke petrol engines, there are no valves and valve operating mechanism. In single cylinder engine, no distributor is required and the gravity fuel feed system needs no fuel pump. There are variations from vehicle to vehicle and engine to engine. Hence tuning job should be carried out according to the construction and working of the engine keeping in view the specifications and instructions of the manufacturer contained in the *servicing manual.*

In diesel engines, fuel is injected into the engine by means of Fuel Injection System. There is no spark-ignition system as in petrol engines. The tuning of diesel engines need the servicing of only fuel injection and cooling systems. As the servicing of cooling system is just like those in petrol engines so let us take up the servicing of F.I. system.

Servicing of F.I. System :

(*i*) Check up leakage or restriction in the fuel feed and return lines. Bleed out the system to expel air. Tighten loose connections.

(*ii*) Remove fuel filter elements, replace or clean them.

(*iii*) Check up injectors. If fuel is reaching the injectors and it is certain that there is no air in the fuel system, any of the injectors fails to give the characteristic *ping* or *squeak*, proceed as under to check up and test faulty injector :—

(*a*) Starting from No. 1 cylinder, slacken off union nut on the injector end of fuel pipe while the engine is running. This prevents fuel being pumped to the injector nozzle. Note the effect on engine. If the cutting of injector has no effect on engine running, the injector is faulty. Repeat this process with each injector one by one, tightening the union of one before proceeding to next.

(*b*) Remove the faulty injector, couple it to its fuel pipe and lay it in such a position so that the spray and holes could be seen and examined. *Keep your hands and face out of the direct line of spray holes lest you are hurt.*

Turn over the engine and watch the spray. If instead of fine mist spray there is dribbling of fuel, the injector is faulty. It should be sent to F.I. service station for proper servicing.

(*iv*) Remove and clean air filter or cleaner.

(*v*) Check up governor.

(*vi*) Check up F.I. pump. Disconnect fuel pipes at the pump. Turn over the engine and watch pumping of fuel at each element of fuel pump. If there is any defect in the fuel delivery, check up delivery valve. If the valve is seating properly, the pump element is defective. In such a case, remove pump and send it to F.I. service station for servicing.

(*vii*) Check up fuel feed pump. Remove sediment bowl. Clean the filter screen, remove sediment and replace defective gasket.

Before proceeding with tune up job in diesel engines, ensure that there is sufficient fuel in the fuel tank. Due to lack of fuel, air will trap into the fuel line and trouble shall breed in. If fuel supply is trouble free, very few items are left over to attend. Tappet adjustment is the only major task in the final adjustment.

15. Engine rating. All engines are rated in horse power.

Horse Power is simply a unit of measurement based on an arbitrary assumption that the average horse can raise 33,000 pounds of weight to a height of one foot or 4500 kg. of weight to a height of one metre in one minute.

The horsepower developed in an engine depends upon many factors, such as compression, revolutions per minute, piston diameter, length of stroke and number of cylinders.

Several types of horse power ratings are used for engines. These are Brake Horse Power, Indicated Horse Power, Friction Horse Power, Taxable Horse Power etc.

Brake Horse Power. The brake horse power (B.H.P.) of an engine is the actual horse power delivered at the flywheel. It is measured by means of dynamometer or Prony Brake.

Indicated Horse Power. It is the rate at which work is done by the expanding gases in pushing the piston down.

The pressures acting on a piston during power stroke are not uniform throughout the stroke. At the start of the stroke, the pressures are at the highest degree and decrease as the piston moves downward. Pressures within the cylinder are recorded by means of *Pressure Indicator* and from the available data, an average or mean effective pressure is determined.

Mean effective Pressure (M.E.P.) is the average of the pressures exerted on the piston during the power stroke minus the

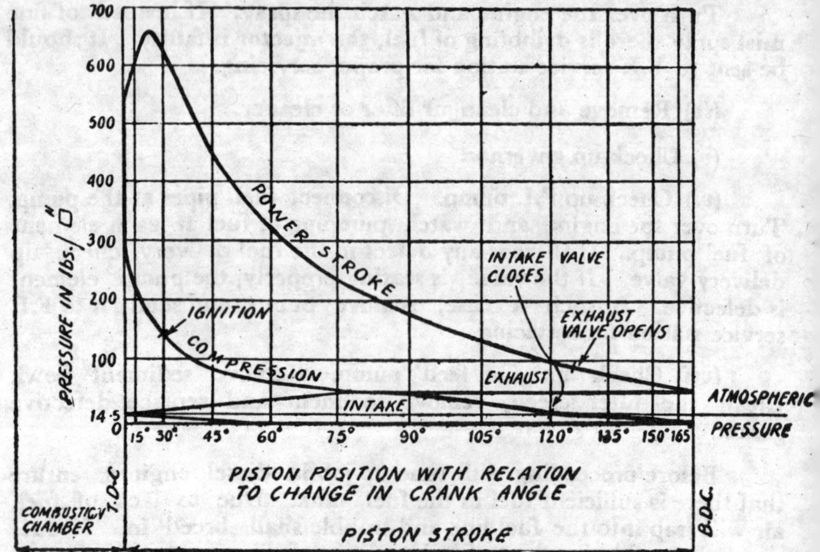

Fig. 5·32. Pressures within cylinder during various strokes.
(Indicator Diagram)

average of opposing pressures encountered on intake, compression and exhaust strokes.

M.E.P. = Positive average Pressure (Average Pressure during Power Stroke)—Negative average Pressure (Average pressure during intake, compression and exhaust stroke)

If the approximate average pressures in the cylinder are —

(*a*) Induction Stroke = 2 lbs/□" (Below atmosphere)
(*b*) Compression Stroke = 35 lbs/□" (Above atmosphere)
(*c*) Power Stroke = 160 lbs/□" (Above atmosphere)
(*d*) Exhaust Stroke = 2 lbs/□" (Above atmosphere)

Then the resultant mean effective pressure = 160 − (35 + 2 + 2)
= 160 − 39 = 121 lbs/□"

Indicated Horse Power could be determined by the following formula :

$$\text{I.H.P.} = \frac{\text{PLANK}}{33{,}000}$$

where
P = m.e.p. in lbs/□"
L = Length of stroke in ft.
A = area of cylinder in sq. in.
N = No. of power strokes per minute
K = No. of cylinders.

or
$$\text{I.H.P.} = \frac{\text{PLANK}}{4500}$$ where m.e.p. is in kg./cm^2
L in meters
A in cm^2

Indicated horse power of an engine is always higher than the actual brake horse power developed by the engine as it does not take into account the power necessary to overcome engine friction. Therefore .

I.H.P. = B.H.P. + Friction Horse power

Friction Horse Power (F.H.P.). The power absorbed by the engine to turn itself over, shear through oil film and to operate the valves and accessories etc. is known as friction horse power.

F.H.P. = I.H.P. − B.H.P.

Taxable Horse Power. It is that horse power on the basis of which tax is charged. This is also known as S.A.E. Horse power which is worked on with the help of the following formula :

$$\text{S.A.E. hp} = \frac{D^2 \times N}{2 \cdot 5}$$

where D=Diameter of cylinder or bore.

N=Number of cylinders.

Engine Torque. Torque is directly related with the horse-power of the engine. It is the turning effect. Power impulse available at the piston is transmitted to the crankshaft through connecting rod. This power impluse acts through the crank arm to develop a torque or turning force at the crankshaft. The amount of torque developed in the engine depends upon the pressure exerted on the pistons and the length of crank arm. Torque is measured in lbs. ft. or kg metres.

Torque=Pushing force × Length of arm

The torque increases with the increase in engine speed up to about 2,500 r.p.m. and then decreases as the engine speed continues to increase. As the engine speed continues to increase, the decrease in torque is due to decreased amount of fuel charge that enters the cylinders because of high engine speeds. The speed or r.p.m. at which the maximum torque is obtained, is determined by the design of the engine.

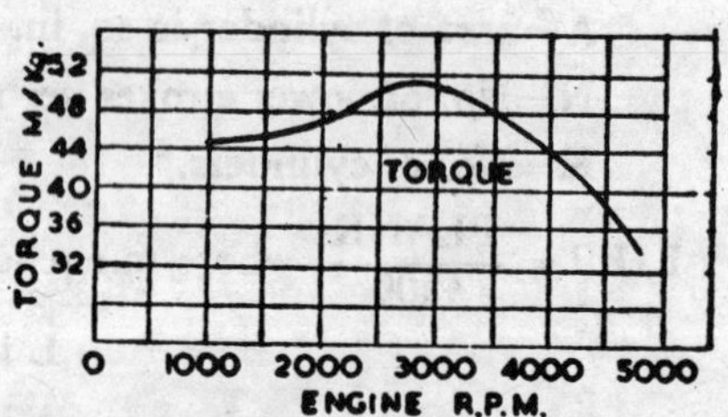

Fig. 5·33. Torque curve of 6·3 Lit V-8 Mercedes Benz Engine Type 600.

16. Engine Efficiency. The ratio of input and output is known as efficiency. In an engine, it is measured in terms of percentages and in relation to fuel induction, heat supplied by the fuel and power produced by the heat supplied by the fuel. With this regard, engine efficiency is known in three ways such as—

(*a*) Volumetric Efficiency.

(*b*) Thermal Efficiency.

(*c*) Mechanical Efficiency.

(*a*) **Volumetric Efficiency.** It is the ratio between the amount of fuel air mixture which actually enters the cylinders and the total piston displacement of the engine.

$$\text{Volumetric Efficiency } (\eta_v) = \frac{\text{Volume of charge at atmospheric temperature and pressure}}{\text{Piston Displacement}}$$

The volumetric efficiency of modern I.C. engines is usually less than 100%. The efficiency of supercharged engines may exceed 100 per cent, as sometimes fuel-air mixture is supplied to the cylinders at pressures higher than atmospheric pressures.

More the volumetric efficiency, more the power shall be developed by an engine. At much higher speeds, volumetric efficiency of the engine falls down due to increase in speed as there is less time for the fuel charge to enter into the cylinders. This affects the torque of the engine. More the volumetric efficiency, more the engine torque.

The following factors tend to decrease volumetric efficiency :

(*i*) High engine speeds.

(*ii*) The degree of throttle opening.

(*iii*) Improper valve timing.

(*iv*) Incomplete scavenging of exhaust gases.

(*v*) Preheated air.

(*vi*) High temperatures in the combustion chamber.

(*vii*) Low atmospheric pressures.

(*b*) **Thermal efficiency**. The thermal or heat efficiency of an engine is the ratio between the power output and the energy contained in the fuel.

Thermal Efficiency (η_t)

$$= \frac{\text{B.H.P.} \times 33{,}000}{\text{weight of the fuel burned per minute} \times \text{heat value of fuel} \times 778}$$

or

$$\frac{\text{B.H.P} \times 4500}{W \times Q \times J}$$

where 4500 kgf. m/min.=one B.H.P.

W=weight of fuel supplied per minute

Q=Heat value of fuel in kcal/kg.

J=427 kgf /kg./°K

(Mechanical equivalent of heat)

All of the heat produced by the fuel is not converted into useful work as most of the heat is lost in cooling system and engine friction and carried away by the exhaust gases. Due to heat lost in engine operation, thermal efficiencies are near about 25 per cent.

(c) **Mechanical Efficiency.** It is the ratio between the power actually delivered at flywheel (B.H.P.) and the power developed within the cylinders of an engine (I.H.P.)

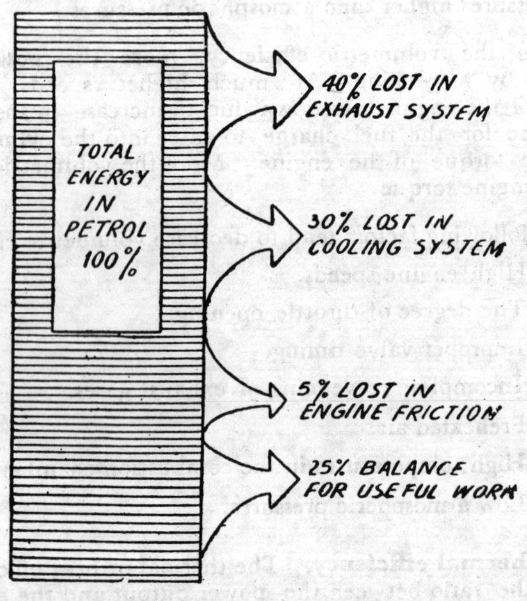

Fig. 5·34. Energy loss in engine.

$$\text{Mechanical efficiency } (\eta_m) = \frac{\text{Brake Horse Power}}{\text{Indicated Horse Power}}$$

The mechanical efficiency of an I.C. engine is approximately 90%. It is never 100 per cent as some of the power is lost in overcoming engine friction.

17. Comparison between Petrol Engine and Diesel Engine.

S. I. or Petrol Engine	*C. I. or Diesel Engine*
1. Spark ignition since engine ignition is by means of a spark produced by the spark plug.	1. Compression ignition engine since ignition takes place due to heat produced by high compression of air.
2. Otto Cycle or Constant Volume Cycle Engine.	2. Diesel Cycle or Constant Pressure Cycle Engine.
3. Petrol is used as fuel.	3. Fuel used is diesel.
4. Mixture of petrol and air is induced during intake stroke.	4. Only air is drawn in during intake stroke.
5. Carburettor for the supply of fuel air mixture and spark plug and distributor or magneto etc. for the supply of spark, are essential accessories of the engine.	5. No carburettor and spark plugs are required but in lieu injectors and fuel injection pump is fitted with the engine.
6. Quality of fuel air mixture is controlled.	6. Quantity of injected fuel is controlled.
7, Spark is provided into the combustion chamber when the piston reaches near compression T.D.C.	7. Fuel is injected into the combustion chamber when the piston reaches near compression T.D.C.
8. Average compression pressure 100 lbs per square inch.	8. Average compression pressure 500 lbs. per square inch.
9. Compression ratio range 5—9 : 1	9. Compression ratio range 14—20 : 1
10. Temperature of compressed fuel mixture 60°C to 80°C.	10. Compressed air temperature 500°C.
11. Thermal efficiency less (about 25%).	11. Higher thermal efficiency—upto 40%.
12. High speed engine.	12. Comparatively low speed.
13. Light in construction.	13. Heavy in construction.
14. Gives radio interference due to electrical ignition system.	14. Does not create radio interference due to the absence of spark ignition.
15. Occupies less space.	15. Occupies more space.
16. Greater risk of fire due to petrol being volatile and highly combustible.	16. Greater security from risk of fire as diesel is non-volatile.
17. Initial cost less.	17. Initial cost more than a petrol engine of same power.
18. Produces less torque.	18. Torque characteristics are better.

18. Comparison between two stroke engine and four stroke engine.

Two Stroke Engine	*Four Stroke Engine*
1. The cycle is completed in two strokes of the piston or in one revolution of the crankshaft.	1. The cycle is completed in four strokes of the piston or in two revolutions of the crankshaft.
2. Provides one power stroke in each revolution of the crankshaft.	2. Gives one power stroke in every two revolutions of the crankshaft.
3. Mostly valveless.	3. Contains valves and valve operating mechanism.
4. Usually air cooled.	4. Mostly water cooled.
5. In petrol engines, fuel air mixture is induced into crank chamber which is required to be air tight.	5. Fuel directly goes into the combustion chamber and has nothing to do with crank chamber.
6. Oil for lubrication of the petrol engine is mixed with petrol in the tank, which goes into the crank chamber as petroil mixture.	6. Oil for lubrication is contained in the oil sump or in a separate oil tank from where it is fed to various parts.
7. Simple in construction having only piston, connecting rod, crankshaft and flywheel as the main working parts.	7. More complicated due to valves and valve operating mechanism, lubrication and water cooling systems etc.
8. Its piston contains a baffle at the crown to help in separating off fresh and used gases in early engines.	8. There is no need of such baffle in the pistons.
9. Mostly single cylinder.	9. Mostly multi-cylinder.
10. Lighter in weight.	10. Heavy in weight.
11. Volumetric efficiency less due to lesser time for induction.	11. Volumetric efficiency more due to more time for induction.
12. Useful gases escape with exhaust gases.	12. Loss of useful gases very rare.
13. Gets hot soon as more heat is produced, combustion taking place in each revolution.	13. Does not get hot soon as less heat is produced, combustion taking place once in every two revolutions.
14. Rate of wear and tear more.	14. Wear and tear rate less due to full-fledged lubricating system.
15. More even torque due to more even power impulses.	15. Turning moment not so uniform.
16. Lighter flywheel fulfils the requirement.	16. Heavier flywheel needed to keep the engine running at uniform speed.
17. More power is produced for the same size engine.	17. Comparatively less power is produced.
18. Initial cost less.	18. Initial cost more.

19. Comparison between 4-stroke single and multi-cylinder engines

Single cylinder engine	Multi-cylinder engine
1. One cylinder.	1. More than one cylinder.
2. Smaller size.	2. Bigger size.
3. Light in weight.	3. Heavy in weight.
4. Lower output.	4. Higher output.
5. Less cost.	5. High cost.
6. Delivers only one power impulse in one cycle or two revolutions of crankshaft.	6. Delivers more than one power impulse in one cycle.
7. Heavy flywheel needed to smooth out torque fluctuation.	7. Lighter flywheel due to better power balance.
8. For getting more power, size of cylinder cannot be increased beyond a certain limit due to dynamic* considerations.	8. Power output can be increased by increasing number of cylinders.
9. Dynamically unbalanced.	9. Dynamically better balanced.
10. Not suitable for cars, trucks, buses etc., due to insufficient power output and vibrations caused by speed fluctuations.	10. Suitable for cars, buses and trucks but not economical for scooters, mopeds, motorcycles etc

*Inertia forces are generated by the accelerating and decelerating reciprocating parts in the engine. By increasing cylinder size, weight of piston assembly and the upper part of connecting rod is also multiplied. This results in the increase of inertia forces to an unacceptable level because they vary as the square of engine speed. If the speed is doubled the inertia force will become four times as great resulting in engine vibration and overloading of crankshaft bearings.

20. Advantages and disadvantages of Diesel engine :

(*a*) **Advantages** :

(*i*) It is economical due to cheap and less fuel consumption.

(*ii*) It has greater security from fire risk.

(*iii*) It needs no ignition system or carburettor.

(*iv*) It is free from radio interference as no high tension current is used in it.

(*v*) It requires less water-proofing.

(*vi*) Fuel is equally distributed to all cylinders in it.

(*vii*) It needs less cooling as it is designed to work at high temperatures.

(*viii*) There are less chances of its overheating as whole of the air drawn into it does not combust but a small amount of air remains unburnt which cools the cylinder up to some extent.

(*ix*) It could be put on load immediately after starting whereas petrol engine needs warming up.

(*x*) There are less chances of fuel to evaporate in it.

(*xi*) In emergency, it could be run on kerosene or lubricating oil.

(*xii*) Its fuel injection equipment is more reliable than the ignition system of petrol engine.

(*xiii*) Its thermal efficiency is more whereas cylinder temperature is less.

(*xiv*) It produces more torque due to its greater thermal efficiency.

(*xv*) It could be kept running for long periods.

(*b*) **Disadvantages :**

(*i*) Its initial cost is more.

(*ii*) It is heavy.

(*iii*) It produces less power in relation to weight.

(*iv*) It is difficult to start during cold.

(*v*) It gives more noise due to diesel knock.

(*vi*) It requires more efficient filters.

(*vii*) Its governors are more complicated.

(*viii*) Servicing of its fuel injection equipment requires specialised repair and machines.

21. **Air pollution and its control.** As explained earlier internal combustion engine converts heat energy into mechanical power to propel the vehicle. Heat energy is obtained from the fuel carried in the fuel tank. Mostly petrol and diesel oil is used as fuel

for automobile engines. These fuels contain hydrogen and carbon in various combinations. During combustion, oxygen combines with hydrogen and carbon to form water (H_2O), carbon monoxide (CO) and carbon dioxide (CO_2). The fuels containing sulphur produce sulphur oxides. Due to combustion heat, some of the nitrogen in the air combines with oxygen forming nitrogen oxides (NO_2). Some of the fuel goes unburnt resulting in smoke and ash.

Petrol being more volatile liquid, evaporates due to heat effect at higher rate. Hydrocarbon (petrol) vapours from fuel tank and carburettor; *crank-case blow-by* containing partly burned air-fuel mixture which blows by the piston rings and into crank chamber ;. exhaust gas constituents consisting of partly burned petrol (HC), carbon monoxide, nitrogen oxides and if sulphur in petrol, sulphur oxides, pollute the air. As a result *smog* is created, the atmosphere becomes dirty and breathing becomes difficult.

Smog is a kind of fog mixed with other substances. The smog covers the cities like a blanket for days together during winter. The heat generated in large cities tends to circulate air within a dome-like shape. It traps smog and holds it over the city affecting the visibility.

Smog along with smoke is the most visible evidence of air pollution. But some atmospheric pollution is not visible until mixed with moisture. Unburnt petrol, carbon monoxide, leaded compounds from leaded petrol and other gases which pollute the air, may not be seen. All these polluting substances are deadly harmful for men, animals and food crops. Smog blurs vision, irritates the eyes, throat and lungs. Eyes tear up, throats get sore, people start coughing and fall sick, making the sick sicker.

Air pollution gives birth to troublesome diseases such as asthma, eczema, emphysema, cardiovascular troubles, lung and stomach cancer. It has damaging effect on food crops and animals and the paint starts peeling from the houses.

The automobile gives off pollutants from the fuel tank, carburettor, crank-case and silencer tail pipe if not controlled. In order to reduce or prevent pollution from these sources, the following measures have been taken in the modern automobiles :—

(*i*) Positive crankcase ventilation.

(*ii*) Vapour recovery system.

(*iii*) Cleaning up the exhaust gas.

(*i*) **Positive crankcase ventilation** (*PCV*). In order to remove *blow-by*, the crankcase must be ventilated. In earlier engines, the crankcase was ventilated by an opening at the front of engine and a vent tube at the rear. The forward motion of the vehicle and rotation of crankshaft caused air to flow through and remove blow-by gases, water and fuel from the crank case, which appear when the engine is cold. The discharge of fuel vapours or partly burnt gases into the atmosphere cause air pollution. For preventing air pollution, the modern engines have a closed system known as *positive crank-case ventilation* (*PCV*) *system*. In this system filtered air from the air cleaner is drawn through the crank-case from where it picks up water, fuel vapours and blow-by gases. The air then flows back to the induction manifold and enters the engine where the unburnt fuel is burned.

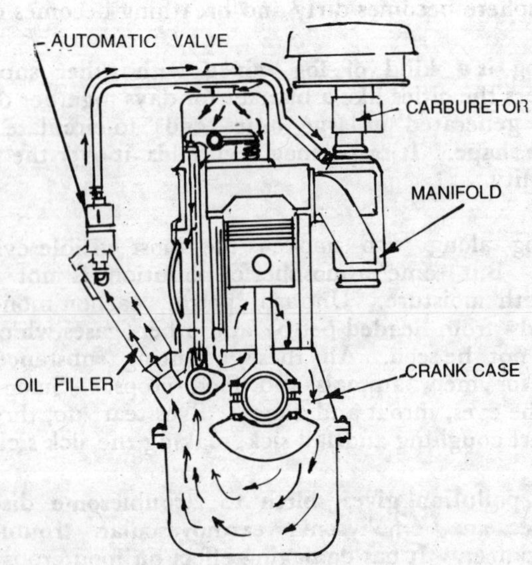

Fig. 5·35. Positive crankcase ventilation system.

Too much air flowing through induction manifold during idling would upset fuel-air ratio resulting in poor idling. In order to prevent this a regulator valve known as *positive crank-case ventilation* (*PCV*) *valve* is placed between crankcase and induction manifold as shown in the figure. This automatic valve allows only a small amount of air to flow through during idling. But as the engine speed increases, fall in manifold vacuum allows the PCV valve to open more to allow more air to flow through the crank-case.

(*ii*) **Vapour recovery system** (*VRS*). In the absence of recovery system, petrol in the tank and carburettor evaporates away

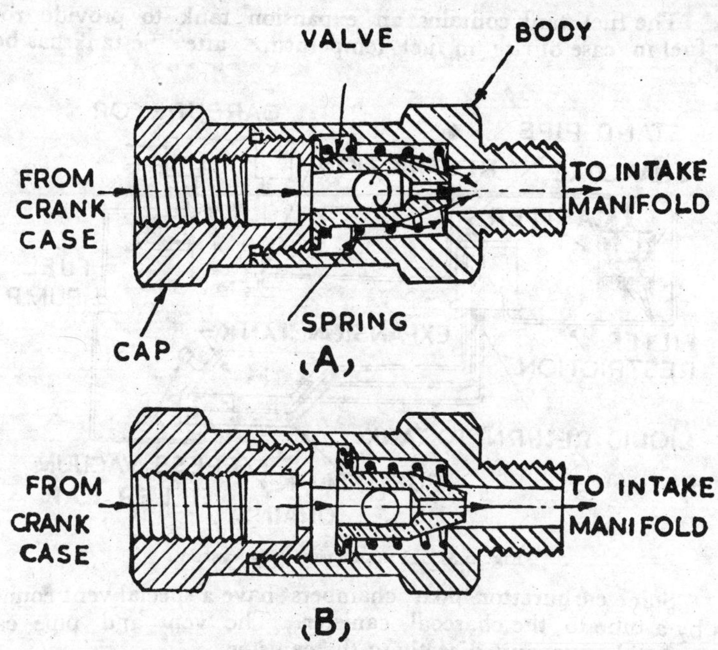

Fig. 5·36. Regulator valve operation in PCV system.
(A) High speed operation (Low manifold vacuum).
(B) Low speed operation (High manifold vacuum).

through the vent holes due to temperature effect. The petrol vapours passing out of the tank and carburettor pollute the atmosphere. A vapour recovery system captures these petrol vapours and prevents them from escaping into air thereby reducing pollution. Almost all the modern cars are equipped with *VRS* under different names such as *Vapour saver system* (*VSS*), *Vehicle vapour recovery* (*VRR*), *Evaporation control system* (*ECS*), and *Evaporation emission system* (*EES*).

In a typical vapour recovery system, a canister filled with activated charcoal is placed between carburettor and fuel tank. Petrol vapours from the fuel tank and carburettor float chamber pass through the canister and are trapped in there. When the engine is started, induction manifold vacuum acts at the canister and the fuel vapours are drawn into the carburettor to mix up with fresh charge.

A *standpipe assembly* between fuel tank and canister is used to separate petrol vapours from the liquid petrol.

The petrol tank contains a sealed cap of pressure and vacuum type as used on the radiator of a sealed cooling system. The cap valve opens when too much pressure develops in the tank. It also opens to admit air as fuel is withdrawn.

The fuel tank contains an expansion tank to provide room for fuel in case of rise in fuel temperature after the tank has been filled.

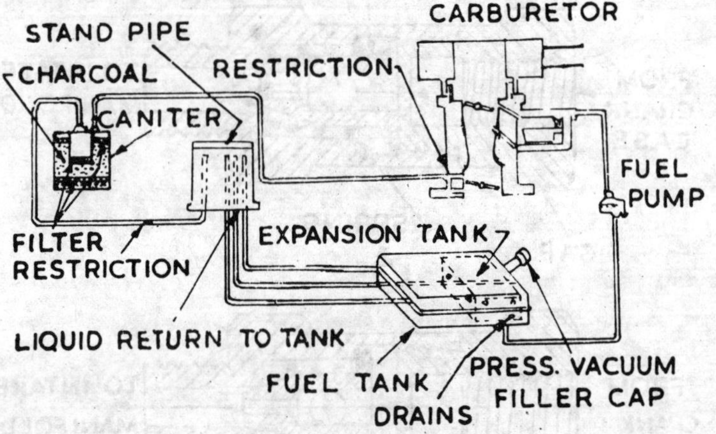

Fig. 5·37. Vapour recovery system.

Some carburettor float chambers have a special vent connected by a tube to the charcoal canister. The vent and pipe carry float chamber vapours directly to the canister.

An insulator is used between carburettor and induction manifold to act as a heat barrier between them so that heat flow from the engine to carburettor is reduced. This is to reduce petrol evaporation at the carburettor.

A fuel return line parallel to main fuel line connects the pressure side of fuel pump to fuel tank to return excess petrol pumped by the fuel pump. This arrangement helps in removing any vapour developing in the fuel pump and keeps the pump cool.

There is slight difference in the construction of V.R.S. of one make of vehicle to another. In *Buick* car, the purge line from canister is connected to the air cleaner snorkel. V.R.S. in *Pontiac* car uses a domed fuel tank. In some cars there is a liquid check valve between fuel tank and charcoal canister. Some canisters contain vacuum signal at their top. *Chrysler Corpn*. has designed a *vapour liquid separator* having overfill limiting valve in place of standpipe assembly. Some Chrysler Corporation cars use crankcase to store petrol vapours from fuel tank and carburettor. When the engine is stopped, petrol vapours from the vapour separator at the fuel tank flow to the crankcase air cleaner from where they flow down into the crankcase. During the same period petrol vapours from the carburettor float chamber flow down into the crankcase. The vapours are two to four times heavier than air and they sink to the bottom of crank-case. When the engine is started, the *positive crankcase ventilation system* clears the crankcase of petrol vapours. The vapours are sucked into the induction manifold from where they are fed to the engine and burned along with the fresh charge from carburettor.

(*iii*) **Cleaning up the exhaust gas.** The most powerful source of air pollution in an automobile is the silencer tail pipe through which exhaust gases from the engine are expelled out. There are the following three ways of cleaning the exhaust gases :—

(*a*) Controlling the fuel-air mixture.

(*b*) Controlling combustion.

(*c*) Treating the exhaust gas.

(*a*) *Controlling the fuel-air mixture.* The fuel air mixture can be controlled by modifying the carburettor to deliver a leaner mixture and faster warming up and quicker choke action. The modern carburettors are fitted with *idle limiter* to fix the idle adjustment. The idle adjustment screw is adjusted and idle-limiter cap is installed at the factory. The cap permits adjustment up to some degrees but will not permit adjustment beyond the lean idle setting permitted by law. The leaner mixtures at idle and part throttle assume a more complete burning of fuel resulting in less hydrocarbon exhausting out or reduced exhaust emissions.

During cold starting, extra rich mixture is required. As soon as the engine starts, the situation changes and the hot exhaust gas circulating around the manifold *heat control valve* begins to heat the ingoing fresh charge through the induction. The modern engines need faster warm up due to which a *thermostatically controlled air cleaner* is used to provide heated air quickly to the carburettor when the engine is cold. This system is known as *heated air system* (H.A.S.). This system improves engine performance after a cold start and during warm up. It makes the engine capable to start and operate satisfactorily when cold even though the idle mixture is lean. Thus leaner mixtures could be used to reduce smog without affecting cold engine performance.

The earlier carburettors were fitted with manually operated choke. Later models contained thermostatically controlled chokes operated by engine heat. The modern carburettors are equipped with *electrically assisted chokes* in which the choke thermostat is subjected to heat from a heating element in the choke and heat from exhaust manifold. This choke acts faster and opens more quickly. This reduces the time period during which the engine works in a choked condition. During choking, a very rich fuel air mixture is fed to the engine and the exhaust gas is loaded with unburnt petrol and carbon monoxide. The electrically assisted choke reduces the length of time during which the exhaust emissions are let into the atmosphere and hence they assist in reducing air pollution.

(*b*) *Controlling combustion.* In an I.C. petrol engine, a mixture of fuel and air is compressed in the combustion chamber and then ignited by a spark from the spark plug. The compressed mixture explodes and provides power thrust at the piston crown to

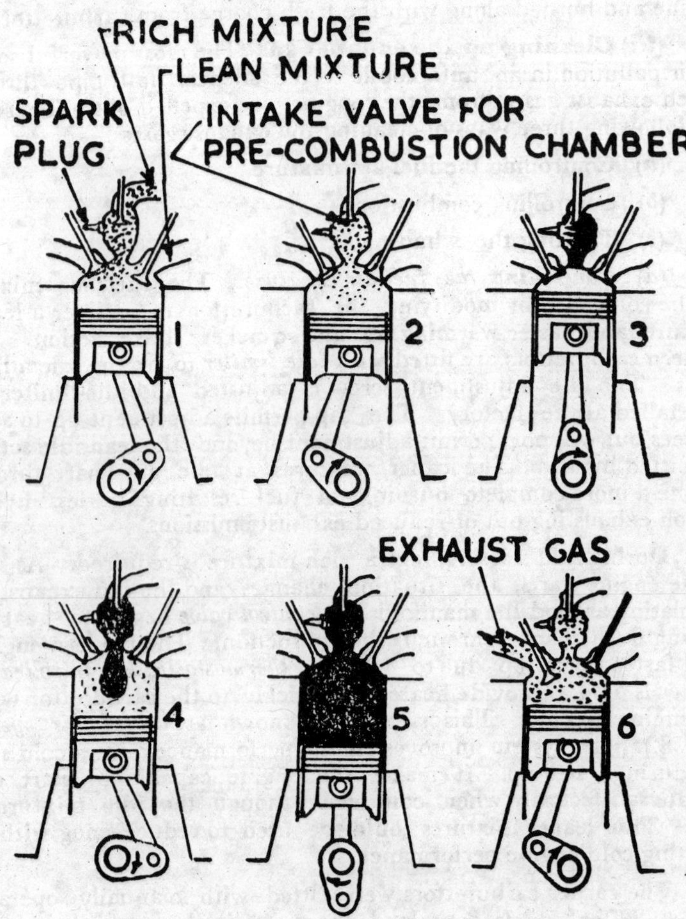

Fig. 5·38. Actions in Honda system. (Stratified charge)

1. Intake stroke.
2. Compression stroke.
3. Ignition.
4. Power stroke.
5. Piston at B.D.C.
6. Exhaust stroke.

drive it down. During combustion, the layers of fuel-air mixture above the relatively cool cylinder head and piston crown do not burn properly. The metal surfaces chill these layers below the combustion point and the unburnt fuel is swept out of the cylinder during exhaust stroke. The unburnt fuel let out in the atmosphere pollutes it. This problem could be tackled by the following methods :—

(*p*) Reducing combustion chamber surface area.

(*q*) Using stratified charge or fuel injection.

(*r*) Increasing combustion temperature.

(*s*) Controlling vacuum advance.

(*p*) *Reducing combustion chamber surface area.* If the ratio between surface area and the volume of combustion chamber is reduced, a lower percentage of unburnt hydrocarbon will be exhausted and hence less air pollution. Change in combustion chamber design affects s/v ratio. *Wedge* type combustion chamber has higher s/v ratio. The *hemispheric* combustion chamber has lower surface area to chill the fuel air mixture and hence lesser air pollution shall result in.

(*q*) *Stratified-charge or fuel injection.* In stratified charge, the rich part of mixture is kept away from the combustion chamber surface. The rich mixture is concentrated in the centre of compressed fuel-air mixture. During combustion, the burning rich mixture spreads outwards and moves into areas where the mixture is lean and harder to ignite. Thus a much leaner mixture could be used with stratified-charge as in *Honda system.* As the fuel is burned more completely so the amount of pollutants is reduced.

In Honda system, a rich mixture is delivered to the precombustion chamber and ignited there from where it streams out into the lean mixture in the combustion chamber and combustion continues. Thus the system assures good burning of the fuel and the polluting gasses—unburnt petrol (hydrocarbon), carbon monoxide and nitrogen oxides are kept at low level.

Fuel Injection in petrol engines improves combustion and reduces hydrocarbon and carbon monoxide in the exhaust as this system measures fuel more accurately. It supplies the same amount of fuel to each cylinder whereas in the carburation system some cylinders could get a richer mixture than others

In petrol injection, fuel nozzles are located in the induction manifold, just at the back of intake valve to inject fuel into the ingoing air. Fuel injection in petrol engines is different from diesel engines. In petrol engine, fuel is injected during suction stroke in the air stream in induction manifold. Whereas in diesel engine, it is injected directly into the combustion chamber near the end of compression stroke.

(*r*) *Increasing combustion temperature.* Increasing of combustion temperature reduces unburnt petrol and carbon monoxide in the exhaust. The combustion of the fuel is improved by the increased temperature but the higher temperature produces more nitrogen oxides (NO_x) which is another problem. The following methods are employed to reduce NO_x :

(*i*) Reduction in *compression ratio* by which top combustion temperatures are reduced resulting in lesser formation of NO_x.

(*ii*) Exhaust gas recirculation or *EGR system* in which a small part of exhaust gas is fed back to the engine that reduces the combustion temperature and lowers NO_x formation.

(*iii*) Additional *valve overlap* to leave more exhaust gas in the cylinders to mix with fresh fuel charge resulting in reduced combustion temperature and lesser NO_x formation.

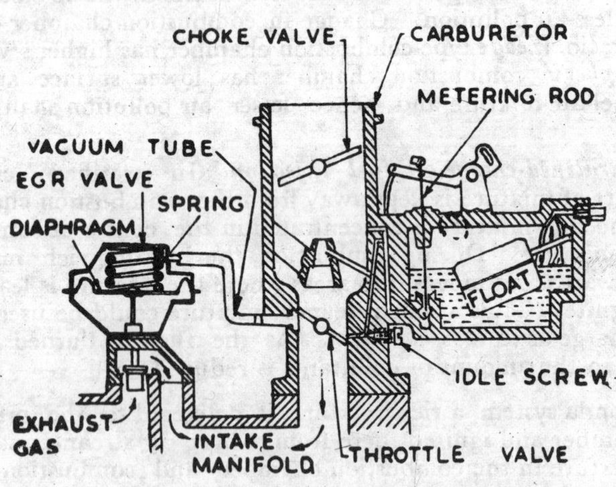

Fig. 5·39. Exhaust gas recirculation (EGR) system.

(*s*) *Controlling vacuum advance*. During part-throttle operation, additional advance in the ignition is provided by the distributor vacuum advance mechanism. The additional advance provides more time for the lesser fuel-air mixture to burn. But this added time also allows the formation of more NO_x. To combat this problem different types of controls have been used to prevent vacuum advance under certain conditions. Some of the vacuum advance control systems are as under :—

(*i*) **Transmission-controlled Spark (TCS) system** as used in *Chevrolet* cars having manual transmissions. This system prevents vacuum advance when the vehicle is operated in reverse, neutral or low gears under which conditions vacuum advance could greatly increase NO_x formation.

(*ii*) **Transmission regulated spark (TRS) systemis** applicable in *Ford* cars employing both manual and automatic transmissions. This system works almost in the same way as TCS system. When the transmission is in the high gear, the solenoid valve is normally open allowing vacuum advance. In the lower gears, the transmission switch is closed, which closes the solenoid valve. When the solenoid valve is closed, vacuum is cut off from the distributor vacuum advance mechanism. Thus there is no vacuum advance.

(*iii*) **Orifice spark-advance control** (OSAC) as used by *Chrysler Corpn.* This system contains a very small hole or orifice which delays by about 17 seconds any change in the application of vacuum to the distributor between idle and part open throttle. Thus there is a delay in vacuum advance until the vehicle is accelerated.

(c) **Treating the exhaust gas.** After the exhaust gases come out of the engine cylinders, they could be treated by *air injection and catalytic converters* to reduce hydrocarbon, carbon monoxide and nitrogen oxide contents. By *air-injection system*, fresh air is blown into the exhaust manifold. The fresh air provides additional oxygen to burn hydrocarbon and carbon monoxide coming out of the engine cylinders.

The air injection pump pushes air through the air lines and air manifold into a series of air injection tubes which are located opposite to the exhaust valves. The oxygen in the air helps in burning unburnt fuel particles in the exhaust manifold. In case of backfire, a check valve prevents any back-flow of exhaust gases to the air pump. There is an air-bypass valve which operates during engine deceleration when manifold vacuum is high. The bypass valve thus diverts momentarily pumped air to the air cleaner to avoid backfire.

The *catalytic converters* convert gaseous pollutants into harmless gases. A *catalyst* is a material which brings chemical change without chemical reaction. The catalyst stands by and encourages two chemicals to react. In HC/CO catalytic converter, the catalyst tends HC to combine with oxygen to form water (H_2O) and carbon dioxide (CO_2). The catalyst in the nitrogen oxide (NO_x) converter splits nitrogen from oxygen, NO_x becoming harmless nitrogen and oxygen. Thus each exhaust is equipped with two catalytic converters, one for HC and CO and the other for NO_x.

QUESTIONS

1. How heat energy is converted into mechanical energy in a conventional engine ?
2. Illustrate the working of 4-stroke petrol engine.
3. Describe with the help of sketches, the working of two stroke valveless petrol engine.
4. Illustrate the working of 4-stroke diesel engine.
5. Describe with the aid of sketches, the working of two stroke diesel engine.
6. Describe the main requirements for the operation of an engine.
7. Describe the main factors which affect power output in an engine.

8. What do you know about the following ?

(*a*) Piston Displacement.

(*b*) Compression Ratio

(*c*) Clearance Volume

(*d*) Ignition Timing

(*e*) Engine Drag

(*f*) Power overlap.

9. Power overlapping results in engine smooth running. Discuss.

10. What is the necessity of engine balancing ? Describe the power balance and mechanical balance.

11. Define the following :

(*i*) I.H.P.

(*ii*) B.H.P.

(*iii*) M.E.P,

(*iv*) F.H.P.

(*v*) Torque.

12. What is the difference between I.H.P. and B.H.P. ? Describe the relationship of F.H.P. with I.H.P. and B.H.P.

13. What is taxable horse power ? How is it calculated ?

14. If a six cylinder engine has 3·5 inch bore what shall be its S.A.E. horse power ? **[29·4 H.P. Ans.]**

15. Write short notes on the following :

(*a*) Volumetric efficiency.

(*b*) Thermal efficiency.

(*c*) Mechanical efficiency.

16. Describe the factors which affect volumetric efficiency of an engine.

17 Compare the following engines :

(*a*) Petrol and diesel engine

(*b*) Two-stroke and 4-stroke engine.

18. Describe the advantages and disadvantages of Diesel Engine.

19. On the basis of following information, calculate I.H.P. of a 4-stroke petrol engine :

(*i*) M.E.P. = 110 lb. □″

(*ii*) Length of stroke = 3·75″

(*iii*) Diameter of cylinder = 3·5″

(*iv*) R.P.M. = 3,000

(*v*) No. of cylinders = 6 **[90·234 H.P. Ans.]**

20. An indicator diagram shows the following average pressure during the different strokes of an engine. Calculate m.e.p. :

(*i*) Suction stroke = 2 lb./□″

(*ii*) Compression = 30 lb./″

(*iii*) Power stroke = 15 lb./□″

(*iv*) Exhaust stroke = 2 lb./□″

[116 lb./□″ Ans.]

21. Explain the path followed by the incoming and outgoing gases in the engine.

22. Show with the help of a sketch, the induction and exhaust system of V-8 Ford engine.

23. Show with the aid of sketches the path followed by the exhaust gases in the exhaust manifold when engine is cold and when engine is hot

24. What is hot-spot and what is its utility ?

25. Illustrate how heat control valve works.

26. Explain the various methods of scavenging employed in two stroke engines.

27. Explain the different types of exhaust silencers used in automobiles.

28. Compare 4 stroke single and multi-cylinder engines.

29. A multi-cylinder engine is better balanced than a single cylinder engine. Discuss.

30. Which of the cylinder arrangements is best from balancing point of view ? Explain why ?

31. Explain the following :

(*i*) Balanced an d unbalanced inertia forces.

(*ii*) Right and left hand crankshafts.

(*iii*) Numbering of cylinders in V-8 engines.

(*iv*) Multi-cylinder engines.

(*v*) In-line vesus V-shape engines.

32. How an automobile engine pollutes the atmosphere ? Which methods are employed to control air pollution ?

33. Explain the following :

(*i*) Positive crankcase ventilation

(*ii*) Vapour recovery system

(*iii*) Exhaust gas recirculation (EGR) system

(*iv*) Catalytic converters

(*v*) Smog.

34. Explain the various methods used to clean exhaust gas for reducing air pollution.

35. What is stratified charge ? Explain the working of stratified-charge engine.

36. Explain the various arrangements employed to crank engines.

37. What is engine tuning ? Explain the tuning process.

6

Engine Lubrication System

1. Lubrication System. It is the system by means of which various engine parts are lubricated to reduce friction and thus ensure their free movement. It provides a film of oil between the moving parts and their bearing surfaces. It avoids direct friction by keeping the parts floating upon the oil film. This enables the parts to work for longer time resulting in longer engine life. Due to decrease in

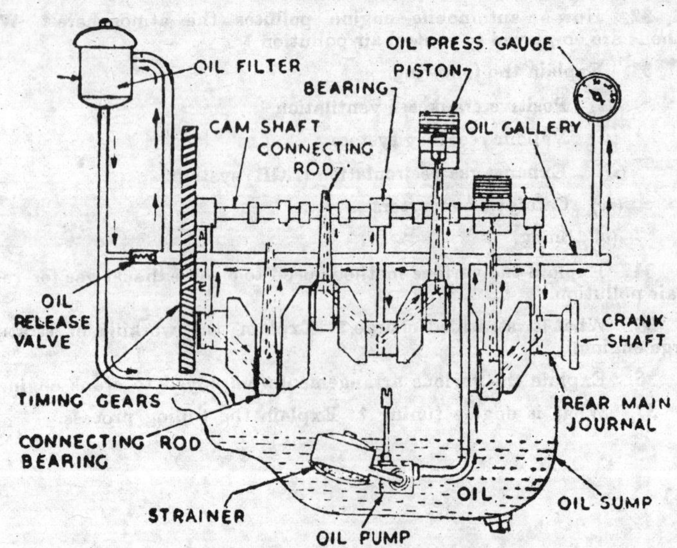

Fig. 6·1. Engine lubrication system.

friction, the engine runs free which results in less power consumption in driving the engine, leading to increase in power output.

2. **Lubricating Oil**. It serves the following purposes in an engine :—

(*i*) Lubricates the moving parts to minimize wear.

(*ii*) Provides an oil film for the floating of parts for free movement so that power losses in friction could be reduced.

(*iii*) Helps in keeping the parts cool.

(*iv*) Absorbs shocks between the bearing surfaces and other engine parts and thus reduces noise and extends engine life.

(*v*) Forms a good seal between piston rings and cylinder walls and thus avoids the leakage of gases down to the crank chamber.

(*vi*) Washes off and carries away dirt, carbon particles and other foreign matter from the passage of moving parts so that these may not create obstruction in the movement of engine parts.

In order to fulfil these objects, lubricating oil should possess the following properties :—

(*i*) It must possess proper *viscosity*.

(*ii*) It must resist oxidation, carbon formation, corrosion, rust, extreme pressures and foaming.

(*iii*) It must have good viscosity at extremes of high and low temperature.

Viscosity. It is a most important property of lubricating oil. It refers to the tendency of oil to resist flowing. Layers of oil must move or slip with respect to each other. Viscosity of oil determines the ease with which this slipping can take place.

Viscosity may be divided into two parts—body and fluidity. *Body* relates to the resistance to oil film puncture or penetration due to the application of heavy loads. The body of oil prevents the load from squeezing out the film between the working parts and the bearing surfaces. This property cushions shock loads, helps in maintaining a good seal between the piston rings and cylinder walls and maintains an adequate oil film on all bearing surfaces under load.

Fluidity relates to the ease with which the oil flows through oil lines and spreads over bearing surfaces. In some respects, the body and fluidity are opposing properties because if an oil is more fluid, it shall have less body.

The temperature affects viscosity. Increasing temperature reduces viscosity. It causes oil to lose body and gain fluidity. Decreasing temperature increases oil viscosity. The oil gains body and loses fluidity.

Viscosity Ratings. Viscosity of oil is determined by means of a device known as viscosimeter. *The viscosimeter* determines the length of time required to flow a definite amount of oil through an opening of a definite diameter with relation to temperature.

SAE (Society of Automotive Engineers) ratings define oil viscosity in two different ways. Winter grade oils are suffixed with 'W' whereas for other than winter, all grades are without the suffix 'W'. Such as SAE 5 W and SAE 30 respectively.

Viscosity Index. Some oils change viscosity to a great extent with the change in temperature. The viscosity index (VI) was adopted to maintain an accurate measure of change in viscosity of a particular oil with temperature change. Originally it ran from 0 to 100. The higher the number, the less the oil viscosity changes with temperature change.

Oil Additives. Any mineral oil, by itself, does not possess all the properties which it should. The oil manufacturing companies, therefore, add a number of additives into the oil during the manufacturing process. The following are the main additives :—

(*a*) Pour point depressants,

(*b*) Oxidation inhibitors,

(*c*) Corrosion and rust inhibitors,

(*d*) Foaming resistance,

(*e*) Detergents—dispersants,

(*f*) Extreme-pressure resistance.

(*a*) **Pour point depressants**. At low temperatures, some oils become so thick that they will not pour at all. Pour point depressants keep the oil fluid at low temperature for adequate lubrication during cold weather operation.

(*b*) **Oxidation inhibitors**. In engines, lubricating oil acting as cooling agent, carries away heat. Sometimes the temperature of oil reaches to a high degree. The oil is well agitated during engine operation for splashing. During this period considerable air is mixed with it. The oxygen in the air tends to combine with oil leading to oxidation. Due to oxidation, the oil breaks down to form harmful substances which result in corrosion and clogging of oil passages in the engine. Oxidation inhibitors are added to lubricating oil to resist oxidation.

(*c*) **Corrosion and rust inhibitors**. At high temperatures, acid may form in the oil which could corrode parts. Corrosion and rust inhibitors, when added to oil keeps away corrosion and rust. Rust inhibitors displace water from the metal surfaces which are coated by oil. They possess an alkaline reaction to neutralize acids formed as a result of combination in the engine.

(*d*) **Foaming resistance**. The churning action in the engine tends the oil to foam leading to overflow. The foaming oil is not able to provide normal lubrication of moving parts. Antifoaming additives known as foaming resistance, are mixed with oil to prevent foaming.

(*e*) **Detergents—dispersants.** Dust deposits owing to oxidation and metal particles in the lubricating oil, gradually reduce the performance of working parts in the engine and lead to more rate of wear. In order to prevent the formation of such deposits detergent additives are added to the oil.

Dispersant additives in the oil do not allow the above-mentioned particles to collect and form large particles which might block oil passage and reduce effectiveness. They keep the particles in a finely divided state.

(*f*) **Extreme pressure resistance.** Lubricating oil in the modern engine is subject to very high pressures in the bearings and valve train etc. Extreme pressure resistance prevents the oil from squeezing out. They react chemically with metal surfaces to form very strong and slippery films.

3. **Types of lubrication system.** Engine parts are lubricated in different ways in different engines. Following are different types of lubrication systems :

(*a*) Petroil system,

(*b*) Splash system,

(*c*) Semi-pressure system,

(*d*) Pressure system,

(*e*) Wet sump-system,

(*f*) Dry sump-system.

(*a*) **Petroil System.** This is the simplest form of lubrication system and is generally adopted in two stroke petrol engines. There is no separate part exclusively meant for lubrication such as oil pump. The lubricating oil is mixed with petrol according to prescribed ratio which is usually 1 to 30, during fuel filling in the tank. When fuel is induced into the crank chamber during engine operation, lubrication particles go deep into the bearing surfaces and lubricate them.

The main drawback of this system being that lubricating oil separates off from petrol if allowed to remain unused for a considerable period. It leads to clogging of passages in the carburettor resulting in engine starting trouble.

(*b*) **Splash System.** In this system, lubricating oil is contained in the oil sump. When the engine operates, oil is splashed in the crank chamber by means of flywheel, crank webs or dippers attached to big ends of connecting rods. The splashed oil spreads over like the mist and runs into minute clearances resulting in their lubrication.

The engines, which are provided with dippers to splash oil, contain turfs below the dippers in the sump. These turfs remain filled

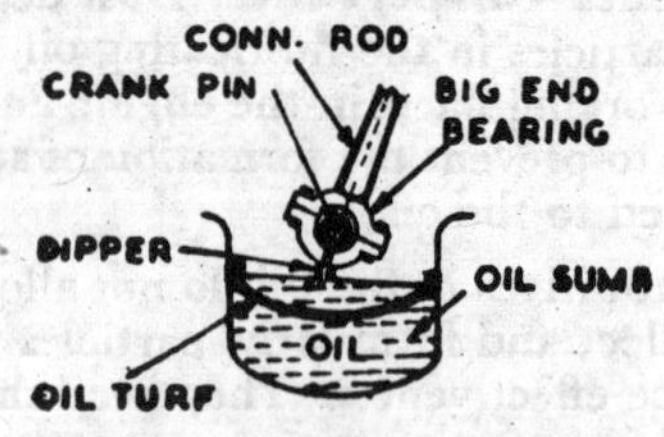

Fig. 6·2 A. Dipper Splash.

with oil to which the dippers dip into and splash oil. This type of arrangement is applicable in Chevrolet engine.

Splash system mostly works in collaboration with pressure feed lubrication system in an engine, some parts being lubricated by splash system whereas some by pressure system.

In four stroke engine, splash system usually lubricates cams on the camshaft, tappets, cylinders, pistons, pins and rings and valve guides, valve stems and springs, inside valve-engines.

(*c*) **Semi-pressure system.** Engine in which some parts are lubricated by splash system and some by pressure system, is said to be having semi-pressure lubrication system. Almost all the four stroke engines are lubricated by this method now-a-days.

(*d*) **Pressure system.** Engine parts are lubricated under pressure feed in this system. The lubricating oil is contained in the oil sump or in a separate tank from where oil is pumped to the engine oil gallery. The oil then flows under pressure to main bearings of camshaft and crankshaft, big end bearings of connecting rods and in certain cases small end bearings of connecting rods and timing gears or sprockets and chains. In overhead vaive engines, rockers, shafts, valve stems, guides and springs etc are also lubricated by pressure feed system.

The lubricating oil flows to various parts of the engine through the specially built-in passages. In order to control the pressure of oil in the system, an oil relief valve is provided which releases the built-in high pressure.

The pressure of oil in the system is indicated by the oil pressure gauge provided at the instrument panel of the vehicle. Oil filters and strainers in the system clear off the oil from dust, metal and other harmful particles.

In certain engines, nozzles are provided in the sump, through which oil under pressure is sprayed on the cylinder walls. This arrangement is provided in *Chevrolet* engine.

(*e*) **Wet sump system.** As mentioned in the description under pressure system, oil for lubrication is contained in the oil sump

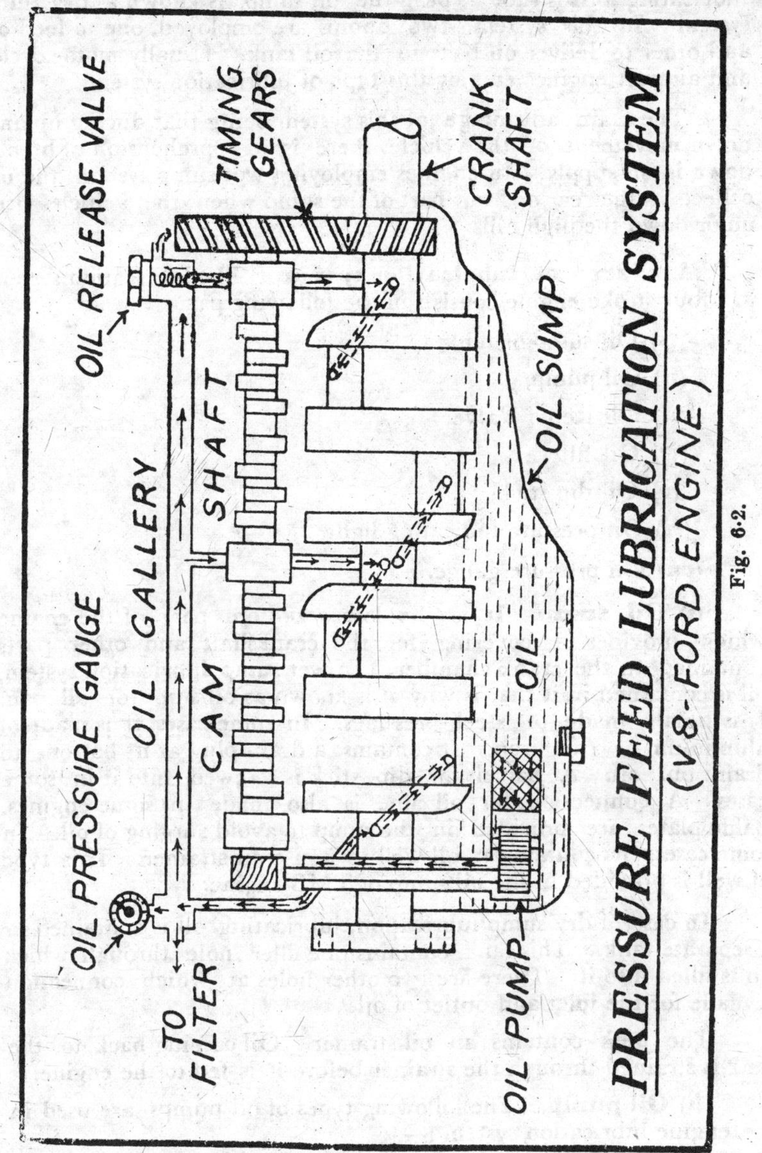

Fig. 6·2.

in most cases. The systems in which lubricating oil is contained in the oil sump, are known as wet sump lubrication systems.

(*f*) **Dry sump system.** In some engines, lubricating oil is carried in separate tanks from where it is fed to the engine. The oil which falls into oil sump after lubrication, is sent back to the oil tank by a separate delivery pump. The lubrication system in which lubricating oil is not kept in the oil sump, is known as dry sump system. In this system, two pumps are employed, one to feed oil and other to deliver oil back to the oil tank. Usually motor cycles and aircraft engines employ this type of lubrication system.

The main advantage of this system being that during up and down movement of the vehicle, there is no apprehension of break-down in oil supply. In engines employing wet sump system, the oil collects in the rear or front part of the sump when the vehicle rises up or down the high hills.

4. **Parts of lubrication system.** The lubrication system in a four stroke engine consists of the following parts :

(*i*) Oil sump or tank,

(*ii*) Oil pump,

(*iii*) Oil Relief Valve,

(*iv*) Oil filter,

(*v*) Oil dip stick,

(*vi*) Oil pressure indicating light,

(*vii*) Oil pressure gauge.

(*i*) **Oil sump.** It is the lower bottom part of the engine which provides a covering for the crankshaft and other parts contained in the crank chamber. In wet sump lubrication system, oil is contained in it that is why it is known as oil sump or oil pan. It is usually made of steel pressings. In some cases, it is made of aluminium or cast iron. It contains a drain plug at its bottom to drain out oil. A barrel for dip stick is screwed into it in some cases. A connection for oil line is also made in some engines. Baffle plates are provided in the sump to avoid surging of oil. In some cases, it contains an oil well to house oil strainer. This type of well is provided in the oil sump of TMB engine.

In case of dry sump lubrication, lubricating oil is contained in a separate tank. This tank contains one filler hole through which oil is filled into it. There are two other holes at which connection is made for the inlet and outlet of oil.

The tank contains an oil strainer. Oil coming back to the tank is strained through the strainer before it is fed to the engine.

(*ii*) **Oil pump.** The following types of oil pumps are used in the engine lubrication system :—

(*a*) Gear type.

(*b*) Rotor type.

(*c*) Plunger type.

(*a*) **Gear type pump**. This type of pump contains two spur gears which rotate in the body. One gear is fixed with the drive shaft whereas the other gear is mounted on a pin. The second gear is free to rotate around its axis *i.e.* pin. In some cases the drive shaft contains a gear through which connection is made with the camshaft from where it gets drive. In some cases, the drive shaft contains a slot at its end which fits the distributor shaft end notch. In this case, the drive from the camshaft is through the distributor.

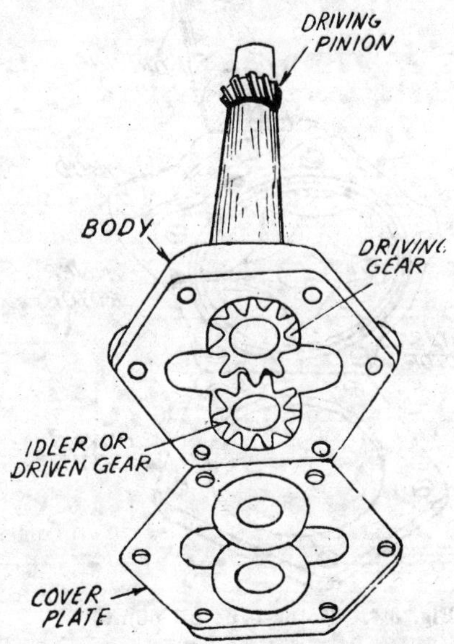

Fig. 6·3. Gear Type oil pump.

The pump contains two holes, one serving as inlet and the other acting as outlet. The inlet hole is connected to the strainer whereas the outlet is connected to the oil gallery. There is a bypass in the body of the pump which contains an oil relief valve to release oil when high pressures are built up.

During fitment with the engine, the pump is filled with oil. When the engine works, the drive gear of the pump drives the idler gear. Since the gears work in oil filled housing so the movement pushes oil out of the chamber. Oil from inlet travels forward to take the place of thrown out oil. This way a circulation is set up due to which oil is drawn from the oil sump and pumped out to the oil gallery.

Gear chamber is kept oil tight so that oil may not leak away. Usually the chamber portion of the pump is kept submerged in the oil of sump.

(*b*) **Rotor type pump.** The construction of this type of pump very much resembles with gear type pump. In rotor type pump, there are two rotors, one inner and the other outer in lieu of gears. The inner rotor is fixed with the drive shaft and gives drive to the outer shaft. When the pump shaft rotates, oil is filled into the segments of outer rotor. Oil is squeezed in the outer rotor due to the movement of inner rotor. Oil under pressure is then pushed out through the outlet hole.

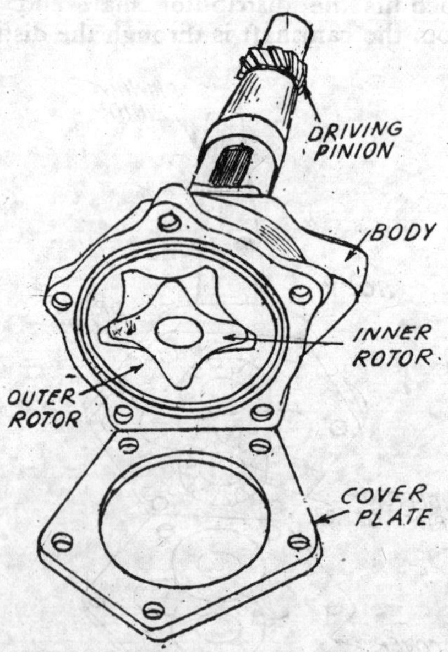

Fig. 6·4. Rotor type oil pump.

(c) **Plunger type pump**. In this type of pump, a plunger works up and down in a barrel. Up and down movement is affected by a cam on the engine camshaft or by an eccentric provided at the crankshaft. Upward movement of the plunger creates suction in the barrel and oil is sucked in from the oil sump or tank, through a non-return valve housed in the pump. When the plunger moves down, oil trapped into the barrel during upward movement, is pumped out through outlet check valve.

(*iii*) **Oil relief valve**. At higher engine speeds, greater pressures are developed in the pressure feed lubrication system. If the high pressures are not released, oil lines may burst or back pressure may develop which will hold the movement of pump and other engine parts. In order to release high pressures built in the system, oil relief valve is employed. It opens when the oil pressure reaches the predetermined limit. This limit is adjusted by increasing or decreasing the tension of spring in the valve.

Oil relief valve contains ball, plunger or flap which is kept against the hole of oil passage. A spring keeps the ball, plunger or flap pushed against the hole. Due to high pressure, the ball, plunger or flap is lifted off the seat and oil goes out through it.

Oil relief valve is located in the oil pump bypass or in the oil gallery.

(*iv*) **Oil filter**. As its name indicates, lubricating oil is filtered through it. Lubricating oil serves several purposes in the engine. During engine operation, metal particles removed due to wear and tear, dirt and other harmful particles are washed away into the oil sump. If oil is not freed from these harmful particles, engine life will decrease as these particles will come in the way of moving parts, thereby restricting their free movement and leading to more friction. Oil filter catches these particles and makes the oil free from them.

Oil filter consists of a round container which is closed at one end, a central tube filtering element, a cap, spring and holding down bolt. The filtering element contains a through hole in the centre into which passes the tube which is fixed in the centre of container.

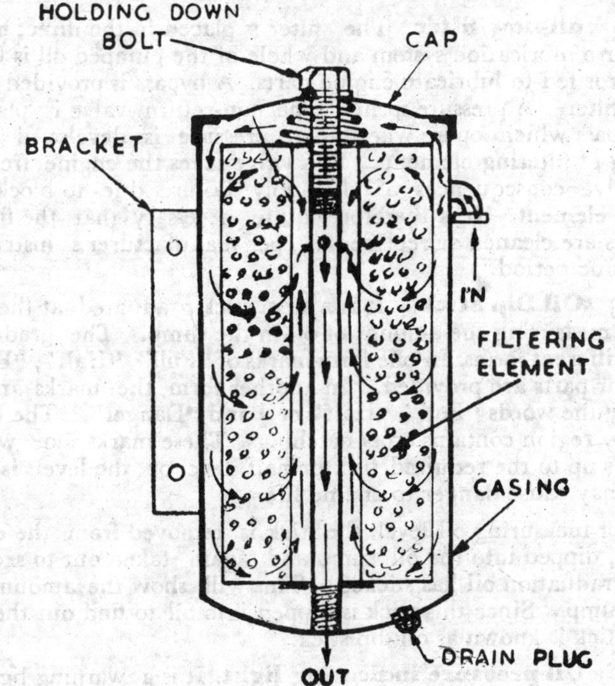

Fig. 6·5. Oil Filter (Bypass type).

After putting in filtering element, the cap is held over the open end of container by means of holding down bolt. The spring is held between the cap and strainer at the centre. This keeps the strainer held down.

When all parts are held in position, the upper end of the tube is closed by the holding down bolt as it is screwed into it. The lower end of the tube acts as outlet for the filter. A tiny hole is provided at the upper part of the tube through which oil from the filtering element enters into the tube for outlet. An inlet connection is made at the upper part of the container, through which oil enters into the filter. This type of filter is placed in the bypass so that regular flow of oil may not be restricted due to clogging of filtering element. This type of filter is known as *Bypass Filter*.

Oil filters are of the following types :

(*a*) Bypass filters

(*b*) Full-flow filters.

(*a*) **Bypass filter.** Its construction has been explained above. Since it is not placed in the direct passage of pressure feed lubrication system so only a part of the oil fed to oil gallery by the oil pump, is filtered. The inlet connection for this oil filter is taken from oil gallery and outlet connection is made at the oil pump.

(*b*) **Full-flow filter.** The filter is placed in the direct passage of pressure lubrication system and whole of the pumped oil is filtered before it is fed to lubricate engine parts. A bypass is provided in this type of filter. A pressure spring type non-return valve is placed in this bypass which opens when high pressure is developed due to clogging of filtering element. This valve saves the engine from the destructive consequences of oil supply failure due to blockade of filtering element. It is therefore quite necessary that the filtering elements are cleaned or replaced as per manufacturer's instructions within due period.

(*v*) **Oil Dip Stick.** It is a steel stick graduated at the front end for measuring the amount of oil in the sump. The graduations are in different forms. In one form marks of "Full", "Half", "Empty" and their parts are provided. In another form, the marks are indicated by the words "Full", "Half" or ½ and "Danger". The danger or empty region contains cross hatchings. These marks show whether the oil is up to the required full or half level or the level is so low which may cause danger to engine life.

For measuring oil level, the stick is removed from the engine, cleaned, dipped into the oil sump and again taken out to see up to which graduation oil has sticked. This will show the amount of oil in the sump. Since this stick is dipped into oil to find out the level, so this stick is known as oil dip stick.

(*vi*) **Oil pressure indicating light.** It is a warning light provided at the instrument panel and indicates the presence and absence of oil pressure in the pressure system It is a small red light which is operated by means of oil pressure operated switch. The switch is located somewhere at the oil gallery. Its connection with the warning light is through the ignition switch. When engine is working and there is sufficient oil pressure in the pressure system, the indicating light switch is open due to oil pressure effect on it and no current flows to the light. During this occasion, the warning light is off. As soon as oil pressure in the pressure system falls down due to any breakdown in the system or engine stoppage, the warning light switch is closed and the light starts glowing. If the engine is not running and the ignition switch is put on, this warning light should glow. If it does not glow, it may be due to some defect in circuit.

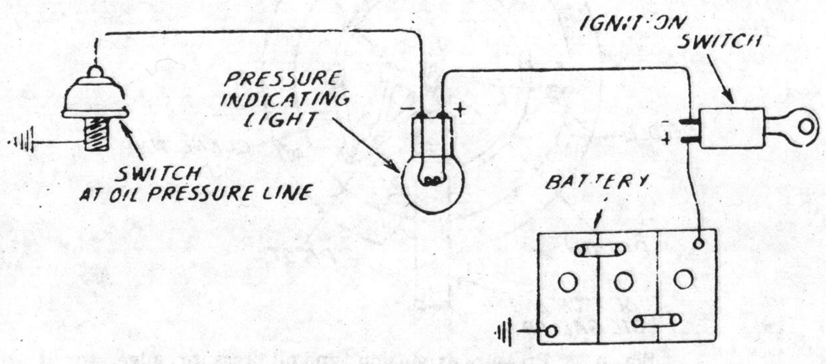

Fig. 6-6.

Pressure indicating light circuit.

(*vii*) **Oil pressure gauge**. As its name indicates, it is a gauge or indicator which records oil pressure prevalent in the pressure system. Oil pressure gauges are of two types as mentioned below :

(*a*) Pressure expansion type,

(*b*) Electric type.

(*a*) **Pressure expansion type**. This type of gauge works with the oil pressure effect. Its connection is made with the oil pressure line through a pipe. This gauge contains a hollow curved tube which is fastened at one end and free at the other. This curved tube is connected with the oil pressure pipe. When the engine is running, oil pressure affects at the curved tube and tends it to straighten out.

This movement is transmitted to the needle of the gauge by means of linkage and gears which are connected through the free end of the tube. The needle moves over the dial of the gauge and registers the amount of oil pressure.

(*b*) **Electric type**. Electrically operated oil pressure gauges are of the following two types :

(*i*) Balancing coil type,

(*ii*) Bimetal-thermostat type.

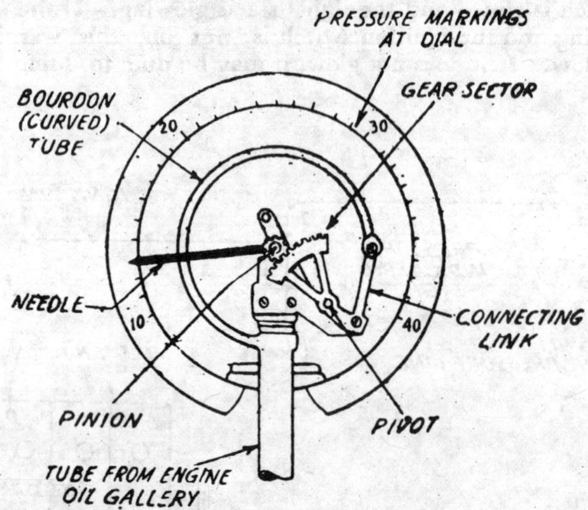

Fig. 6·7. Pressure expansion type oil pressure gauge.

(*i*) **Balancing coil type**. It consists of two separate units—engine unit and gauge unit. The engine unit is located at the oil pressure line of the engine and the gauge unit is mounted at the instrument panel. The effect of oil pressure at the engine unit is transmitted to the gauge unit through an electric circuit.

The engine unit consists of a variable resistance and a moveable contact which moves from one end of the resistance to the other in accordance with the varying oil pressure acting against a diaphragm in the unit. Due to increase in oil pressure, the diaphragm moves inwards, tending the contact to move along the resistance so that more resistance is affected in the circuit between engine and gauge units. This reduces the amount of current that could flow in the circuit.

The gauge unit consists of two coils which balance each other. When ignition switch is turned on, current from the battery flows through the two coils. This produces magnetic field which acts on the armature to which needle is attached. When the resistance of engine unit is high due to high pressure in the oil pressure system, the current flowing through the left side coil also flows through the right side coil. Thus the armature is pulled to the right and the needle indicates high pressure as it moves to the right side across the dial. But when the oil pressure in the engine begins to fall down, the resistance of the engine unit drops. Thus the more current flowing through the left side coil passes through the engine unit. Since less current is flowing through the right side coil so its magnetic field is weaker. As a result, the left side coil pulls the armature towards it and the needle moves anticlockwise to indicate low oil pressure.

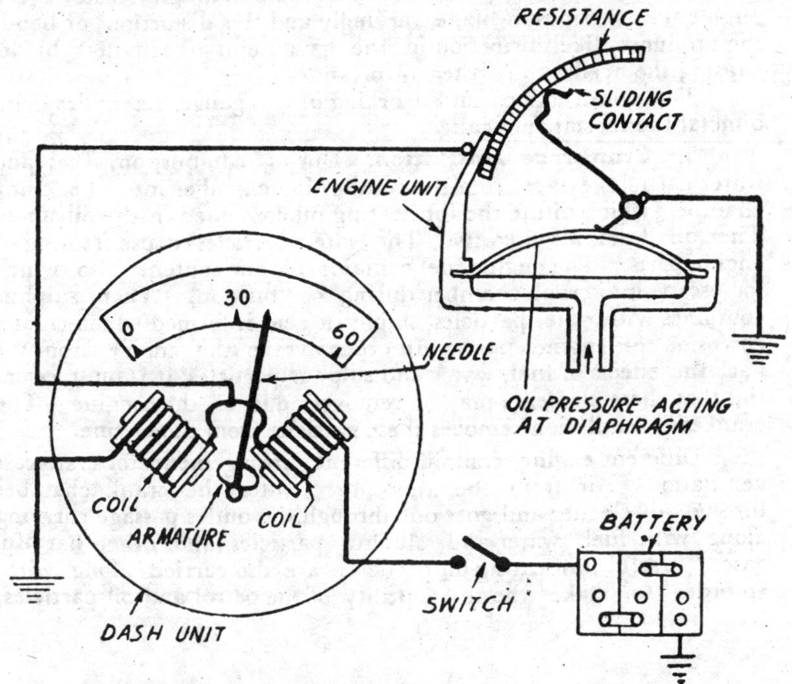

Fig. 6·8. Electric circuit of Balancing Coil type oil pressure gauge.

(ii) Bimetal-thermostat type. This type of oil pressure gauge too consists of an engine unit and a gauge unit. Each unit contains one thermostat blade having a heating coil. The coils of both the units are connected in series through the ignition switch to the battery.

The engine unit contains a diaphragm at which oil pressure reacts. The movement of the diaphragm in or out, imposes more or less bending on the thermostat blade. When ignition switch is turned on, current flows through the heater coils. When the engine unit blade is hot enough, it bends farther and the contacts are separated off. The blade then cools down and the contacts are closed. Again the blade is heated and the points reopen. This action continues as long as ignition switch is on. In the same time, the blade in the gauge unit is heated and bends relatively. Movement of this blade is transmitted to the gauge needle through linkage. The needle thus moves clockwise to indicate oil pressure increase in the lubrication system.

When there is no or less pressure in the oil line, there is no or less bending effect at the thermostat blade. There is relative effect at the gauge unit blade. If the oil pressure is less, the gauge unit blade bends only a little and the needle points towards low side. Thus the varying oil pressure on the diaphragm distorts the engine unit thermostat blade varyingly and this distortion or bending produces like distortion in the gauge unit thermostat blade, causing the needle to register oil pressure.

The construction and working of this gauge resembles with bimetal thermostat fuel gauge.

5. **Crankcase ventilation**. During combustion, fuel and water particles escape from the combustion chamber into the crank chamber, which dilute the lubricating oil contained in the oil sump. The oil loses its viscosity. The water particles cause rusting of engine parts. The engine fuel contains sulphur contents also which too escape into crank chamber during combustion. When sulphur combines with water particles, sulphuric acid is formed which creates corrosion of engine parts. In order to save the engine from the harmful effects of fuel, water and sulphur particles, it is most essential that these elements may be removed out of the engine. The crankcase ventilation removes these particles from the engine.

Different engines contain different arrangements for crankcase ventilation. Air from the atmosphere enters the crank chamber through air cleaner and goes out through the outles passage carrying along with fuel, water and sluphur particles and other harmful gases. Useful lubricating oil particles are also carried along with. In order to make the best utility of the petrol and oil particles,

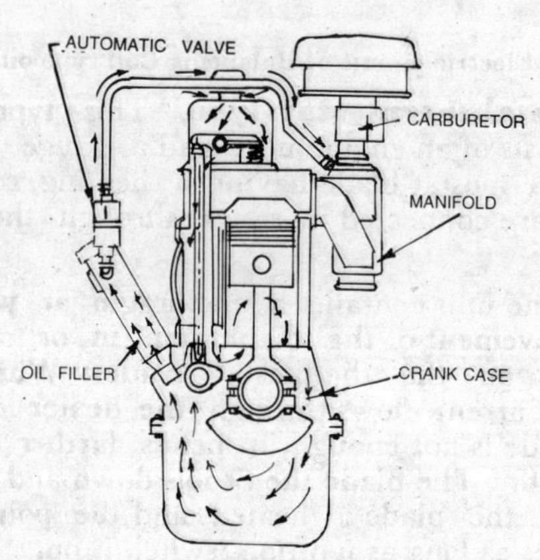

Fig. 6·9. Positive Crankcase Ventilation System.

positive crankcase ventilation outlet is connected with the engine manifold so that useful gases may go into the combustion chamber along with fresh charge where power is produced out of them.

A regulator valve is arranged between the induction manifold and positive crank case ventilation outlet. This is a check valve which controls the excessive flow of oil particles into the manifold. The valve contains a stepped circular piece having a plunger in the centre, which are held in the housing by a coil spring. When manifold suction increases, the valve is closed and the suction of oil vapours from the crankcase is stopped. When the manifold suction falls down, tension of the spring overcomes the valve seat and opens it.

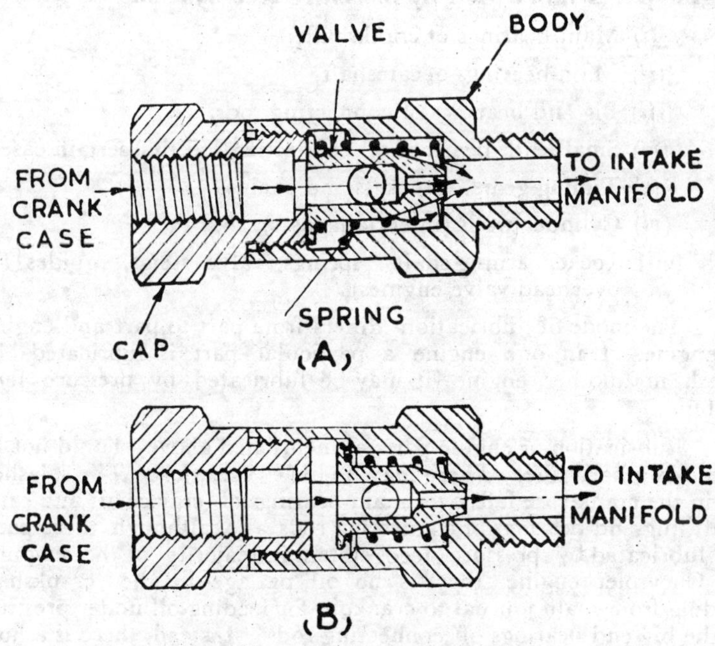

Fig. 6·10, Regulator valve operation in PCV system.
(A) High speed operation (Low manifold vacuum).
(B) Low speed operation (High manifold vacuum).

6. Lubrication system for the conventional 4 stroke engine. Lubrication is fed to the various parts of the engine by different ways. Some of the parts are lubricated by the splash system and some by pressure feed system. *Upper cylinder lubrication* is by means of oil particles which are induced into the cylinders with the fuel air mixture. The oil particles are carried out by the air which passes through oil bath air cleaner for reaching the carburettor to form fuel air mixture. In diesel engines too, the air which is drawn into the cylinders comes through the air cleaner carrying along with oil particles for upper cylinder lubrication.

Engine parts lubricated by splash system :

(*i*) Cylinder walls,

(*ii*) Pistons, pins and rings,

(*iii*) Cams on the camshaft,

(*iv*) Tappets,

(*v*) Valve stems, springs and guides (in side-valve-engines) ; Push rods.

(*vi*) Small end bearings of connecting rods (in some cases),

(*vii*) Distributor or oil pump drive gear.

Engine parts lubricated by pressure feed system :

(*i*) Main bearings of crankshaft,

(*ii*) Main bearings of camshaft,

(*iii*) Big end bearings of connecting rods,

(*iv*) Small end bearings of connecting rod (in certain cases),

(*v*) Timing gears, sprockets and chains,

(*vi*) Cylinder walls (in certain cases),

(*vii*) Rocker arms, shafts, springs, valve stems, guides (in overhead valve engines).

The mode of lubrication differs from part to part and engine to engine. If in one engine a particular part is lubricated by splash, in another engine it may be lubricated by pressure feed system.

Lubrication of engine parts as mentioned above should not be taken as a hard line. There may be little variation. The splashed oil in the crank case falls over main bearings of crankshaft and camshaft, big-end bearings of connecting rods also, although these parts are lubricated by pressure feed system in majority of the engines. In Chevrolet engine, there is no oil passage in the crankshaft leading from main journal to crankpin for feeding oil under pressure to the big-end bearings of connecting rods. Instead, there is a hole in the cap of big end. When the dipper dips into the turf in the oil sump, the hole in the big end cap is filled with oil. From here the oil goes inside the big-end bearing.

It is, therefore, concluded that the majority of the automobile four stroke engines employ semi-pressure lubrication system which consists of splash and pressure systems and the different parts are lubricated differently.

QUESTIONS

1. What role lubricating oil plays in the working of an engine ? What characteristics lubricating oil should possess ?

2. Describe the following :

(*i*) Viscosity.

(*ii*) Viscosity Ratings.

(*iii*) Viscosity index.

3. What are oil additives ? What is their necessity ? Which are the various additives in common use ? Describe any one of them.

4. Describe the following :

(*a*) Pour point depressants.

(*b*) Foaming resistance.

(*c*) Detergents-dispersants.

5. Which are the various systems employed in the engine lubrication ? Describe with the aid of a sketch semi-pressure system.

6. What is the difference between wet sump and dry sump lubrication system ? What are the advantages of one over the other ?

7. Which are the different types of oil pumps employed in the engine lubrication ? Describe with the help of a sketch thé construction and working of any one type of oil pump.

8. Write notes on the following :

(*i*) Oil sump.

(*ii*) Oil relief valve.

(*iii*) Oil dip stick.

(*iv*) Oil pressure indicating light.

9. Which are the different types of oil filters ? Explain with the aid of a sketch any one type of oil filter.

10. What is the difference between gear type and rotor type pump ? Describe the construction and working of rotor type pump.

11. Describe the following :

(*i*) Plunger type pump.

(*ii*) Full flow filter.

(*iii*) Balancing coil type oil pressure gauge.

(*iv*) Crankcase ventilation.

12. What is pressure indicator ? What is the difference between pressure indicating light and oil pressure gauge ?

13. Draw the circuit diagram of pressure indicating light and explain it.

14. Which are the various types of oil pressure gauges ? Describe with the help of a sketch the construction and working of pressure expansion type gauge.

15. Which are the different types of electric type oil pressure gauges ? Describe any one of them with the help of a sketch.

16. Describe the lubrication system as employed in majority of the conventional four srtoke engines. Which parts of the engine are usually lubricated by which way ?

17. Explain crank case ventilation system.

7

Engine Cooling System

1. Cooling system. It is the system by means of which engine is kept at working temperature.

2. Necessity of cooling system. During combustion in the engine, temperature in the cylinder rises up to 4,000°F. A large amount of this heat is absorbed by the cylinder walls, cylinder head, pistons and valves due to which there is great rise in their temperature. If this absorbed heat is not dissipated, the combustion chamber surfaces would become red hot, the valves would burn and warp and the various parts of the engine would expand enormously, resulting in the seizure of pistons in the cylinders, bearings on the journals and pins and scoring of cylinder walls. In an I.C. engine, about 20% of the latent heat produced during combustion passes through the cylinder walls into the cooling system. The cooling system is required to absorb and dissipate the excess heat conducted on cylinder walls, cylinder head, pistons and valves etc. and thus to keep the engine temperatures within safe limits.

3. Types of engine cooling systems. There are two types of cooling systems applicable in automobile engines. These are as under :

(*a*) Air cooling system.

(*b*) Liquid or water cooling system.

(*a*) **Air cooling system.** This type of cooling system is mostly employed in light engines used in motor cycles and scooters. Air cooled engines contain fins or ribs on the outer surfaces of the cylinders and cylinder heads. These fins provide more area for air contact resulting in better radiation of heat. The heat produced by the combustion of fuel passes through the cylinder walls and cylinder head to the fins, from where it is dissipated into the surrounding atmosphere.

The air cooled engines require the circulation of large volumes of air over and past the fins. In motor cycles, flow of air is achieved by their forward motion. In case of cars, the air is thrown over the engine by a fan or blower built into the engine flywheel. A shroud or cowling often encloses the engine to control the flow of air over the engine. Baffles are provided near the cylinders to deflect air through the fin area.

(*b*) **Liquid or water cooling system.** In liquid or water cooling engines, heat from the cylinders is transferred to the liquid or water contained in the jackets surrounding the cylinders, combustion chambers and valve ports etc. Through the water or liquid, heat is carried into the radiator where it is conducted to its fins for radiation to the passing air. Forward motion of the vehicle and an engine driven fan accelerate the circulation of air over the radiating surface.

This system serves the following two purposes :–

(*a*) Cools the engine and saves it from overheating.

(*b*) Warms the engine to bring it to the working temperature for efficient and economical working.

Cooling is done by circulating water or liquid in the jackets around the hot spots of the engine, *i.e.*, cylinders, combustion chambers, exhaust valve ports.

In winter season, the temperature of the engine is raised to the working temperature through a thermostatically controlled cooling system. In this system, hot water is kept in circulation inside the water jackets. The water in the jackets obtains heat from the cylinders due to conduction. The hot water keeps the cylinders warm. The water is not allowed to flow into the radiator till it attains 140°F temperature at which the thermostat valve starts to open and provides passage for the flow of water towards the radiator for cooling down. Hence water cooling system warms the engine too, to bring it to the working temperature.

Water cooling system is of the following two types :—

(*i*) Thermosyphon system,

(*ii*) Pump circulation system.

(*i*) **Thermosyphon system.** This system works on the principle that hot water being lighter, rises up and cold water being heavier, falls down. It depends upon gravity to circulate water or liquid in the system. In this system radiator is kept above the bottom level of water jacket. This system contains no pump to keep the circulation of water regular. When water in the jackets becomes hot as a result of engine combustion, it expands and therefore becomes lighter. The hot water rises up and the cold water comes to take its place. The hot water flows to the radiator through the upper connection where it is cooled down by the air passing through the core. The cold water being heavier, moves downwards to take the place of displaced hot water. This way a circulation is set up due to which heat is carried away from the engine.

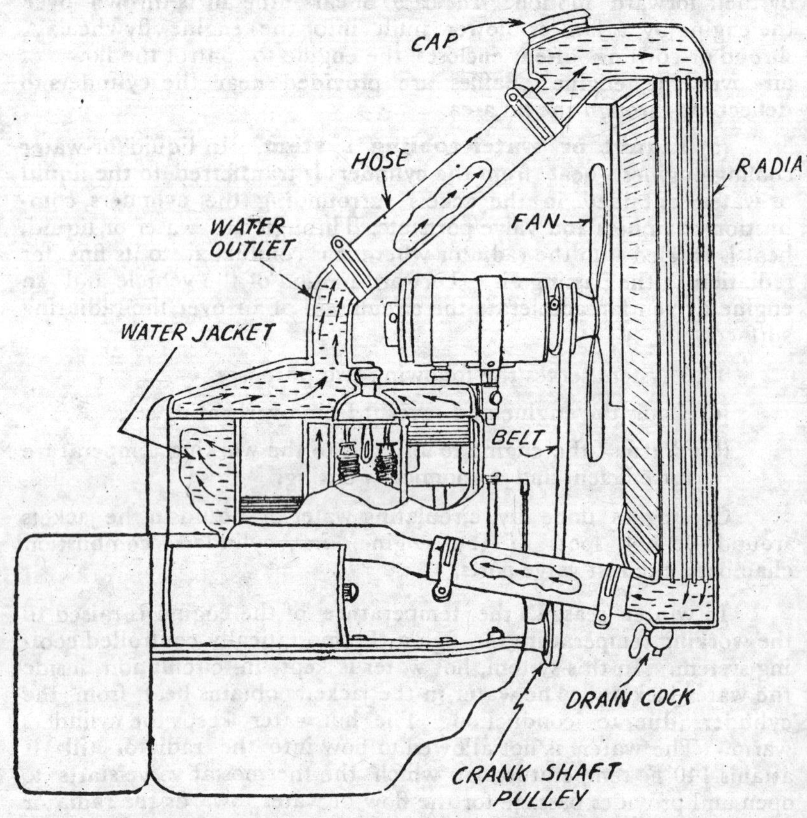

Fig. 7·1. Thermo-syphon water cooling system.

Radiator and an engine driven fan are two main parts in this system, which play an important part in keeping the engine cool.

(*ii*) **Pump circulation system.** The pump or forced circulation system very much resembles in construction with the thermo-syphon system. This system makes use of a centrifugal pump to circulate water. In certain engines, water distributing tubes or nozzles are inserted in the cylinder block or head to direct the coolest water towards the exhaust valve ports.

The following are the main parts of this system :—

(*a*) Radiator, (*b*) Radiator cap, (*c*) Overflow Tank, (*d*) Fan, (*e*) Water pump, (*f*) Thermostat valve, (*g*) Temperature gauge or indicator.

(*a*) **Radiator.** The purpose of the radiator is to cool the water received from the engine. The radiator consists of three parts—

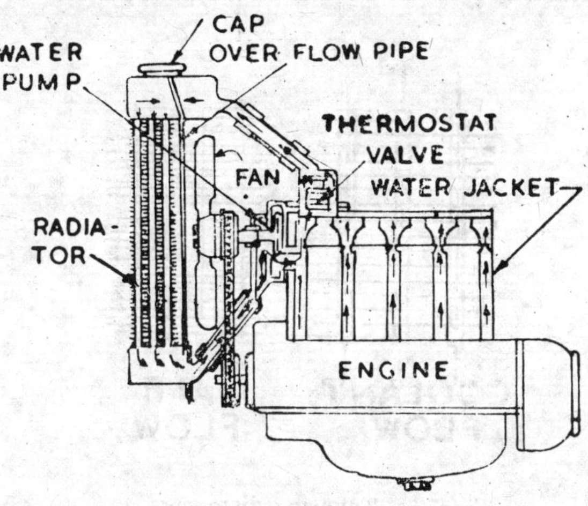

Fig. 7·2. **Water Cooling system (Pump circulation).**

upper tank, core and lower tank. Upper and lower tanks are joined by the core. The upper tank contains a neck through which water is filled into it and to the side of which an overflow pipe is connected. A connection is also provided for the inflow of hot water from the engine into it. The lower tank again contains connection for the outflow of water. A drain cock is provided at the bottom of lower tank to remove water from the radiator. The core is in between the upper and lower tanks. Hot water flows down to the lower tank through the core during which process heat from the water is transferred to the passing air by the core.

The radiator is usually made of copper or brass in order to provide rapid dissipation of heat. The radiators can be divided into following classes with regard to the construction of their core :

(*i*) Tubular,

(*ii*) Gilled tube,

(*iii*) Honeycomb or cellular.

Tubular. It is a tube and fin type radiator which consists of a series of long tubes extending from upper to lower tank. The fins are copper or brass strips of fine thickness through which pass the tubes at right angles. Air passes around the outside of the tubes, between the fins, absorbing heat from the water when it passes through them.

(ii) **Gilled tube.** In gilled tube radiator, each tube contains individual fins surrounding it. Rest of its construction is like the tubular radiator as it is also tube and fin type radiator.

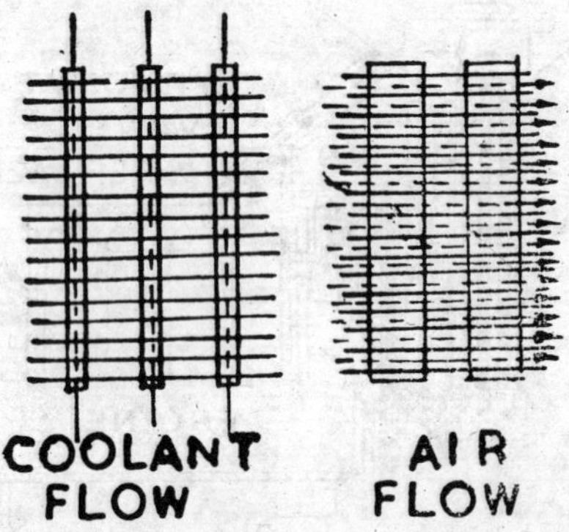

Fig. 7·3. Tubular radiator core.

(iii) **Honeycomb or cellular.** A cellular radiator contains a large number of individual air cells which are surrounded by water. It is made up of a large number of narrow water passages formed by pairs of thin metal ribbons soldered together along their edges, running from upper to lower tank. The water passages are separated by air fins of metal ribbon which provide air passages between the water passages. Air passes through these passages from front to

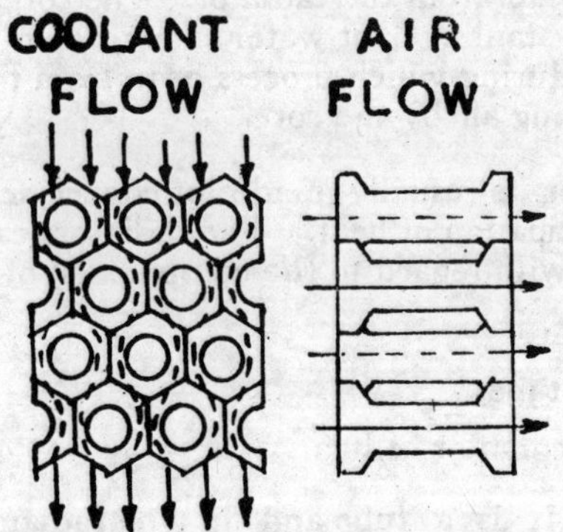

Fig. 7·4. Honeycomb radiator core.

back, absorbing heat from the fins which in turn absorb heat from the water flowing downward through the water passages.

The construction of this type of radiator is like the beehive that is why it is known as Honeycomb. In tubular radiator, if one tube is clogged, the cooling effect of the entire tube is lost. In the cellular radiator, the clogging of any passage results in loss of small part of the total cooling surface.

(*b*) **Radiator Cap**. It provides a cover for the filler hole through which water is filled into the radiator. There are two types of caps in common use. One is the simple cap which acts only as a cover so that water may not flow out through the filler hole. The other type of cap contains a double check valve through which cooling contents go in or out to the overflow tank. This type of cap is known as *pressure cap*. By the use of this cap, cooling efficiency is increased ; evaporation is prevented and water losses due to surging when the vehicle is suddenly stopped, are avoided.

At sea level where atmospheric pressure is higher (about 15 lb/□″), water boils at higher temperature (about 212°F). At higher altitudes, where atmospheric pressure is less, water will boil at lower temperature. Higher pressures increase the temperature required to boil water whereas lower pressures lower down boiling temperature. The use of a pressure cap on the radiator increases air pressure within the cooling system and the water is circulated at higher temperatures without boiling. The water enters the radiator at a higher temperature and the difference between the water and air temperature is greater. Heat is thus transferred from water to air more quickly which results in improvement in cooling efficiency.

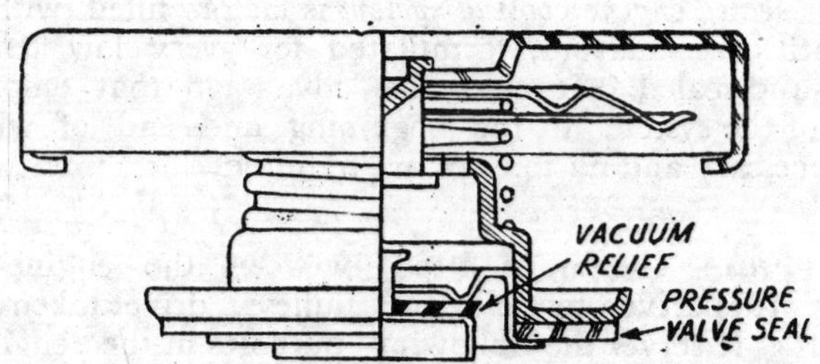

Fig. 7·5. Radiator pressure cap.

The double check valve in the pressure cap consists of a blow off valve and a vacuum valve. The blow off valve is held against the seat by means of a calibrated spring. The spring keeps the valve held close so that pressure is produced in the cooling system. If the pressure reaches the predetermined value for which the system is designed, the blow off valve is raised off its seat and the excessive pressure is released through the overflow pipe.

The vacuum valve prevents the formation of a vacuum in the cooling system when the engine is put off and begins to cool down. When vacuum is formed inside the cooling system, atmospheric pressure from the outside causes the vacuum valve to open, admitting air into radiator. In case overflow tank is connected with the radiator overflow pipe, coolant collected therein runs back into the radiator along with air.

(*c*) **Overflow tank.** It is a small tank which collects the coolant after it flows out of the radiator during higher pressure. It is connected with the overflow pipe of the radiator. It contains a small curved tube at the top through which atmospheric pressure acts in the tank.

During higher pressure, when blow off valve in the pressure cap opens, the contents of the radiator run out and are collected by the overflow tank. There the coolant vapours condense down. When the vacuum valve in the pressure cap opens due to vacuum created in the cooling system during cooling down of the contents at engine shut off, the coolant runs back into the radiator from the overflow tank, thereby making up the deficiency of coolant and maintaining proper level in the cooling system.

Overflow tank is necessary when anti-freeze solution is used as it saves the costly solution to run waste.

The sealed engine cooling system is factory filled with a permanent antifreeze mixture, formulated for very low temperatures (—35°C) and sealed. It offers the advantage that emptying and flushing of the system at the beginning and end of winter is no longer necessary and no top-ups are required.

(*d*) **Fan.** The fan is fitted between the engine and the radiator. It is driven by belt and pulleys, drive taken from the crankshaft. It serves the following purposes in the engine cooling system :

(*i*) It ensures regular flow of air through the radiator fins and thus increases its heat radiating ability.

(*ii*) It throws fresh air over the outer surface of the engine due to which heat conducted is taken away and thus it increases efficiency of the cooling system.

The fans vary in number of blades which range from two to eight. In cold countries or during winter season, there is more problem of keeping the engine warm than to cool down. Fan is

usually removed during extreme cold. Even a canvas is spread over the radiator core or radiator shutters are employed to keep it warm.

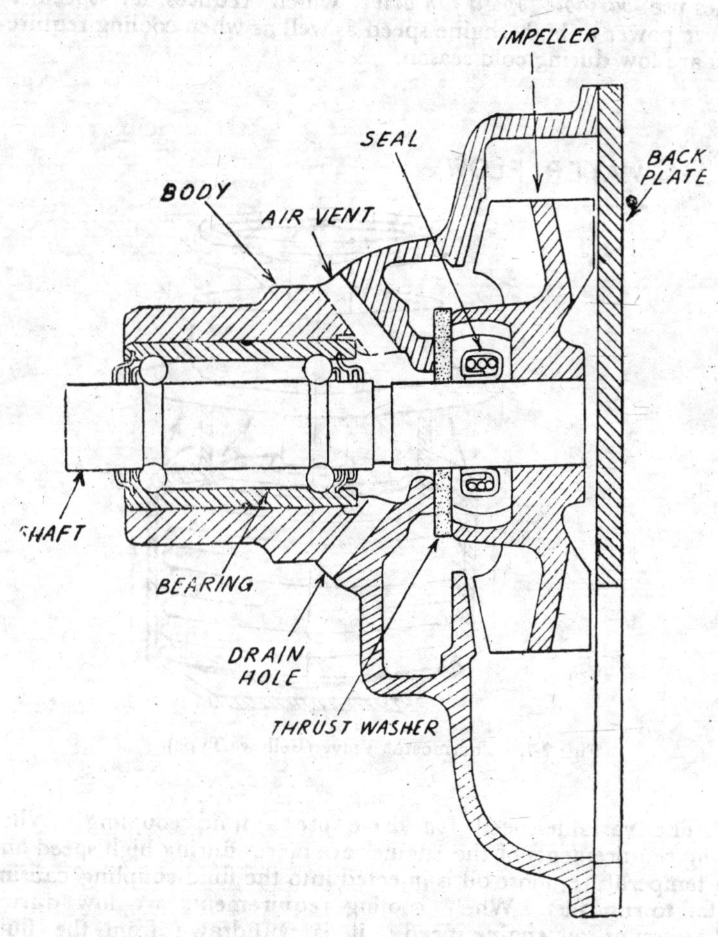

Fig. 7·6. Water Pump.

Removal and replacement of the fan leads to unnecessary headache. In order to overcome this difficulty, *automatic fans* have been developed and employed in cars. Automatic fans are operated by different ways, electromagnetically or thermostatically. When engine

temperature is low or falls down, the drive to the fan is cut off and when the temperature rises up, the fan begins to rotate. Electro-magnetically controlled fans are employed in Fiat 2300 cars. Automatic fans of Kenlowe make are quite popular in U.K. Many engines use *variable speed fan drive* which reduces fan speed to conserve power at high engine speed as well as when cooling requirements are low during cold season.

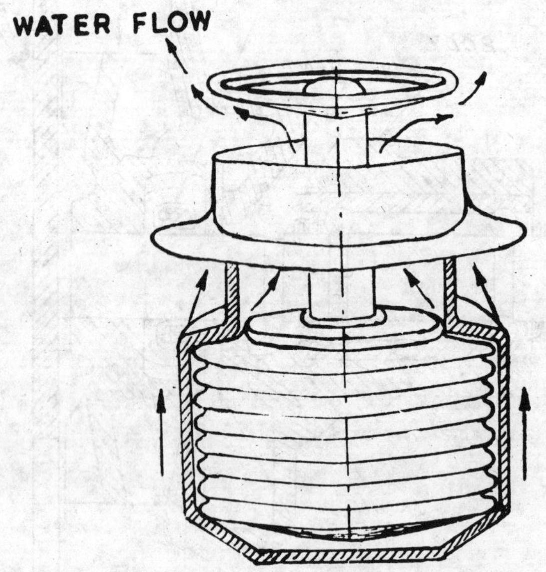

Fig. 7·7. Thermostat Valve (Bellows Type).

The variable-speed fan drive uses a fluid coupling. When cooling requirements of the engine are more during high speed and high temperature, more oil is injected into the fluid coupling causing the fan to run fast. When cooling requirements are low during cold season or low engine speed, oil is withdrawn from the fluid coupling due to which the fan speed drops off. The oil flow to the fluid coupling is controlled thermostatically.

(*e*) **Water Pump**. Usually the water pumps are of impeller type and are fitted at the front end of the cylinder block, between the block and the radiator.

The impeller pump consists of a housing with a water inlet and outlet and an impeller. The impeller is a sort of flat disc or rotor having series of flat or curved blades or vanes and is fixed over the pump shaft. When the impeller rotates, the water between the blades is thrown outward by centrifugal force and is forced through the pump outlet into the engine water jackets The pump inlet is connected to a connection at the lower tank of radiator through a hose. The water from the radiator is drawn into the pump to replace water forced out through the outlet.

A plain bearing supports the impeller drive shaft and a seal prevents water leakage out of the pump. The pump is driven by a pulley and V-belt, the drive being provided by a pulley at the crank-shaft.

(*f*) **Thermostat Valve.** It is a kind of check valve which opens and closes with temperature effect. It is fitted in the engine water outlet. It helps in raising the temperature of coolant during cold-season which warms the engine to bring it to the operating temperature.

The thermostat valves are designed to open at certain fixed temperatures and do not provide passage for the engine water to flow into the radiator until the temperature of the coolant reaches that degree. Usually thermostat valve is open at nearabout 170°F.

Thermostat valves are of the following three different types :

(*i*) Bellows types,

(*ii*) Sleeve type,

(*iii*) Butterfly type.

The bellows type valve contains a poppet valve attached to metallic bellows held in a cage The bellows are filled with liquid such as acetone or alcohol. The liquid evaporates due to increase in temperature and the internal pressure tends the bellows to expand resulting in the rising of valve off its seat. When the temperature falls below the specified at which the valve is designed to open, the valve closes down due to squeezing of bellows.

Sleeve and **butterfly** type thermostat valves differ in construction from bellows type. Instead of liquid, they are powered by a wax pellet which expands with increasing temperature to open sleeve or butterfly valve.

A bypass is provided for the engine through which water keeps circulating between engine outlet and water jacket when thermostat valve is closed. A small valve is located in the bypass which is forced open by the water pressure during thermostat valve shut off.

(*g*) **Temperature indicator or gauge.** It is desirable that the operator should know the temperature at which the engine is working so that he may stop it and look into the defect if the temperature reaches the danger region. Temperature indicator or gauge records the engine temperature at which it is working and warns operator to stop the engine before serious damage occurs.

Temperature indicators or gauges are of the following two types :

(*i*) Vapour pressure type,

(*ii*) Electric type.

(*i*) **Vapour pressure type.** This gauge works due to vapour pressure affecting in it. It consists of two units, temperature sender unit and gauge unit. The temperature sender or bulb is a small tube closed at one end and connected to the gauge unit through a fine tube or pipe at the other end. It contains liquid inside it. Due to temperature effect, the liquid begins to evaporate, creating pressure which is transmitted to the gauge unit through the fine pipe.

The gauge unit contains a curved tube, one free end of which is closed and linked with the gauge needle. The other end of the tube is fixed. At this end, connection of the pipe coming from temperature sender is made.

The temperature sender is fitted inside the engine water jacket whereas the gauge unit is provided at the dash board. The construction of the gauge unit is just the same as that oil gauge. The only difference being that oil pressure affects the curved tube in

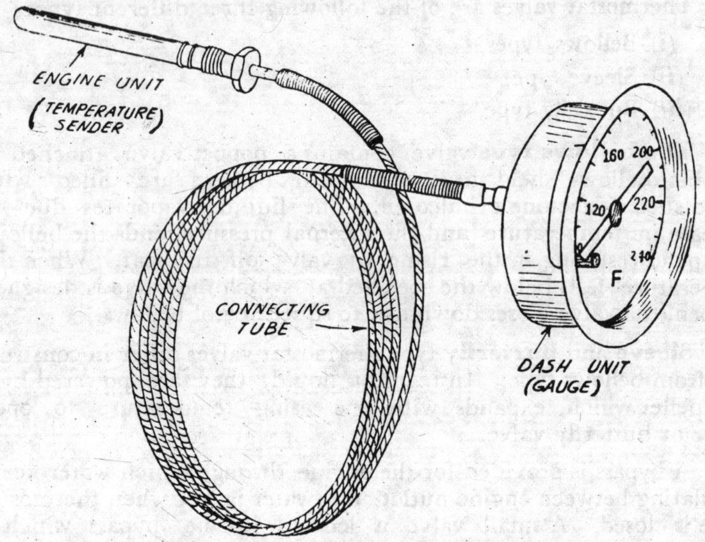

Fig. 7·8. Vapour-pressure type temperature-gauge.

oil pressure gauge whereas liquid pressure acts at the curved tube in temperature gauge to move needle across the gauge dial.

When temperature affects the temperature sender, the inside liquid which is usually alcohol or acetone, begins to evaporate and requires more volume to withstand. The vapour pressure acts the curved tube in the gauge unit and tends it to straighten up. This movement is conveyed to the gauge needle through linkage and gears which are connected to the free end of curved tube. The needle moves across the gauge dial and records the engine temperature.

(*ii*) **Electric type**. Electrically operated temperature indicators are of the following two types :

(*a*) Balancing coil type,

(*b*) Bimetal thermostat type.

Both these types of gauges operate in the same manner as electric type oil gauges. Both these types of gauges consist of engine unit and gauge unit. Gauge unit is quite identical with oil pressure gauge.

The engine unit changes resistance with temperature in such a way that at higher temperatures, it has less resistance and will thus pass more current. More current passing through right side coil of balancing coil type gauge, attracts the gauge needle due to increased magnetic pull and the needle indicates higher temperature.

In bimetal thermostat type gauge, the temperature of cooling water is directly imposed on the engine unit thermostatic blade. A

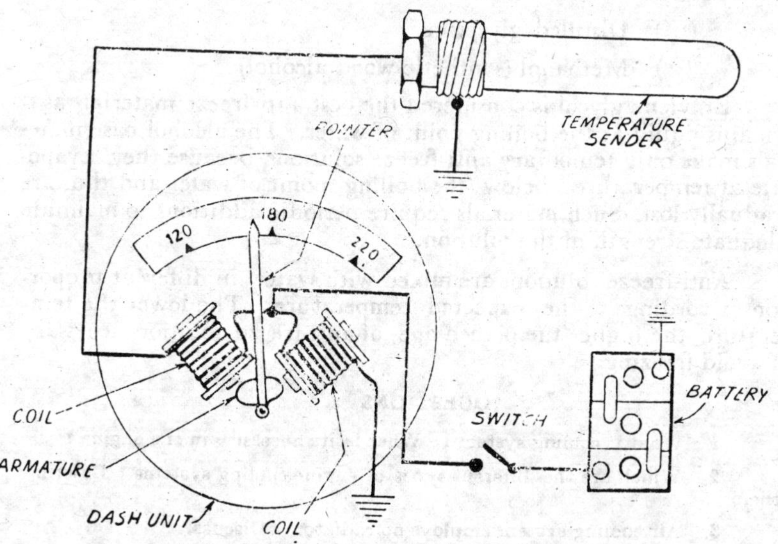

Fig. 7·9. Balancing coil type temperature gauge.

low temperature, most of the blade heating comes from the electric current. Due to flow of more current, the gauge unit blade distorts considerably to indicate low temperature. As the temperature increases, less heat from current flow is required to bring the engine unit blade up to working temperature. Due to flow of less current, the gauge unit shows a higher temperature.

4. **Anti-freeze solutions.** In order to prevent the cooling water from freezing down, some chemicals known as anti-freeze solutions, are mixed up with water. If water freezes in the engine, the resulting expanding force is sufficient often to crack cylinder block and radiator. Anti-freeze solutions added and mixed with cooling water prevent its freezing.

Anti-freeze solution should possess the following characteristics :

(*i*) It should mix readily with water.

(*ii*) It should circulate freely.

(*iii*) It should prevent freezing of the mixture at the lowest possible temperatures.

(*iv*) It should not damage the cooling system.

(*v*) It should not lose its anti-freezing properties after extended use.

The following are common anti-freeze solutions :—

(*i*) Denatured alcohol,

(*ii*) Ethylene glycol,

(*iii*) Distilled glycerine,

(*iv*) Methanol (synthetic wood alcohol).

Ethylene glycol is considered the best anti-freeze material as it remains liquid at the boiling point of water. The alcohol base materials make only temporary anti-freeze solutions because they evaporate at temperatures below the boiling point of water and thus are gradually lost. Such materials require periodic additions to maintain adequate strength of the solution.

Anti-freeze solutions are mixed with water in different proportions according to the expected temperature. The lower the temperature, the higher the percentage of anti-freeze solution necessary to avoid freezing.

QUESTIONS

1. What is cooling system ? What is its necessity in the engine ?

2. Which are the different types of engine cooling systems ? Describe them.

3. Air cooling system employs no radiator. Discuss.

4. Which are the different types of water cooling systems ? Describe them.

5. Illustrate the working of thermosyphon cooling system.

6. What is the difference between thermosyphon cooling system and pump circulating system ? Describe the advantages of one over the other.

7. Pump circulating system is the most popular water cooling system. Discuss.

8. Describe with the aid of a sketch, pump circulating system.

9. Which are the different types of radiators ? Describe them.

10. Describe the following :—

    (*i*) Cellular radiator.

    (*ii*) Pressure cap.

    (*iii*) Overflow tank.

    (*iv*) Thermostat valve.

11. What is the difference between tubular and honeycomb radiators ? Describe the construction of any one of them.

12. Which are the different types of thermostat valves ? Describe the construction and working of any one of them.

13. Which are the different types of temperature indicators ? Describe any one of them.

14. Describe the construction and working of vapour pressure type temperature gauge.

15. What are anti-freeze solutions ? Which properties they should possess ? Which is the best solution and why ?

8

Fuels and Fuel Systems

1. Fuels. The most common fuels used in automobile engines are diesel and petrol or gasoline. Other fuels used in I.C. engines are alcohol, alcohol-gasoline blends, benzol, liquid petroleum gas and methanol.

(*a*) **Petrol.** Petrol or gasoline is a colourless liquid obtained from crude petroleum by means of distillation and cracking process. It has two main properties, volatility and anti-knock value.

The volatility of any liquid is its vaporizing ability. It is usually determined by its boiling point. Petrol or gasoline is a mixture of hydrocarbon compounds, each having its own boiling point. Petrol for automobile engines has boiling point range from about 100°F to 400°F.

The volatility of petrol affects ease of starting, length of warming up period and engine performance during normal operation. Petrol should, therefore, remain a liquid until it enters the air stream in mixing chamber of the carburettor to form fuel air mixture. At this time, it should quickly vaporize and mix uniformly with the air in correct proportions.

Too high percentage of fuel volatility causes *vapour lock* in the fuel line, fuel pump and carburettor. As a result of vapour lock, the fuel supply to the engine is not regular and the engine stops working.

Anti-knock value denotes the ability of fuel to burn without causing detonation or knocking. The tendency towards detonation is overcome by adding certain compounds such as tetraethyl lead in petrol.

The ability of a fuel to resist detonation is measured by *octane* rating. The octane rating of a fuel is determined by matching it with the mixtures of normal heptane and iso-octane in a test engine

The octane rating is in octane number and is known as ONR. A fuel of low octane number knocks easily whereas a high octane petrol is highly resistant to knock.

Oil refineries add several *additives* to the petrol to improve its quality. The following are the main additives :—

(*i*) **Oxidation inhibitors** which prevent the formation of gum while the petrol is in storage.

(*ii*) **Metal deactivators** which protect the petrol from the harmful effects of certain metals that mix up in the refining process in the fuel system of the vehicle.

(*iii*) **Rust resistors** which protect the fuel system of the vehicle.

(*iv*) **Ice-resistors** which combat carburettor icing and freezing of fuel line.

(*v*) **Detergents** which keep the carburettor clean.

(*vi*) **Phosphorus compounds** which combat surface ignition and fouling of spark plugs.

(*vii*) **Dye** which give colour to the fuel for identification.

Combustion of gasoline fuel. Combustion is the rapid combination of fuel with oxygen to produce heat. Petrol is a hydrocarbon. When a hydrocarbon burns, hydrogen and carbon combine with oxygen of the air. The combination of hydrogen and oxygen form water which usually appears as a vapour in the exhaust gas. The carbon combines with oxygen forming carbon dioxide. The chemical equation of combustion for petrol is as under :—

$$C_8H_{18}+12{\cdot}5O_2 = 8CO_2+9H_2O$$

Where C_8H_{18} represents chemical formula for petrol and consists of 8 volumes of carbon and 18 volumes of hydrogen. The equation shows that one volume of petrol requires 12·5 volumes of oxygen for combustion and the products of combustion consist of 8 volumes of carbon dioxide and 9 volumes of water steam. This means that :—

Petrol+air or fuel air mixture (after combination)
= carbon dioxide+water

In the engine, the fuel never burns completely due to which carbon monoxide is formed. Carbon monoxide is a compound of carbon and oxygen which has not received enough oxygen to form carbon dioxide. Engines which run on rich mixture (having less air) have higher percentage of carbon monoxide in their exhaust gases.

(*b*) **Diesel Fuels.** Diesel fuel varies from highly volatile jet fuels and kerosene to the heavier furnace oil. Automobile engines use fuels between these two extremes. Diesel oil is relatively light, having a rather low viscosity and proper cetane number.

Cetane number indicates the ease with which the fuel ignites. In other words, it denotes the *delay* between the time the fuel is injected into the cylinder and ignited by the hot air. If a diesel fuel is low in cetane number, it will ignite with relative difficulty and at a relatively high temperature. If the fuel is of high cetane number, the fuel will ignite with relative ease or at a relatively low temperature. The lower the cetane number, the higher the temperature is required to ignite the fuel and there is more likelihood of the fuel to knock. Fuel having lower cetane number will not ignite quickly and will tend to accumulate. When ignition starts, there will be a *combustion or diesel knock* as the fuel present burns suddenly. If cetane number is sufficientty high, the fuel will ignite and begin to burn as soon as the injection of fuel spray starts. Thus, there will be no knock due to even rise in combustion pressure.

(*c*) **Liquefied petroleum gas (LPG).** It is a mixture of gaseous petroleum compounds (principally propane and butane), together with smaller quantities of similar gases.

Chemically, *LPG* is similar to petrol as it consists of a mixture of compounds, of hydrogen and carbon. However, it is more volatile and at usual atmospheric temperatures it is a vapour. Special types of carburettors are required for this fuel.

LPG is compressed and cooled down to bring to it to the liquid stage for storing and transporting purposes. Owing to its pressure, it is stored in strong tanks.

LPG is made of surplus material in the oil fields that is why its cost is low. It has a high octane value. It is a dry gas and does not create carbon in the engine and does not cause dilution of engine oil. Cold weather starting is easy with this fuel.

2. **Fuel System.** This is the system by means of which fuel is supplied to the engine according to its requirements.

3. **Types of Fuel Systems.** Different ways have been adopted to supply fuel to the engine from time to time. The fuel is carried in a tank located at a suitable place in the vehicle. From the tank, it is fed to the carburettor or injection pump by different ways. The following are the different ways by means of which fuel is fed to the engine :—

(*a*) Gravity feed system,

(*b*) Autovac feed system,

(*c*) Pressure feed system.

(*a*) **Gravity feed system**. In this type, the fuel tank is placed at a higher level than the engine. Fuel from the tank flows through the fuel line towards the carburettor or injection pump by its own weight (gravity) and hence the system is known as gravity feed system. In gravity systems, the fuel tank was placed beside or above the dash board which being nearer to the engine, was a great risk of fire. Due to this drawback, this system has lost its popularity in the vehicles. However, this system is still employed in motor cycles and scooters due to its simplicity.

(*b*) **Autovac feed System**. This system uses engine manifold vacuum to draw fuel from tank to autovac tank from where it

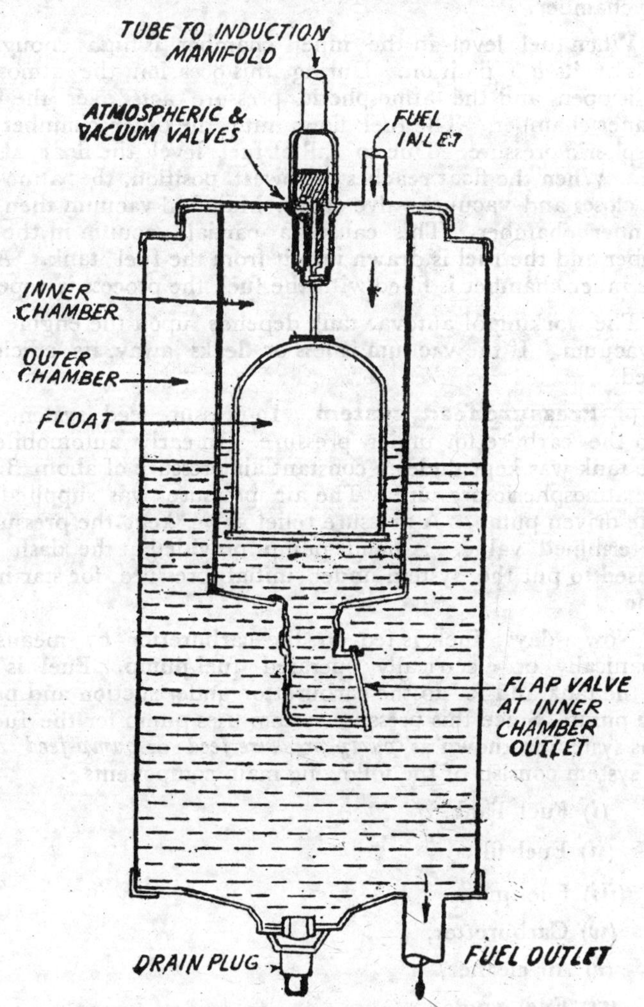

Fig. 8 1. Autovac tank.

is fed to the carburettor by gravity. The fuel tank is placed away from the engine at some suitable place in the vehicle. To autovac tank as shown in the diagram, is located above the engine level.

The autovac tank contains two chambers, the outer enclosing the inner one. The outer chamber acts as a reservoir to feed fuel to carburettor. It is always under pressure and contains an outlet connection and a drain plug at its bottom. The inner chamber contains a float which operates the vacuum valve and the atmospheric valve contained in vacuum line. Fuel line from the main fuel tank is connected at the inner chamber. A flap type valve is placed in the outlet of inner chamber which opens in the outer chamber.

When fuel level in the inner chamber is high enough, the float is at its top position. During this occasion, the atmospheric valve is open and the atmospheric pressure acts over the fuel in the inner chamber. The fuel flows into the outer chamber under atmospheric pressure. Due to fall of fuel level, the float also falls down. When the float reaches its lowest position, the atmospheric valve closes and vacuum valve opens. Manifold vacuum then affects the inner chamber. This causes a partial vacuum in the inner chamber and the fuel is drawn into it from the fuel tank. As soon as the inner chamber is filled with the fuel, the process is repeated.

The working of autovac tank depends upon the engine manifold vacuum. If the vacuum is less or leaks away, its efficiency is marred.

(c) **Pressure feed system.** In pressure feed system, fuel is fed to the carburettor under pressure. In early automobiles, fuel in the tank was kept under a constant air pressure of about 3 lb/□″ above atmospheric pressure. The air pressure was supplied by an engine driven pump. A pressure relief valve kept the pressure at a predetermined value. A hand pump provided at the dash board, was used to put the system under initial pressure for starting the engine.

Now-a-days, fuel is fed to the carburettor by means of a mechanically or electrically operated fuel pump. Fuel is drawn from the tank and fed to the carburettor under suction and pressure of the pump. Since this pressure system uses pump for the fuel feed so this system is known as *pump pressure feed* or *pump feed system*. This system consists of the following main components :

(*i*) Fuel Tank,

(*ii*) Fuel filter,

(*iii*) Fuel pump,

(*iv*) Carburettor,

(*v*) Air cleaner,

(*vi*) Fuel gauge.

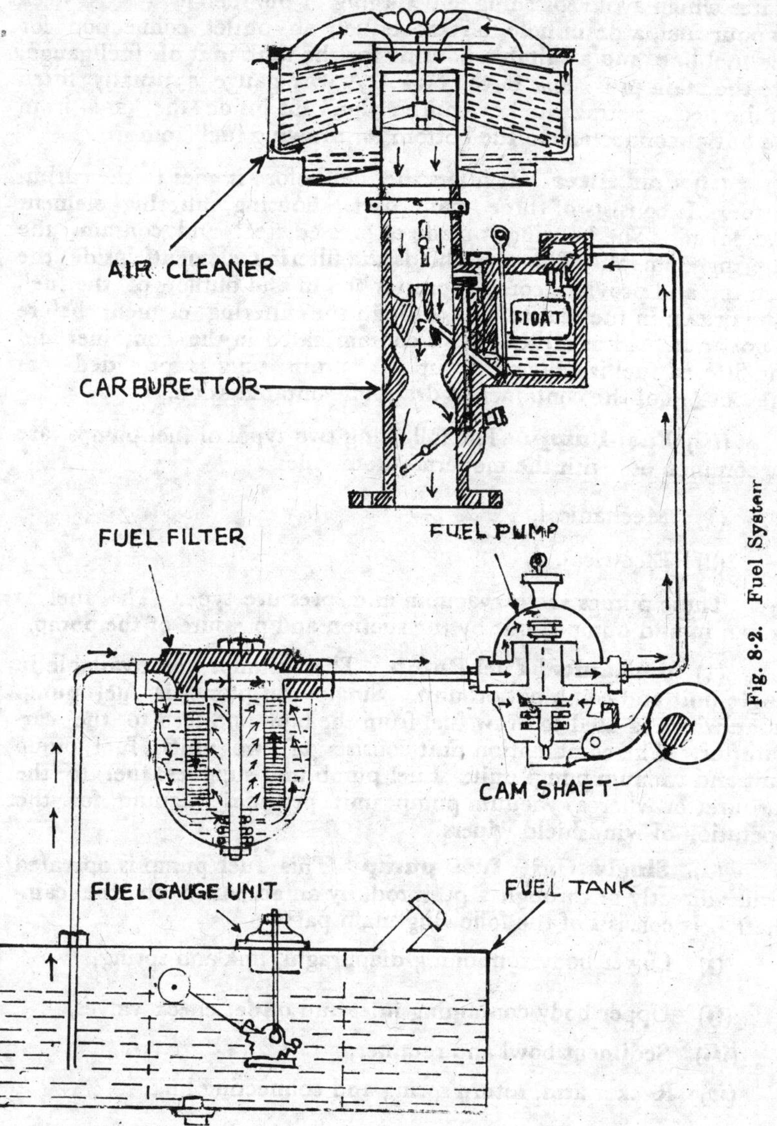

Fig. 8·2. Fuel System

(*i*) **Fuel Tank.** In cars, the fuel tank is usually placed below the luggage boot. In some case it is located below the driver's seat. In heavy vehicles, it is carried by the side of chassis frame. In certain cases, there are two fuel tanks, one on each side of the frame.

The fuel tank is made of steel metal and is usually of rectangular or barrel shaped. The tank is reinforced by means of baffle plates which avoid splashing and surging of the fuel in it. A neck to pour fuel, a drain hole to remove fuel, an outlet connection for the fuel line, and a suitable housing for the tank unit of fuel gauge, are the main provisions in the tank. A wire gauge is usually fitted in the neck to strain fuel. An iron pipe runs inside the tank from the outlet connection to the bottom for drawing fuel from it.

(*ii*) **Fuel filter.** It filters the fuel before it goes to the carburettor. It consists of three main parts, housing, filtering element and cover. The housing is made of pressed steel and contains the filtering element. The cover holds the filtering element inside the housing and provides connections for the in and outflow of the fuel. The drawn in fuel has to pass through the filtering element before it goes out. Thus all the dirt is contaminated in the container and the filtered fuel is delivered out. A drain plug is provided at the bottom of the container to drain off contamination.

(*iii*) **Fuel-Pump.** The following two types of fuel pumps are in common use with the modern automobiles :

(1) Mechanical,

(2) Electrical.

These pumps are of vacuum and pressure type. The fuel is drawn in and pumped out by the suction and pressure of the pump.

(1) **Mechanical Fuel Pump.** These pumps are available in single unit and combination unit. Single unit pump is fuel pump alone which is used to draw fuel from the tank and feed to the carburettor. The combination unit consists of two units, fuel pump unit and vacuum pump unit. Fuel pump unit supplies fuel to the carburettor whereas vacuum pump unit provides vacuum for the operation of windshield wipers.

(*a*) **Single Unit fuel pump** This fuel pump is operated either directly or through a push rod, by an eccentric on the camshaft. It consists of the following main parts :

(*i*) Lower body containing diaphragm, link and spring,

(*ii*) Upper body containing inlet and outlet check valves,

(*iii*) Sediment bowl and retainer,

(*iv*) Rocker arm, return spring and connecting link.

The various parts have been shown in the diagram of conventional fuel pump which is most popular.

When camshaft rotates, rocker arm of the fuel pump is moved back and forth. The eccentric on the camshaft pushes the rocker arm back towards the body of the fuel pump. When the eccentric moves off, return spring pushes the rocker arm forward towards the engine camshaft. When the rocker arm is pushed back by the eccentric, diaphragm in the pump is pulled down, which creates suction in the main chamber. During this occasion, inlet valve

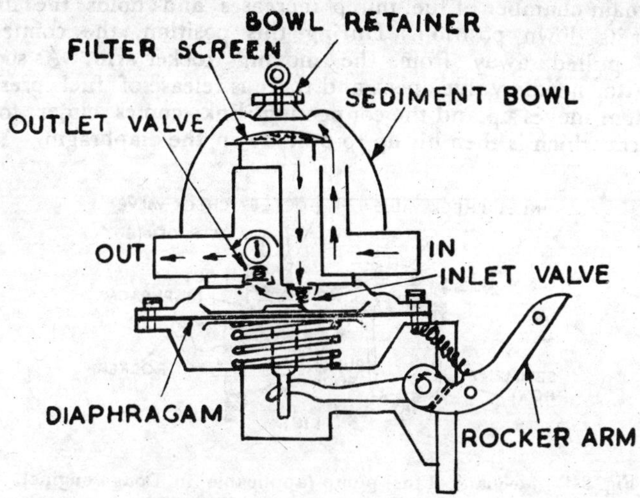

Fig. 8·3. Fuel Pump (Mechanical).

opens due to suction effect and fuel rushes into the main chamber from the inlet chamber. When eccentric pressure on rocker arm is released, the diaphragm moves upwards due to the tension of the spring placed below it. At this time the outlet valve opens and the fuel is pumped out of the fuel pump.

The working of the fuel pump can be summed up as below :

When eccentric pushes rocker arm :

(*i*) Diaphragm is pulled down through the connecting link,
(*ii*) Suction or vacuum is created in the main chamber,
(*iii*) Inlet valve opens,
(*iv*) Fuel rushes into the main chamber,
(*v*) Outlet valve remains closed.

When pressure of eccentric is released from the rocker arm :

(*i*) Rocker arm moves forward,
(*ii*) Diaphragm rises up,
(*iii*) Inlet valve closes and outlet valve opens due to fuel pressure,
(*iv*) Fuel is pumped out of the main chamber.

Suction and pumping effect is repeated again and again by the up and down movement of the diaphragm caused by rocker arm. Fuel from the tank is drawn and pumped to the carburettor by this action of the pump.

When needle valve in the float chamber of carburettor is closed and no more fuel could go into the carburettor, fuel pressure in the main chamber of fuel pump increases and holds the diaphragm at its down position. During this position, the connecting link is pulled away from the moving rocker arm. As soon as carburettor needle valve opens and there is release of fuel pressure, diaphragm moves up and the connecting link comes closer to the ocker arm which is then hit up to pull down the diaphragm.

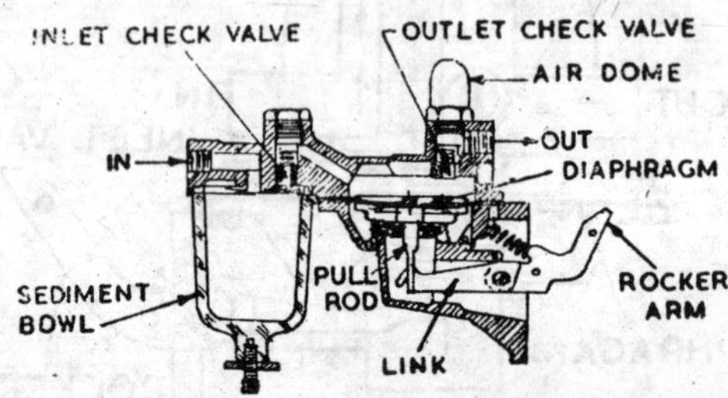

Fig. 8·4. Mechanical fuel pump (applicable in Dodge engine).

Single unit fuel pump employed in Dodge engines differs in design, although the working is the same. The difference could be well judged from its diagram which shows the various parts.

(*b*) **Combination unit fuel and vacuum pump.** As explained already, this fuel pump consists of two units, fuel pump unit and vacuum pump unit. The upper unit is vacuum pump and the lower unit is fuel pump or *vice versa.* Both the units are operated by the same rocker arm. The construction of the vacuum pump is similar to the construction of fuel pump. Inlet of the vacuum pump is connected to the wiper machine and outlet is connected to the induction manifold through pipe line.

When vacuum is caused by the downward movement of diaphragm in the vacuum pump, inlet valve opens and this vacuum operates wiper machine by drawing in air. When the diaphragm rises up, exhaust or outlet valve opens and the air drawn from the wiper-machine is pumped out to induction manifold.

When the wiper motor is not being used, engine manifold vacuum holds the diaphragm against its actuating spring. The diaphragm therefore, cannot make a complete stroke.

(2) **Electrical fuel pump**. Fuel pumps which are operated electrically are of the following two basic types : (*i*) *The suction type* which draws fuel in a similar manner as mechanically operated pumps, (*ii*) *The pusher type pump* which is located in the bottom of fuel tank and pushes fuel to the carburettor.

(*a*) **Suction type pump.** Suction type electrically operated fuel pump functions upon the same principle of pulling and pushing effect of diaphragm. The only difference being that in this pump, up and down movement of the diaphragm is affected electrically whereas in mechanical fuel pump, this movement is through mechanical means.

The electrical fuel pump of suction type consists of the following parts :

(*i*) Lower main body containing bellows or diaphragm, link, contact breaker solenoid and cover.

(*ii*) Upper body containing inlet and outlet check valves.

(*iii*) Sediment bowl and strainer.

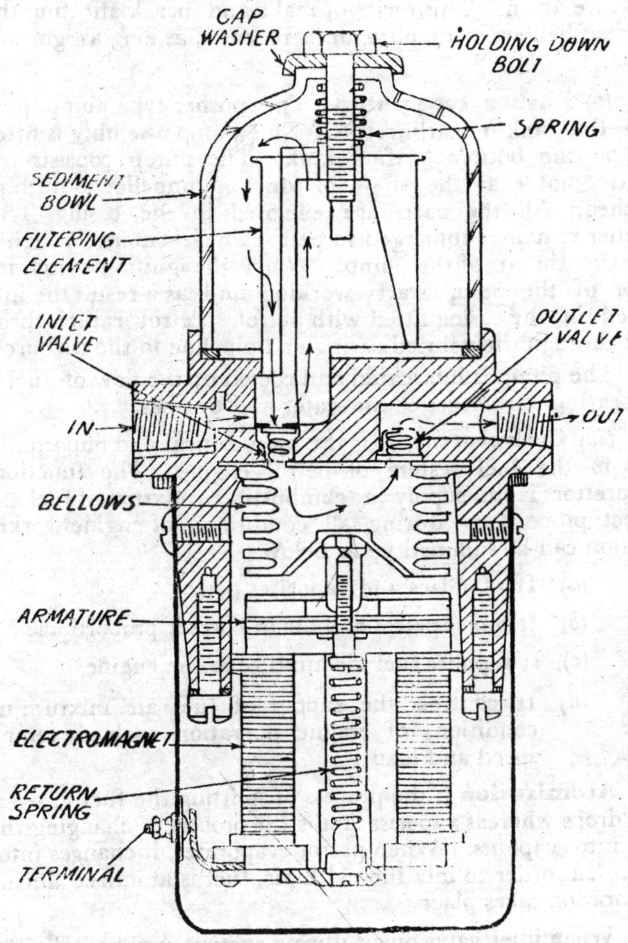

Fig. 8·5. Electrical Fuel Pump (Auto-pulse).

The metal bellows or diaphragm is operated by a solenoid. When ignition switch is turned on, current from the battery flows to the windings of solenoid and the armature is pulled down which results in the extending of bellows or downward movement of diaphragm. This produces suction or vacuum in the main chamber due to which fuel is drawn into the main chamber through inlet valve. As the armature reaches its lower limit of travel, the contact points are separated off which cut off the electrical circuit. The return spring then pushes the armature up and the bellows are contracted or diaphragm moves up. This results in the pumping of fuel out of the fuel pump main chamber through the outlet valve. As the armature reaches the upper limit of its travel, it closes the contact points and the solenoid is again energized to pull down the armature again. This series of make and break affecting the movement of bellows or diaphragm, is repeated as long as ignition switch is on.

(*b*) **Pusher type pump**. The pusher type pump pushes fuel from the tank to carburettor. The pump assembly is fitted inside and on the bottom of fuel tank. The pump consists of a small electric motor at the shaft of which an impeller to push petrol, is attached. All the parts are enclosed in the casing. The pump impeller remains submerged in petrol which enters through a screen into the throat of the pump. When the ignition switch is put on, motor of the pump starts working and as a result the impeller is rotated in the casing fitted with petrol. Petrol is then thrown outward through the voluted casing and pipeline to the carburettor.

The pump is lubricated and cooled by the flow of fuel around the bearings, armature commutator and brushes.

(*iv*) **Carburettor**. It is the most critical and important component in the fuel system of petrol engine. The function of the carburettor is to supply a combustible mixture of fuel and air in correct proportions during all conditions of engine working. Its function can be summed up as below :

(*a*) It atomizes and vaporises petrol,

(*b*) It mixes petrol and air in correct proportions,

(*c*) It supplies fuel air mixture to the engine

(*d*) It controls the supply of fuel air mixture under all conditions of engine operation such as temperature, speed and load.

Atomization is the process of splitting the fuel into atoms or fine drops whereas vaporisation is the process of changing the liquid fuel into vapours. When petrol evaporates, it changes into vapour state. In order to mix fuel with air, fuel is atomized during which evaporation takes place.

When inlet valve opens during suction stroke, fuel air mixture is drawn into the cylinder from the carburettor. This process is known as supply of combustible fuel air mixture. The amount of fuel supply determines the speed of the engine. More the fuel supply to the engine, more the power developed and more the speed of the engine. The flow of fuel air mixture into the engine is controlled by the throttle in the carburettor.

The carburettor is supposed to supply fuel air mixture in correct proportions under different conditions of temperature, speed and load on the engine. The strength of fuel air mixture is varied to suit engine operating conditions by changing the fuel air ratio. Mixing of more fuel with air forms *rich mixture* and decreasing the amount of fuel in the same amount of air makes the *mixture lean*. *Metering rod* in the carburettor measures the amount of fuel flowing into mixing chamber whereas the choke orifice measures the amount of air flowing into the mixing chamber. The flow of air into the mixing chamber depends upon engine suction, throttle valve opening, venturies in the carburettor and atmospheric pressure apart from choke orifice.

Higher the engine suction, more the vacuum produced within the cylinders and more the fuel air mixture is drawn in. More the throttle opening, more the fuel air mixture that goes into the cylinders. More the atmospheric pressure, more the air goes into the carburettor during engine suction. More effective venturi in the carburettor, more the velocity of air that goes in to form fuel air mixture.

For normal operating conditions, the best economy is obtained by a mixture of one part of fuel and 16 to 17 parts of air by weight. For quick acceleration and maximum power, a somewhat richer mixture having a ratio of 1 to 12-13 is required. For idling, a somewhat richer of 1 to 14-15 ratio is needed. Similarly, during cold starting, an extremely rich mixture having a ratio of one part of petrol and 9 to 10 parts of air is required.

Principle of working. The working of carburettor resembles with that of a simple mosquito oil spray gun. In order to understand the working of carburettor, let us first of all understand the working of mosquito oil spray gun.

The spray gun consists of two parts : the air pump and the oil tank fitted with a small nozzle. The pumped air passes over the mouth of small nozzle, one end of which is immersed in oil tank, nozzle being keeping the same level of oil as that in the tank. The passing stream of air over the mouth of the nozzle containing oil, takes along with oil particles which are split up and evaporated as air spreads them over, forming spray. Thus the spray provided by the oil spray gun is a mixture of oil and air.

The carburettor works in the same way to form a mixture of fuel and air. Instead of a separate pump, the piston in the cylinder acts as a pump and creates suction during its downward movement. During suction stroke, a pressure difference betwaen the pressure inside the cylinder and outside the carburettor is caused due to suction. Due to this pressure difference, air from the atmosphere rushes through the air cleaner into the carburettor mixing chamber where it passes over the mouth of fuel loaded nozzle placed in the venturi. This nozzle is connected with the float chamber on the other end. The float chamber acts as small fuel tank. As air goes in, the fuel is drawn out of the nozzle which is split up and surrounded by air forming fuel air mixture.

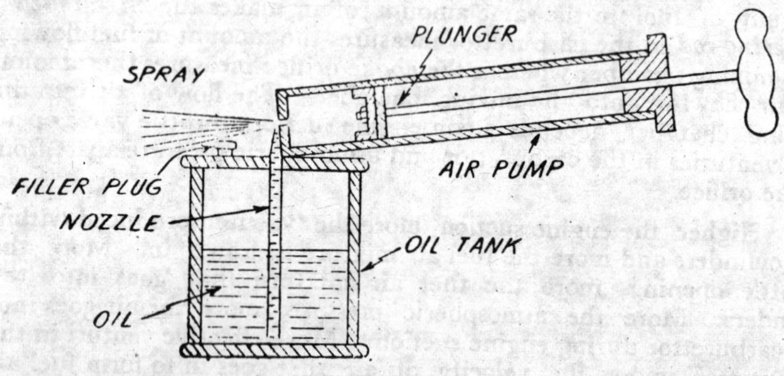

Fig. 8·6. Spray gun.

Venturi system. In order to have an even flow of fuel from the nozzle, venturi system is adopted in the carburettor which works on the following principle :

"If the pressure is decreased, the velocity increases and if the velocity is decreased, pressure increases."

A venturi or constriction which keeps less diameter at the centre and more at the ends, is provided in the carburettor mixing chamber to increase suction within the carburettor.

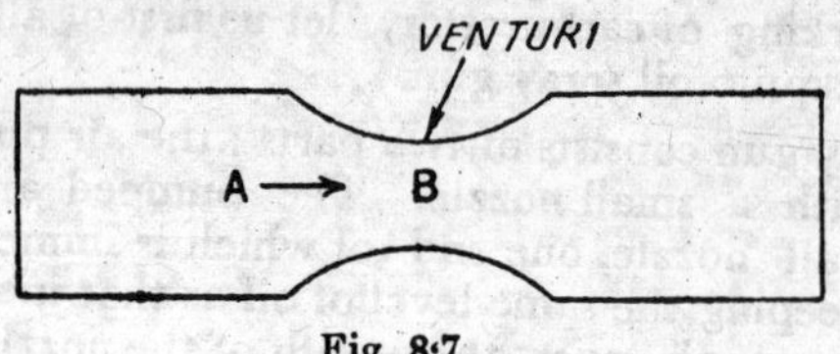

Fig. 8·7.

Pressure at B is lower than at A.

Velocity at B is higher than at A.

The outlet of the fuel nozzle is placed in the venturi due to which fuel is drawn out of the nozzle and mixes with the passing air forming fuel air mixture.

The difference between the pressure affecting at fuel nozzle in the venturi and the pressure on the surface of fuel in the carburettor float chamber causes the flow of fuel. More the pressure difference more the velocity of the fluid. Pressure and velocity can be affected by altering the diameters of the venturi tube through which the fluid passes.

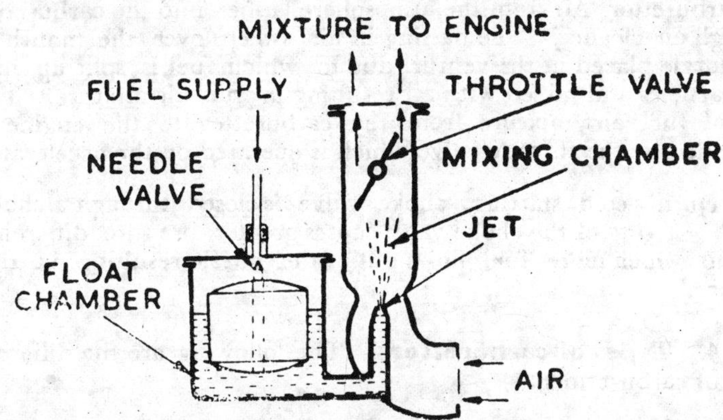

Fig. 8·8. A simple curburettor.

Simple plain tube carburettor. A simple carburettor consists of the following two main parts :

(*a*) Float chamber containing float, needle valve, metering jet and metering rod.

(*b*) Mixing chamber containing main nozzle, throttle and choke valves.

(*a*) **Float Chamber.** It is a small tank or bowl incorporated in the carburettor in which the flow of fuel from the line, is regulated by means of float controlled needle valve.

The metering rod, if installed, measures the amount of fuel going out from the float chamber to the main nozzle. It works in the metering jet placed at the junction of main nozzle in the float chamber.

(*b*) **Mixing Chamber.** It is long barrel attached to the float chamber, which contains venturi. The outlet of main nozzle opens at an angle in the venturi. It contains two butterfly type valves, one fitted in the air horn or upper part of the barrel, known as *choke valve* and the other installed in the carburettor outlet known as *throttle valve*.

The choke valve controls the flow of air into the mixing chamber and the throttle valve controls the flow of fuel air mixture from the carburettor to the engine.

Working. The main nozzle keeps the same level of fuel as contained in the float chamber since one end of the nozzle is connected with the float chamber. This is upon the principle that the liquids keep equal levels.

When engine intake valve is open during the suction stroke, the carburettor barrel is in line with the cylinder. Downward movement of the piston creates suction or vacuum in the cylinder. Suction leads to pressure difference between the cylinder and outside the carburettor. Air from the atmosphere rushes into the carburettor through air cleaner. The passing in air sweeps over the mouth of fuel nozzle placed in the venturi due to which fuel is split up into fine particles and mixed with air resulting in fuel air mixture. The flow of fuel air mixture from the carburettor to the engine is controlled by the throttle valve which is operated by the accelerator.

During cold starting, choke valve is closed through a choke cable. Closing of the choke valve causes greater pressure difference due to which more fuel flows out of the nozzle resulting in rich mixture.

4. **Types of carburettors**. The following are the different types of carburettors :—

(*a*) **With regard to draft** :

(*i*) **Down-draft carburettor** in which fuel air mixture goes downward towards the induction manifold. It is fitted above the induction manifold. The gravity assists in it for the flow of fuel.

(*i*) **Up-draft carburettor** in which the fuel air mixture flows upward towards the induction manifold. It is fitted below the induction manifold. Liquid fuel cannot enter into the engine if it overflows.

(*iii*) **Side-draft carburettor** in which the mixture flows horizontally towards the manifold. It is fitted by the side of induction manifold.

(*iv*) **Semi-down draft carburettor** ; a mixture of down draft and side draft carburettors.

(*b*) **With regard to the working construction** :

(*i*) **Constant choke carburettors** in which the orifice area is constant and the pressure difference is varied by means of venturies. Carter, solex and zenith are constant choke carburettors.

(*ii*) **Constant vacuum carburettor** in which the orifice area is varied to meet the changing demand and the pressure or depression is kept constant. The vacuum or suctional effort at all the throttle positions and engine speeds remains constant. S.U. (Skinner Union) carburettor is an example of this type.

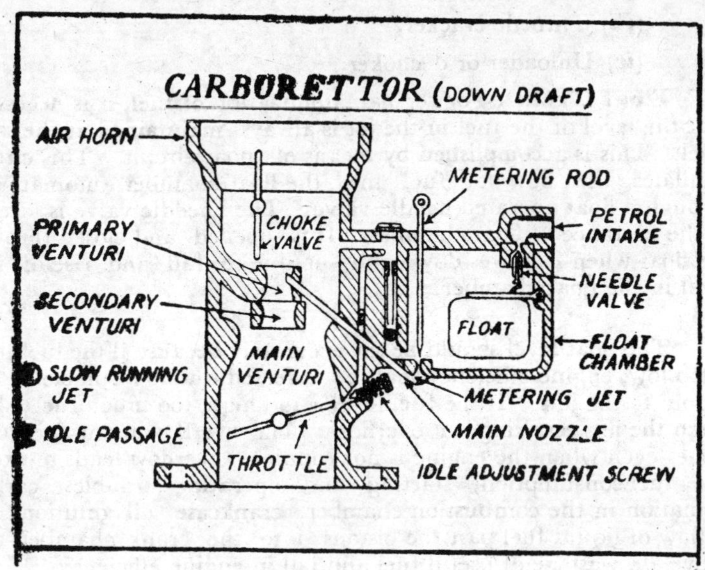

Fig. 8·9.

(*c*) **With regard to number of units.**

(*i*) **Single barrel carburettor** containing single barrel or only one throat.

(*ii*) **Dual or multiple barrel carburettor** containing two or more throats or outlets or barrels. Two or more carburettors combined into one unit.

5. Circuits and controls in the carburettor. The carburettor is subject to supply suitable mixture of fuel and air during all conditions of engine operation To accomplish this, the carburettor must be provided with additional circuits and controls. The following are the main circuits and controls are required by the carburettor for its proper operation :—

(*a*) **Circuits :—**

(*i*) Float circuit.

(*ii*) Idle circuit.

(*iii*) Low speed circuit.

(*iv*) High speed circuit.

(*v*) Pump circuit.

(*vi*) Coke circuit.

(*b*) **Controls :—**

(*i*) Throttle.

(*ii*) Choke.

(*iii*) Throttle cracker.

(*iv*) Unloader or dechoker.

Float Circuit. For regular atomization of fuel, it is necessary that the level of the fuel in the jet is always maintained at the same level. This is accomplished by means of float circuit. This circuit regulates the flow of fuel into the float chamber automatically through a float operated needle valve. The needle valve is located in the entrance of float chamber. It is opened and closed through the float when it moves down and up due to fall and rise of fuel level in the float chamber.

The float level should be set accurately because if the fuel level is too low, engine efficiency shall be affected due to insufficient fuel supply to the jets. If the fuel level is too high, too much fuel shall reach the jets resulting in overflow. The overflowing of fuel continues even when the engine is not running. Overflow leads to excessive fuel consumption, starting and operating troubles, carbon formation in the combustion chamber, crankcase oil dilution due to flow of liquid fuel past the pistons into the crank chamber and above all, wastage of useful fuel and fall in engine efficiency.

The float chamber contains a vent to admit outside atmospheric pressure. In some carburettors, the vent opens into the air horn on which air cleaner is installed. The air has to pass through the air cleaner before it enters the float chamber. This results in equalized pressure on the fuel in float chamber and on the flow of air through carburettor air horn. This is known as balancing. Carburettors provided with this type of arrangement are called *balanced* carburettors. In the balanced carburettors, restrictions due to accumulation of dirt in the air cleaner do not affect the fuel-air ratio.

During extremely hot weather, the fuel in float chamber vaporizes and leads to vapour lock. In the balanced carburettor, the vapours formed in the float chamber pass through the vent tube into the air stream in the air horn where they mix up with the air. This helps in eliminating vapour lock but enriches the fuel air mixture unnecessarily.

Idle Circuit. Fuel flows from the float chamber towards the mixing chamber through the *idle circuit* when the throttle valve is fully closed. During closed throttle, the engine suction acts at the tiny hole known as *idle port*, provided in the mixing chamber wall below the throttle disc. Discharge of fuel air mixture is at this hole. This idle passage is shown in the diagram. Air from the air horn enters into the idle passage through a hole and bypass provided in the upper part of the mixing chamber. The air sweeps over the slow running jet, contained in the passage connected with floa chamber. This passage contains the same level of fuel as in the float chamber. The fuel particles are sweeped into the idle passage by the

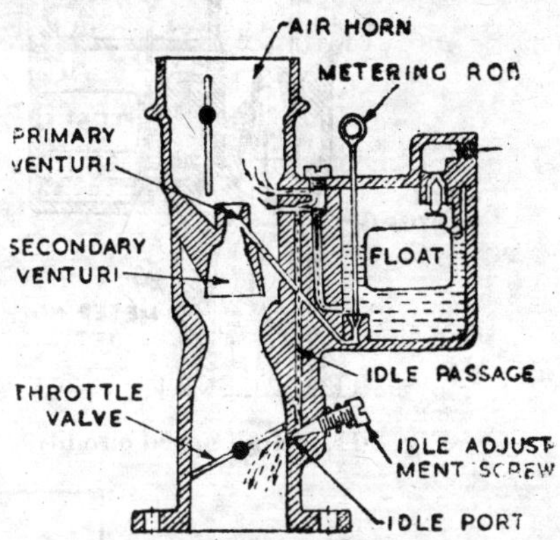

Fig. 8·10. Idle circuit.

ingoing air. The air entering through the bypass helps in pushing down the heavier fuel particles. The discharge of fuel air particles is at idle port from where they travel into the engine.

The idle port-hole contains a conical seat for the spring loaded idle mixture-adjusting screw. By turning the screw clockwise, the amount of fuel discharge from the idle port is decreased while turning the screw anti-clockwise, the discharge rate is increased.

The idle circuit supplies fuel air mixture to the engine during its idle speed when throttle valve is closed. In the absence of idle circuit, the engine would stop with the closing of throttle valve.

Low Speed Circuit. During low engine speed or slow running of the engine, the throttle valve is partly opened and the air from the air horn passes down by the sides of throttle disc. The throttle is operated by means of accelerator pedal. When the accelerator pedal is pressed down, throttle valve is opened, metering rod rises up and the pump plunger moves down in a 'Carter' like carburettor. When the metering rod rises up along with the opening of throttle valve, fuel flows into the main nozzle through the metering jet. The sweeping in air takes along with fuel particles from the main nozzle, the mixing of which results in fuel air mixture.

High Speed Circuit. The fuel flows through the same passage as in the low speed circuit and the operation is similar with the exception that the metering rod is raised high and throttle valve is wide open. So, proportionately, more fuel air mixture is supplied by the carburettor through this circuit.

In order that the correct mixture of fuel air may be supplied through this circuit during all speeds and throttle openings, *compen-*

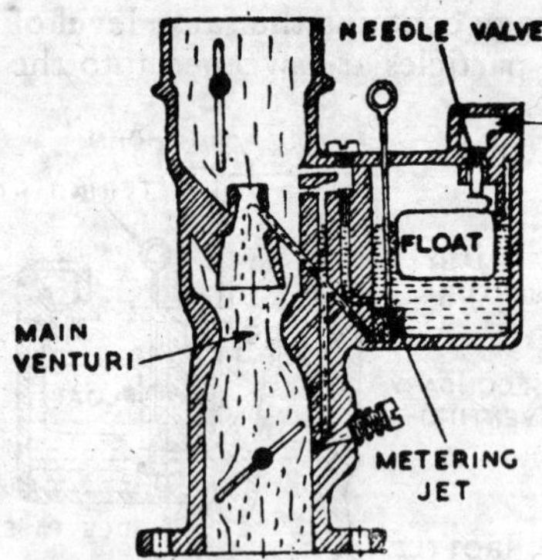

Fig. 8·11. High speed circuit.

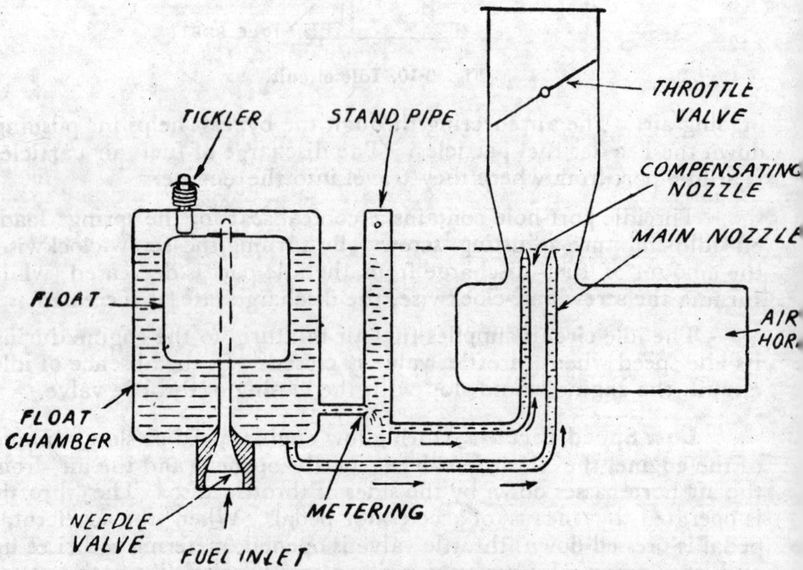

Fig. 8·12. Carburettor employing compensating nozzle parallel to main nozzle.

sating or ***air bleed*** arrangements are provided in the carburettors. In the ***compensating*** arrangement, an additional nozzle is provided which discharges fuel at the constant rate. In combination, the two nozzles supply a constant mixture.

The main nozzle gets fuel directly from the float chamber whereas the compensating nozzle is fed through a standpipe. The upper end of standpipe is open to the atmosphere. The fuel enters the standpipe through a metered passage from the float chamber. The discharge rate of fuel from the compensating nozzle is constant as the flow of fuel into the standpipe is constant due to the constant level in the float chamber. At high engine speed, the compensating nozzle delivers less fuel than at low speed. This way, it compensates the tendency of main nozzle to supply a rich mixture at high speed or wide open throttle.

In *air-bleed* arrangement, air bleed holes are provided in the upper section of main nozzle at a point below the level of fuel in the nozzle. Introduction of air into the air bleed holes reduces the surface tension of fuel and assists it to flow at low pressures. Air bleed also restricts the flow of the fuel through the main nozzle during high vacuum. Flow of fuel at low pressure and at low rate, controls the fuel air ratio and offset the tendency of increasing the richness of mixture due to increasing air velocity. Thus by means of air-bleed arrangement, the increased richness of mixture is compensated and the fuel-air ratio is maintained at a constant rate.

For maximum engine speed and power, additional fuel must be supplied. The *economizer valve* or *power jet* and the metering rod help in providing additional fuel for maximum power. The power jet is provided in Ford carburettors. It is operated by the engine manifold vacuum. It is a diaphragm or plunger type non-return valve. It opens to permit the required amount of fuel to pass through the main nozzle when engine vacuum is less. During idle speed or low speed, when throttle valve is fully or partly opened, the engine vacuum keeps the diaphragm or plunger pulled down resulting in closed position.

In some carburettors, the power valve is controlled by a vacuum actuated piston assembly which operates in accordance with the throttle opening. When throttle valve is closed, a high manifold

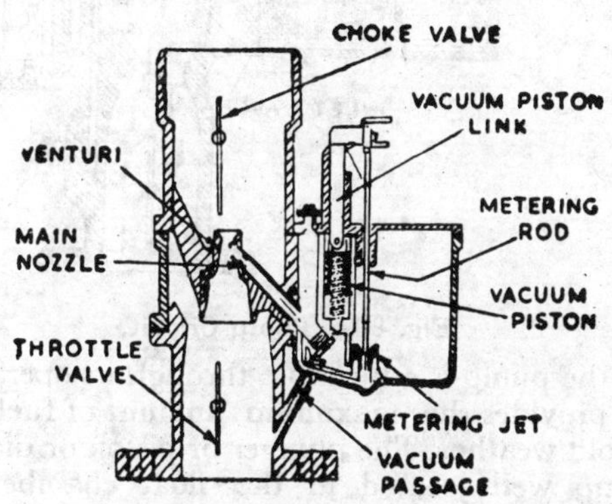

Fig. 8·13. Vacuum piston operated metering rod.

vacuum is acting on the vacuum-controlled piston assembly. The piston assembly is pulled up in its cylinder against the tension of a spring due to vacuum resulting in the closing of valve. When the throttle valve is opened for high speed operation, the manifold vacuum drops down and the springs at the piston assembly overcomes the piston and moves it down in its cylinder. As a result the power jet is opened and additional fuel is supplied into the high speed circuit.

The *metering rod* also assists in supplying additional fuel for more power during high speed circuit. The metering rod contains several steps or tapers on its lower end and operates in a calibrated metering jet which supplies fuel to the main nozzle. When in the high speed, throttle valve is wide opened, the metering rod is raised high and more fuel is permitted to flow through the circuit. Metering rods are also operated by means of a vacuum controlled piston as discussed above and shown in the diagram.

Pump Circuit. When the throttte valve is opened quickly for rapid acceleration, the fuel mixture tends to becom lean due to the fact that fuel is heavier than air. The fuel lags behind the flow of air, resulting in a lean mixture. In order to keep the fuel mixture of proper strength for more speed, additional fuel is injected directly into the air stream by means of a pump.

The pump is operated by the accelerator and is related with throttle valve. The pump is of plunger or piston or diaphragm type.

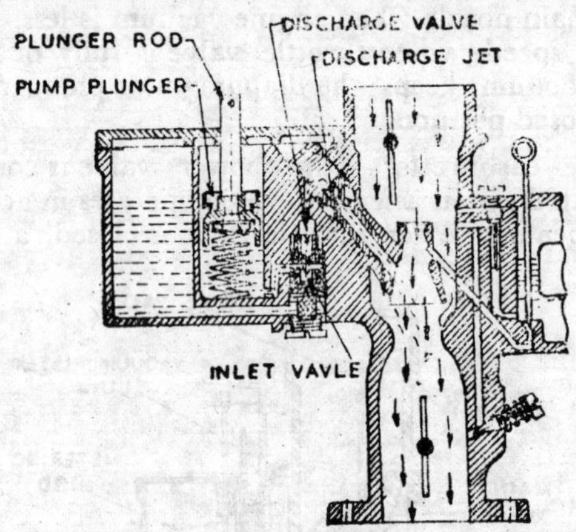

Fig. 8·14. Pump Circuit.

The stroke of the pump is adjustable through its operating link. The longest stroke provides the maximum amount of fuel and is usually used during cold weather. The plunger or piston or diaphragm works inside the pump well located in the float chamber. The pump system consists of the following parts :—

(*i*) Pump plunger, or piston or diaphragm with link and spring.

(*ii*) Inlet and outlet check valves.

(*iii*) Pump jet.

When the accelerator pedal is released and throttle valve is closed, the plunger moves upward in its cylinder and the fuel is drawn in due to vacuum caused therein, through the inlet check valve from the float chamber. When the accelerator is pressed down for more speed, the plunger moves down and the fuel is pumped out through the outlet valve and pump jet into the mixing chamber.

The discharge action of the pump is prolonged or delayed so that it may continue the discharge of fuel into the air stream for several seconds. This action of the pump is known as *delayed action.* This is necessary because it requires a few seconds for the high speed circuit to begin to deliver sufficient fuel through the main nozzle.

Choke Circuit. Rich mixture is required for cold starting. The rich mixture is obtained by the application of choke. When choke valve is closed, the air enters into the mixing chamber through a tiny hole, or a light spring loaded poppet valve in the butterfly disc or by the sides of the choke disc in case it is elliptical. The closing of choke valve causes greater pressure difference between the pressure affecting inside the mixing chamber and the pressure outside the carburettor. Pressure above the choke disc is much more than the pressure acting on the other side as the air passage decreases due to the closing of disc.

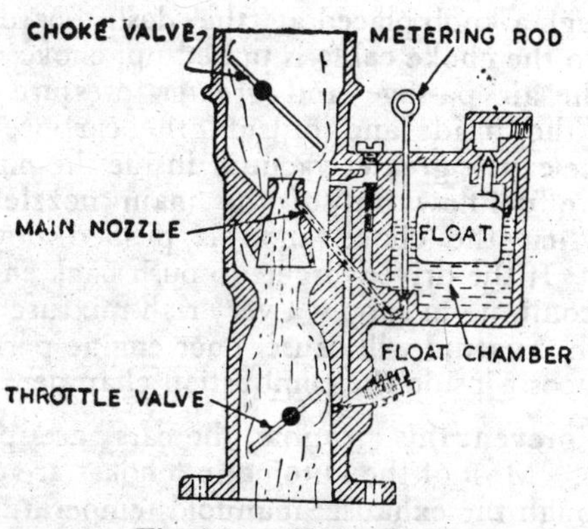

Fig. 8·15. Choke circuit.

The decreased air passage increases the air velocity and decreases the pressure inside the mixing chamber. The varying difference between the atmospheric pressure affecting the float chamber and the decreased pressure inside the mixing chamber, causes the fuel to flow rapidly towards the mixing chamber which results in rich fuel air mixture.

Some carburettors are provided with special passages which carry hot exhaust gases or hot engine water around the carburettor to keep it warm during cold weather to avoid icing in the carburettor.

Throttle Control. As mentioned already, supply of fuel air mixture into the engine is controlled by means of throttle. The throttle is a butterfly or sleeve or disc type valve which opens or closes the passage for the flow of fuel air mixture into the engine. Butterfly valve is quite common. Sleeve valve is used in 'Amal' carburettors applicable to motor cycles and disc valve is used in small carburettors just as one fitted with Vespa scooter.

The throttle valve is located in the throat of carburettor and is operated by the accelerator. It is connected with the accelerator pedal or twist grip by means of linkages or cables. More the throttle opening, more the engine speed and less the throttle opening, less the speed of the engine.

Opening of the throttle valve is adjusted by means of a stopper screw known as slow running adjustment screw.

Choke Control. Choke in the carburettor helps in providing rich mixture for starting a cold engine. During cranking, vacuum affecting in the mixing chamber is insufficient to draw adequate fuel flow for starting. To produce sufficient fuel flow during cranking, the choke is installed in the carburettor. The choke is a butterfly type valve which is placed in the air horn. It is operated mechanically or automatically. In mechanically operated choke, the butterfly valve is opened and closed by means of a cable and lever which is operated through a knob placed at the dash board. When the knob connected to the choke cable is pulled up, choke valve is closed which restricts the air passage and greater pressure difference is caused between the inside and outside the carburettor. Greater pressure difference causes greater vacuum inside the mixing chamber due to which more fuel flows out of the main nozzle resulting in rich mixture. When the choke knob is pushed down, the choke valve is opened. If the driver forgets to push back choke knob, the carburettor will continue to supply a very rich mixture to the engine. The excessive rich mixture will cause poor engine performance and lead to carbon deposit inside the combustion chamber.

In order to prevent this trouble, the cars are provided with automatic chokes. Most of the automatic chokes are operated thermostatically through the exhaust manifold temperature and induction manifold vacuum. The thermostatically controlled choke consists of a thermostatic spring and a vacuum piston or diaphragm which are linked with the choke valve. The thermostatic spring is made up of two different metal strips which are welded together and formed into a spiral. Due to difference in expansion rates of the two metals, the spring tends to wind or unwind with the changing temperature. During low temperature, spring is moved up and keeps the choke valve in the closed position. When the engine is cranked, a rich mixture is supplied to the engine. As the engine starts, movement of air through the air horn causes the choke valve to open slightly. Also the induction manifold vacuum acts upon the piston which is pulled outwards to open the choke valve further. When manifold vacuum falls down during high engine speed, the piston is pushed back due to which choke valve is closed when the engine temperature is low.

The thermostatic spring is enclosed in a housing which is connected to the exhaust manifold through a small tube. The exhaust gas circulates over the thermostat spring and warms it. Warming of the spring unwinds it which results in the choke valve to start opening. When operating temperature is reached, the thermostatic spring unwinds enough to keep open the choke valve up to full extent. When the engine is switched off and cools down, the thermostatic spring again winds up to close the choke valve.

In some carburettors, hot water from the engine cooling system is used in place of exhaust hot gas from the exhaust manifold, to operate the thermostat spring. The operation of vacuum operated diaphragm in lieu of piston is quite similar.

Throttle Cracker. When the engine is cranked for starting, the throttle valve must be opened or cracked slightly so that enough air may pass it. This necessity is fulfilled by a special linkage between the self-starter and the throttle linkage. When self-starter is operated, the throttle valve is opened slightly.

Unloader or Dechoker. When an engine does not start immediately, prolonged cranking will result in carburettor overflow. The mixture no longer remains vapour and goes to the engine in the liquid state which is difficult to explode. In order to overcome this problem, special linkage is provided between the throttle and choke levers, which holds the choke valve open when the accelerator is pushed to the floor board. Then, when the engine is cranked again, the excessive petrol is cleared off as air enters the manifold and cylinders.

Constant Vacuum Carburettor. In this carburettor, the orifice area is varied to meet the varying demand of carburettor and the pressure difference is kept constant. S.U. carburettor is of this type. This carburettor contains only one jet in which a tapered needle slides up and down. The needle is carried by a piston which is contained in a housing and moves up and down due to engine suction. The movement of the piston varies the size of the jet through the needle in accordance with the quantity of air flowing past it.

The correct level of petrol is maintained in the jet due to the flow of petrol from the float chamber. The passing in air sweeps over this jet and takes along with fuel particles to form fuel air mixture which is fed to the engine. The movement of the taper needle carried by the piston is controlled by the suction disc which is cast with the piston. The underside of suction disc is open to atmospheric pressure through the hole in the flange. The upper side of the disc is open to the engine suction through a passage from the underside of the piston. As engine suction increases, the difference between the pressure on the upper and lower sides of the disc will

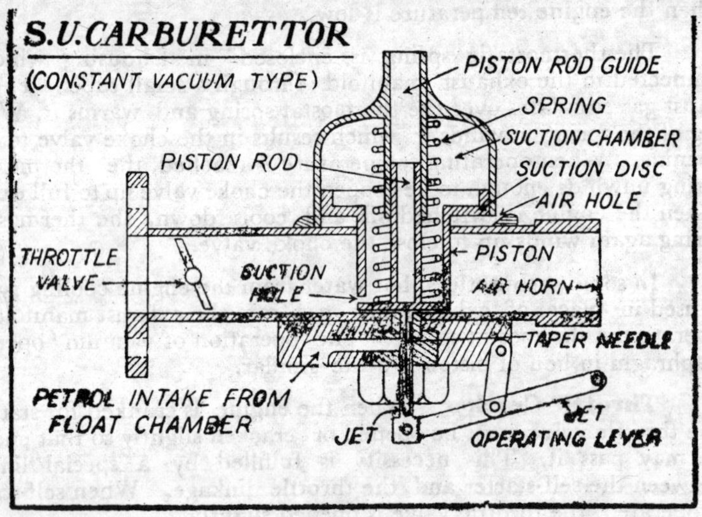

Fig. 8·16

increase, causing the disc to rise. Thus the needle is withdrawn from the jet due to which more fuel flows out. When the engine is stationary, pressure above and below the disc is the same and as a result the piston moves to its lowest position, closing the jet through the needle.

For starting purposes, the jet itself is moved downward from the needle by means of the jet lever. The jet lever is operated by a knob controlled cable attached at dash-board. When the knob is pulled, the jet moves down and the fuel flows from it to form fuel air mixture for starting. When the knob is pushed down, the jet returns to its normal position.

Solex Carburettor. M/s. Carburettors Ltd. Madras, are manufacturers of Solex carburettors in India. These carburettors are employed in Fiat/Premier, Standard Herald, Ambassador cars, Jeep C J B and other Indian vehicles.

Solex carburettors have been in use since long in many cars of foreign make. The basic feature of this carburettor is the straight forward design of the spraying assembly as in other improved carburettors. The main jet supplies petrol to the spraying well which contains a perforated emulsion tube, at the upper end of which is filled a calibrated air-correction jet. As the engine speed rises and petrol level in the well falls, an increasing amount of air is drawn through the holes in the emulsion tube, automatically weakening the mixture and providing necessary correction at high speeds.

Solex C 32 PBIC carburettor is of downdraft type. It contains progressive action starting device which ensures prompt starting from cold and regular running of engine at low speeds when still cold. The starting device consists of a flat-disc valve having different size holes. These holes connect the starter petrol jet and starter air jet sides to the passage which opens below the throttle valve. The disc valve is operated through a lever connected to control knob at dash board.

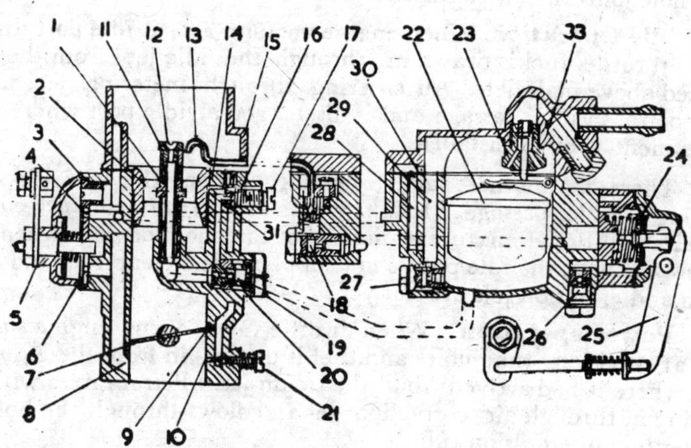

Fig. 8·17. Solex C 32 PBIC Carburettor.

1. Spraying well.
2. Venturi.
3. Starter air jet.
4. Hole in starter disc.
5. Starter lever.
6. Hole in starter disc.
7. Throttle disc.
8. Passage to starter device.
9. Transition orifice.
10. Idle port.
11. Emulsion tube.
12. Air correction.
13. Accelerator pump injection tube.
14. Calibrated orifice.
15. Body.
16. Idle speed air metering jet.
17. Idle speed jet.
18. Accelerating pump jet.
19. Main jet holder.
20. Main jet.
21. Idle adjustment screw.
22. Float.
23. Needle valve.
24. Pump diaphragm.
25. Accelerating pump operating lever.
26. Non-return valve.
27. Starter petrol jet.
28. and 29. Wells.
30. Air inlet orifice from bowl for staking device.
31. Idle well.

Operation of starting device. The fuel flows through the starter petrol jet and fills two wells above the jet. During closed throttle, engine suction acts at the starting passage and draws in fuel from the starter jet through the two holes in starter valve disc, the opening of which varies according to the position of operating lever. This fuel combines with air entering through hole above petrol well and starter air jet and forms a very rich mixture. When the engine is running steadily and operating lever is pushed back to the intermediate position, the amount of fuel coming through petrol jet is reduced because the wells go empty in the meantime. The mixture becomes still leaner because air entering through the well hole joins the air streams.

Idle Operation. When engine vacuum acts at idle port during closed throttle, fuel is drawn in through the idle jet from the well located above main-jet. Air entering through metering jet mixes with petrol in idle passage and is discharged at idle port where idle-adjustment screw is installed.

There is another small opening above the throttle disc leading to idle passage. It is known as *transition-orifice* which supplies additional mixture at minimum throttle openings when due to low vacuum, the idle port is not supplying sufficient fuel and both main and emulsion jets are not yet *primed*.

Power operation. When throttle valve opens, engine suction acts at emulsion tube holes and fuel is drawn in from the emulsion well where it had arrived from the main jet. Petrol mixes with air coming in through air correction jet and flows through the holes at the bottom of emulsion tube.

Accelerating pump. As in other carburettors, the accelerating pump injects extra petrol needed to step up acceleration. The diaphragm type pump is actuated by the accelerator through a lever secured on throttle shaft. The diaphragm is held in position by a compression spring. When accelerator is released, fuel is drawn into the pump reservoir through a non-return valve. When accelerator is pressed down, pump diaphragm moves against the tension of spring and pumps out fuel through the pump jet. The fuel is sprayed into venturi through curved injector tube. The size of jet determines injection speed.

Float circuit. Like other carburettors, float operated needle valve controls the flow of fuel into the float chamber.

As is clear from the above explanation, Solex carburettor, like others, contains float, choke, idle, main and pump circuits.

Zenith carburettor. This carburettor is quite popular with British cars and is available in different forms and models. Up-draft and down-draft Zenith carburettors are quite common. Like other types of fixed choke carburettors, the Zenith employs various jets, air bleeds, an accelerating pump, choke for cold starting, float, idle or slow running, acceleration, low and high speed circuits.

Zenith 30 VIG11 downdraft carburettor consists of mixing and float chambers. Flow of fuel into the float chamber is controlled by float operated needle valve as in other carburettors. Emulsion block is attached with one side of float chamber which covers the mixing chamber open side. By unscrewing only two top screw bolts, the float chamber along with emulsion block, can be removed easily. Main and compensating jets are contained at the bottom of float chamber. Slow running and pump jets, plunger pump, non-return valves are all accessible after the float chamber is removed.

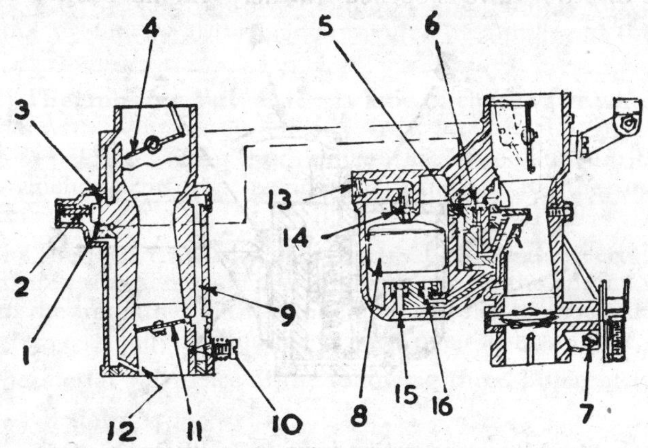

Fig. 8·18. Zenith carburettor (30 VIG-11).

1. Diaphragm in economy valve.
2. Economy valve spring.
3. Restricted passage.
4. Choke strangler.
5. Capacity well.
6. Slow running jet.
7. Throttle stop screw.
8. Float.
9. Idle passage.
10. Idle adjustment screw.
11. Throttle valve.
12. Suction port.
13. Fuel inlet.
14. Needle valve.
15. Main jet.
16. Compensating jet.

Carburettor air horn contains choke strangler (flap) to create pressure difference for drawing in more fuel during cold starting. Throttle valve is housed at the lower part of mixing chamber and controls the flow of fuel air mixture into the engine. An idle port opens below the tip of throttle valve. Fuel from the float chamber flows to the idle passage having air hole at the top. During closed throttle, fuel air mixture is supplied through the idle circuit which contains a spring loaded conical screw known as idle adjustment screw.

Upon pressing down accelerator, fuel is pumped into the air stream in mixing chamber through the pump circuit. The additional fuel ensures a smooth and progressive gateway from slow running. As the throttle valve is opened further, engine suction acts at the

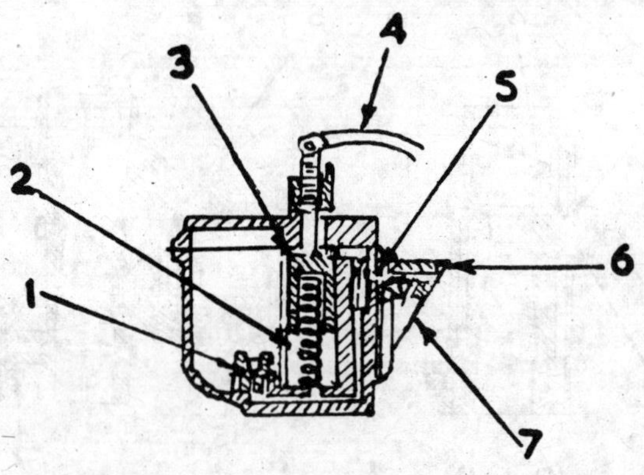

Fig. 8·19. Accelerating pump in Zenith Carburettor.

1. Non-return valve.
2. Pump well.
3. Piston.
4. Operating link.
5. Pump jet.
6. Beak.
7. Emulsion block.

spraying beak in emulsion block and petrol is drawn out from the main and compensating jets and from the capacity well. The compensating jet is connected directly to the capacity well through a passage. During small throttle openings, the well is partly filled with petrol which is drawn out during acceleration to provide temporary enrichment. An air bleed provided at the capacity well help in pushing out fuel from main and compensating jets due to cross-connection between them.

This carburettor contains a diaphragm type economy valve which is fitted at the side of carburettor. There are two holes on the inner side of diaphragm. The small hole provides a permanent air bleed to the capacity well for compensation. The second hole is larger and is normally closed by a plate valve at the centre of diaphragm. The opposite side of diaphragm is subject to induction manifold vacuum through a passage which runs from a point in the carburettor throat below the throttle to the diaphragm housing. When throttle is partly closed and manifold vacuum is high, the diaphragm is pulled back against spring pressure, uncovering the larger hole. This results in the supply of more air to the capacity well and weakening the mixture.

(*v*) **Air Cleaner.** It serves the following purposes :

(*a*) It screens and filters air entering the carburettor to form fuel air mixture.

(*b*) It helps in decreasing wear rate in the engine as it cleans the air entering the carburettor from dust and dirt particles which otherwise could act as an abrasive to damage the engine.

(*c*) It acts as a silencer to decrease noise of air rushing into the carburettor.

(*d*) Oil bath type air cleaner serves as a trough of oil for upper cylinder lubrication.

There are two main types of air cleaners in common use. These are as under :

(*a*) Dry air cleaners

(*b*) Oil bath type air cleaners.

(*a*) **Dry air cleaners.** This type of air cleaner consists of three main parts : housing, cleaning element and cover. The cleaning element is usually made of multi-wire netting and is contained in the housing with top cover or cap on it. Air from the atmosphere enters from below the cap border, passes through the cleaning element and goes down to the carburettor through the outlet provided at the bottom of housing. The dust and dirt particles from air entering the carburettor are held by the filtering element and thus the air is cleaned. These types of air cleaners are mostly used in motor cycles and scooters.

(*b*) **Oil bath air cleaners.** The main difference between dry and oil bath air cleaners being that in the latter case, oil is contained in the housing. Air which enters the air cleaner has to pass or dip into the oil before it enters the carburettor. The direction of incoming air is reversed and directed over the surface of oil bath. This results in a large portion of dust to be retained in the oil bath. The air then passes through the cleaning element which is usually coppermesh screen, before it goes out of the carburettor. Thus the air entering the carburettor is saturated with oil for *upper cylinder lubrication.*

Some air cleaners are directly mounted over the carburettor whereas in most cases these are placed away from the carburettor to decrease overall height of the engine. In such cases air from the air cleaners is fed to the carburettor through specially designed pipes which are known as silencers. These help in reducing the noise of air rushing into the carburettor a great deal.

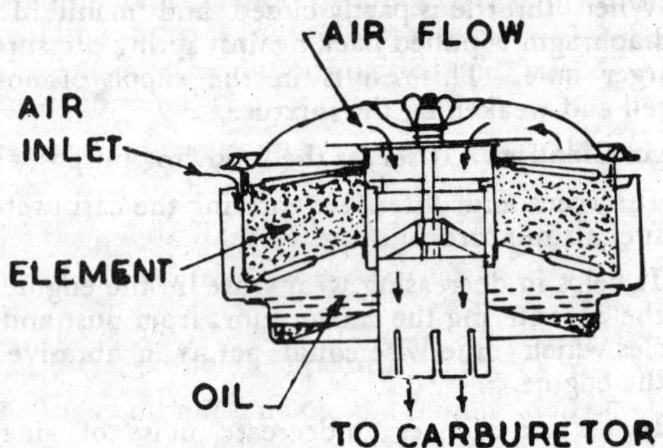

Fig. 8·20. Air clearner (oil bath type).

(*vi*) **Fuel Gauge.** It indicates the amount of fuel in the fuel tank.

The fuel gauges can be classified as under :

(*a*) Mechanical Fuel Gauges,

(*b*) Electrical Fuel Gauges.

(*a*) **Mechanical Fuel Gauge.** This type of fuel gauge is located direct at the fuel tank. It consists of the dial, needle, lens, cap and casing, cork or copper or brass float, link and gears. The float is kept inside the tank and connected with the gauge unit through link and gears. The float rises up and down with the level of fuel contained in the tank. The movement of the float is transmitted to the gauge needle through link and gears. When the float rises up due to more level of fuel in the tank, the gauge needle moves across the dial to the right side and registers the amount of fuel in the tank. As the fuel level falls down in the tank, the gauge needle moves backward to the left due to the downward movement of float and shows the amount of fuel accordingly. The gauge dial is marked differently in different gauges. Mostly the markings indicate full, half, empty and their parts.

This type of gauge is becoming obsolete as the driver has to come down from his seat to read the fuel gauge. This type of gauge is applicable to Foden heavy duty diesel vehicles.

(*b*) **Electrical Fuel Gauges.** These gauges are of the following two types :

(*i*) Balancing coil type,

(*ii*) Bimetal thermostat type.

These gauges operate in the similar manner as electric type oil pressure gauges. The construction too resembles very much with one another. The only difference being that resistance at the tank unit is caused to alter due to the movement of a float which rises up and down with the level of fuel in the tank.

(*i*) **Balancing coil type**. The tank unit in this gauge contains a sliding contact which slides back and forth on a resistance as the float rises up and down in the fuel tank. This alters the amount of electrical resistance which is offered by the tank unit. As the tank empties, the float falls down and the sliding contact moves to reduce the resistance.

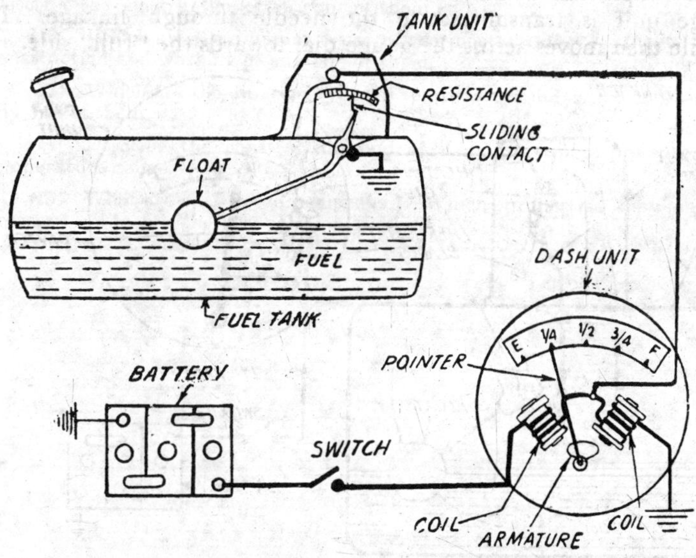

Fig. 8·21. Balancing coil type fuel gauge circuit.

The dash unit contains two coils. When ignition switch is turned on, current from the battery flows through both the coils due to which magnetic field is set up. The magnetic flux acts on the armature to which the gauge needle is attached. When the tank is filled with fuel, float is in the up position which results in high resistance at the tank unit. During this position, current flows from the left side to right side coil due to which the armature is pulled to the right and the needle deflects to the full side of the dial. When fuel level in the tank begins to fall down, the resistance of the tank unit drops resulting in the flow of more current of left side coil through the tank unit. Since less current is flowing through the right side coil so its magnetic field is weak due to which the left side coil pulls the armature towards it and the needle deflects towards the empty or left side of the dial.

(*ii*) **Bimetal thermostat type**. Each unit in this gauge contains a thermostat blade having heating coil. The coils are connected in series through the ignition switch to the battery. The float in the fuel tank actuates a cam which imposes more or less bending on the thermostat blade in the tank unit. When the float is up due to high level of fuel in the tank, the cam puts a considerable bend in the blade. Upon putting on ignition switch, current flows through the heater coils. Due to heating of blade in the tank unit, it bends further resulting in the separation of contacts. The blade then cools and the points are closed. Heating of the blade again starts and the points reopen. This action continues as long as ignition switch is on. Heating and bending of blade in the gauge unit is affected in the same proportion as that of tank unit. Movement of blade in the gauge unit is transmitted to the needle through linkage. The needle then moves across the gauge dial towards the "full" side.

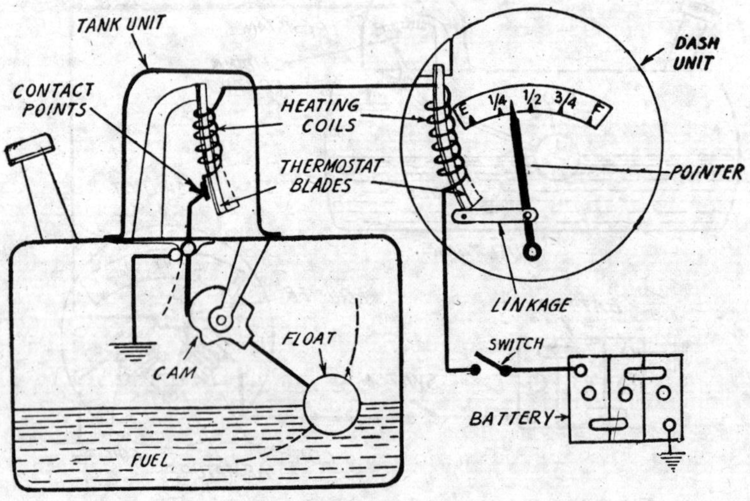

Fig. 8·22. Thermostat fuel-gauge circuit.

When there is no or less fuel in the tank, the float is down and the cam bends the tank unit thermostat blade up to a small limit. Consequently, only a small amount of heating is enough to bend the blade any more. Accordingly, the dash unit blade bends only a little due to which the needle points towards "empty" side.

4. Supercharger. It is a type of blower or compressor which accelerates the flow of fresh charge into the engine cylinders. It is located between the carburettor and induction manifold. Racing cars are usually fitted with superchargers for more power. The supercharger increases volumetric efficiency of engine. It pushes in more volume of fresh charge into the cylinders. More the fuel charge more the power.

Supercharged engine will have a higher overall compression which will increase the tendency toward detonation in petrol engines. When a supercharger is used on an automobile engine, fuel of higher octane number is needed to overcome detonation.

The following are the different types of superchargers :

(*a*) Roots type,

(*b*) Centrifugal type,

(*c*) Vane type.

(*a*) **Roots type**. It consists of two rotors each having two, three or more lobes. The shafts of both rotors are interconnected through gears and operate on the same speed. It resembles with gear type oil pump in operation There is a slight clearance between the rotors and the surrounding housing.

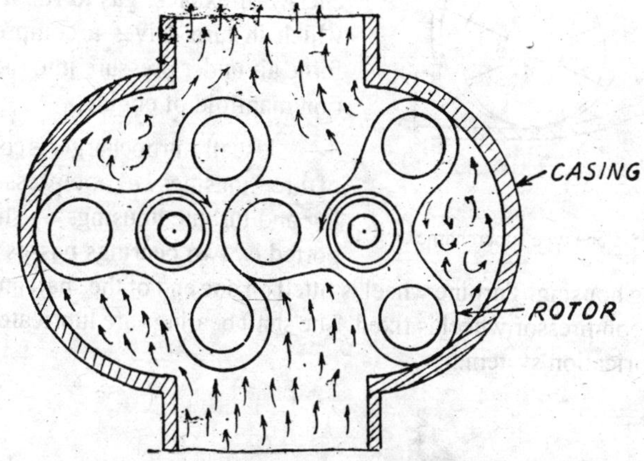

Fig. 8·23. Roots supercharger.

When the rotors rotate, air enters the housing and passes between the lobes of the rotors and the housing. The air is then forced out through the outlet into the cylinders. This supercharger is known as positive type.

(*b*) **Centrifugal type.** It consists of an impeller which rotates at a high speed inside a housing. Air enters at the centre of

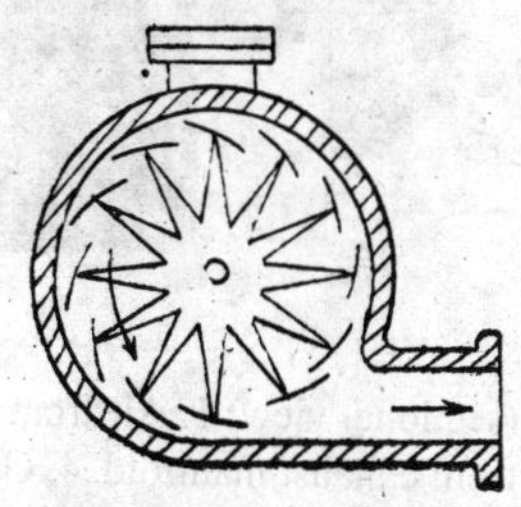

Fig. 8·24. Centrifugal supercharger.

housing and is thrown off rapidly by the impeller blades. This supercharger requires accurately balanced impeller, the blades of which should be strong enough so that centrifugal force at high speed may not cause them to stretch and strike with the housing.

This type of supercharger depends upon centrifugal force to increase the pressure of air and is usually known as non-positive type.

(*c*) **Vane type.** This is a positive type supercharger. Its vanes are so designed as to crowd the air in a smaller space. The air blades slide in slots or grooves in the hub of the supercharger.

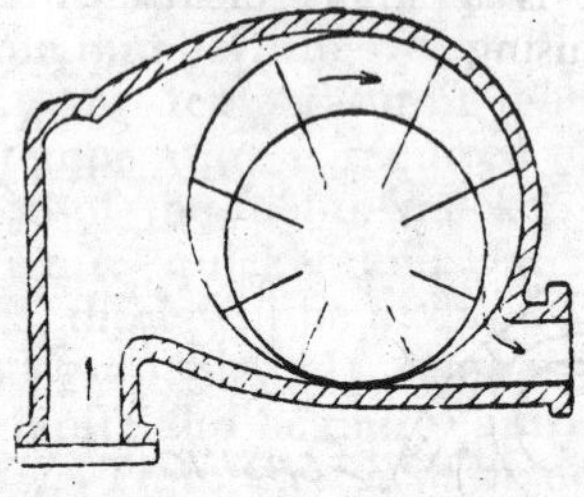

Fig. 8.25 Vane type supercharger.

5. Turbocharger. It is a supercharging device which uses thermal energy of exhaust gas to run a turbine which in turn drives a compressor to force air under pressure into the induction manifold of engine.

Usually turbocharger is composed of three housings *i.e.*, compressor, bearing and turbine housings. A shaft supported by two bearings passes through the bearing housing. Turbine wheel is fitted on one end of the shaft and on the other end compressor wheel is fixed. The shaft bearings are lubricated by the engine lubrication system.

Fig. 8.26. Turbocharger (sectional view) 1. Air from filter. 2. Air to induction manifold. 3. Gases from exhaust manifold. 4. Gases to exhaust pipe.

Gases from engine exhaust manifold enter the turbine housing, push round the turbine wheel and run into exhaust pipe.

The turbine sets in motion the compressor wheel which draws air from filter and pushes into induction manifold. Hence the turbocharger acts as a super charger.

QUESTIONS

1. Which are the different fuels used in modern automobiles? Describe them.

2. Discuss the combustion of gasoline fuel.

3. Which different additives are added to petrol for improving its quality? Why?

4. What do you know about the following?

(*i*) Octane rating (*ii*) Volatility (*iii*) Anti-knock value
(*iv*) Diesel knock (*v*) Cetane Number (*vi*) Detonation

5. Which are different types of fuel feed systems? Explain them.

6. Illustrate the working of autovac tank.

7. Describe the pump feed system.

8. Illustrate the construction and working of mechanical fuel pump.

9. Describe with the aid of a sketch, the construction and working of electrical fuel pump.

10. Describe the construction of fuel tank.

11. Describe the action at fuel pump when float chamber of carburettor is fully filled.

12. What is carburettor? What part is plays in the working of spark ignition engine?

13. Illustrate simple carburettor and describe how fuel mixture is prepared by it.

14. On what principle the carburettor works? Discuss.

15. What is venturi system? Discuss.

16. Which are the different types of carburettors? Describe the construction and working of any one of them.

17. Illustrate the construction and working of constant vacuum carburettor.

18. Which are the various circuits in a carburettor? Describe them.

19. Which are the various controls in a modern carburettor? Describe.

20. Illustrate the working of idle and pump circuits.

21. What do you know about the following?

(*i*) Delayed action (*ii*) Idle passage (*iii*) Air bleed
(*iv*) Compensation (*v*) Throttle cracker (*vi*) Unloader

22. Which are the different types of air cleaners? Illustrate oil bath types air cleaner.

23. Which are the different types of fuel gauges? Illustrate the construction and working of any one of them.

24. Illustrate the construction and working of balanced coil type fuel gauge.

25. What is a supercharger? Which are the different types of super charger? Describe them.

26. Illustrate the working of solex carburettor.

27. Explain Zenith carburettor.

28. Compare Solex and Zenith carburettors.

29. Explain Turbocharger.

9

Fuel Injection System

1. Fuel Injection system, is the system which injects fuel into the combustion chamber of a diesel engine when the piston reaches near compression Top Dead Centre (T.D.C.).

This system must fulfil the following main requirements :

(*i*) Measure and control the correct quantity and rate of fuel injected.

(*ii*) Fully atomize the fuel into fine particles, so that it is distributed properly in the combustion chamber.

(*iii*) Develop pressures well in excess of the combustion chamber pressures.

If the amount and rate of fuel being injected is not timed, controlled and measured, it will result in uneven running of engine, leading to vibrations. The atomised fuel mixes up with hot compressed air inside the combustion chamber at high pressure. So, the injection of fuel should take place at a well high pressurc in excess of this pressure. The injection should take place at the correct time according to the set firing order of the engine. In certain engines *pre-combustion chambers* are provided, into which the fuel is injected, where it swirls and then enters into the main combustion chamber. This provides better turbulence of the fuel.

2. Parts of fuel injection system. The following are the main parts of a typical fuel injection system :

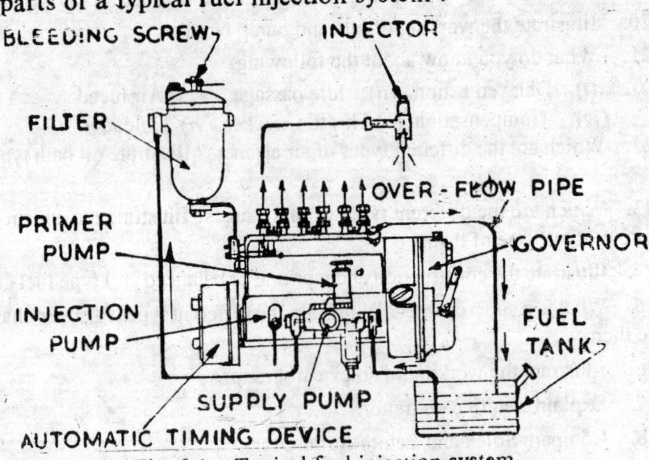

Fig. 9·1. Typical fuel injection system.

(i) **Fuel tank** serves as the main reservoir of fuel. It has two separate pipe lines, one serving as feed (supply) line, whereas the other as return-line. The return-line returns excess fuel from the injectors to the tank.

(ii) **Fuel filter.** A small amount of dirt can ruin the working parts. It is, therefore, quite essential that the fuel should be well filtered before it is fed to the fuel injection pump and injectors.

(iii) **Fuel feed or transfer pump.** This pump is used to draw fuel from fuel tank and supply onward to the fuel injection pump. Gear type fuel transfer pump is similar in construction and operation to the gear type oil pump.

A primary pump is also used in conjunction wlth fuel feed pump, to prime fuel. before starting. It is operated manually through a plunger.

(iv) **Fuel injection pump.** This pump supplies, under extremely high pressure, equal quantity of fuel at equal intervals, to the injectors, according to the engine firing order.

(v) **Injector.** It is a device with fine holes in its nozzletip which atomizes and injects fuel into the combustion chamber.

3. **Types of fuel injection.** Fuel injection can be classified as under :

(a) Air injection, *(b)* Mechanical or solid injection.

(a) **Air injection.** In this system, a blast of air from an external source, forces a measured amount of fuel into the engine cylinder. This was the original method as used by Dr. Rudolf Diesel in his first engines.

This type of injection lengthened the injection period, with the result that the pressure did not rise during combustion. Also volume of liquid fuel delivered out in each cycle was an extremely small one which lengthened the injection period.

(b) **Mechanical or solid injection** is the system mostly adoped now-a-days and employs a fuel injection pump, which pumps fuel under high pressure, through the injector, into the engine eylinder. The spray of fuel which is injected by the injector, comes out of the fine hole or holes in the injector nozzle. Thus its essential components are,

(i) Fuel Injection Pump. *(ii)* Injector.

(i) **Injection Pump** is fitted by the side of the engine. In certain cases, there is a separate or individual pump for the individual injector.

A single unit pump contains a camshaft, which operates ali the pumps contained in that unit, depending upon the number of cylinders in the engine. Each pump measures the amount of fuel to be injected into its respective cylinder and delivers it to the injector. The injection pump plunger is lifted by a cam on the camshaft. The plunger always makes a full stroke, moving in the barrel. The amount of fuel pumped per stroke can be varied by turning the plunger in the barrel by the governor action through the rack, which meshes with the gear segment on the bottom of pump plunger. This rack is operated through the accelerator too.

The operation of an injection pump, as the plunger makes a stroke, has been shown in figures A, B and C.

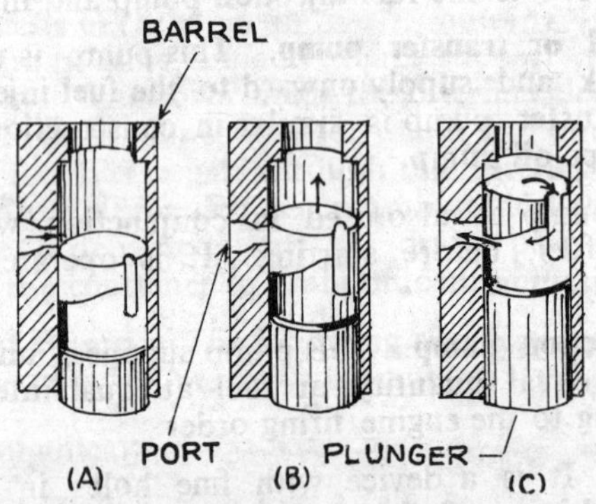

Fig. 9.2. Working of a fuel injection pump :
A—Port uncovered; Fuel enters barrel
B—Port covered; Injection begins
C—Port uncovered; Injection finishes.

Figure 'A' :

(*i*) The plunger is at its down position.

(*ii*) Inlet port is uncovered.

(*iii*) Fuel flows into the space above the plunger through the slot and into the recess around the plunger.

Figure 'B' :

(*i*) The plunger has started its upward storke.

(*ii*) The inlet port is covered.

(*iii*) The fuel is trapped and will be forced through the delivery valve, fuel line and injector into the cylinder, as the plunger moves upward.

Figure 'C' :

(*i*) The plunger has moved up, until the port is uncovered by the recess in the plunger.

(*ii*) The fuel finds chance to escape back, through the port, into the fuel gallery in the pump.

The recess in the pump plunger forms a helix around the upper end of the plunger. Rotating of the plunger affects the quantity of fuel injected, as shown in figures P, Q and R.

Figure 'P' :

(*i*) The plunger has been rotated into the shut-of position.

(*ii*) The slot connecting the top of the plunger with the recess, is in line with the port.

(*iii*) No fuel can be trapped and injected in this position.

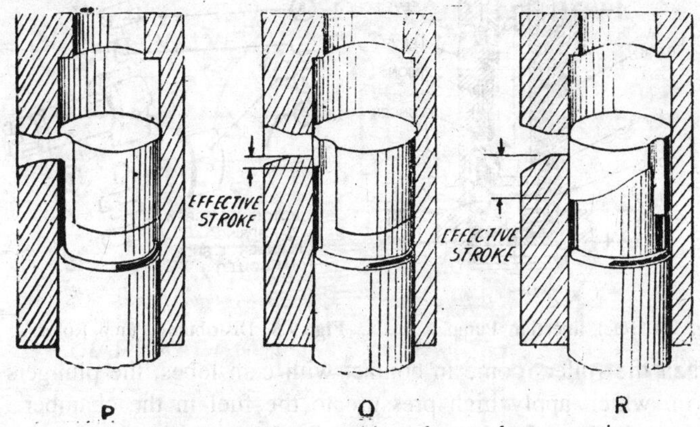

Fig. 9.3. Different positions due to plunger rotation.
P = Shut off Q = Idling R = Full load

Figure 'Q' :

(*i*) The plunger has been rotated into the idling position.

(*ii*) The narrow part of the plunger formed by the helix, will cover the port for only a short part of the stroke.

(*iii*) This position permits only a small amount of fuel to be injected per stroke.

Figure 'R' :

(*i*) The plunger has been rotated into the full load position.

(*ii*) The wide part of the plunger formed by the helix, covers the port for a longer part of the stroke.

(*iii*) A larger amount of fuel to be injected per stroke is permitted in this position.

The various parts of separate unit type pump have been shown in Fig. 9.4.

Rotary Distributor Pump. Like Ignition Distributor, in petrol engines, this pump pushes out fuel to the injectors in diesel engine, as the rotor in the distributor pump rotates. The pump rotor includes a pair of cam-rollers and plungers. When the rotor rotates, the rollers roll on the inner surface of cam. They move in and out, as they roll over the cam lobes. Moving out of rollers causes the plungers to move out, resulting in increase in the size of internal chamber and flow of fuel into the chamber.

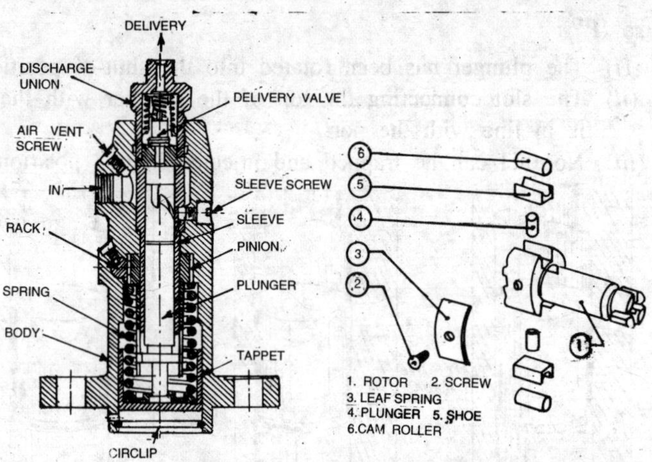

Fig. 9.4. Fuel Injection Pump. **Fig. 9.5.** Distributor Pump Rotor.

When the rollers come in contact with cam lobes, the plungers are pushed in, which apply high pressure to the fuel in the chamber. The fuel is forced out through an opening in the rotor.

Opening in the rotor indexes with stationary openings in the outer shell of the pump. Each opening is connected with the injector through separate pipes. When the rotor opening indexes with the stationary opening, high pressure surges of fuel are pushed into the injectors, according to injection (firing) order of the engine.

4. Setting of Fuel Injection Pump. It should deliver fuel in equal quantity and at equal interval, failing which the engine shall run with jerks and jolts. The pump should be calibrated and phased, on a proper Test Bench.

Calibration : Measurement of quantity of fuel, delivered by each element of an injection pump, at a fixed control setting and adjusting these settings, so as to deliver equal quantities of fuel per stroke of each plunger to all the cylinders, is known as *calibration.*

Adjustment is usually made by changing the relative positions of control sleeve of each pump element. Clockwise movement of control sleeve increases and anticlockwise decreases the quantity of fuel delivered, on a right-hand driven pump and vice versa.

Phasing : The testing and adjustment of an injection pump for the accuracy of its firing-interval is known as *phasing.*

In phasing, the exact firing or critical point is spotted where injection starts at the closure of spill port, at the beginning of upward stroke of the plunger. For this the pump is installed on a Test Bench; fuel line connection is made and air is bled out from the supply system and pump. After that, proceed as under :-

(*i*) Detach No. 1 delivery valve-holder. Remove only valve and place it safely away in a tray containing some clear diesel oil.

(*ii*) Replace delivery-valve-holder alone and turn on fuel.

(*iii*) Move the control rod to the maximum delivery position.

(*iv*) Rotate the pump shaft in the direction of rotation, until the plunger is at the bottom of its stroke and that the fuel flows freely through the delivery valve orifice.

Further rotation of the pump shall diminish fuel-flow which will finally cease as the ports are closed. This *point of port-closure*, which is the beginning of effective part of pumping stroke, is the exact *firing point.*

(*v*) Mark the 'firing point', at the graduated disc of Test Bench.

(*vi*) Repeat the above process on the other elements of the pump, one by one, in the sequence of firing order.

(*vii*) Rotate the camshaft through the required angle *i.e.*, 360°/Number of elements, and see that the point of port-closure coincides. The tappet should be adjusted for correct phasing. Decrease the length of tappet if firing is early and increase if late.

(*viii*) Fit back the delivery valves and make a final check. The delivery valves should not be mixed up, as these are not interchangeable. Tappet clearance (free lift) at the end of plunger stroke, should also be checked up. (normal clearance is = 0.3 mm)

5. Injector. It is a kind of valve through which fuel under high pressure is injected into the combustion chamber in the form of fine spray. When fuel is pumped into it, needle in its nozzle is lifted off the seat under high pressure and the fuel comes out through fine hole or holes, forming spray. It consists of the following three parts :

(*i*) **The nozzle tip,** containing the orifice or jet, which directs the fuel stream into the combustion chamber.

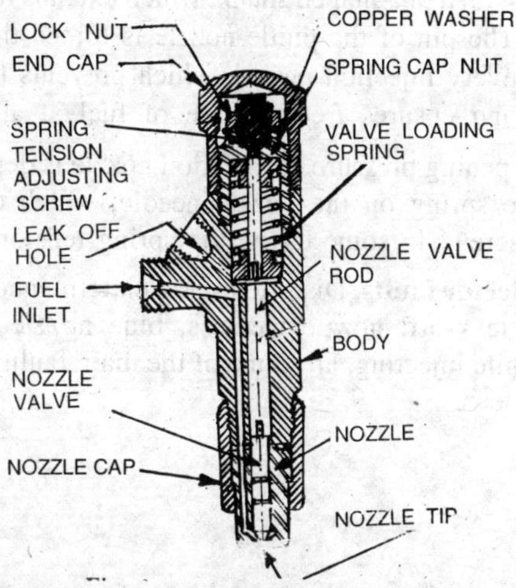

(*ii*) **The body,** which houses the component parts.

(*iii*) **The holder,** which holds the injection nozzle unit with the cylinder head.

Injectors are classified according to the type of nozzle tip fitted to them. The following are different types of nozzle tips employed in injectors :

(*i*) **Single hole** which contains only one hole.

(*ii*) **Multiple hole** having more than one hole.

(*iii*) **Circumferential** which contains a disc placed beneath a single hold orifice. It breaks the fuel stream into a thin flat sheet and thus distributes the highly atomized fuel to all parts of the cylinder.

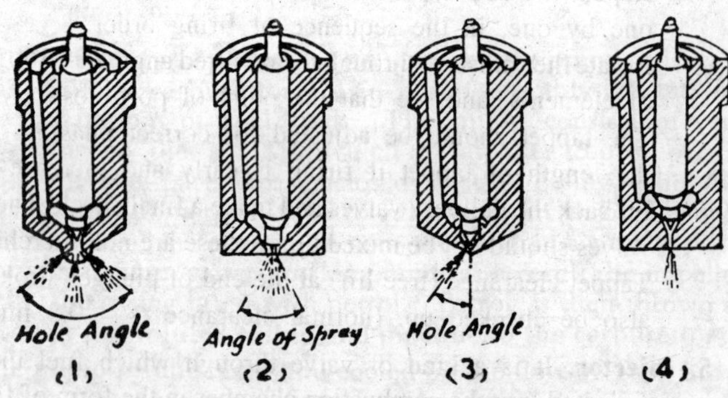

Fig. 9.7. Types of injector nozzles.

1. Multiple holes. 2. Pintle.
3. Single side holes. 4. Single central hole.

(*iv*) **Pin or pintle orifice.** In this case, the plunger valve is fitted with a pin-shaped shank, which extends through the orifice-hole. The pin of the pintle-nozzle is forced through the orifice after every injection period, which prevents the hole from clogging and ensures free passage of fuel at all times.

The opening pressure of injection nozzle is determined by the tension of pressure spring on the nozzle needle, which can be adjusted by an adjusting screw. In some cases, the spring tension is adjusted by shims.

6. Injector faults. Distorted spray pattern, incorrect opening pressure, excessive leak-off, nozzle wetness, blue nozzle body, non-buzzing of injector while injecting, are some of the main faults. Their probable causes are as under :

(*a*) **Distorted spray pattern :**

(*i*) Nozzle valve damaged, (*ii*) Nozzle body holes choked (*iii*) Carbon on needle valve.

(*b*) **Incorrect opening pressure :**

(*i*) Needle valve seized, (*ii*) Adjusting nut slackened, (*iii*) Nozzle body holes chocked.

(*c*) **Excessive leak off :**

(*i*) Loose nozzle cap nut, (*ii*) **Slack needle valve.** (*iii*) Foreign matter on mating faces of injector at nozzle body.

(*d*) **Nozzle wetness :**

(*i*) Sticking needle valve, (*ii*) Carbon on needle valve.

(*e*) **Blue nozzle body :**

(*i*) Defective cooling, (*ii*) Incorrect installation.

(*f*) **Non buzzing of injector while injecting :**

(*i*) Distorted nozzle cap nut, (*ii*) Binding needle valve, (*iii*) Leaking valve seat.

7. **Governors.** The engines must either maintain a specific constant speed, regardless of load-variations, or if they operate within a specific speed-range, the minimum and maximum speed-limits must not be exceeded. For this, the quality of the fuel must be metered, in accordance with the torque required. In diesel engines, this is achieved by controlling the quantity of the fuel injected by shifting the control rack of the injection pump to control fuel quantity. To prevent the engine from stalling or from exceeding the permissible maximum speed, both idling and maximum speed have to be limited automatically. All these requirements are met by the governors automatically.

There are two following types of governors in common use :—

(*a*) Centrifugal or mechanical (Flyweight type), (*b*) Pneumatic.

(*a*) **Centrifugal Governors.** These are driven by the engine. As the engine speed increases, the flyweights of the governors move outwardly, under the effect of centrifugal force, when that fecor overcomes the tension of the governor springs. When the engine speed decreases, the centrifugal force drops down until it is finally overcome by the tension of the governor-springs and the flyweights move inwardly again.

The movement of flyweights is transmitted to the control rod through bell-cranks, the transmission-rod and the fulcrum-lever. When the engine speed increases, the control rod is moved in the direction of less fuel. Fuel delivery is thus reduced and the engine speed controlled. When the engine speed is reduced, the process is reversed. By using springs of correct size and by adjusting their initial tension appropriately, the governor can be made to maintain speed automatically.

(*b*) **Pneumatic Governors** are operated by the vacuum created by the airflow in the induction manifold. Its two main parts are :

(*i*) The venturi control unit, (*ii*) The diaphragm unit.

The venturi control unit is mounted on the engine induction pipe at its inlet end. The oil-filter is attached to this unit. The throat of the body forms a venturi, in which a butterfly-valve is placed. The connection for vacuum line is made at the venturi. The butterfly-valve is connected to the accelerator pedal, through a control-lever and the linkage.

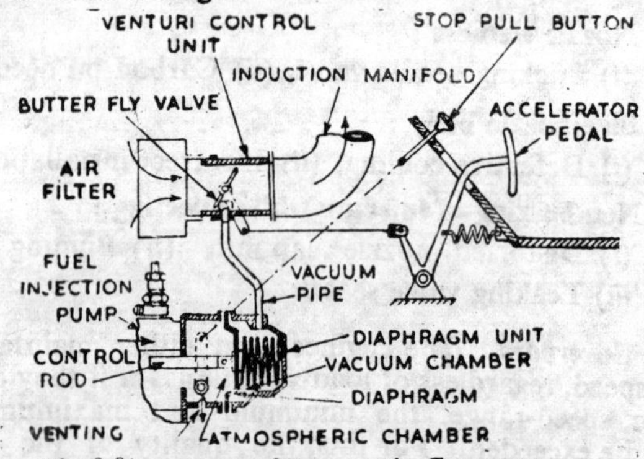

Fig. 9.8 Layout of Pneumatic Governor.

The diaphragm unit is mounted on the one end of the fuel injection pump. It is sub-divided, by a leather-diaphragm, into two chambers: atmospheric and vacuum chambers.

The atmospheric chamber is always under atmospheric pressure. The vacuum chamber is connected to the venturi through a pipe. The diaphragm is connected with control-rod through a pin. The diaphragm is under pressure of the governing spring which tends to move the diaphragm and thereby the control-rod towards the "full" position. The control-rod is thus forced by the governing spring to the full-load position, when the engine is stopped.

8. **Timing Device.** The timing device advances the injection by turning the pump camshaft forward to a certain degree of angle, in relation to engine crankshaft. By means of this device, the commencement of injection can be advanced by up to 10 degrees. Advancing is needed for improving the engine's efficiency during high speed and low compression pressure.

The timing devices are of two types : hand-operated and automatic. In *hand operated timing device*, desired advance is obtained by shifting a lever. When the lever is moved, the position of the sliding collar is changed in a linear direction, through an appropriate linkage and a link fork, which are connected with the lever. The sliding collar carries along with a steep-threaded nut which turns, giviug movement to the coupling connected to the pump shaft, thereby turning the shaft forward on rotation, in relation to drive shaft.

The automatic timing device depends upon engine-speed, as centrifugal force is employed to advance injection. Its main parts are : coupling flange, housing, driving flange, bearing cover, two flyweights and two helical springs. The flyweights have a curved surface, the curvature of which is different, according to required degree of advance.

The driving and driven couplings are inter-connected by two helical springs. When engine-speed increases, the centrifugal force

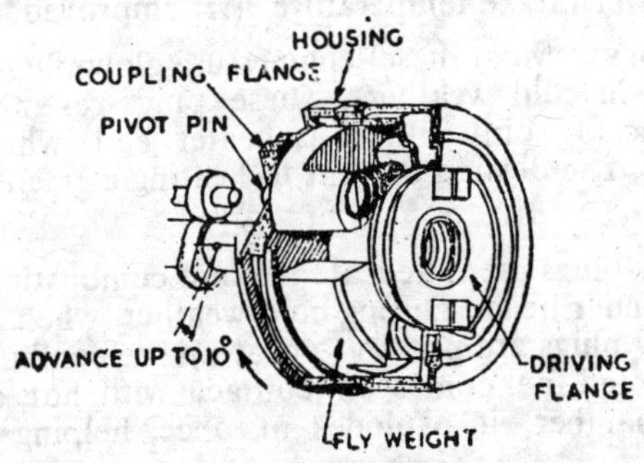

Fig. 9.9 Automatic Timing Device.

causes the flyweights to move outward, resulting in the curved surfaces of the flyweights to slide towards the pin of driving flange. By this, a pull is exerted on the pivot pins of coupling flange, in such a way that the coupling-flange, along with camshaft, advances in relation to driving-flange. The commencement of fuel delivery in each pump element is, therefore, advanced accordingly. When the engine-speed falls down, the centrifugal force, affecting the flyweights, also decreases. The flyweights are contracted inward by the helical springs and the advance is reversed accordingly.

9. **Petrol injection**. Robert Bosch in Germany developed the first petrol injection system for Mercedes racing cars. Petrol injection is different from diesel injection system. In petrol injection, spray of fuel is injected in the induction manifold near inlet valve during suction stroke when intake valve is open. The injection has not to work against the compression pressure, as in diesel engines; rather the petrol engine acts as a vacuum pump, to draw-in injected fue . The injection can, therefore, be made at lower pressure. That is why this system is known as *low pressure system*.

Petrol injection is of following two types :

(*a*) The constant or dribble system,

(*b*) Timed or intermittent system.

(*a*) **Constant or dribble system.** In this system a very small stream of fuel runs constantly into the induction manifold while the engine is running. This does not need to be timed.

(*b*) **Timed or intermittent system.** In this system a, slant of fuel is supplied on the suction stroke of each cylinder. This system is timed with the crankshaft in the same way as the timing of ignition distributor.

10. Advantages of Petrol Injection. Following are the main advantages of petrol injection :

(*i*) More power, (*ii*) Higher torque.

(*iii*) Quicker cold starting. (*iv*) Faster warming up.

(*v*) No necessity of manifold heat to warm ingoing fuel.

(*vi*) Lower intake temperature. (*vi*) Improved fuel economy.

Glow plugs. Most diesel engines use glow-plugs for easy starting, specially in cold weather. These plugs resemble with spark-plugs. In placeof points, there is heater coil which glows when switched on. The desired current to heating element is supplied by the battery.

The glow-plugs are filted at the pre-combustion chambers in the engine cylinder head. During cold weather, when air temperature is low the glow-plugs are put on to heat up pre-combustion chambers When the injected fuel comes in contact with hot air in the pre-combustion chamber, it explodes at once, helping in the ignition process.

QUESTIONS

1. What is fuel injection system ? Which requirements should it fulfill and why ?

2. Describe with the help of a suitable sketch, a typical fuel injection system.

3. Which are the different types of fuel injection ? Explain them.

4. Illustrate the construction aud working of a fuel injection pump. Show how the quantity of fuel is controlled.

5. Which are the different types of injector nozzle tips ? Explain.

6. Illustrate the construction and working of injector.

7. Which are the different types of governors ? Explain them.

8. What is timing device ? What is its necessity ? Describe a band operated timing device.

9. Explain the automatic timing device.

10. What is petrol injection ? How it differs from diesel injection ? Narrate its advantages.

11. Describe the following :

(*i*) Dribble injection system. (*ii*) Intermittent injection system.

(*iii*) Pneumatic governer. (*iv*) Pintle orifice nozzle.

(*v*) Air injector (*vi*) Calibration (*vii*) Phasing

12. Explain the usual injector faults and their probable causes.

13. What is a Rotary Distributor pump Explain, with specia emphasis on its rotor.

14. Explain glow-plugs

10

Electrical Systems—I

1. Electricity in Automobile. Electricity is used for a number of purposes in the automobile. By means of electric current different lights are put on, horn is blown, engine is cranked, spark is provided, battery is charged and numerous other components such as radio, clock, trafficators, fan, heater, wiper machines etc. are put into working order.

There are two sources for the generation of electric current in an automobile : storage battery and generator or dynamo.

2. Storage Battery. It is an electro-chemical apparatus which supplies regular current due to chemical action inside it. It stores energy in the chemical form that is why it is known as storage battery. It is a lead storage cell which consists of a hard rubber partitioned into different sections, one for each cell. The cells are made of lead grid plates into which a lead paste is pressed. The positive plates which are brown in colour, have lead peroxide while the negative plates which are grey in colour, have pure lead paste. Each cell of the battery is made up of alternate positive and negative plates separated by wooden or plastic or rubber or glass separators. All the like plates in each cell are connected by a lead strap, soldered to a lead lug. There is always one more negative plate than the positive one in each cell as the outside plate in each cell is a negative plate. All the cells are connected in series by means of lead straps and are sealed with covers by means of sealing compound. Holes are provided in each cell to fill and unfill electrolyte or distilled water. The filler holes are covered with plugs which contain fine holes as air vents.

The electrolyte or battery solution is made up of two parts of pure sulphuric acid added to approximately four parts of distilled water. The electrolyte of a fully charged battery usually contains 31 per cent sulphuric acid by weight or about 21 per cent by volume.

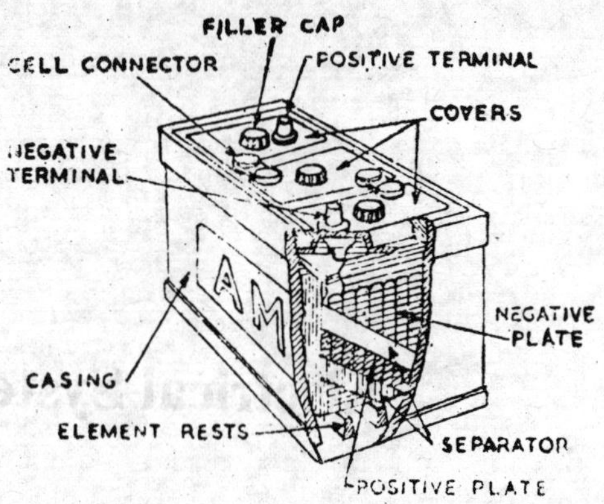

Fig. 10·1. Storage Battery.

3. **Chemical action in the battery**. Electrical energy is converted into potential chemical energy during charging and during discharge, this potential chemical energy is re-converted into electrical energy.

During discharge, hydrogen is formed at positive plate and oxygen at negative plate. The hydrogen reduces lead peroxide to lead oxide and forms water in this process. The oxygen at negative plate oxidizes the metallic lead to lead oxide. The lead oxide present at both plates reacts with sulphuric acid forming lead sulphate on both plates and releasing more water. The chemical equation is as under :—

Lead oxide + Sulphuric acid = Lead sulphate + Water
(PbO) (H_2SO_4) $(PbSO_4)$ (H_2O)

or $$PbO + H_2SO_4 = PbSO_4 + H_2O$$

Water formed during discharge reduces specific gravity of electrolyte and it indicates discharged state of cell.

During charge, D.C. current flows in the opposite direction through the electrolyte resulting in giving off hydrogen at negative plate and oxygen at positive plate. Hydrogen at negative plate displaces lead from lead sulphate in the form of spongy metallic lead and takes its place by combining with sulphate to sulphuric acid.

The lead from lead sulphate on positive plate combines with oxygen formed at the plate and some oxygen from water to form lead peroxide. Hydrogen from water combines with sulphate to form sulphuric acid. Formation of sulphuric acid at both positive and negative plates increases specific gravity of electrolyte and restores the battery to its original value. After charge, lead sulphate $(PbSO_4)$ + Water (H_2O) = Lead oxide (PbO) + sulphuric acid (H_2SO_4).

When the cell is full charged, chemical action ceases and gassing occurs due to liberation of oxygen and hydrogen at the plates. This is an evidence of completion of charge.

The following are typical ranges of specific gravity for a cell in various stages of charge with respect to its ability to crank the engine at 80°F :—

Specific Gravity Ranging		*Stage of Charge*
From	*to*	
1,220	1,230	100% charged
1,200	1,210	75% „
1,175	1,185	50% „
1,150	1,160	25% „
1,125	1,135	Very little useful capacity
1,100	1,110	Discharged

Lead acid battery

Fully discharged									*Fully charged*			
Positive Plates (+)		*Electrolyte*		Negative Plates (—)	Positive Plates (+)		*Electrolyte*		Negative Plates (+)
Lead sulphate ($PbSO_4$)	+	water ($2H_2O$)	+	Lead sulphate ($PbSO_4$)	Lead peroxide (PbO_2)	+	Sulphuric acid ($2H_2SO_4$)	+	Spongy lead (Pb)

Charging——→

←——Discharging

Effects on chemicals in the battery during

	Discharge	*Charge*
1. Sponge lead	Decreasing	Increasing
2. Lead peroxide	Decreasing	Increasixg
3. Sulphuric acid	Decreasing	Increasing
4. Water	Increasing	Decreasing
5. Lead sulphate	Increasing	Decreasing

4. Battery rating. The amount of current that a battery can deliver, depends upon the total area and volume of active plate material, and the amount and strengh of the electrolyte.

Storage batteries are rated, as below, as per accepted standard.

20-Hour rating in Ampere-Hour. This indicates the lightning ability of the battery. A fully charged battery is brought to a temperature of 80°F (27°C) and is discharged at a rate equal to 1/20 of the specified 20-hour capacity in ampere-hours. For example, a 6 volt battery, rated by the manufacturer at 100. Ampere-Hour capacity, would be dlscharged at 1/20 of 100 or at 5 amperes, until the terminal voltage falls to 5. 25 vols. Thus

Ampere-Hour Capacity = No. of hours required for discharge of a battery) (Rate of discharge in amperes)

Typical 20 hour rating discharge, showing hours of discharge at 20-Hour rate (Battery temperature 80°F (27°C) are shown in the graph shown in Fig. 10·2,

5. **Battery efficiency**. This is ability of battery to supply current. It depends upon temperature and rate of discharge. At low temperature, sulphuric acid cannot work so actively on plates and thus chemical action is highly reduced. So the battery cannot supply high current for long time. At high discharge rates, chemical action takes place only on the surface of plates due to short time. But during slower rates of discharge, there is sufficient time for the chemical activities to penetrate deep below the plate surfaces and act on the active material.

When a battery is discharged slowly, it is far more efficient than when it is discharged rapidly. High rates of discharge do not produce as many ampere-hours as low rates of discharge. Hence lower the discharge rate, higher the battery efficiency and higher the discharge rate, lower the efficiency.

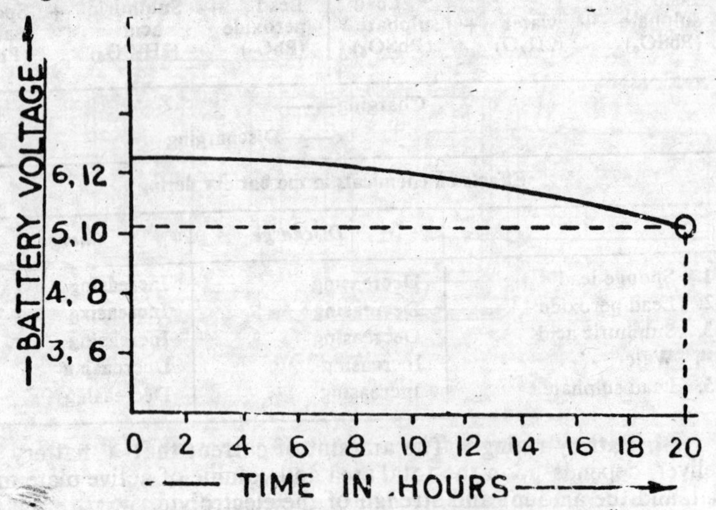

Fig. 10·2 20-Hour rating discharge

The following table shows the approximate battery efficiencies in relation to battery temperatures ;

The efficiency of a battery can be determined either as the quantity or the energy efficiency.

$$\text{Quantity efficiency }\% = \frac{\text{Ampere hour output}}{\text{Ampere hour input}} \times 100$$

Battery temperature (Degrees Centigrade)	*Battery efficiency (Per cent)*
26·7	100
0	65
—17·8	50

$$\text{Energy efficiency } \% = \frac{\text{Watt hour output}}{\text{Watt hour input}} \times 100$$

Battery output is dependent upon the rate of discharge. The quantity efficiency may vary from 75 to 95% and as low as 50% if discharge rate is very high. For finding energy efficiency, average voltage during charge and discharge is taken into consideration. Owing to difference in cell voltage on charge and discharge, energy efficiency is much lower than quantity efficiency.

For a 75Ah (ampere hour) battery, energy efficiency is 86% for normal discharge and 66% for higher discharge rate.

6. **Battery charging**. There are the following three different ways of charging batteries :—

(*a*) Constant current charging,

(*b*) Constant voltage or potential charging,

(*c*) Quick charging.

(*a*) **Constant current charging**. In this method, the current input to the battery is adjusted to the value as recommended by the battery manufacturer. The charging is continued until the battery is gassing freely and there is no rise in gravity for two hours.

Constant current chargers usually employ a rectifier to convert A.C. to D.C. and a rheostat to adjust current flow to the battery.

The undermentioned procedure should be followed to charge batteries with constant chargers :—

(*i*) Remove filler plugs, check electrolyte level and add distilled water if the level is low.

(*ii*) Connect battery to charger. If *more than one battery is to be charged, connect the batteries in series* up to the capacity of charger.

(*iii*) switch on the charger and adjust charging rate.

(*iv*) Check gravity of batteries by means of *hydrometer* after every two hours until there is no further rise in specific gravity.

(*v*) If a battery boils violently or overheats, remove it from the charging line and check up the defect.

(*b*) **Constant voltage or potential charging**. In this method, the charging voltage is kept at a constant value. As the battery approaches a charged condition, it resists to the charging current which tapers off gradually. When the battery is fully charged, the current input has been reduced to few amperes. In this method, the battery temperature remains within limits. If the temperature of battery electrolyte rises high, battery resistance shall fall down and the battery shall be damaged due to overcharging. In such a case the battery should be removed from the charging line to avoid damage.

Constant voltage charger consists of a motor-generator set in which an A.C. motor drives a D.C. generator. The generator is usually rated at 7·5 volt for 6-volt batteries or 15-volt for 12-volt batteries.

The undermentioned procedure should be followed to charge batteries with constant potential charger :—

(*i*) Remove filler plugs.

(*ii*) Check electrolyte and add distilled water if the level is low.

(*iii*) Connect battery to the charger. ***The batteries should be connected in parallel if more than one battery is to be charged.***

(*iv*) Switch on the charger and adjust the *voltage* to specified value.

(*v*) Keep watch on battery boiling and temperature as there is possibility of overheating in this method. If the battery boils or overheats, remove it from the line.

(*vi*) Stop charging when the gravity shows no further rise after an hour of charging.

(*c*) **Quick charging**. In this method the battery is charged at a high rate as much as 100 amperes for a short time of about 30 to 40 minutes. The battery is brought to a fair state of charge before it reaches at harmful high temperature.

Quick chargers or boosters supply current to the battery at high rate and work on the theory that the battery will be brought up to a charged condition before excessive battery temperatures are reached. But as a rule, the quick chargers cannot bring a battery up to *full* charge in a short time. The battery can be substantially charged or *boosted* but for bringing a battery to a fully charged condition, the charging cycle must be completed by charging at a low or normal rate.

The following precautions should be kept in mind at the time of charging batteries by means of quick charger :—

(*i*) Check the gravity and height of electrolyte. If the gravity is uniformly low and there is sufficient electrolyte in all cells to take gravity reading, it is probably safe to begin the charge.

(*ii*) Observe the colour of electrolyte carefully. If it is discoloured with brownish sediment which is from positive plate, don't charge the battery with quick-charger. Quick charging will stir up sloughed-off material in the sediment chambers which will settle on top of negative plates resulting in short circuit. Such a battery should be always charged at low rate.

(*iii*) Don't quick charge a battery if the gravity readings vary more than 25 points between cells.

(*iv*) Don't quick charge a badly overcharged or sulphated battery.

(*v*) After a few minutes' charge, check up colour of electrolyte in each cell. Stop quick charging if electrolyte discoloured.

(*vi*) Check cell voltages by means of *cell tester* after every few minutes. Stop quick charging if the difference between cell voltages is more than 0·2 volt.

(*vii*) Stop quick charging if battery temperature exceeds 50°C.

(*viii*) Avoid sparking while using cell tester because a spark or flame around the battery could ignite the gases given off by the battery during charge. Sparking at the battery could lead to a serious explosion and fatal accident.

7. **Overcharging and its effects**. During battery charging, water in the electrolyte is split up into hydrogen and oxygen which gather at negative and positive plates. Hydrogen combines with sulphate (SO_4) and results in sulphuric acid. Oxygen penetrates into positive plates to form lead peroxide (PbO_2). The battery is fully charged when all the sulphate has been extracted from the plates and all the material in positive plate has been converted into lead peroxide. After full charging, if charging current continues to flow into battery, the battery becomes overcharged.

During overcharge, oxygen and hydrogen find nothing to combine with and they bubble up through the electrolyte and pass out through the cap holes, since hydrogen and oxygen are constituent parts of water, so water is rapidly lost from the battery cells. As a result, the electrolyte level falls down the plate tops which are exposed to air. The dry plates get hard and lose their ability to take a charge.

Due to loss of water, the acid at the lower parts of plates becomes more thick or concentrated owing to which the plates deteriorate more rapidly. The upper parts of separators are also exposed to air due to fall in electrolyte level. Wood separators dry out and crack at the top and lose their insulating properties. They are charred with strong acid at their lower parts and result in short circuit between the plates.

Considerable heat is generated during overcharging due to which a good amount of water is lost from the battery due to evaporation. More the heat, higher the evaporation. Greater the water loss, lower the electrolyte level and more the plate and separator upper parts exposed to air leading to higher damage.

The high temperatures developed in the battery cells during overcharging cause the plates to buckle. The buckling plates crush the separators and lead to short circuit between the positive and negative plates. Thus the battery is totally ruined.

During overcharging, some of the oxygen gathering at the positive plates attacks the grid structure of plates causing them to oxidize. This results in swelling of plates resulting in pushing up of cell covers. Too much swelling causes the upper edges of positive plates to be pushed against the negative plate straps resulting in direct short circuit and battery failure.

8. Checking battery state of charge. There are the following two general ways of determining battery state of charge :

(*a*) By measuring the *strength* of electrolyte ;

(*b*) By measuring the voltage of battery cells under various battery-load conditions.

(*a*) **By measuring the strength of electrolyte.** The strength of electrolyte or the amount of sulphuric acid present in the electrolyte can be measured by means of a *hydrometer* which checks

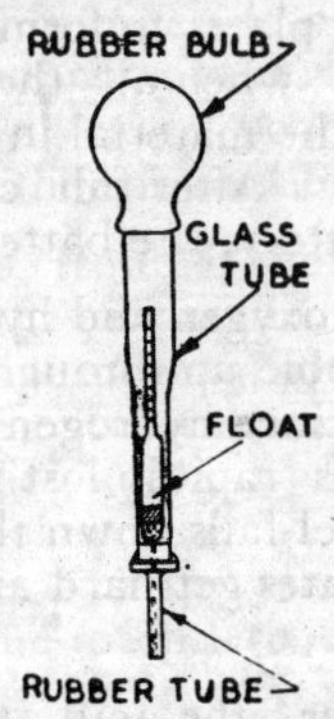

Fig. 10·3. Hydrometer.

the specific gravity of electrolyte. The specific gravity reflects the percentage of sulphuric acid present in the electrolyte and thus the battery state of charge.

The electrolyte of a fully charged battery consists of about 39% acid and 61% water by weight. When the battery is discharged, the percentage of acid and water is about 15 and 85 respectively. Sulphuric acid is 1·835 times heavier than water. The float in the hydrometer according to acid percentage, sinks more or less deeply into the sucked in electrolyte from the battery cell. Thus the hydrometer measures specific gravity directly on the basis of which battery state of charge is ascertained.

Actual battery performance cannot be checked with a hydrometer as the temperature has far-reaching effect on specific gravity of battery. The table showing state of charge in relation to specific gravity of electrolyte may not prove true in Simla during winter although it is correct at Bombay. Many batteries are manufactured to have a gravity of not more than 1·280 at full charge while batteries for hot climates may have a top gravity 1·225. It is therefore necessary to take into account temperature while measuring gravity to determine accurately the state of battery charge.

Many hydrometers have a temperature table and a thermometer incorporated in them to make proper correction in the gravity reading with reference to table showing state of charge.

The gravity of electrolyte changes about 0·004 for every 10°F. Referring to table based on 80°F, we must add or subtract 0·004 for every 10°F above or below 80°F respectively. For example, if the gravity is 1·240 at 100°F, add $2 \times 0{\cdot}004$ or 0·008 and the corrected reading shall be 1·248. If gravity is 1·210 at 40°F, subtract $4 \times 0{\cdot}004$ or 0·016 and the corrected reading shall be 1·194.

The following precautions should be kept in view while measuring specific gravity by means of hydrometer :—

(*i*) While taking specific gravity readings, do not suck in too much electrolyte into the hydrometer. The float should be suspended freely in the electrolyte.

(*ii*) Hold the hydrometer vertically and take the reading at eye level.

(*iii*) Replace the electrolyte in the cell from which drawn.

(*iv*) If a cell does not contain sufficient electrolyte to permit hydrometer reading, add distilled water but do not take reading until the battery has been charged for an hour at least for mixing water with electrolyte.

(*v*) Be careful lest electrolyte falls on clothes or body resulting in damage or skin burns. In case electrolyte is spilled during gravity measurement, flush it away with water. Baking soda may also be used to neutralise spilled electrolyte.

(*b*) **By measuring the voltage of battery cells.** The

voltage of battery can be measured under different conditions and by means of different instruments. Generally the following methods are used :—

(1) Light load test

(2) High discharge test.

(1) **Light load test.** This test is carried out by applying light load to the battery and measuring cell voltage after stabilizing the battery with a heavy load. Proceed as under for light load test.

(*i*) Examine the battery for broken case or covers, electrolyte level and unusual smell in the cells. (The defective battery should not be put under this test.)

(*ii*) Add distilled water if electrolyte level is low.

(*iii*) Put load on battery by switching on starting motor for about three seconds in "off" ignition position.

(*iv*) Turn on head-lights (low beam) and after one minute note down voltage of each cell by means of a voltmeter.

If the voltage of all the cells is 1·95 volts or more and the difference between the highest and lowest cell voltages is less than 0·05 volt, the battery is *well charged and in good condition. If the cell voltages are above and below 1·95 volts* and the differences between the highest and lowest cell voltages are less than 0·05 volt, *the battery is in good condition but needs charging.*

If the voltage of all the cells is below 1·95 volts, the battery is *discharged.* The battery is *defective* if there is a difference of 0·05 volt or more between the highest and lowest cell voltages, even if one cell gives 1·95 volts more reading.

(2) **High discharge test.** Under this test the cell voltages are measured while the battery is placed under heavy load as during engine cranking. In place of cranking motor (self-starter) load, a tester having a heavy variable resistor, adjustable to provide the specified load, is also used.

Before undergoing this test see that electrolyte level in all the cells of battery is correct and the battery has no apparent defect. The battery should be at least half charged and must be between 60° and 100°F.

A cell tester or voltmeter having heavy resistance fixed with prods is used to check up voltage of individual cell. As the prods are pressed over the cell terminals, the resistance puts the cell under high discharge. The voltmeter shows cell voltage at the same time.

Cadmium-tip test. Chrysler Corporation (U.S.A.) uses a special tester known as *Cadmium-tip battery-cell analyzer* to find out the charge state of battery. The tester contains two number cadmium tips, one red and the other black. For finding battery state of charge, the cadmium tips are inserted into the electrolyte of

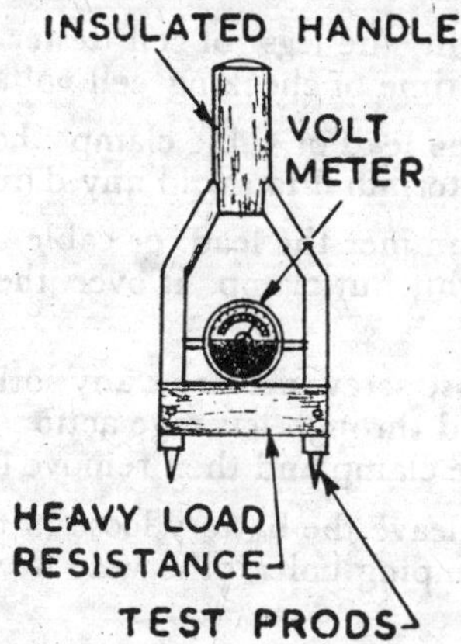

Fig. 10·4. Cell Tester.

neighbouring cells after removing filler plugs. The red probe is put into the cell which contains positive terminal and black probe is inserted into the next cell. After noting down meter readings, the probes are moved to the cells 2 and 3 and so on. The meter shows state of battery charge directly on its scale, the red and green bands indicating recharge and safe conditions respectively.

9. Battery care. (*i*) Keep the battery clean. Moisture on battery top arrests dust. If it is not cleaned off periodically, the battery shall be discharged slowly across the battery top.

(*ii*) Avoid accumulation of corrosion around the terminals. As a result of overfilling or overcharging, electrolyte escapes from the battery cells and spreads over the battery top due to which the terminals are sulphated and corroded. Periodic cleaning of the battery top and terminals will prevent corrosion to progress further.

(*iii*) Smear petroleum jelly or vaseline on the battery terminals to act as anti-corrosion paste.

(*iv*) Avoid overfilling and overcharging.

(*v*) Don't let electrolyte level fall below the top of battery plates but keep the level atleast ¼″ above the plates.

(*vi*) Don't use tap or ordinary water to top up battery level but use distilled water otherwise the battery shall be sulphated due to hard water.

(*vii*) Keep the filler plugs tight and ensure that vent holes in them are clear for the gases to escape.

(*viii*) Avoid loose connections at the battery otherwise the terminals shall be damaged due to arcing whenever there is any vibrator.

(*ix*) Hold tight the legs of cell tester over the terminals to avoid arcing at the time of checking cell voltage.

(*x*) Loosen the lead or cable clamp thoroughly before lifting it from the battery terminal to avoid any damage.

(*xi*) Don't hammer the lead or cable clamp over the battery terminal to hold tight but clamp it over the terminal through the clamping bolt.

(*xii*) Don't use screw driver or any other lever to disconnect battery cable or lead through leverage action but unscrew the clamping bolt, loosen the clamp and then remove lead.

(*xiii*) Don't leave the battery loose in the car cradle but hold it tight through clamping bolts otherwise it will bounce around its carrier.

(*xiv*) Don't let the battery stay unused and unchecked for a long time. If the battery is to be stored for enough long time, drain out electrolyte, wash the cells and store it in a cool dry place after cleaning and coating anti-corrosion paste over the terminals.

If the battery is to be kept unused for a short period, check it at least once a week and see that electrolyte level in the cells has not fallen down and corrosion has not settled over the terminals. Keep the battery clean and electrolyte level up and put it on charge if gravity or voltage falls down. The unused battery should be recharged once a month.

Chemical activity continues in a battery up to some extent even if it is not connected to a circuit and delivering current. This chemical action is known as *self-discharge* which varies with temperature and the strength of electrolyte. The higher the temperature and stronger the electrolyte, the faster the self-discharge.

(*xv*) Never put the batteries one above the other as the weight on the bottom batteries might compress down the plate assemblies resulting in their collapse and ruin.

(*xvi*) While removing battery from the vehicle, first disconnect the grounded terminal cable from the battery to reduce the possibility of any short circuit resulting in sparking. During installation, connect the insulated terminal first and then the grounded terminal.

(*xvii*) While installing battery in the vehicle ensure that negative terminal is grounded in *negative earth return system* and positive terminal is grounded in *positive earth return system*. The negative terminal post of the battery is smaller in diameter than the positive terminal post. The polarity could be checked by switching on ignition or lights. The ammeter should read on the *discharge* (—) side. If the ammeter shows *charge* (+), the battery is wrongly connected.

(*xviii*) Don't connect a cable having defective insulation, broken strands, badly corroded or defective clamp with the battery.

(*xix*) Be careful in removing and installing battery in the vehicle. If the steel chain of your, watch or steel bangle in your hand or a spanner connects the two terminals of cell, the metallic chain, bangle or spanner will heat up. They will not damage your hand alone but will spoil the battery too.

(*xx*) The battery contains acid which is injurious to skin and clothes. Hence beware in handling them.

10. Earth polarity of battery. Formerly negative terminal was used to be earthed in automobiles. Later on *positive earth system* was adopted by so many manufacturers due to the reasons that :

(*i*) it requires low voltages to provide spark at the spark plug.

(*ii*) it minimises corrosion at battery terminals and other connections in the system.

(*iii*) it reduces wear of spark plug points and distributor cap electrodes.

With the introduction of A.C. generators (Alternators) and transistorised ignition system, the automobile manufacturers are shifting back to the original negative earth system.

11. Generator. It is a machine which converts mechanical energy into electrical energy. When rotated, it generates electricity by electro-magnetic induction. In automobiles, it is driven by the engine through a belt pulley or sprocket and chain. It consists of a body or frame, pole shoes, field coils, armature and brushes as shown in the diagram. Automobile generators are known as shunt (parallel) wound generators, because their field coils and armature are connected in parallel.

The soft iron pole shoes are attached to the cylindrical housing made of steel. The pole shoes form the cores for the field windings and become south and north poles of electromagnets when current flows through the field windings. The number of pole shoes are either two or four.

An armature is placed inside the housing and is driven on bearings. The armature consists of a slotted core which is made up of a number of soft iron laminations mounted on a shaft. The armature windings are coiled in the slots of the core and are insulated from the core. The windings consist of a number of coils, each coil being composed of a number of turns of the wire. The ends of the coils are soldered to the copper segments from one another, and mica is placed between them.

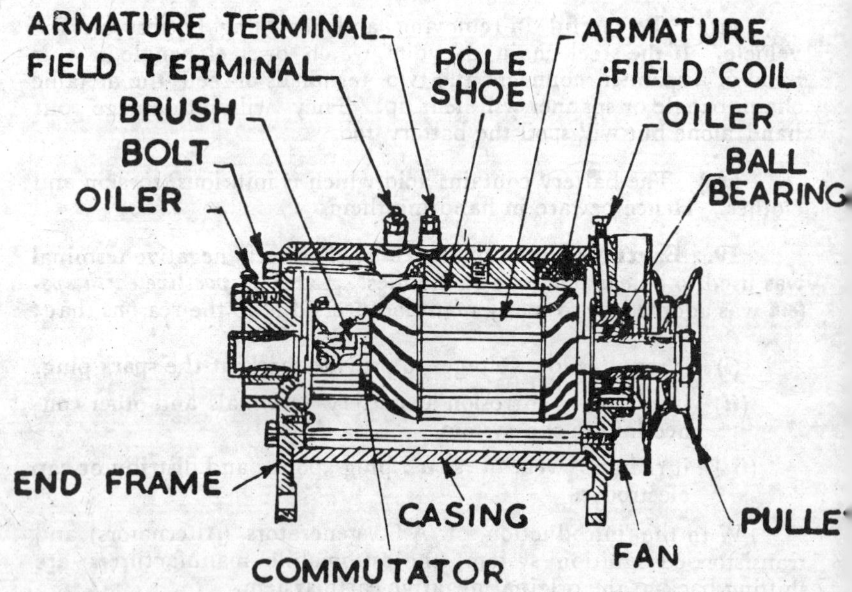

Fig. 10·5. Generator.

For making contact with the commutator, while the armature revolves, carbon brushes are employed in the generator. The brushes are supported in brush holders and are held against the commutator by springs. They convey the current from the armature to the field coils or to the external connected load. One each ground brush and armature brush is provided in all generators. Three brush generators have an additional adjustable field brush.

Two or four field coils are used in automobile generators. The coils are wound in such direction as to give an alternate north and south magnetic polarity to the inner faces of the pole shoes.

The field windings consist of coils of wire wound around the soft-iron pole shoes fitting in the housing. The flow of current through the field windings magnetizes the pole shoes, resulting in a magnetic field across the shoes.

The D.C. generator which is mostly used in automobiles, works on the following principle :

"When a coil of wire is moved through a magnetic field, current will be induced or generated in the coil."

The amount of current induced in a conductor depends upon the strength of magnetic field, the number of wires, cutting the field and the speed with which they pass through the field.

In an automobile generator, the armature is rotated by the engine through its magnetic field which is set up by the field winding wound around the pole shoes attached to the housing or casing.

When the armature is first rotated, only small amount of residual magnetism in the magnet poles is available but this is sufficient to generate a small current in the armature windings which in turn energizes the field poles resulting in high output. An automatic switch or cut out is used to control the output of the generator as it continues to increase with the increase in speed or the armature. Without some form of regulation, the generatof would overheat and burn up.

In recent years, alternating current (A.C.) generators oı *alternators* have started taking place of direct current (D.C.) generators in automobiles. The D.C. generator rotates the conductors in a stationary magnet whereas A.C. generator or alternator rotates a magnetic field so that stationary conductors may cut the moving lines of force. This way the voltage is generated in the stationary coils. This has the advantage that heavy currents do not have to pass through brushes and collector rings.

An alternator consists of two major parts : a stator in which the voltage is generated and a rotor which provides the magnetic field. The rotor revolves within the stator and the electricity is generated.

The battery, ignition system, lights and other electrical components in an automobile cannot use A. C. current produced by the alternator as they are D.C. units. Some arrangement, therefore, must exist to rectify or change A.C. output to D. C. This problem is solved by the rectifiers. Instead of a commutator by means of which alternating current generated within the armature is changed into direct current as it leaves the brushes, a rectifier converts alternating current into direct current.

Rectifier is a device which permits current to flow through it in only one direction. Different types of rectifiers are in use but in automobile alternators, silicon-diode rectifiers are mostly employed.

12. Types of D.C. generators. D.C. generators are classified according to the number of brushes employed in them. Following are the main different types :

(*a*) Two-Brush generators,

(*b*) Three-Brush or third-brush generators,

(*c*) Four-Brush generators.

(*a*) **Two-Brush generators.** In these types of generators, the field receives its current from the armature brush and the full voltage of the generator is applied to it. With the increase of generator voltage, the strength of the field current increases. This, in turn, further increases the output of the generator and so on.

These generators are classified according to the polarity of the generator and according to whether the field circuit is internally or externally grounded. Generators may have a positive or negative polarity and it must be the same as that of the rest of the electrical system.

(*b*) **Three or third Brush generators.** In three or third brush generators, both terminals of the field windings are not connected directly to the brushes but instead, one end of the field winding is connected to the third brush while the other end is connected to the ground. The position of the third brush is adjustable and its position controls the voltage applied to the circuit. Nearer the third brush to the armature brush, more the current in the field circuit and higher the output of generator. The farther the third brush positioned away from the armature, the less the field current and the lower the output of the generator.

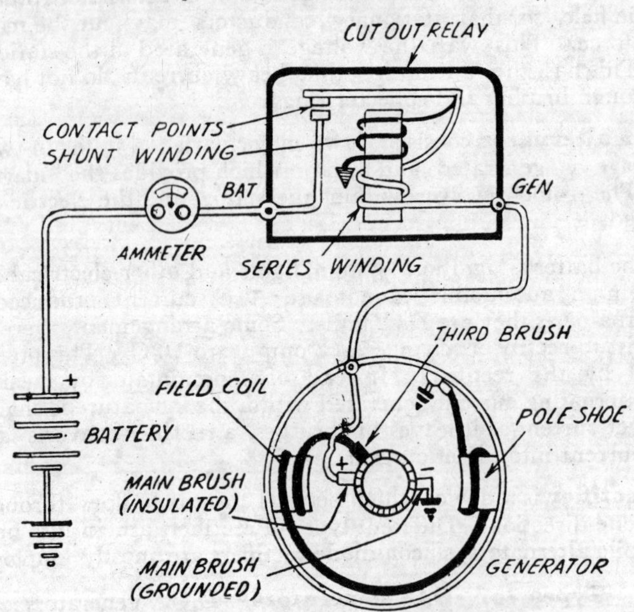

Fig. 10·6 Third brush generator circuit.

(*c*) **Four-Brush generators.** These are shunt wound generators which contain four brushes and four field coils. The construction and operation of these generators is similar to those of two-brush generators. In these generators, there are four brushes instead of two, for greater output. The field coils are connected directly to the main brushes.

The shunt wound generators have taken the place of third-brush generators due to the fact that the output is limited at low speeds in third-brush type generators.

13. Electrical systems. In an automobile, electric current flows through different systems in order to fulfil the different objects of cranking the engine for starting, sparking the compressed fuel air mixture in combustion chambers for ignition, charging the battery, lighting the lamps, blowing the horn and operating the numerous other electrical components. The following are the main electrical systems in an automobile :

Group 'A' **Group 'B'**

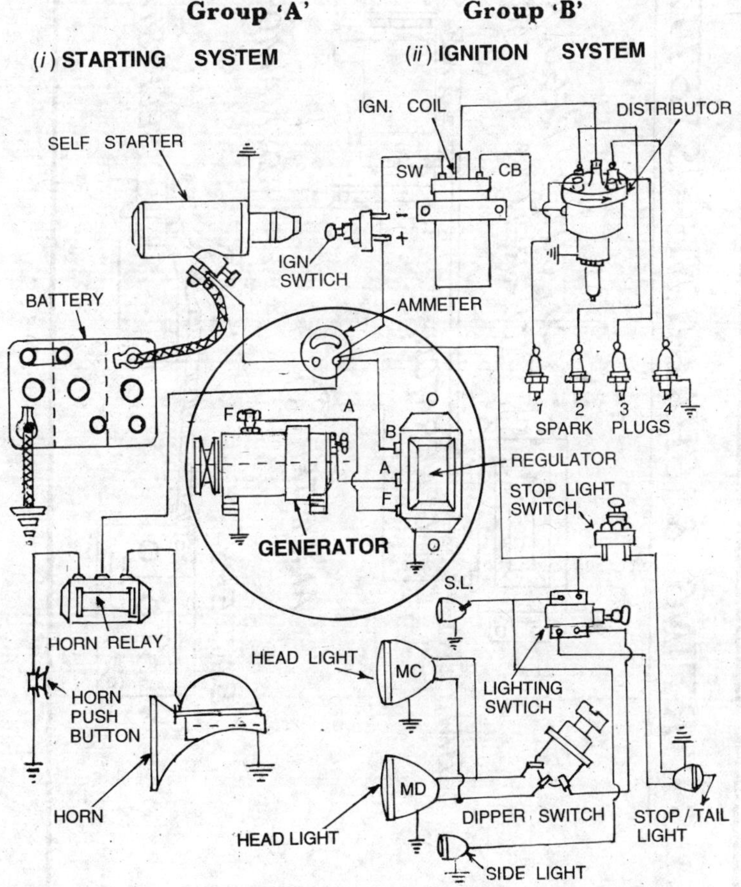

Fig. 10·7. Electrical system.

(*ii*) Charging system
(*iii*) Lighting system
(*iv*) Miscellaneous system

I. Starting system

Starting system. It is the system by means of which an engine is cranked for starting purposes. Although engine is cranked

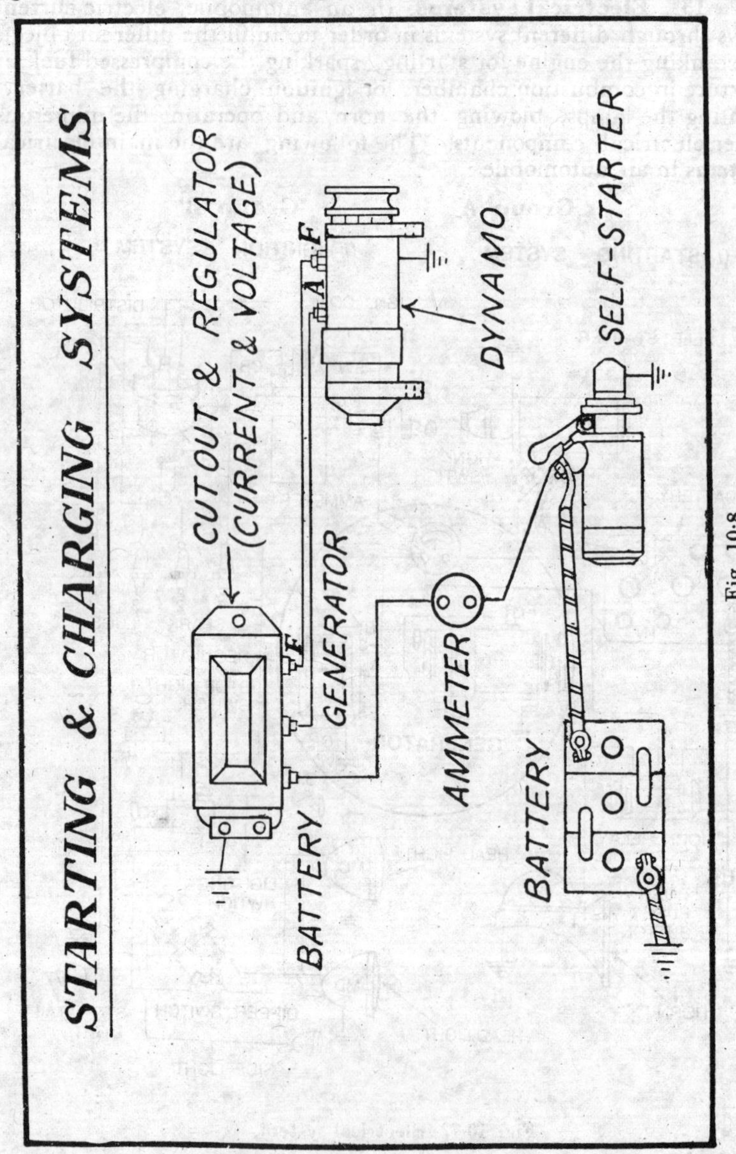

Fig. 10·8

by different methods such as by hand crank or inertia starters, compressed air starters or cartridge-type starters, auxiliary engine known as starting engine etc., yet this system is confined to engine cranking by means of a starting motor or self-starter.

This system consists of the following main components

(*i*) Starting motor or self-starter,

(*ii*) Storage battery,

(*iii*) Starter switch.

(*i*) (*a*) **Starting motor or self-starter.** It is a low voltage D.C. motor incorporating a gear-drive mechanism which converts the electrical energy received from the storage battery, into mechanical energy. When the circuit between the starting motor and the storage battery is completed through the starting switch, the starter cranks the engine. The starter drive gear is engaged with the ring gear at engine flywheel, manually or automatically. Once the engine starts, the starter is automatically disengaged or the starter armature is prevented from overspeeding until disengagement is affected.

The starting motor consists of the following two main units *i.e.*, (*a*) motor unit, (*b*) drive unit.

(*a*) **The motor unit.** It consists of a cylindrical housing, field coils, pole shoes, armature and brushes etc. Steel pole shoes are securely attached to the inner surface of steel housing. The number of pole shoes varies from two to six but mostly there are four pole shoes. The pole shoes hold the field coils and become magnetized when current flows through the coils. The armature revolves between the pole shoes with a slight clearance.

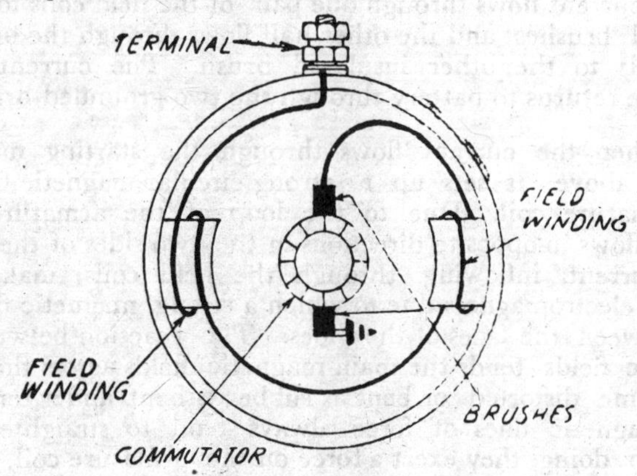

Fig. 10·9. Electric circuit in cranking motor.

The armature consists of a slotted iron core, shaft, commutator and armature windings. The windings are made of heavy flat copper strip for carrying safely the heavy current which passes through them. The windings consist of a number of coils, each coil having single loop. The sides of the coils fit into the slots of the armature core and are insulated from the core. The coils are connected to the commutator in such a way so that the current from the brushes may flow through all the coils at one time. This creates a magnetic field around each conductor due to which a repulsion force is produced all around the armature and the armature is rotated.

Field coil windings are also of heavy flat copper wire to carry heavy current. The coils on adjacent pole shoes are wound in opposite directions giving their respective pole shoes a north and south magnetic polarity. The starter housing acts as a return circuit for the magnetic lines of force. The field coils are connected together in series with the armature. One end is connected with starter terminal and the other to the insulated brushes. In case the starter is having four brushes, two opposite brushes are insulated from the end frame and are connected with the field coils. The other two brushes are earthed to the end frame through the commutator and armature conductors.

The starting motor works on the following principle :

"Unlike magnetic fields attract each other and like magnetic fields repel each other."

When the starter switch is put on, current from the battery flows to the starting motor. In four pole, four brush starting motor, the current entering the motor is divided into two halves as the field windings are paired off at the starter terminal. One half of the current flows through one pair of the field coils to one of the insulated brushes and the other half flows through the other pair of field coils to the other insulated brush. The current from the armature returns to battery through the two grounded brushes.

When the current flows through the starting motor as explained above, it sets up a strong circular magnetic field around the armature coil. Due to the loop of the armature coil, the current flows in opposite directions in the two sides of the coil. The same current, inflowing through the field coils, makes the pole shoes as electromagnets due to which a strong magnetic field is created between the faces of the shoes. The reaction between the two magnetic fields tends the main magnetic fields across the pole shoes to become distorted or bent as rubber is bent under tension. The bent magnetic lines of force always tend to straighten out and while so doing, they exert a force on the armature coil, causing the armature coil to rotate between the pole shoes. Thus the repulsion force causes the starting motor armature to revolve. The greater the current flowing through the armature and field coils, the greater the repulsion force exerted on the armature coils and greater the torque produced by the starting motor.

(*b*) **Drive Unit**. The drive unit or drive mechanism of the starting motor, transmits the torque developed by the starting motor to the engine flywheel for cranking engine as starter armature revolves.

The ratio between the drive gear on the starting motor and ring gear on the engine flywheel is about 1 to 12 as drive gear has few gear teeth as compared to the number of gear teeth on flywheel ring gear. Thus a gear reduction is provided which enables the torque developed by the starting motor to turn the engine over cranking speeds. The starting motor drive mechanism permits quick disengagement of drive gear from the flywheel ring gear or the drive gear from the armature shaft in order to save the armature from ruin due to centrifugal force developed at engine speed.

The drive mechanism is of the following two types :—

(*i*) Inertia drive,

(*ii*) Over-running clutch type.

(*i*) **Inertia-drive**. This type of drive depends upon the inertia of the drive gear or pinion to produce meshing. *Inertia* is the property which all things have that resists any change in motion. When the drive gear is not rotating, it resists any force which tends to set it in motion.

Bendix drive is known as inertia drive. In this drive the drive pinion is mounted loosely on a threaded sleeve. The sleeve is fastened to the armature shaft by means of a heavy spiral Bendix spring. This spring absorbs the shock caused during meshing.

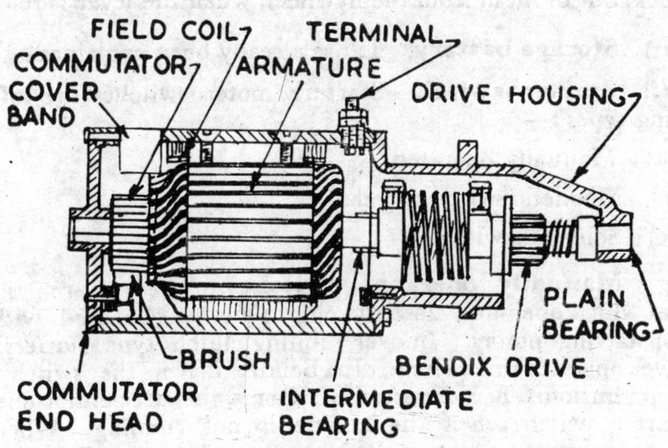

Fig. 10·10. Self-starter, Bendix drive type.

When the starter switch is pushed on, the armature begins to rotate. The inertia of the drive gear on the Bendix drive prevents it from turning, while the threaded sleeve on which it is mounted revolves with the armature shaft. This tends the drive gear to be thrust forward into mesh with the flywheel ring gear.

When the engine starts, the flywheel rotates at a higher speed than the starter. The engine flywheel tends to drive the starting motor but because the drive gear is mounted on the threaded sleeve, so it is put out of mesh from the flywheel.

(*ii*) **Over-running clutch type**. In this type a shift lever is employed to shift the drive gear and over-running clutch mechanism on the armature shaft into and out of mesh with the engine flywheel. The shift lever is operated manually by means of a lever or pedal or by an electrically operated solenoid control.

The over-running clutch is located between the drive gear and sleeve attached to the splined armature shaft. It consists of the outer shell which contains hardened steel rollers about four in number, fitted into notches and the pinion and collar assembly. The notches are not concentric but are smaller in the end opposite to the plunger springs. When the armature and the shell begin to rotate, the pinion is momentarily stationary. This causes the rollers to rotate into the smaller sections of the notches where they jam tight. The drive pinion now rotates with the armature, cranking the engine. After the engine starts, it spins the pinion faster than the armature. The rollers are then rotated into the larger sections of the notches where they are free. This happens due to difference in speed between the engine flywheel and armature shaft. This allows the pinion to spin independently or over-run the remainder of the clutch. A return spring on the shift lever pulls the drive gear back, out of mesh from the flywheel, when the lever is released.

(*ii*) **Storage battery**. It has already been explained.

(*iii*) **Starter switch**. Starting motor switches are of the following types :—

(*a*) Manually operated,

(*b*) Magnetic switch or relay,

(*c*) Solenoid switch.

(*a*) **Manually operated switches**. These are simple switches which open and close the circuit and can be used with all types of starting motors. In over running clutch type starters, the shift lever operates this switch. In Bendix drive, the switch is a push type button. In some starting systems, the accelerator operates the starter switch when the engine is not running. When the engine is running, a latch device operated by a vacuum control holds the starting motor control mechanism away from the accelerator.

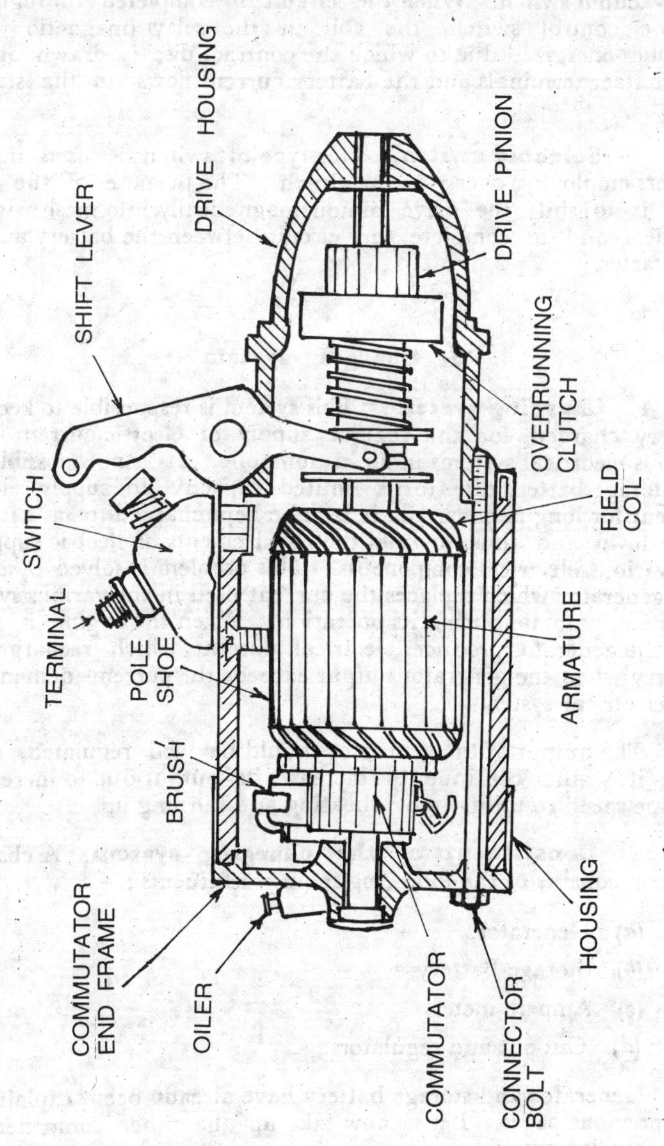

Fig. 10·11. Self-starter ; over.running clutch type

(*b*) **Magnetic switch or relay.** In Bendix type self-starters, simple starting motor relays are used. These are controlled either by means of a remote control having push button type switch or a vacuum switch. When the circuit is completed through the remote control switch, the coil in the relay (magnetic switch) becomes energized due to which the contract disc is drawn against the contact terminals and the battery current flows to the starting motor.

(*c*) **Solenoid switch.** This type of switch is used in self-starters employing over-running clutch. The purpose of the solenoid is to shift the drive pinion magnetically into mesh with the flywheel and to complete the circuit between the battery and the self-starter.

II. Charging system

1. **Charging system.** This system is responsible to keep the battery charged for the regular supply of electric current to the various electrical systems in an automobile. It is not possible for the storage battery due to its limited capacity, to supply electric current for long periods. So it is to be kept charged to save it from run down and thus to fulfil the requirements of electric supply to the various electrical components. This problem is solved by means of a generator which replaces the current used in the various systems of the automobile during its operation. When the engine is working, the generator produces electrical current which recharges the battery when the generator output exceeds the current demands of the electrical systems.

The output of the generator should be well regulated, otherwise it would continue to increase its output due to increase in engine speed resulting in overheating and burning up.

2. **Constituents of the charging system.** A charging system consists of the following main constituents :—

(*a*) Generator,

(*b*) Storage Battery,

(*c*) Ampere-meter,

(*d*) Cut out and regulators.

Generator and storage battery have already been explained in the previous pages. Let us now take up the other components of the charging system.

(*i*) **Ampere-meter.** It is an instrument fitted on the dash board, by means of which rate of current flow is measured. For the continued operation of an automobile, both a properly working generator and a fully charged battery are essential. The ammeter keeps the driver constantly informed about the performance of the generator and battery. It indicates the direction of flow of electric current. When the current is being used from the battery, the ammeter needle indicates on the negative (—) side and while the battery is being charged by the generator, the needle indicates on the positive (+) side of the gauge the strength of the current flowing into the battery in amperes.

The ampere meter offers practically no resistance to the flow of current and is always connected in series. Ammeters commonly used in automobiles are of the following types :—

(*a*) Moving Vane type,

(*b*) Loop type.

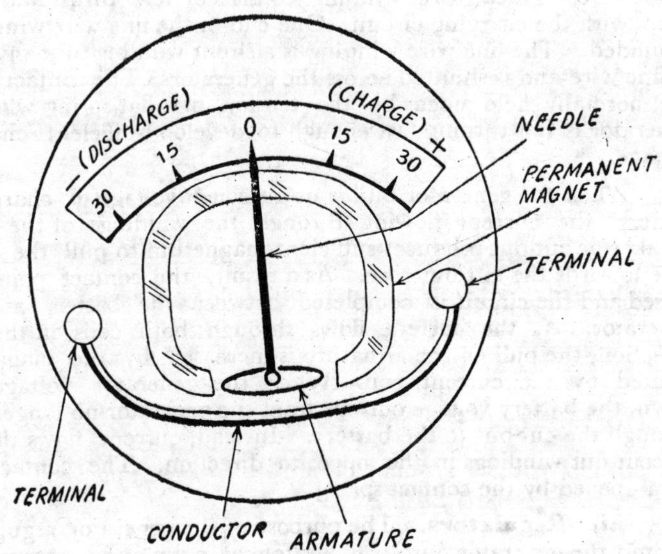

Fig. 10·12. Ammeter (Moving Vane Type).

(*a*) **Moving Vane type.** In this type of ammeter, a permanent magnet, an armature and needle assembly are provided in its frame. The magnet, which is attached to the frame, holds the armature on which the needle is mounted. When no current is flowing, the magnet holds the needle through the armature to zero. As current flows through the armature in either direction, a magnetic field is built up around the armature. This overcomes the effect of permanent magnet, thus providing a reading proportional to the strength of current flowing through it.

(*b*) **Loop type.** This type of ammeter contains a magnet and needle assembly having a wire assembled in it. When the current flows through the wire, a magnetic field is set up around it. This tends the needle to move in proportion to the strength of the magnetic field.

(*ii*) **Cut-out.** Cut-out or circuit breaker is an automatic switch which opens and closes the circuit between the generator and the battery. It prevents the battery from discharging through generator when engine is stopped or is operating at such a low speed that the generator is not producing enough voltage to charge the battery.

It consists of an electromagnet and an armature having a contact breaker. The electromagnet contains two types of windings, one of fine wire and the other of thick wire wound around a single core. Thick wire winding consists of few turns and is in series with the charging circuit. One end of the fine wire winding is grounded. The fine wire winding is a shunt winding of many turns of fine wire and is shunted across the generator. The contact points are normally held open by the tension of a flat spring when the generator is not turning fast enough to develop sufficient charging voltage.

When the generator builds up sufficient voltage to charge the battery, the current flowing through the windings of the circuit breaker or cut-out produces sufficient magnetism to pull the armature towards the cut-out core. As a result, the contact points are closed and the circuit is completed between the battery and the generator. As the current flows through both coils in the same direction, the pull on the armature is increased by the magnetism created by the current coil. When the generator voltage falls down, the battery voltage current from the generator no longer flows through the cut-out to the battery. Instead, current flows through the cut-out windings in the opposite direction. The contacts are then opened by the contact spring.

(*iii*) **Regulators.** The purpose of the generator regulator is to limit the generator output to a safe maximum and to permit the output to vary according to the requirements of the battery and the connected electrical load of other electrical systems.

Apart from cut-out relay, the charging system employs a voltage regulator and a current regulator.

Voltage regulator. It controls the maximum voltage of the generator and prevents it from exceeding a predetermined value which is fixed by setting the regulator.

The voltage regulator consists of a shunt winding and a flat steel armature mounted on a hinge above the winding core. The regulator does not operate when battery charge rate is low. The generator output which is determined by speed and battery condition, increases to the determined value. Battery voltage increases as the battery approaches a charged condition. There is then an increasing magnetic pull from the shunt winding on the armature. When the regulating voltage is attained, the magnetic pull is sufficient to overcome the spring tension. Thus the armature is held away from the winding core. The armature is pulled down and the points are separated off. This inserts resistance into the generator field circuit so that the generator output and voltage are reduced. The magnetic strength of shunt winding is reduced by the reduction in generator voltage. As a result, the armature is pulled up by the armature spring tension, the points close and the generator output and voltage increase.

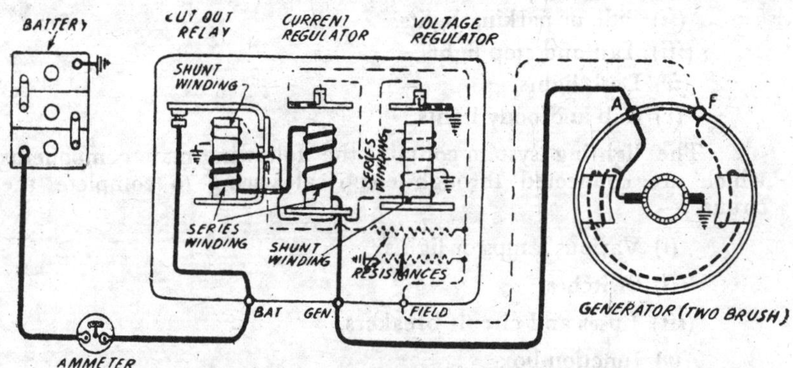

Fig. 10·13. Wiring circuit in generator regulator. (Delco Remy G.M.).

Current regulator. It controls the maximum amperage output of the generator and prevents it from exceeding 30 to 35 amperes which is obtained by setting it. This prevents the generator from damage due to overload. The construction of current regulator is similar to that of a voltage regulator except that the winding consists of a few turns of heavy wire through which the full output of the generator passes. When the generator output reaches the rated maximum, the current passing through the regulator winding is sufficient to overcome the armature spring tension and separate contact points. This inserts the resistance into the generator field circuit, causing the generator output to fall. The magnetic pull of the regulator winding is also reduced due to fall in generator output. The contact points are then no longer held open. As soon as the points close, the generator field circuit is directly grounded and the output rises This cycle is repeated very rapidly resulting in the current regulator to limit the current output of the generator to the rated value.

The current and voltage regulators never operate during the same time. The current regulator operates when the load requirements are high and the battery is low. It then prevents the generator output from exceeding its safe maximum. The voltage is not sufficient to tend the voltage regulator to operate.

When the load requirements are reduced or the battery begins to come up to charge, the line voltage increases to a sufficient value to cause the voltage regulator to operate. At this occasion, the generator output falls down below the value required to cause the current regulator to operate. All regulation is then done by the voltage regulator.

III. Lighting system

Lighting system. This system is responsible to glow various lights in an automobile. The main lights are as under :—

(*i*) Head lights,

(*ii*) Side or parking lights,

(*iii*) Tail and stop lights,

(*iv*) Dash lights,

(*v*) Cab and body lights.

The lighting system contains the following main components which are connected through electrical wiring to complete the circuit :—

(*i*) Various lamps or lights,

(*ii*) Switches,

(*iii*) Fuses and circuit breakers,

(*iv*) Junction box.

2. **Lamp**. An electric lamp consists of a cover, reflector, bulb, holder and lens.

(*i*) **Bulbs**. Small incandescent bulbs are used in most lamps. The bulbs are filled with a gas and are designed to operate on a low voltage. The bulbs contain single or double filaments and one or two contacts at the base accordingly. One end of the filament is attached to a contact in the base while the other end is grounded to the metal base of the bulb. Bulbs used in automobile lights vary from 1/2 candle-power to 50 candle-power.

(*ii*) **Reflectors**. The reflectors are employed in the lamps to collect the light rays and distribute them outward in the form of a beam. They control the direction of light beam and increase the intensity of light.

The reflector is made of either of brass with a silvered reflecting surface or of glass with a polished aluminium surface. It has a fixed focal length or focal point which is the point at which the filament of the bulb is located to throw light beams straight ahead and at the greatest intensity.

(*iii*) **Lens**. The main light lens contains flutes and prisms built into them which help in bending and distributing the light emitted by the reflector into the desired beam pattern to cover the road.

3. Head lamps or lights. The modern head lights are fitted with sealed beams which consist of a permanently sealed glass lens and reflector containing two filaments—upper and lower. The sealed beam is housed in a suitable cover to form the complete lamp.

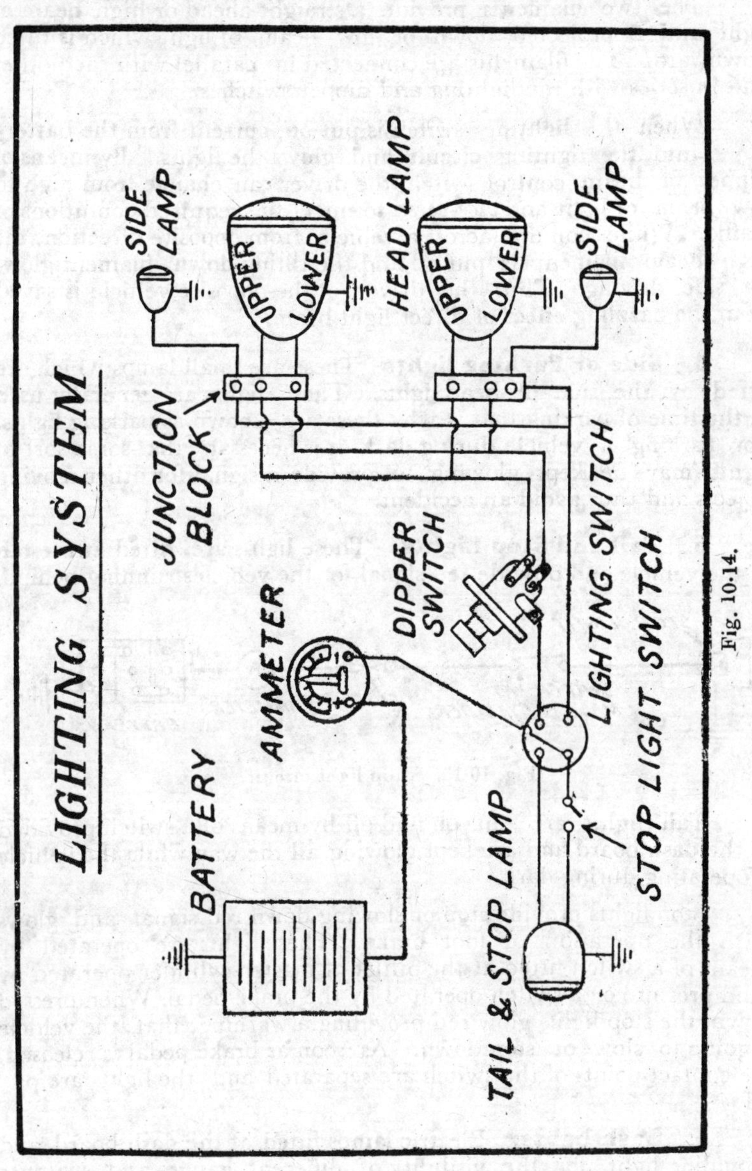

Fig. 10·14.

The sealed beams are of two types—glass back and metal

back. In the glass back type, the reflector forms a hermetically sealed single piece bulb which is filled with an inert gas. The metal back type sealed beam contains a bulb which is soldered to the reflector. The lens is clamped on to provide a tight seal.

The two filaments provide a straight ahead or high beam of light and a projected down or low beam of light which is tilted downward. The filaments are connected in parallel with each other and in series with the lighting and dipper switches.

When the lighting switch is put on, current from the battery flows into the lighting circuit and glows the lights. By means of dipper or beam control switch, the driver can change from high to low beam of light and *vice-versa* to meet the required conditions of traffic. While coming across a vehicle from opposite direction, the high beam filament is put off and the tilting down filament glows for safe driving. Thus the driver of the opposite vehicle is saved from the dazzling effect of direct light beam.

4. Side or Parking lights. These are small lamps which are fitted by the side of head lights. These lights are generally used at the time of parking that is why these are known as parking lights. For parking a vehicle during dark, it is necessary that some sort of lights may be kept glowing to provide a signal for other moving objects and thus avoid an accident.

5. Tail and Stop Lights. These lights are fitted at the tail of the vehicle and provide red signal for the vehicles running behind.

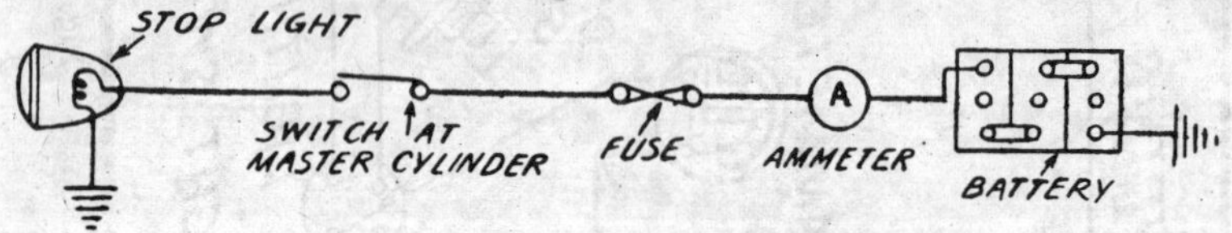

Fig. 10·15. Stop light circuit.

Tail lights are put on and off by means of a switch provided at the dash board and are kept glowing all the way while the vehicle is operating during dark.

Stop lights provide stop or slowing down red signal and glows with the operation of foot brake. These lights are operated by means of a switch fitted at the outlet of master cylinder operated by fluid pressure or a switch operated by the brake pedal. When pressed down, the stop lights glow red providing a warning that the vehicle is going to slow or stop down. As soon as brake pedal is released, the contact points of the switch are separated and the lights are put off

6. Dash lights. Electric lamps fitted at the dash board and providing light for the visibility of different gauges and switches etc. are known as dash lights.

7. **Cab and body lights.** These are the lights which are fitted inside the cab and body for the convenience of the driver and passengers.

8. **Switches.** A switch opens and closes an electrical circuit. Various important switches used in the lighting system of an automobile are as under :—

(*i*) Main Lighting switch,

(*ii*) Dipper switch,

(*iii*) Stop light switch.

(*i*) **Main light switch.** Mostly it is either pull-push or rotary type switch. Usually it contains three positions : 'Off', 'On' for side, tail and dash lights and 'full' for head, side, tail and dash lights.

(*ii*) **Dipper switch.** It is a rotary disc and plunger or rotary contact and push-pull type switch by means of which dimming of head light is done. When the plunger button is depressed, it turns the rotary contacts one position within the switch. The button is spring loaded and when released, automatically returns to the reset position.

(*iii*) **Stop light switch.** When operated direct by the brake pedal, it is an 'off' and 'on' type mechanically operated switch. In case the vehicle is employing hydraulic brake system, this switch is operated by the hydraulic fluid pressure which acts on it.

9. **Fuses and circuit breakers.** Fuses used in automobiles consist of a glass-enclosed strip of alloy metal having a low melting point. When a short circuit or overload occurs in a circuit, the fuse burns out and opens the circuit so that no further damage may occur.

In certain cases *circuit breakers* are employed in place of fuses in the electrical system to prevent an excessive amount of current from flowing, which would damage the wiring, gauges and other electrical components. Circuit breakers are current limiting relays and are meant for breaking the circuit when an excessively high current flows through the circuit due to overload or short circuit. Vibrating and lock out type circuit breakers are in common use.

Vibrating type circuit breaker. In this type an abnormal condition in the circuit causes the circuit-breaker to vibrate, intermittently cutting off the flow of current in the circuit.

Lock-out type circuit breaker. In this type, the contact points are held open by a separate winding connected between the contact points. A small current continues to flow through the winding but the electrical equipment is put out of order.

Thermal type circuit-breaker. It consists of two contacts mounted on a bimetal strip in the circuit. When a short circuit occurs, the excessive current heats the bimetal, causing it to bend and open the contacts. The circuit is opened till the bimetal cools after which the contacts again close.

10. **Junction box.** As its name indicates it is a box which serves as junction for the various electrical lines. Usually fuses are provided in the junction box to avoid the damaging effects of overload and short circuits.

IV. Miscellaneous system

This system contains the following main circuits :

(*a*) Horn circuit,
(*b*) Trafficator circuit,
(*c*) Wiper motor circuit,
(*d*) Speedometer circuit.

(*a*) **Horn circuit.** This circuit consists of the following main components ;—

(*i*) Electric Horn,
(*ii*) Horn Relay.
(*iii*) Switch.

(*i*) **Electric Horn**. It provides a warning sound to other road users. Most of the modern horns contain a diaphragm which is vibrated to produce sound waves, by means of an electromagnet. When the electromagnet is energized, it pulls on an armature which is attached to the diaphragm. Movement of the armature flexes the diaphragm and opens the contact points due to which circuit is broken. As the circuit breaks, electromagnet loses its magnetism and the flexed diaphragm regains its original position. The contact points are closed then. The cycle again starts which is repeated again and again resulting in continuous vibrating of the diaphragm which provides sound. Tone of the horn much depends upon the movement and stiffness of the diaphragm. A resistance or condenser is connected in series with the magnet winding and across the contact points for avoiding arcing at the points.

(*iii*) **Horn Relay**. It is connected in the horn circuit to provide a better direct connection between the horn and battery. This way, the voltage drop in the wiring from battery to horn is eliminated and higher voltage is available for operating the horn with better performance.

The horn relay consists of a winding on a core above which an armature is placed. The armature contains a contact point which comes in contact with a stationary point.

When horn push button is depressed, circuit from battery to the horn relay is completed and the relay points are closed. As shown in the diagram, when the contacts are closed, current flows direct from the battery to the horn.

(iii) **Switch.** Push button or ring type switch is usually used in the horn circuit.

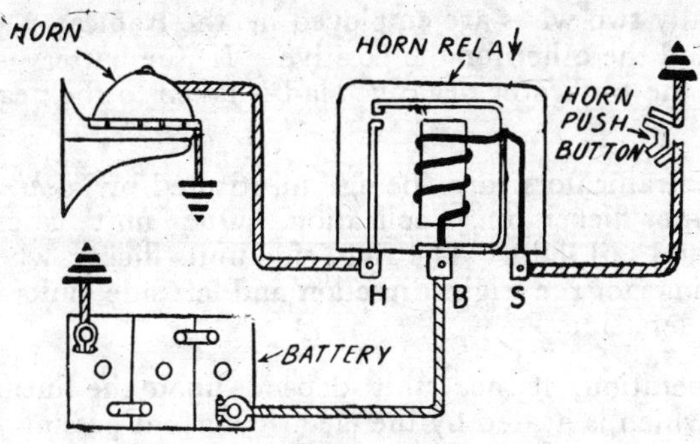

Fig. 10·16. Horn circuit.

(*b*) **Trafficator Circuit.** This circuit controls the operation of trafficators or direction indicators provided one on each sides of the car. This circuit consists of two trafficators and an operating lever type switch placed in the steering wheel. The switch has three positions, *i.e.*, 'off', 'right' and 'left'. When the switch lever is moved

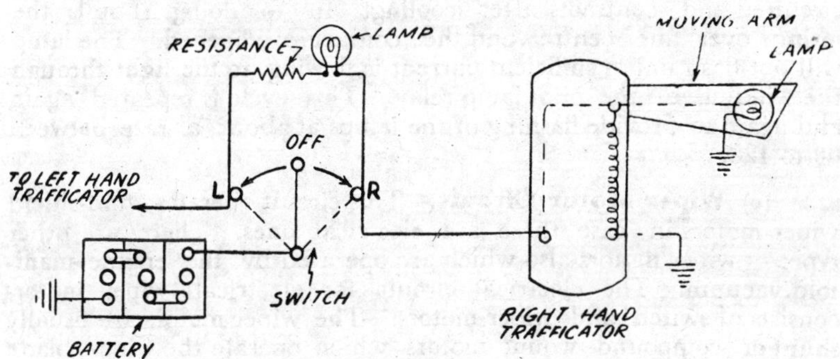

Fig. 10·17. Trafficator circuit.

to either 'right' or 'left' position, trafficator of that side is raised up horizontally which indicates the intention of the driver to turn to that direction.

The trafficator consists of a solenoid operated arm. The arm contains a bulb which illuminates when the arm is raised up. The sides of the arm are fitted with transparent red sheets so that when bulb illuminates inside the arm, it may provide red signal for attracting the attention of other road users.

The solenoid and bulb are in parallel with single pole wiring. The arm is pivoted at the top. A link connects the arm to an iron plunger which is operated by the solenoid. When the trafficator switch is turned on, the solenoid is energized and pulls the plunger downwards, resulting in the raising of arm to its horizontal position and at the same time completing the circuit through the bulb.

Usually two wires are employed in the trafficator, one for the negative and the other for the positive. If the battery negative is grounded, the trafficator negative lead is taken to the nearest ground point.

The trafficators are being substituted by *flashing indicators* which flash or flicker on application. One unit is employed in each side and tail light. The right side units flicker when the vehicle is turning to the right direction and left side units flash while turning to left side.

In operation, flasher unit depends upon the linear expansion of a wire which is heated by the electric current passing through it. This wire controls the movement of a spring blade having contact which provides flashing action.

When current flows into the unit through resistance wire and indicator lamp relay, the wire heats up and expands. This releases the tension on the spring which closes the contacts, thereby applying full voltage to the lamp through the windings of pilot lamp relay. While the contacts are closed, the resistance wire is short circuited and contracts after cooling. In so doing, it pulls the spring over the centre and the contacts are opened. The lamp will not flash unless sufficient current is flowing to the light through the windings of the pilot lamp relay. This cycle is repeated again and again to provide flashing of the lamps at about a rate between 60 to 120.

(*c*) **Wiper Motor Circuit**. This circuit operates wind-shield wiper motors in case these are electrical ones. There are other types of wiper motors also which are operated by the engine manifold vacuum. The electrical circuit for electrical wiper motors consists of switch and wiper motors. The wiper motors are usually shunt or compound wound motors which operate the wiper blade through an arm over the wind-shield.

(*d*) **Speedometer circuit**. This circuit operates electrically driven speedometer which records the speed of the vehicle at which it runs. The speedometer unit contains an odometer also which registers the distance travelled by the vehicle.

There are mechanically operated speedometers also which are operated by a flexible cable or shaft. Electrically driven speedometers are driven through an electrical sending unit.

An *electrical speedometer* consists of a sending unit and speedometer head. The sending unit is fitted at the transmission and speedometer head is mounted on the dash board. The sending unit generates the current required for speedometer drive. Therefore this circuit is not completed through the battery. The sending unit contains a revolving magnet and armature. When the magnet rotates around the armature, current proportional to the speed at which it is driven, is produced. The current generated by the sending unit is transmitted to the speedometer head through wiring, which causes speedometer needle to indicate speed of the vehicle at which it is running. The speedometer head contains a speed cup assembly which is connected to the needle. The current generated by the sending unit flows through resistors, compensator and armature, causing a rotating field. The rotating field drags the speed cup and needle shaft clockwise against the tension of hair-spring employed in the head and the vehicle speed is indicated.

QUESTIONS

1. Which are the different sources from which electrical energy is drawn in an automobile ? Describe any one of them.

2. Show with the help of a sketch the construction of a storage battery. Explain the chemical action in it.

3. How a battery is rated ? What is 20-hour rating ? Explain.

4. What is generator ? How it produces electrical current ?

5. What is the difference between D.C. generator and an alternator ?

6. Describe the construction and working of a third brush generator.

7. Which are the various types of D.C. generators employed in automobiles ? Describe them.

8. What does a starting circuit consist ? Show this system through a diagram.

9. Which are the various electrical systems employed in an automobile? Describe the function of each system.

10: Which are the different drive mechanisms employed in self-starters? Explain them.

11. Write short notes on the following :

(*i*) Electrolyte.

(*ii*) Shunt wound generator.

(*iii*) Over-running clutch.

(*iv*) Bendix drive.

(*v*) Solenoid switch.

12. Describe the construction and working of one type of self-starter.

13. What role is played by the charging system in an automobile How output of the generator is regulated ?

14. Draw the circuit diagram of a charging system.

15. Describe the following :

(*i*) Ammeter.

(*ii*) Cut-out.

(*iii*) Current regulator.

(*iv*) Voltage regulator.

16. Draw wiring diagram of lighting system.

17. Describe the construction of a sealed beam.

18. Which are the various lights employed in an automobile ? Describe them.

19. Draw a circuit diagram for head lamp through the dipper switch.

20. What do you know about the following :

(*i*) Fuse.
(*ii*) Dipper switch.
(*iii*) Circuit-breaker.
(*iv*) Stop light switch.
(*v*) Junction box.

21. Which are the different types of circuit breakers ? Describe them.

22. Describe the following :

(*i*) Electric horn.
(*ii*) Trafficator.
(*iii*) Horn relay.
(*iv*) Flashing indicator.
(*v*) Speedometer.
(*vi*) Wiper motor.

23. Explain the following :

(*i*) Battery efficiency.
(*ii*) Battery self-discharge.
(*iii*) Negative and positive earth system.
(*iv*) Battery overcharging and its effects.

24. Explain the various methods of battery charging and precautions required.

25. Explain the different methods for checking battery state of charge.

26. Describe the hints in battery care.

11

Electrical Systems—II

(Ignition System)

1. **Ignition system.** This is the system by means of which spark is provided in the petrol engine to ignite the compressed fuel-air mixture. This system supplies high-voltage surges (of as much as 20,000 volts) of current to produce spark at the spark plug gap. The spark is provided at the exact time in the various cylinders according to the firing order of the engine.

2. **Types of ignition system.** There are two main types of ignition systems as employed in petrol engines. These are as follows : (*a*) Battery Coil Ignition System and (*b*) Magneto Ignition System. These in turn are divided as below :

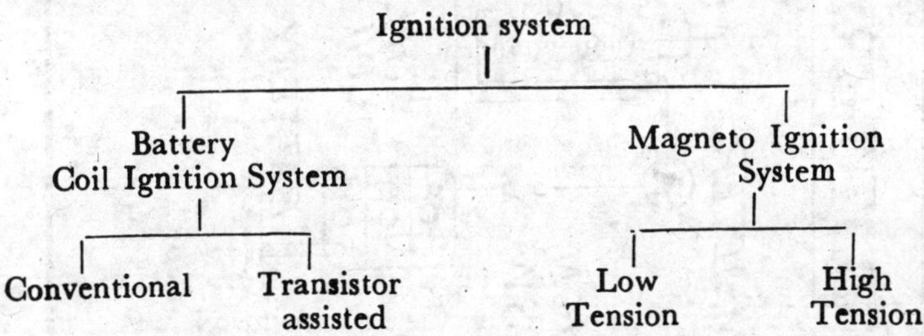

3. **Battery Coil Ignition System.** In battery coil ignition system, electric current is obtained from the storage battery. An induction coil is employed to transform low tension current into high tension current, through a make and break mechanism operated by the distributor in multicylinder engines. When a break is provided in the primary circuit, high voltage surge is produced by the induction

coil, which flows through the secondary circuit to the spark plug gap and produces spark. The break in the primary circuit is affected at

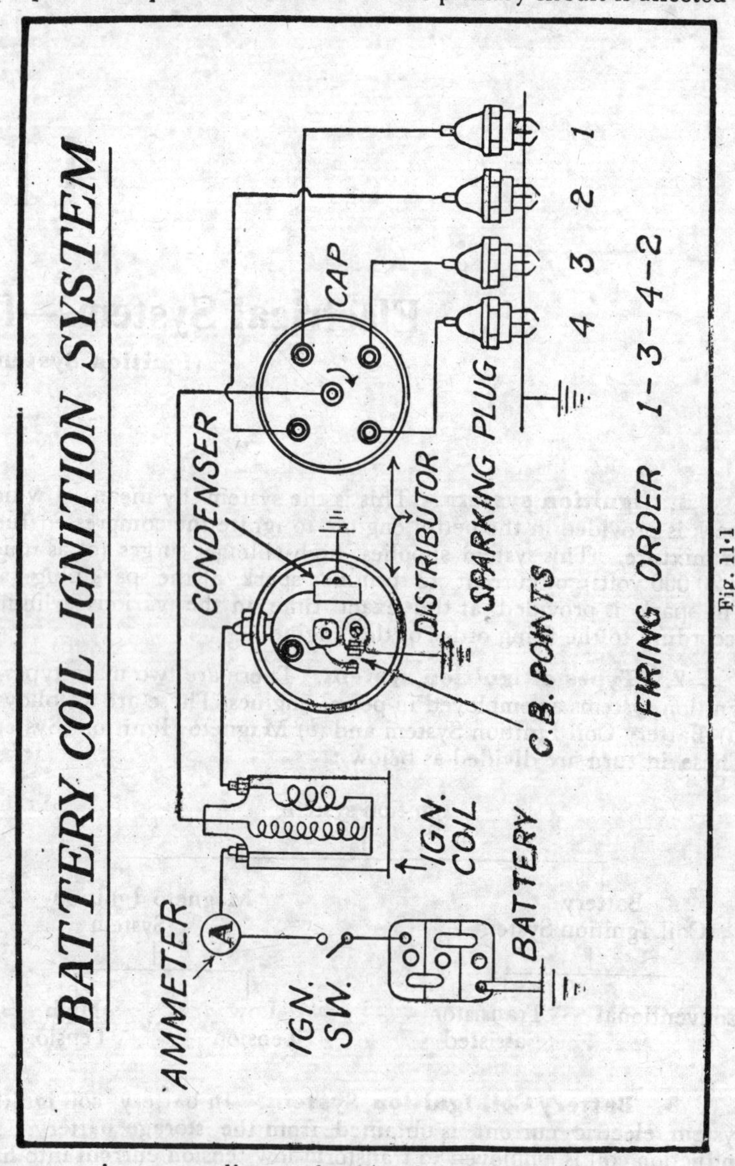

Fig. 11·1

the exact time according to load and speed of the engine. The breaker points are caused to open early or late through automatic advance and retard mechanisms provided with the distributor. The distributor distributes the high voltage surge to the different spark plugs for producing spark in the combustion chambers as per order of firing in the engine.

4. Conventional battery coil ignition system. This system consists of primary and secondary circuits. In *primary circuit*, current flows from the battery through the ammeter, ignition switch, primary windings of ignition coil, contact breaker points and condenser and then through earth return back to the battery.

When there is break in the primary circuit due to opening of breaker points, electrical induction takes place into the secondary windings of the ignition coil and a high voltage surge is produced. The high tension current then flows from the ignition coil through high tension lead to central point of distributor cap from where it reaches the tip of rotor. From the rotor, the high voltage current then goes to the spark plug points through the distributor cap terminal and high tension lead. When the high voltage surge jumps across the spark plug gap for completing the secondary circuit, a spark is produced which ignites the compressed fuel air mixture in the cylinder.

This system contains of the following main parts :

(*i*) Storage battery,

(*ii*) Ignition switch,

(*iii*) Ammeter,

(*iv*) Ignition coil,

(*v*) Distributor,

(*vi*) Vacuum advance mechanism,

(*vii*) Spark plugs.

(*i*) **Storage Battery.** It is source of electrical energy in the ignition system and supplies electric current for the primary circuit. Its construction has already been explained in the previous chapter.

(*ii*) **Ignition switch.** It is usually a locking switch by means of which ignition system and ultimately engine is kept under lock and key.

Ignition circuit is closed when the key is put in and turned in the lock. The lock is opened and circuit is closed resulting in the flow of current from battery to primary circuit of the ignition system. When ignition lock is closed, the electrical circuit opens and the engine stops working as the ignition is put off.

(*iii*) **Ammeter.** It is a gauge which indicates the flow of electrical current through the system.

(*iv*) **Ignition coil.** It is a type of induction coil which steps up low voltage current of 6 or 12 volts to high voltage current up to about 20,000 volts so that it may jump the spark plug gap. A high voltage is required because the fuel air mixture between the two electrodes of spark plug provides a high resistance to the flow of current. The voltage or pressure of the current must go up very high for pushing the current from the central electrode to the outside casing of spark plug.

Ignition coil consists of two types of windings—primary and secondary, wound around a soft iron lamination core. The primary winding is made up of relatively thick wire than the secondary winding which is of fine wire. The ratio between the number of turns of primary and secondary windings is about 1 to 100. Usually secondary winding is wound around the soft iron core and the primary winding surrounds secondary. The purpose of the *core* is to concentrate the magnetic field.

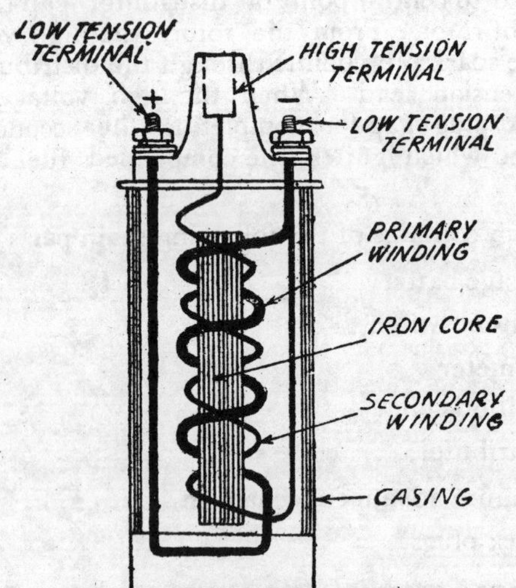

Fig. 11·2. Can type ignition coil.

There are usually two types of ignition coils—can and core types. In *can type coils*, the core assembly is placed in a steel casing which is fitted with a cap of moulded insulating materials. Terminals are provided in the cap. In many coils, the windings are immersed in oil to improve insulation and reduce the effects of moisture. The coils are hermetically sealed to prevent them from absorbing moisture.

In *core type ignition coil*, the primary winding is wound around the iron core made of laminations forming double C with air gaps at the outside of secondary winding. The secondary winding is around the primary winding. This assembly is contained in an insulating material casing which contains terminals.

Working. When ignition switch is turned on and the distributor contact points are closed, the current flows through the primary windings of the ignition coil. A magnetic field is set up around the coil windings. When there is break in the primary circuit due to opening of contact breaker points, the magnetic field collapses. Rapidly collapsing magnetic field induces high voltage in both the primary and the secondary windings. Current flowing into primary windings runs towards contact breaker points and is absorbed by the condenser placed across them and the circuit is completed through earth return. Current of high voltage flowing through the secondary windings is carried from the ignition coil to distributor cap through a high tension lead for onward transmission to spark plug.

Vibrating type ignition coil. In early days of automobile history, this type of coil was quite popular with the automobiles. Popular model 'T' Ford was employing this type of coil.

In this coil, primary winding is wound around the iron case and the secondary winding is above the primary. A set of contact points is connected in series with the primary winding. The contact points are placed in the coil in such a way so that magnetic field from the core may pull the vibrating point away from the stationary point resulting in make and break in the primary circuit. A condenser is placed across the contact points. The whole assembly is contained in a casing made of insulating material.

Working. When current flows through the primary windings, magnetic field is set up which attracts the armature carrying one of the contact points. This causes the contact points to separate and open the primary circuit. When there is break in the primary circuit, magnetic field collapses and high voltage is induced in the windings. The armature is then released and the points close. The primary circuit is again established and the current starts flowing through the primary windings. This cycle is repeated again and again. Each time the magnetic field is built up and collapses, high voltage current is induced in the secondary windings which is further carried to the spark plug to produce spark.

(*v*) **Distributor.** It serves the following purposes in the ignition system :—

(*i*) To operate make and break mechanism.

(*ii*) To distribute high tension current to the proper spark plugs at the correct time.

The distributor consists of a housing, a drive shaft with breaker cam, a governor, a breaker plate with contact points and a condenser, a rotor and a cap. The shaft is driven by the engine cam-shaft directly or indirectly through the oil pump drive shaft. The drive shaft is rotated at half the engine speed in four stroke engines. When the shaft moves, it opens and closes the contact breaker points. The rotor is mounted over the breaker cam which is carried by the drive shaft. The breaker cam usually contains the same number of lobes as the number of cylinders in the engine.

Contact Breaker. It is the mechanism by means of which make and break is affected in the primary circuit of the ignition system for obtaining high tension current.

It consists of a plate and arm having platinum points through which electrical connection is made or separated. The plate is fixed

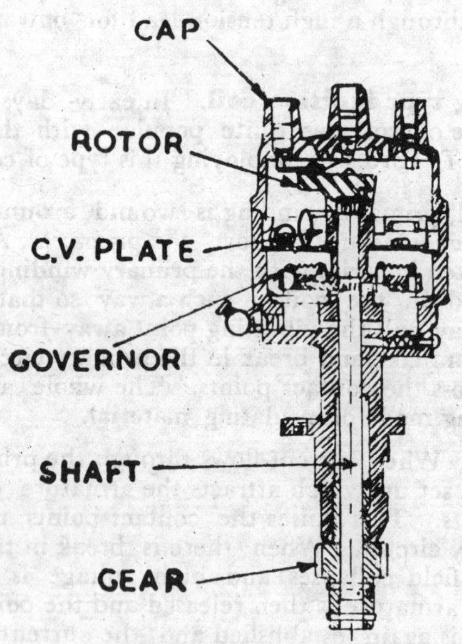

Fig. 11·3. Distributor.

and earthed with the distributor main plate carrying the contact breaker points. The breaker arm is insulated from the distributor shaft breaker cam. It contains fibre lobe which comes in contact with the breaker cam. The breaker arm is operated by the breaker cam. The arm is connected with L.T. wire carrying current from primary windings of ignition coil. The primary circuit is completed through the contact breaker points. The circuit is broken when the points are separated off by the movement of breaker arm which is affected by the breaker cam rotating with the distributor shaft.

Condenser. It is a device which avoids arcing at the contact breaker points at the time of break in the primary circuit and thus saves them from pitting or burning away.

It is made of alternate sheets of metal foils and insulation such as mica or special paper and is housed in a cylindrical casing which is hermetically sealed.

The condenser is fitted parallel to the contact breaker points and provides a reservoir for the current induced in the primary circuit at the time of break. It also gives more rapid break to the circuit when C.B. points open which results in a higher voltage in the secondary circuit.

Rotor. It is an insulated arm which contains a brass tip connected with spring steel leaf at the top. It rotates with the breaker

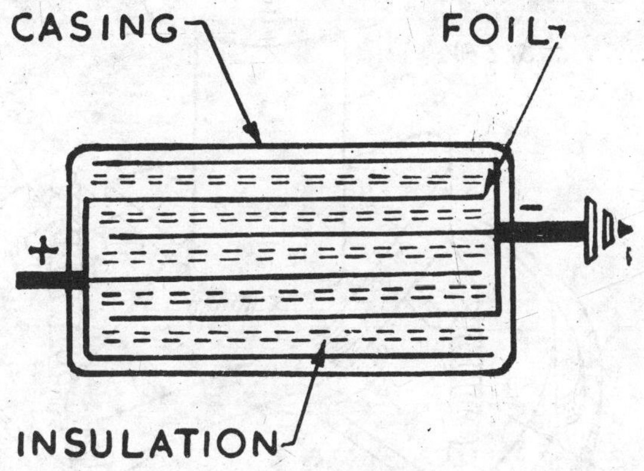

Fig. 11·4. Condenser.

cam at distributor drive shaft and provides connections from the central point to the different segments of the distributor cap by rotation.

Distributor cap. It is a bakelite cap which covers the distributor at the top and provides connections for the high tension leads.

Governor or Centrifugal advance mechanism. It is a centrifugal governor which automatically advances and retards ignition according to load and speed of the engine for its efficient working.

It consists of two flyweights and springs contained in the distributor housing, beneath the main plate which carries contact breaker points. The flyweights are carried by the distributor drive shaft through a base plate which is fixed with the drive shaft. The flyweights are fitted on the pins over the base plate. The pins act as fulcrum. The flyweights are attached to the breaker cam through the springs. The breaker cam is connected with the distributor drive shaft through the springs, flyweights and the base plate, otherwise breaker cam is free to move over the end pin of distributor drive shaft.

The flyweights are thrown outward against the tension of the springs due to centrifugal force developed by the rotating distributor

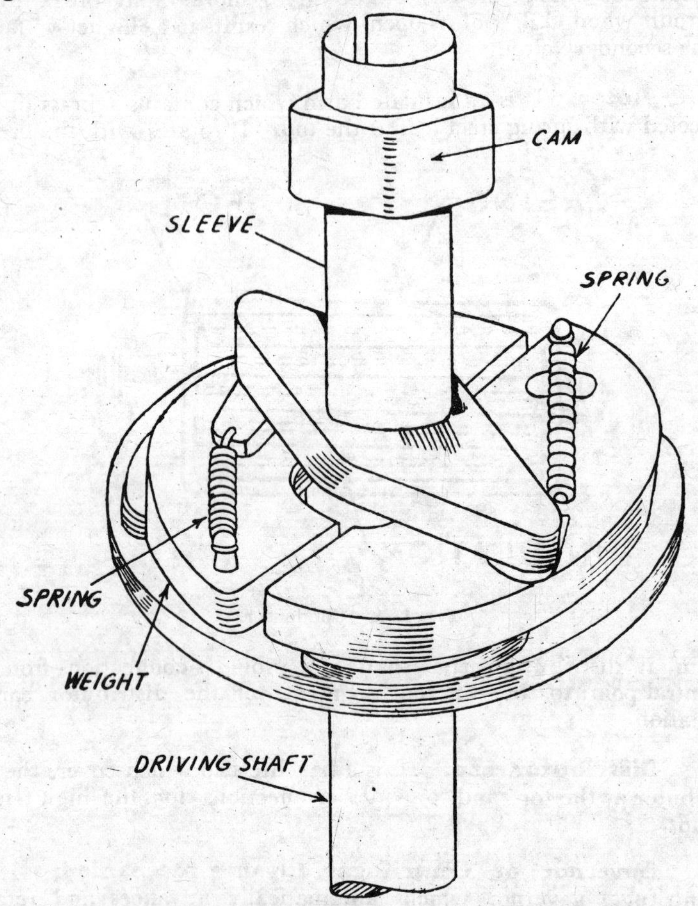

Fig. 11·5. Centrifugal Governor (Advance Mechanism).

shaft. The outward movement of flyweights tends the breaker cam to an advanced position. In some distributors, breaker plate is rotated around the axis of distributor shaft. Ignition advance due to this mechanism is subject to the centrifugal force developed by the distributor shaft. The faster the speed of the distributor shaft, the greater the centrifugal force and greater the movement of fly-weights resulting in greater ignition advance.

(*vi*) **Vacuum advance mechanism**. During part throttle operation, a small amount of fuel air mixture is sucked in and the compression pressure is low in the engine. With lower pressures, the mixture does not burn as quickly as it should. In order to obtain maximum efficiency under such conditions. the spark should be advanced more than that obtained by the centrifugal governor during low engine speed. The additional ignition advance is obtained by means of vacuum advance mechanism.

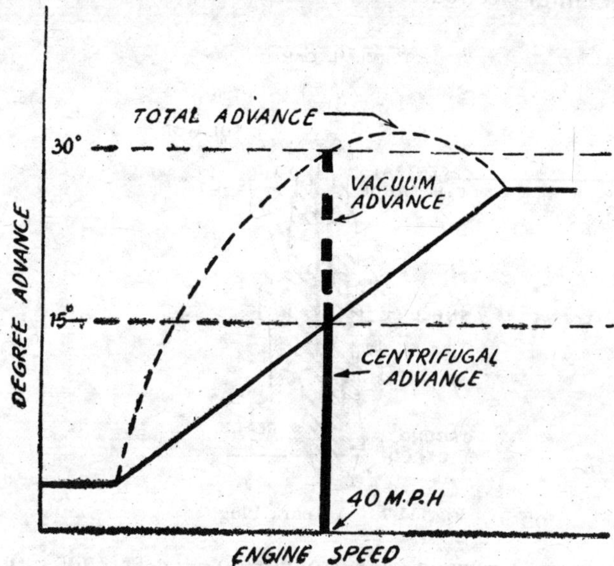

Fig. 11·6. Centrifugal and Vacuum advance curves.

This mechanism contains a spring loaded diaphragm which is connected with the distributor by means of a link and clamp. The diaphragm is contained in a housing, the spring loaded side of which is air tight and is connected with the carburettor by means of pipe line. Connection at the carburettor is made at a point on the atmospheric side of fully closed throttle so that engine suction does not affect the diaphragm during idling speed when additional advance is not required.

When the throttle is opened, suction of the engine affects the diaphragm through the suction line which causes the diaphragm to deflect. The movement of the diaphragm is transmitted to the distributor through the linkage. In some cases, the distributor is rotated, on its mounting while in other cases, the breaker plate is moved which results in ignition advance.

When distributor body is rotated opposite to the rotation of rotor, ignition is advanced while when the body is rotated in the direction of rotation of rotor, ignition is retarded.

Spark plugs. Spark plugs are fitted at the engine combustion chambers to provide spark for the ignition of compressed fuel air mixture contained therein. A spark plug consists of the following main parts :

(i) Metal base,

(ii) Ceramic insulator,

(iii) Central electrode.

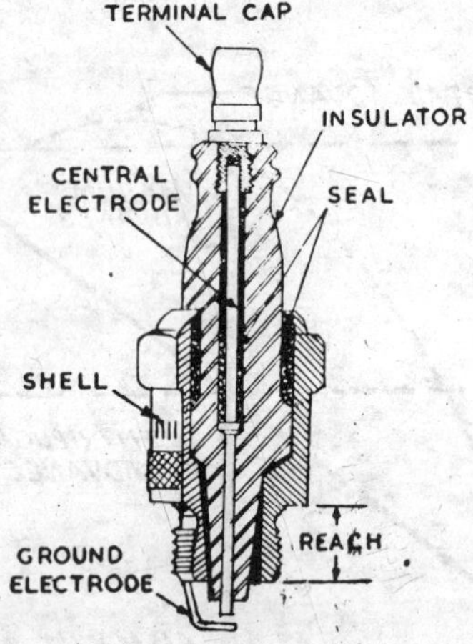

Fig. 11·7. A Spark Plug.

The central electrode is surrounded by the ceramic insulator which is sealed in the metal base. A small tip is welded with the screwed end of the metal base and provides a gap between the central electrode and the tip. The complete assembly known as spark plug is screwed into the cylinder head of the engine and the lower end thus reaches the dome of combustion chamber. High tension lead from the distributor is connected at the outer end of central electrode through a screwed cap.

When high voltage surge travels over the central electrode, it has to cross or jump over the gap of spark plug. While so doing, a spark is liberated for the ignition of fuel air mixture in the combustion chamber. Shape of the electrodes forming gap, the conductivity of the gases in the gap, temperature, pressure, and fuel-air mixture ratio existing in the gap affect the voltage required to jump a certain gap which differs from plugs to plugs in different engines.

Spark plugs are available in different sizes. The most common sizes are 10 mm., 14 mm. and 18 mm. This indicates the diameter of screwed portion of metal base.

According to the construction, spark plugs are of two types : *permanently sealed* and *split type*. Permanently sealed spark plugs cannot be split up into different parts unless their seal is broken. The split type spark plugs can be split up into two parts ; metal base and central electrode with insulation. Both these parts are held together through a screwed holding down cap and are sealed by means of a gasket. In some spark plugs there is more than one tip attached to the metal base.

The spark plugs are required to work under varying engine conditions of load and speed. A spark plug must be capable to work in varying temperatures, mixture ratios, and fuel qualities apart from engine load and speed.

A spark plug can work satisfactorily only if the surface of the insulator nose along with the tip of central electrode maintains a temperature between 500°C to 850°C under all operating conditions of the engine. If the temperature of the insulator nose is lower than 500°C for a certain minimum time, carbon and oil particles can settle down on its surface to such a degree that initially the spark is weakened and finally all the current flows through the lower resistance on the insulator surface instead of jumping over the electrode gap. This results in the spark plug fouling and ignition failure.

If the temperature of the insulator nose or tip of the central electrode rises above 850°C, combustion of the fuel air mixture will be started by these yellow glowing portions of the plug before the ignition spark is provided by the spark plug. This results in *pre-ignition* (ignition starting before the scheduled ignition time), and drop of engine power. The engine life is also shortened due to abnormal mechanical stresses on crankshaft and connecting rod bearings and overheating of exhaust valves.

The spark plugs are classified according to their *heat value* which is determined by measuring the temperature of the core nose. **Higher the heat value**, colder the plug and lower the heat value, hotter the plug. According to heat value, the spark plugs are of two types : cold plugs and hot plugs.

Cold spark plugs. Spark plugs having short insulator nose are known as cold spark plugs because their heat value is high. These plugs remain high in the combustion chamber and their insulator noses are not heated up sufficiently by the combustion gases as heat is transferred quickly from insulator nose to cylinder head through the metallic base of the plug. Cold spark plugs are generally used in heavy loading and high speed operation where high temperatures are encountered.

Hot spark plugs. If heat absorbing surface of the insulator nose of a spark plug is large due to long nose, the spark plug is known as hot plug. Hot spark plugs have low heat value. Lower speed, intermittent loading and colder operating conditions of the engine require a hotter plug.

4. Transistor assisted ignition system. The output of the conventional ignition system depends upon the following factors : (*a*) the amount of current in the primary circuit and (*b*) the period the contact points remain closed.

The period of C.B. points remaining closed becomes increasingly short as the engine speed increases. The amount of current

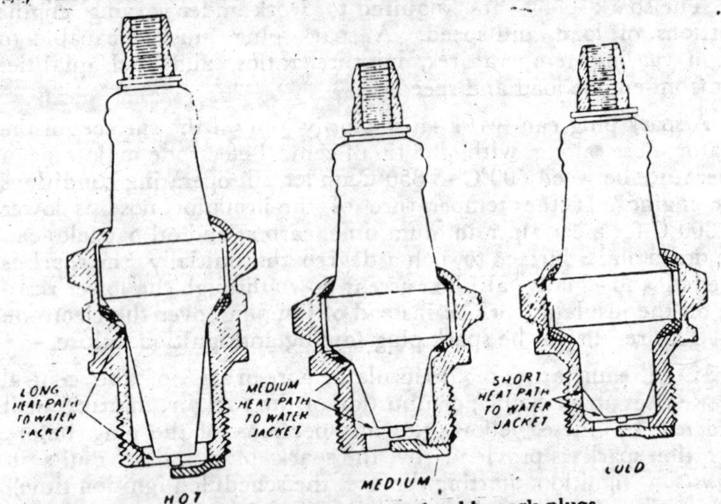

Fig. 11·8. Different hot and cold spark plugs.

that can be interrupted by the conventional tungsten contact points is about 5 amperes for reasonable contact life. The transistor assisted ignition system removes these drawbacks of conventional ignition system and gives easier starting, with better running at high speeds.

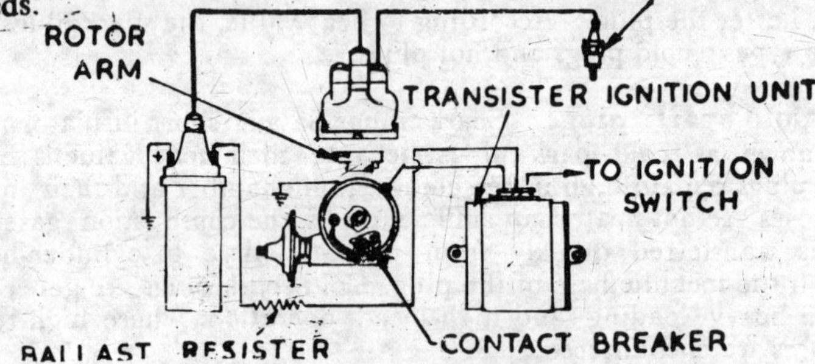

Fig. 11·9. Transistor assisted ignition system.

In transistor assisted ignition system, a special high voltage transistor relieves the contact breaker points from the duty of making and breaking the primary circuit. They only do the light duty of turning the transistor 'on' and 'off' at the proper time in relation to engine timing. Thus arcing at the contact points and the need for a condenser are eliminated. The controlled current is thus increased. It can be reduced also to such a low value that the life of the contact breaker points is greatly extended.

The transistor assisted ignition system contains a transistor ignition unit, in addition to other components of conventional ignition system, except condenser. These units can be easily installed

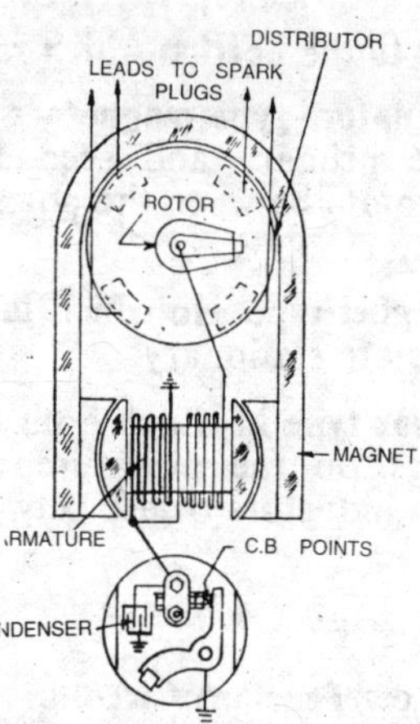

Fig. 11·10. Magneto ignition system.

into the conventional ignition system, to convert them into transistor assisted ignition system.

5. **Magneto Ignition system,** does not require storage battery to provide spark in the engine. Magneto, in the system, itself generates current, and distributes it to the engine cylinders. This system is of the following two types :

(*a*) **Low tension magneto ignition system**, in which a low voltage is stepped up to a high voltage, by means of a separate ignition coil.

(*b*) **High-tension magneto ignition system,** in which the high tension magneto itself produces voltage of sufficient value.

The main advantages of this system are :

(*i*) They do not require any battery or other source of current.

(*ii*) The intensity of the generated voltage does not decrease with the engine speed but increases. The only drawback is that at low speed, it does not provide current of high intensity.

6.. **Magneto.** It is a self-contained device, which generates current, steps it up to high voltage and distributes it to the various cylinders of the engine at the correct time. It works on the principle that "Electricity is produced by revolving a coil of wire, in a magnetic field" produced by permanent magnets.

7. **Classification of Magnetoes.** Magnetoes can be classified differently as below :

(*a*) **With regard to the type of current produced** :

(*i*) **Low-tension magnetoes** which produce a low voltage and require an external coil to step up voltage.

(*ii*) **High tension magnetoes** which incorporate a coil for stepping up low voltage to high voltage.

(*b*) **With regard to the portion which revolves :**

(*i*) **Rotating armature type magnetoes** in which the armature carrying the primary and secondary windings and the condenser, rotate between the poles of a stationary horse-shoe magnet.

(*ii*) **Rotating magnet type** in which the magnet revolves and the windings are stationary.

(*iii*) **Polar inductor type** in which both the magnet system and the windings on the armature core are stationary, but the soft iron inductors rotate between the poles of armature core.

8. Magneto circuits :

(*a*) **Primary (Low-Tension) Circuits.** One end of the primary winding is earthed inside the magneto and the other end is conneted, internally or externally to one contact point. The circuit is completed through the other earthed contact point.

(*b*) **Secondary (High Tension) Circuit.** The current generated passes from the carbon brush to the distributor rotor. It jumps the gap to the distributor segment, which is connected through H.T. lead to the spark plug. It provides spark across the plug gap and returns to the other earthed end of the secondary winding through the spark plug shell, engine and magneto body.

9. High-Tension Magnetoes, generate high tension electricity within itself and distributes it to the various spark plugs.

The armature, along with primary and secondary windings, revolves in the magnetic field set up by the permanent magnets. As the windings or coils pass through the magnetic field, current is generated and when it reaches its maximum, the C.B. points, which are fitted in the primary circuit, open. At this, a current is induced in the secondary windings. Because number of turns of secondary windings are more than the primary windings, so the voltage is multiplied. The end of the secondary winding is connected to a collector ring from where the high tension current is collected by the high tension lead through a carbon connection and fed to the distributor for onward transmission to the spark plugs.

A Safety gap, across the terminals of secondary winding and the ground protects the secondary winding insulation from excessive voltage effects, which would occur due to incomplete circuit, if a high tension lead is disconnected from a spark plug.

10. Disadvantages of rotating armature type magneto :

(*i*) Centrifugal force affects the windings.

(*ii*) There are many rubbing contacts to collect the current from the moving armature.

(*iii*) Only two sparks per revolution are produced.

11. Rotating magnet type magnetoes. These magnetoes are used in small I.C. engines and are known as *flywheel magnetoes.* In this type, the magnet is located on the outer rim of the fly wheel which revolves around the stationary ignition coil, condenser ar l C.B. point assembly. As the ends of the magnet pass by the pole pieces, an alternating magnetic flux is set up through the ignition coil and current is generated in the primary circuit during the period that the C.B. points are closed. When the primary currrent is at its highest peak, the C.B. points are opened through cam. When there is break in the primary circuit, the magnetic field collapses, which in turn induces a high voltage in the secondary voltage.

12. Comparison between Battery-Coil Ignition System and Magneto Ignition System.

Battery Coil Ignition system	*Mogneto Ignition system*
1. Batttery is must.	1. No battery needed.
2. Current for primary circuit is obtained from the battery.	2. The required electric current is generated by the magneto.
3. A good spark is available at spark plug at low speed.	3. During starting, quality of spark is poor due to low speed.
4. Starting of engine is easier.	4. Engine starting is rather difficult.
5. Impossible to start engine when battery is discharged.	5. No such difficulty as battesy is not required.
6. Efficiency of the system falls with the falling of spark intensity as engine speed rises.	6. The intensity of spark keeps on improving as the speed goes on increasing. The efficiency of the system thus improves as the engine speed rises.
7. Mostly emplyed in gasoline cars, buses and trucks.	7. Used on high speed engines provided in racing cars, motor cycles, scooters, etc.
8. Occupies more space.	8. Occupies less space.

13. Ignition Timing. When the C.B. points open, high voltage surge flows from the ignition coil to the spark plug to produce spark. Connecting of distributor with the engine camshaft after ascertaining the position of C.B. points in relation to the piston position in the cylinder is known as *setting of ignition timing*. When piston in cylinder number one of the engine is near compression top dead centre and the timing mark either on the flywheel rim or crank pulley is against the pointer, the position of contact breaker points is fixed as "*ready to open*". This method fixes the initial ignition timing.

Ignition timing must be accurately set so that the spark occurs in the combustion chamber at the correct time. Ignition being started early shall result in back pressure at the piston while the piston is still rising up due to inertia of the flywheel. This shall set up a noise in the engine known as "*pinking*". If the spark occurs too late, this will lead to engine overheating and loss in power output.

The initial timing would not be suitable for different engine speeds at different loads. At higher speeds, the ignition should be started earlier, as the time left for ignition is short. The ignition timing should, therefore be altered to suit the varying conditions of engine operation, using automatic advance and retard mechanisms.

The main factors which affect the ignition timing of an engine are :

(*i*) **Engine load**. At the time of less load on the engine combustion is slower and more ignition advance is required ; and vice versa.

(*ii*) **Engine speed**. At low speed and load, combustion is slower and more ignition advance is required. At low speed and full load, combustion is faster and less ignition advance is needed. At high speed and full load, combustion is slower and more ignition advance is required.

(*iii*) **Quality of fuel**. If the fuel is of high octane value, the combustion shall be slower and more ignition advance is needed and vice versa.

(*iv*) **Compression Pressure or ratio**. If the compression pressure or ratio is low, combustion shall be slower and more ignition advance is required and vice versa.

(*v*) **Fuel-air mixture ratio**. If the fuel-air ratio is high or the mixture is lean, combustion shall be faster and less ignition advance is required and vice-versa.

(*vi*) **Engine temperature**. When the engine is cold, combustion is slower and more ignition advance is required and vice versa.

(*vii*) **Cylinder bore**. Larger the bore, slower the combustion and more the ignition advance required.

Apart from the above factors, shape of the combustion chamber, location of spark-plug, fuel-distribution to the cylinders, carbon deposit in the combustion chamber, volatility of the fuel, throttle opening and combustion rate of the fuel also affect the ignition timing of the engine.

14. Firing Order. It is the order in which the different spark plugs initiate firing in different cylinders of the engine. H.T. leads from the distributor are connected with the spark plugs at the different cylinders, according to the firing order of the engine. Firing order differs from engine to engine. Probable firing orders for different engines are as under :—

(*i*) **Three Cylinder engine** : 1-3-2

(*ii*) (*a*) **Four Cylinder engine.** (*a*) 1-3-4-2
(in line) (*b*) 1-2-4-3

(*b*) **Four Cylinder horizontal apposed engine :**
(Volkswagen engine) 1-4-3-2

(*iii*) **Six cylinder in line engine.** (Cranks in 3 pairs)
(*a*) 1-5-3-6-2-4, (*b*) 1-4-2-6-3-5, (*c*) 1-3-2-6-4-5,
(*d*) 1-2-4-6-5-3.

(*iv*) **Eight cylinder in line engine.** 1-6-2-5-8-3-7-4.

(*v*) **Eight Cylinder V-shape engine.**
(*a*) 1-5-4-8-6-3-7-2 (*b*) 1-8-4-3-6-5-7-2
(*c*) 1-6-2-5-8-3-7-4 (*d*) 1-8-7-3-6-5-4-2
(*e*) 1-5-4-2-6-3-7-8.

Cylinder No. 1 is taken from front of the in-line engines whereas in V-shape, front cylinder on right or left bank is considered cylinder No. 1 for fixing H.T. leads, according to engine firing order.

15. Electronic Ignition system is also known as *Breakerless Ignition System*. In this system, there is no contact-breaker (C.B.) point, cam and sliding block etc. Owing to wear of these parts in conventional ignition system, the gap between the points continually decreases. This results in gradual change in the dwell (cam-angle) and ignition-timing, which affect engine performance badly. At higher engine speeds, *floating* and *bouncing* of movable breaker point arm causes the dwell decrease, which reduces the time to flow current in primary circuit, resulting in lower output voltage of ignition coil and poor spark at the plug. As C.B. point gap decreases it becomes far more difficult for the points to break the circuit. Due to wear of rubbing block, the cam has to travel farther, before it starts to open the points, resulting in more and more retarded ignition-timing. Sometimes the C.B. point gap becomes so small that the engine misfires and shortly after, refuses to run.

In electronic ignition, distributor cam, breaker plate, C.B. points and condenser, have been replaced by an armature (trigger wheel, reluctor etc.) and an electronic control module (amplifier), It creates magnetic pulses, which trigger an *electronic control unit* to create high voltage in a special coil.

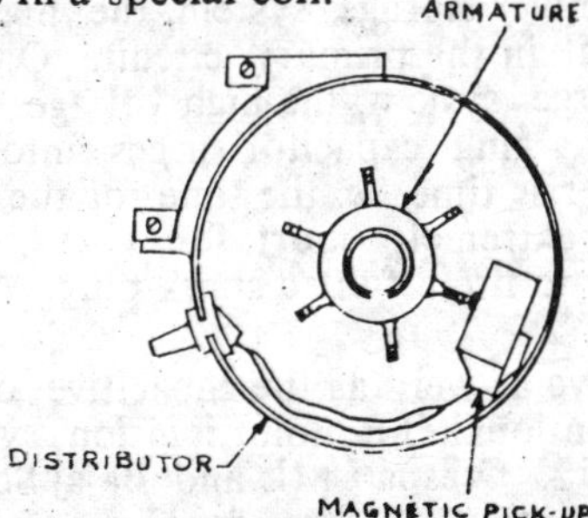

Fig. 11·11. Position of armature and magnetic pick up in the distributor.

When ignition switch is turned on, current flows through the primary windings of ignition coil. As each tooth (leg) of revolving armature approaches the magnetic pick up, it induces a voltage which signals the *electronic control module* (amplifier) to turn off the primary circuit of ignition coil.

A timing circuit in the *control module* turns the circuit 'on' again, after collapsing magnetic field in the ignition coil. The collapsing magnetic field induces high voltage in the secondary windings of ignition coil. Thus high voltage is produced, each time the magnetic field in ignition coil is built up and collapses, resulting in spark through spark plug for igniting compressed fuel air mixture in the engine.

In scooters, the *magnetic pick up* is positioned on the stator plate of flywheel magneto. The function of C.B. point is taken over by the magnetic pick up and that of cam is performed by the extended pole-shoes and specifie positioning of magnetic pick up, which decides *ignition timing*.

In this system, output voltage is higher (upto 35,000 V) owing to which wider plug gap is maintained, resulting in lesser troubles.

16. Capacitive Discharge (Electronic) Ignition System (C.D.I. System). The conventional breaker point ignition system is known as *inductive storage system* because energy stored in the magnetic field of ignition coil is used to create the spark. The same concept, used with a transistor for switching the primary circuit current is called *transistor-inductive storage system*. Here an ignition coil is used as a high voltage transformer. A large capacitor is used for the storage of energy. The capacitor is charged to about 300 Volts and when it is triggered, it discharges its energy into primary windings of the ignition coil, which in turn steps up the voltage to above 30,000 Volts in the secondary windings, to create blue fat spark at spark plug.

In order to switch energy from the capacitor to ignition coil, a solid state switch, known as *thyristor* is employed, which is triggered 'on' by a voltage impulse from magnetic pick-up coil.

In an inductive storage system, the high voltage is produced when there is break in the primary circuit. On the other hand, in a capacitive discharge system, the high voltage is produced when the current from discharging capacitor surges into primary windings of ignition coil. The rise time *i.e.* the time for the high voltage to reach its full value, is extremely short for C.D.I. system. Hence, this system is capable to fire a fouled spark plug.

The inductive as well as the capacitive discharge systems are better than conventional C.B. point ignition system as shown by the curves in Fig. 11·12. Vespa LML and Bajaj Scooters employ C.D.I. system, because output voltage does not fall too much at higher speeds. It is almost uniform. In the case of breaker point ignition the level of output voltage falls down as the engine speed increases.

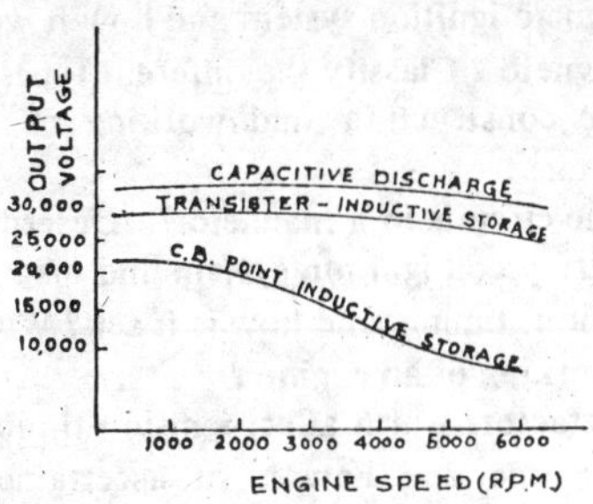

Fig. 11·12. Ignition coil maximum output voltage versus engine speed.

17. Advantages of Breakerless Electronic Ignition System (E.I.S.) :

(*i*) Periodic checks of ignition timing, are no longer required, once the basic timing has been set.

(*ii*) No change in the ignition timing owing to absence of traditional mechanical devices such as C.B. points, cam, sliding block etc. which are exposed to wear and tear.

(*iii*) Produces a higher tension peak (maximum output voltage) which is attained in a very short time and with a very limited discharge, resulting in regular engine running, even with dirty/foul spark plug and wider plug gap.

(*iv*) Better starting with cold engine.

(*v*) Longer life of spark plug, due to less electrode wear.

(*vi*) Less possibility of arcing at spark plug.

(*vii*) Longer life due to stable timing.

QUESTIONS

1. What is an ignition system and which are its different types ? Discuss.

2. Describe the battery coil ignition system with the help of a suitable diagram.

3. Describe the different methods of obtaining automatic ignition advance. Discuss its necessity.

4. Describe the construction and working of one popular type of ignition coil.

5. Describe the following :–

   (*a*) Ignition swich. (*b*) Vibrating type ignition coil.
   (*c*) Vaccum advance and retard mechanism.
   (*d*) Contact Breaker. (*e*) Rotor.
   (*f*) Condenser.

6. Describe the construction and working of ignition distributor.

7. What is a spark plug ? What role it plays in the working of ignition system ?

8. Which are the different types of spark plugs ? Describe them.

9. What is heat value of a spark plug ? How is it determined ?

10. What is a hot plug and cold plug ? What is the effect of too hot and too cold plugs in the working of an engine ?

11. Describe the transistor assisted ignition system.

12. What is magneto ignition system and how it works ?

13. What is magneto ? Classify the different types of magnetoes.

14. Describe the construction and working of high tension rotating armature type magnetoes.

15. Which are the circuits in a magneto ? Deseribe them.

16. Compare battery coil ignition system and magneto ignition system.

17. What is ignition timing and how is it set ? What is the effects of too early ignition in the working of an engine ?

18. Describe the factors which affect ignition timing in an engine.

19. What is firing order and how is it ascertained ? Describe the firing orders for the following engines :—

(*ii*) Four cylinder in the line engine.
(*ii*) Volkaswagen engine.
(*iii*) Six cylinder in line engine.
(*iv*) Eight cylinder in line engine.
(*v*) Eight cylinder-V-shape engine.

20. Whst do you know aboui the following

(*a*) Pinking (*b*) Split type spark plug.
(*c*) Magnetic field. (*d*) Ammeter.
(e) Pre-ignition.

21. Explain electronic ignition system.

22. What isC.D.I. system ? Explain.

23. Define disadvanıages of contact-breaker point ignition system.

24. Compare breaker point and breakerless ignition systems.

25. Point out advantages of electronic ignition systems.

26. Explain the followings :—

(*i*) Inductive storage system.
(*ii*) Capacitive discharge system.
(*iii*) Rise time.
(*iv*) Bouncing and rolling of breaker arm and their effect in engine performance.
(*v*) Spark plug gap size with respect to electronic and conventional ignition systems.

12

Transmission System

1. **Transmission system.** It is the system by means of which power developed by the engine is transmitted to the road wheels to propel the vehicle.

2. **Transmission system requirements.** Since power developed by the engine is to be transmitted to the road wheels so the transmission system of an automobile should be in accordance with the engine characteristics which are as under :

(*a*) External energy is applied to crank the engine for starting.

(*b*) The maximum torque produced by the engine is small as compared to steam engine or traction electric motor of the same maximum horse power.

(*c*) It is comparatively high speed engine.

For starting an engine, it is to be cranked manually through a hand crank or through a self-starter for which electrical energy is obtained from the storage battery. The external energy applied to crank the engine for starting shall not be much and with which it will not be possible to crank the engine if any load is applied to it. Hence the engine should be free from any load so that engine could be cranked by the external energy applied to it.

The engine is to be connected to the transmission line for transmitting the power to the wheels. Therefore, there should be such type of connection between the engine and the rest of transmission system by means of which engine could be connected and disconnected with the transmission line, smoothly and without shock so that the vehicle mechanism is not damaged and the passengers do not feel inconvenience.

The torque developed by the engine is transmitted to the drive wheels through the transmission system to propel the vehicle. When the torque reaches the road wheels, it acts as tractive effort or propulsive force to drive the wheels. While moving off from rest or

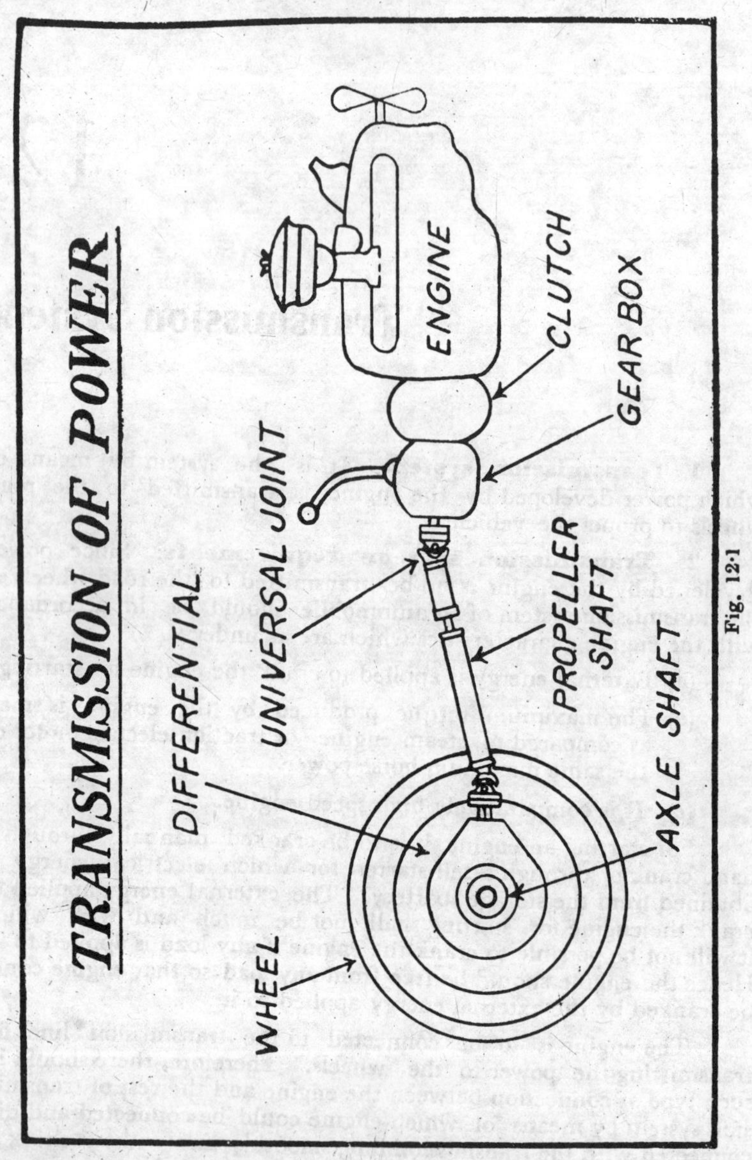

Fig. 12·1

climbing hills or running through rough or slippery roads, a large tractive force is required for rapid acceleration. This requires that considerable leverage should be introduced between the engine and the driving wheels so that the small torque available at the engine could be multiplied to produce a large tractive effort. The leverage thus applied should be variable to meet the different conditions of load, speed and road.

Since the engines employed in automobiles are high speed, so a speed reduction is necessary as the automobiles are not generally expected to travel much more than 60 m.p.h. If the road wheel diameter is 30 inches, the maximum speed of the wheels will be about 700 r.p.m. If the maximum speed of the engine is 3500 r.p.m., the ratio between the engine speed and road wheel speed would be 5 to 1. This means that a gear reduction to this value shall be needed. This ratio varies with weight of the vehicle and the size of engine installed in it.

The nature of transmission system is affected by the type and drive of the vehicle, location of the engine and suspension system in the vehicle. Transmission system for a four wheeler shall be different from that of a two or three wheeler. In a four wheeler, the power usually flows to the driving wheels in one line to which the wheels are provided at a perpendicular axis. Here the problem shall be to change the path of power flow at right angles. In addition to this, some arrangement must exist by means of which the outer wheel should travel more distance than the inner wheel within the same time for negotiating a turn or moving over upheaval road.

If the power is to flow to all the wheels of the vehicle, transmission requirements would be different than that if the power is to be transmitted to only two wheels. In four or six wheel drive vehicles, a power transferring device is required to transmit power in the opposite direction.

The location of the engine has major effect on the transmission system. It differs from vehicle to vehicle. Engine location in different vehicles is as under :

(*a*) Engine at the front.

(*b*) Engine at the rear.

(*c*) Engine in the centre.

The drive of the vehicle changes in relation to engine location. If the engine is located at the rear and the drive is given to the rear wheels, transmission arrangement would be different to those vehicles in which the engine is at the front and drive is given to the rear wheels. If the engine is at the front and drive too is given to the front wheels alone, transmission shall be of different nature.

Suspension system of the vehicle also affects the nature of transmission system. How the driving thrust, torque reaction and brake effects are to be carried to the frame, changes the design of suspension system and this in turn affects the transmission arrangement. If the vehicle is having independent suspension at all the four wheels, transmission requirement shall be different.

Mostly the rear suspension contains conventional arrangement. The drive too is given to the rear wheels in most cases. When the road springs flex due to road irregularities, the drive axle to which the drive wheels are attached, also moves back and forth. This action requires that the drive line should increase and decrease in length as per requirement. At the same time the drive axle moves down or is raised up and down due to road pits and falls and the angle of drive line is changed at which it is transmitting power.

In view of the above discussion, the transmission should fulfil the following requirements :

(*a*) It must provide a means of connection and disconnection of engine with the rest of power train without shock and smoothly.

(*b*) It must provide a varied leverage between the engine and the drive wheels.

(*c*) It must provide a means to transfer power in the opposite direction.

(*d*) It must enable the power to be transmitted at varied angles and varied lengths.

(*e*) It must enable speed reduction between engine and the drive wheels in the ratio of about 5 to 1.

(*f*) It must enable the power flow to be diverted at right angles.

(*g*) It must provide a means to drive the driving wheels at different speeds when needed.

(*h*) It must bear the effects of torque reaction, driving thrust and braking effort effectively.

The above requirements are fulfilled by the following main units of the transmission system :

(*i*) **Clutch**. It provides a smooth means of disengagement and engagement between the engine and the remainder of transmission system.

(*ii*) **Gearbox**. It provides varied leverage to the drive wheels.

(*iii*) **Transfer case**. It helps in diverting power flow in the opposite direction.

(*iv*) **Propeller Shaft and Universal Joints**. Propeller shaft enables the drive wheels to move back and forth due to flexing of road springs. It enables the power to be transmitted to the road wheels at varied lengths and varied angles owing to slip joint and universal joints contained in it.

(*v*) **Final Drive.** It diverts the power at right angles towards the driving wheels and provides a final gear reduction between the engine and the driving wheels.

(*vi*) **Differential.** It allows the outer wheel to run at greater speed than the inner one while negotiating a turn or moving over upheaval road.

(*vii*) **Torque Tube.** In torque tube drive, all the driving thrust, braking effort and torque reactions are taken by the torque tube.

(*viii*) **Road Wheels.** The road wheels provide last leverage to the engine power to act as tractive effort at them. More the diameter of a road wheel, more the leverage it would provide.

3. Type of transmission systems :

(*a*) Electrical and electromagnetic transmission system.

(*b*) Hydraulic transmission system.

(*c*) Mechanical transmission system.

Electric transmissions have been used for motor vehicles but nowadays they have become obsolete. Hydraulic transmissions are now becoming common in the shape of fluid coupling, torque converter, automatic transmission etc. Mechanical transmissions can be divided into the following classes :—

(*a*) Clutch, gearbox and live axle transmission.

(*b*) Clutch, gearbox and dead axle transmission.

(*c*) Clutch, gearbox, axleless transmission.

(*a*) **Clutch, gearbox and live axle transmission.** Majority of the vehicles employ this type of transmission system. Power deve-

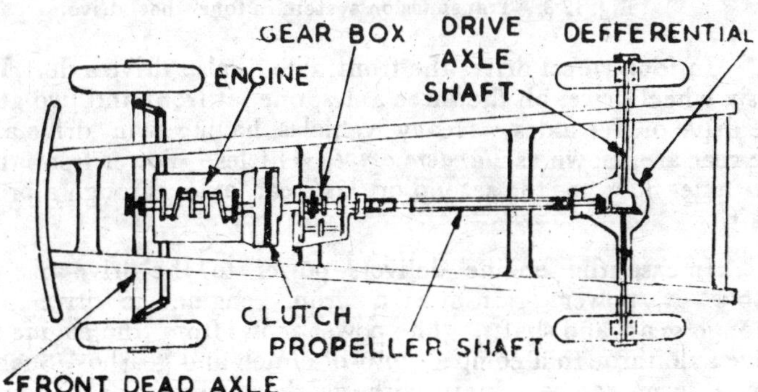

Fig. 12·2. Power flow in two wheel drive (4×2).

loped by the engine flows to the driving wheels through clutch, gearbox, propeller shaft, universal joints, final drive, differential, and half shafts. The live axle acts as a drive axle and houses final drive, differential and half shafts. Clutch assembly is attached to engine flywheel which acts as a driving member for the clutch. The gearbox is bolted to the clutch housing. The drive axle is placed at right angle to the side members of chassis frame through road springs. The wheels are mounted over the ends of the drive axle where brake drums are connecting the outer ends of axle shafts. Drive pinion of the drive axle is coupled with the rear end of the propeller shaft. The front end of the propeller shaft is connected with the gearbox output shaft coupling. The type of arrangement is applicable in two wheel drive vehicle.

In the case of four wheel drive, a transfer case is located next to the gearbox from where the power flows to the front and rear wheels through separate propeller shafts. In certain cases the transfer case is bolted to the gearbox and power flow from gearbox to transfer case is through gears and shafts. In some cases, the transfer case is located away from the gearbox and is then quite independent unit. The power then flows from gearbox to the transfer case through a small propeller shaft.

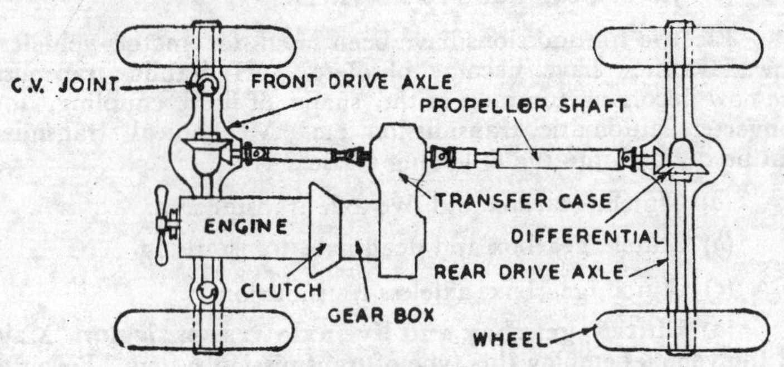

Fig. 12·3. Transmission system in four wheel drive.

In four wheel drive, the front axle is also drive axle. In case of six wheel drive, all the three axles, one at front and two at rear, are drive or live axles. Heavy vehicles having twin drive axles at the rear are known as *tandem drive* vehicles. Power from the first rear axle flows to the second or last rear axle through a propeller shaft.

In case the engine delivers power to the drive axle placed below it, power transmission from engine to drive axle is through gears and shafts. The power flows from the engine to the drive axle through a compact unit of clutch and gearbox. Such axles which form compact unit with gearbox, clutch and engine are known as *power packed axles*. Volkswagen and D.K.W. cars contain this type of arrangement which dispenses with propeller shaft. Absence of propeller shaft gives chance to build deep body.

To sum up, this type of transmission can be split up into the following classes:—

(*i*) Engine at front and drive to rear wheels on a live axle (two wheel drive);

(*ii*) Engine at front and drive to all the four wheels mounted on front and rear live axles (four wheel drive);

(*iii*) Engine at front and drive to front live axle and two rear live axles (six wheel drive);

(*iv*) Engine at front or rear and power transmission to front or rear drive axle placed below the engine. (Power pack axles).

(*b*) **Clutch, gearbox and dead axle transmission.** This type of transmission uses chain and sprockets to transmit power from

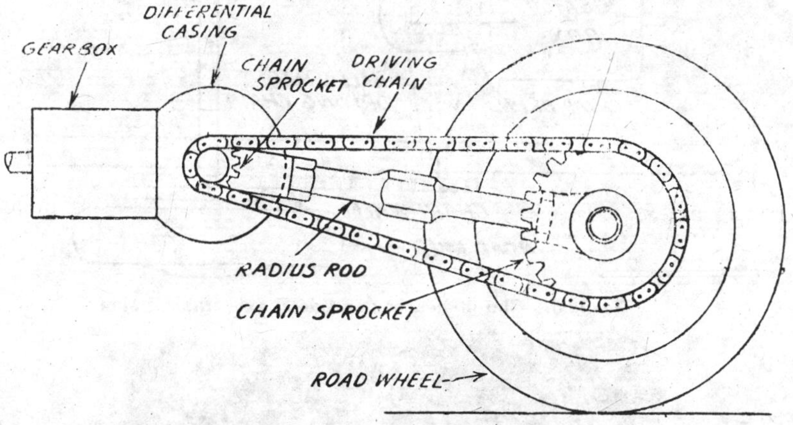

Fig. 12·4. **Chain drive.**

the transmission unit to dead axle over which the driving wheels are mounted. This arrangement is known as *chain drive* and was quite popular with the early motor vehicles.

In a four wheeler, as shown in the sketch, power developed by the engine flows through the clutch, gearbox, final drive, differential, half shafts, drive sprockets, and then through chains to sprockets fixed at the dead axle which rotates like a shaft to move the driving wheels. In this arrangement, gearbox, final drive, differential, half shafts and drive pinions are housed in a suitable housing which is attached direct to the gearbox. The dead axle over which the drive wheels are mounted, is rotated through the chain drive from transmission unit drive sprockets.

In a three wheeler, the chain drive either drives the rear dead axle similar to the chain drive applicable to four-wheeler or the drive is given to the front single wheel through chain and sprockets. The latter arrangement is applicable in Bajaj Tempo, three wheeler. In this arrangement, the drive goes direct from the gearbox to the front single wheel as in certain auto cycles.

In motor cycles and scooters, being two wheelers, no differential is needed. The machine is turned round a corner by turning

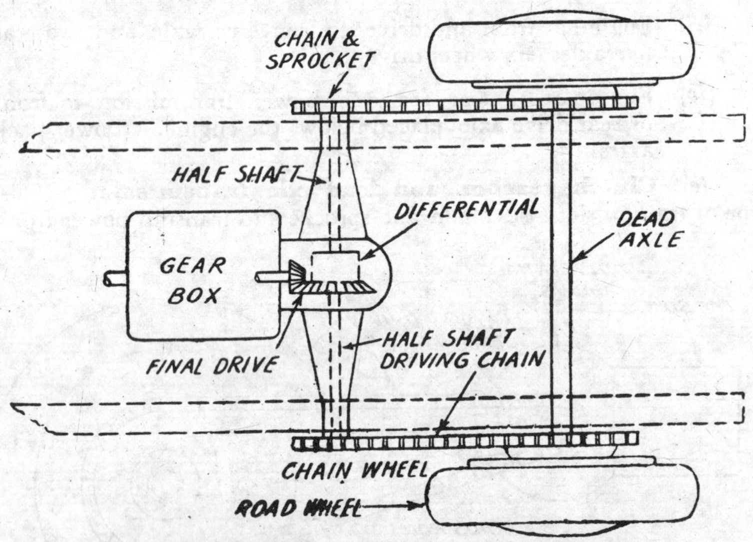

Fig. 12·5. Clutch, gearbox and dead axle transmission.

Fig. 12·6. Power flow in rear independent suspension through carden shaft.

the front steered wheel assisted by tilting the motor cycle to the direction of turn. There are certain compact units in which engine and transmission are enclosed to form a single one unit as in "Vespa" scooter.

Final gear reduction in the chain drive is obtained by putting bigger size sprockets on the dead axle.

(c) **Clutch, gearbox and axleless transmission.** This type of drive is applicable to the vehicles employing independent suspension for the driving wheels. The differential housing is attached to the chassis frame. Drive from the differential to the individual stub axles of the driving wheels is given through carden shafts having universal joints at their ends. The power from the gearbox flows to the differential through the propeller shaft and universal joints.

4. **Drive to the vehicle.** The torque developed by the engine is transmitted to the wheels by means of transmission system. The torque at the driving wheels gives rise to a propulsive force between those wheels and the road. This force is known as *tractive force.*

When the drive wheels begin to rotate, the whole drive axle unit moves forward and pushes the vehicle with it. This thrust of the axle to the chassis frame drives the vehicle and is known as *driving effort.*

When the brakes are applied to a moving vehicle, the drive axle will tend to drag behind and the mass of vehicle to continue in motion. The tendency to hold a moving vehicle is known as *braking thrust.*

When the wheel rotates in one direction, the drive axle housing tries to rotate in the opposite direction due to the fact that action and reaction is equal and opposite. A twisting effort acts at the axle housing opposite to the rotation of wheel. The twisting effort thus applied to the axle housing is known as *torque reaction* or *rear end torque.*

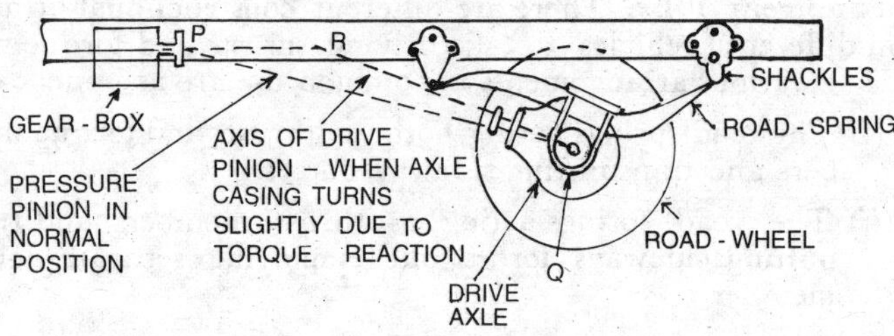

Fig. 12·7. Torque reaction : Hotchkiss Drive.

There are two main different ways to take driving thrust,

braking effort and torque reaction in different vehicles. These are as under :—

(*a*) Hotchkiss Drive,

(*b*) Torque Tube Drive.

(*a*) *Hotchkiss Drive*. This is that type of drive in which twisting effect is taken by the road springs. The driving thrust, braking efforts and torque reactions are imparted from the drive wheels to the chassis frame through the road springs. The road springs, thus, have to bear the effects of different forces in addition to their springing action effect.

(*b*) *Torque Tube Drive*. In this type of drive all driving and braking efforts and all torque reactions are taken by the torque tube. In this arrangement, the road springs are relieved of all strain except deflection due to road irregularities.

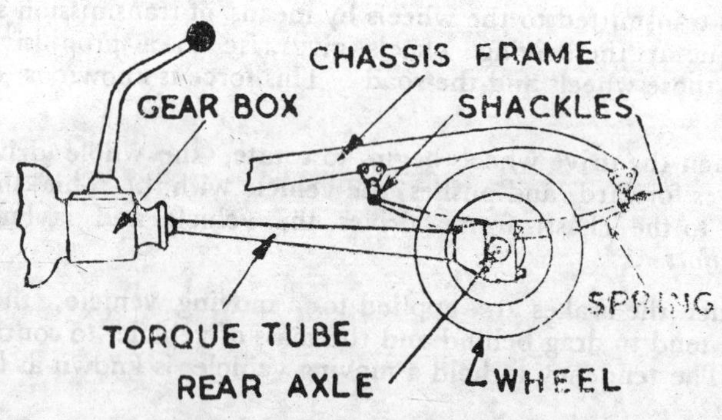

Fig. 12·8. Torque Tube Drive.

The torque tube is a tubular member which encloses the propeller shaft. It is rigidly attached to the drive axle housing. Its front end forms a ball and socket joint and is anchored to a cross member of the chassis frame where gearbox output coupling is attached to the propeller shaft.

5. **Constructional Arrangements to take driving thrust and torque reactions**. There are different constructional arrangements in different vehicles to take driving thrust and torque reactions. The various arrangements in common use are as under :—

(*i*) The road springs acting both as thrust and torque members and transmitting sideways forces.

(*ii*) The road springs acting as thrust members and transmitting sideways forces but employing separate torque member.

(*iii*) Torque reaction and driving thrust taken by separate members and the road springs dealing with sideways forces.

(*iv*) The road springs transmitting only the weight of the body and torque reaction, driving thrust and sideways forces being taken by separate members.

The above arrangements are made by providing appropriate connections between the drive axle and the chassis frame. These connections must be capable to deal with the following actions :—

(*a*) The weight of the body,

(*b*) The torque reaction,

(*c*) The driving thrust,

(*d*) The sideways forces.

The connections between the drive axle and the chassis frame take the weight of the body. In addition to the torque reaction and driving thrust, they deal with the sideways forces which are transmitted from the body to the wheels. These connections should be such so that various components are not involved in undue stresses.

(*i*) *Road spring acting as torque and thrust members.* The arrangement in which road springs act as torque and thrust members is known as "Hotchkiss Drive". This is the simplest arrangement and is most widely used in the four wheelers. In this arrangement, the drive axle is connected with the chassis frame through road springs. The road springs are suspended with the chassis frame through shackles and pins. The front end of the road springs is directly connected with the chassis frame through a pin around which this end could move. The rear end of the leaf spring is connected with the frame hanger through a shackle and pins. The shackle provides the ability to the spring to oscillate. When there is any road bump at the road wheel, the spring extends in length and contracts back when the bump goes away. Due to this effect, the spring swings back and forth. The shackle helps the spring in this action.

The drive axle is connected with the gearbox by means of a propeller shaft which is provided with universal joints at both ends for angularity movement. The propeller shaft contains a slip joint which allows the shaft to increase in length when the drive axle is pushed backward while the leaf springs flatten up.

When the leaf spring assumes its initial curvature, it decreases in length. The drive axle then moves forward thereby pushing the main rear part of propeller shaft inside the splined front part resulting in decrease in length.

When the vehicle is unloaded, the road springs keep the chassis raised up. When the vehicle is loaded, and the load falls over the road springs, they flatten away. Similarly when a wheel falls into a pit, the road spring is pulled down and when a stone comes below the wheel, it is raised high. All the effects are borne by the road springs. During all these different conditions, the angle through which the propeller shaft is to rotate, does not remain the same. It varies with the varying conditions. Moreover, the propeller shaft is fitted in the inclined state due to the fact that the gearbox level is higher than that of drive axle to which the propeller shaft is connected at both ends. Universal joints provide the ability to the propeller shaft to transmit power at varied angles of rotation.

In this type of arrangement, the torque applied by the axle housing to the road springs tends to lift the front ends of the springs. At the same time, it tends to lower the rear end of the road springs. The springs flex a little to permit a slight amount of housing rotation. The road springs thus absorb the rear end torque of twisting effort and thereby act as torque members.

The road springs offer considerable resistance to overcome torque reaction. The driving thrust is taken to the chassis frame through the front ends of the road springs. Thus the road springs act as thrust members.

At the time of torque reaction, the road springs flex a little to permit a slight movement to the axle housing. The condition gives rise to the necessity to provide two universal joints, one on each end of the propeller shaft. When the axle casing turns slightly, the axis of drive pinion at final drive is raised up and no longer passes through the centre line of the front universal joint. If there is no universal joint at the rear end of the propeller shaft, it would bend down.

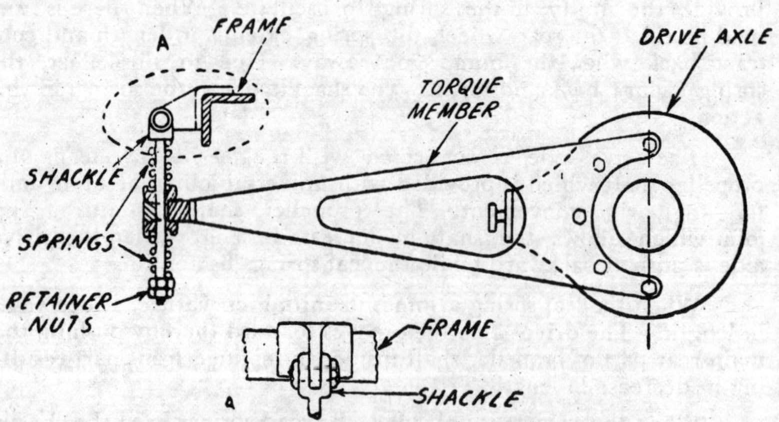

Fig. 12·9. Hotchkiss drive plus torque member.

(*ii*) *Road springs acting as thrust members but employing torque member*. (The Hotchkiss Drive plus torque member).

This arrangement is similar to Hotchkiss Drive. With the addition that a torque member is included in it, the road springs do not take torque reaction as it is taken by the torque member.

In this arrangement, the road springs do not require to be rigidly bolted to the drive axle housing. Instead, the road springs are bolted to separate spring seats which are free to turn on the axle casing. All the other arrangement of spring suspension with the frame and propeller shaft attachment is similar to that of Hotchkiss Drive.

The torque member is rigidly bolted to the axle housing at the rear end coupled to the chassis frame at the front end through a shackle. The shackle allows the front end of the torque member to move slightly back and forth. This is necessary due to the fact that the axle does not move up and down in a true circle when road springs flex.

(*iii*) *A separate torque and thrust member and the springs being separate members.* This type of arrangement is known as Torque Tube Drive. As explained earlier, a rigid tube known as torque tube, surrounds the propeller shaft in torque tube drive. The torque tube is rigidly connected with the axle housing at its rear end. The front end is coupled to the transmission through the frame cross member. During torque reaction when the axle housing tries to rotate, it tends to bend the tube but the tube resists the effort.

The road springs contain shackles at both ends. They just swing away when the driving thrust or torque reaction affects them. The driving thrust is carried to the chassis frame through the torque tube which also takes the torque reaction. The road springs absorb effects of sideways forces acting between body and road wheels.

The propeller shaft employs only one universal joint at the front end and there is no necessity of universal joint at the rear end because the centre line of final drive pinion always passes through the centre of spherical cup of front end coupling of torque tube.

Where leaf springs are employed in the suspension of an automobile, sideways forces are absorbed by them. It is not feasible to deal with sideways forces when coil springs, torsion bars or air springs are used in lieu of leaf springs. In such cases transverse radius rods are employed to do this job. There is one or more rods which are connected with axle housing on one end and chassis frame or body on the other end. The transverse radius rods are nearabout parallel to the axle. The joints at the ends of radius rods are movable to allow for the relative motions which occur.

(*iv*) *Road springs bearing only weight of the body ; Torque reaction, driving thrust and sideways forces being taken by separate members.* This is the arrangement in which three radius rods are employed. It contains two rods parallel to propeller shaft and one wishbone type rod fitted at the centre of drive axle. The rods are connected with the axle housing on one end and frame cross member on the other end through movable joints. The wishbone member takes the sideways forces. The driving thrust and torque reactions are taken by all the three radius rods.

6. **Units of Transmission System**. The transmission system consists of the following main units :

(*a*) **Clutch unit** consisting of clutch assembly or fluid coupling.

(*b*) **Transmission unit** consisting of gearbox, transfer case, over drive, free-wheeling device, torque converter, etc.

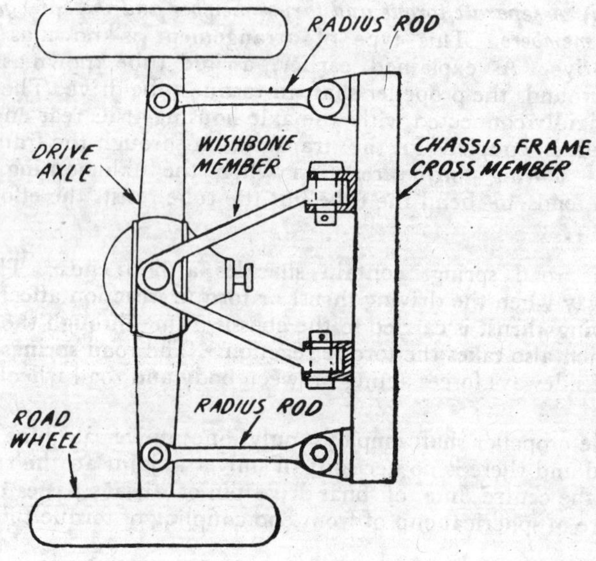

Fig. 12·10. Use of three radius rods.

(*c*) **Drive-line unit** consisting of propeller shaft, and universal joints or sprockets and chains.

(*d*) **Driving Axle unit** consisting of final drive, differential, half shafts, carbon-shafts, etc.

QUESTIONS

1. What is transmission system ? What are its requirements and how they are fulfilled ?

2. Why a different type of transmission is required in an automobile having petrol engine than a steam engine ?

3. Which are the different types of transmissions ? Explain.

4. Which are the different forms of mechanical transmission? Explain any one of them.

5. Explain the following :—

   (*a*) Axleless transmission.

   (*b*) Dead axle transmission.

   (*c*) Driving thrust.

   (*d*) Torque reaction.

   (*e*) Braking effort.

6. Which are the different ways to take driving thrust and torque reactions is automobiles ? Explain.

7. Explain the Hotchkiss Drive and Torque Tube Drive.

8. What other role is played by the road springs in addition to bearing weight of the body ?

9. Describe the arrangement in which——

   (*a*) Road springs act as torque and thrust members.

   (*b*) Road springs bear only weight of the body, torque reaction, driving thrust and side ways forces are taken by separate members.

10. Which are the main units of transmission system ? What do they consist of ?

13

Clutch

1. Clutch. It is a mechanism by means of which engine is connected or disconnected from the rest of transmission.

2. Types of clutches :

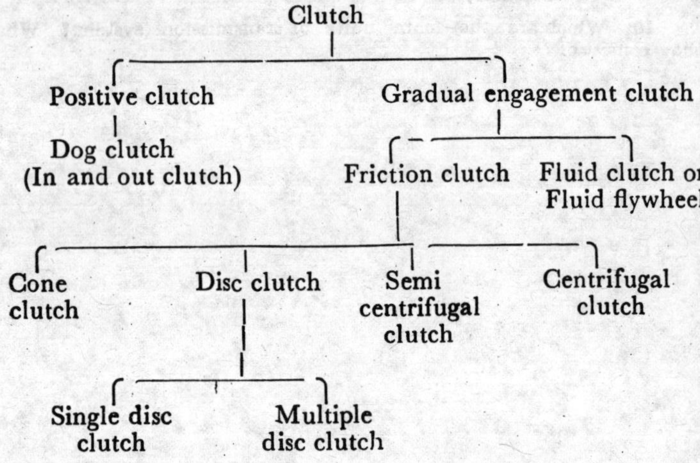

Clutches enable the rotary motion of one shaft to be transmitted to a second shaft having the same axis, when required. They could be classified as follows :

(*a*) Positive clutches.

(*b*) Gradual engagement clutches.

(*a*) **Positive clutches.** These are "in and out" type clutches by means of which the driving shaft could be coupled in or out with

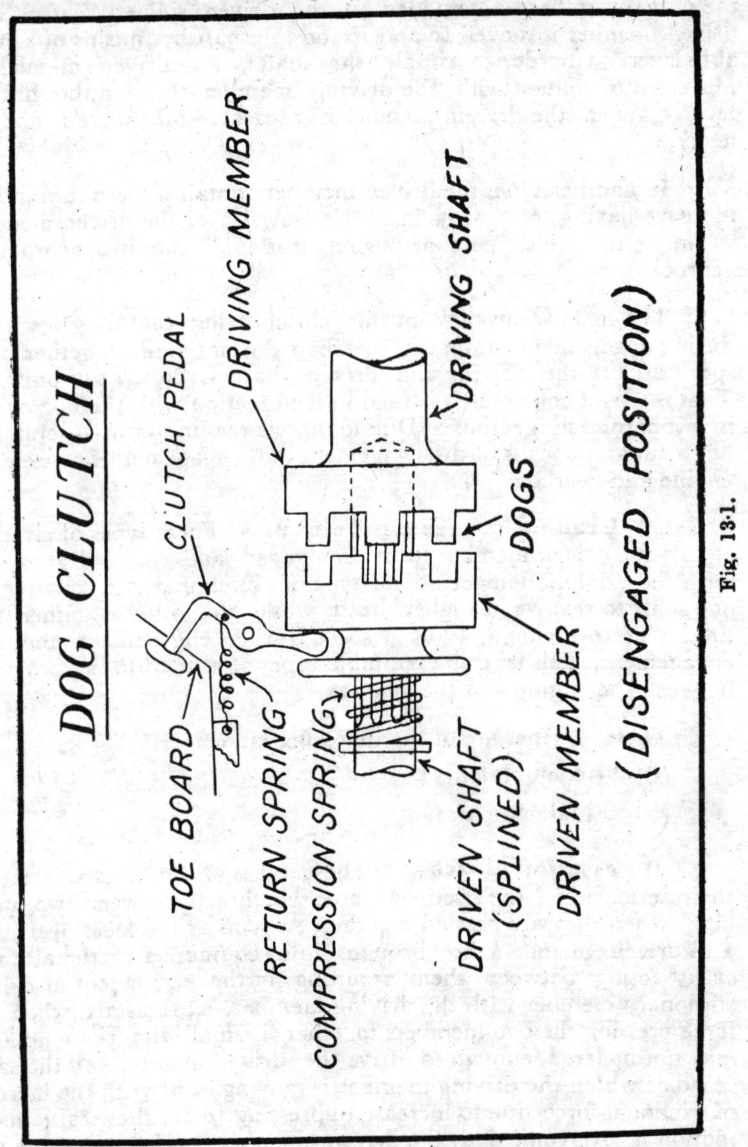

Fig. 13·1.

the driven shaft at will. These clutches are also known as dog clutches and are widely employed in gearboxes to engage one gear with the fitted other on the shafts having the same axis. The synchronizing units in the gearbox employ dog clutch. The progressive type gearbox widely uses dog clutches to engage or disengage gears.

This type of clutch uses two members having dogs or notches at their facings. One member is fixed with the driving shaft and the other member is provided on the driven splined shaft. The driven member is moved to and fro on the shaft by means of a suitable lever. In order to couple the shafts, the driven member is moved into contact with the driving member through the shifting lever. When the driven member is separated off, there is disconnection.

In another type, the driven member contains teeth around it. A sleeve having teeth at its inside, is moved over the driven member to couple it. This type of arrangement is applicable in synchromesh gearboxes.

The main drawback of this clutch being that it gives noise during engagement as the dog members do not mesh together until speed of both the driving and driven shafts is brought to uniform. That is why a cone clutch is used in conjunction with the dog clutch in synchromesh gearbox. Due to above reason, positive clutch is not a suitable means of disengagement and engagement between the engine and gearbox.

(*b*) **Gradual engagement clutches**. These types of clutches enable the driven member to be disengaged and engaged gradually with the driving member. This type of clutch makes it possible for one shaft to revolve at higher speed while the other is either stationary or revolving at a lower speed during engagement and disengagement. This is quite suitable type of clutch to be provided between the engine and the gearbox.

These clutches are of the following kinds :

(*i*) Friction clutches,

(*ii*) Fluid clutches.

(*i*) **Friction clutches**. These types of clutches depend for their action upon the frictional force which acts between two members when they are pressed together. As soon as the faces of driving and driven members are brought into contact, a frictional force starts acting between them resulting in the movement of driven stationary member with the driving member. At the start, when the force pressing the two members together is small, the frictional force may not be large enough to drive the driven member at the same speed at which the driving member is moving. But with the increase of frictional force due to increase in pressing force, the resistance to motion is overcome and the driven member begins to rotate gradually at the same speed at which the driving member is moving. When the driving and driven members are moving at the same speed and there is no slip between them, the clutch is considered to be fully engaged.

In friction clutches, the pressure is exerted by means of coil springs. The springs tend to keep the clutch always engaged and when it is desired to disengage the clutch, the spring pressure is released.

Engine torque is transmitted to the gearbox through the clutch. The torque which a clutch can transmit, depends upon the coefficient of friction, spring force and the contact surfaces.

Friction clutches are of the following main types :

(*a*) Cone clutch,

(*b*) Disc clutch,

(*c*) Semi-centrifugal clutch,

(*d*) Centrifugal clutch.

(*a*) **Cone clutch.** It consists of two cones having leather facings. The cones are known as male and female ones. The female cone is fixed with the driving shaft whereas the male member is splined on the driven shaft. The driven shaft is supported in the driving shaft having the same axis. The spring pressure keeps the driven member in contact with the driving member and the clutch is considered to be engaged when there is no slip between the male and female cone. In order to disengage clutch, the driven member is moved back through leverage, against the tension of the spring.

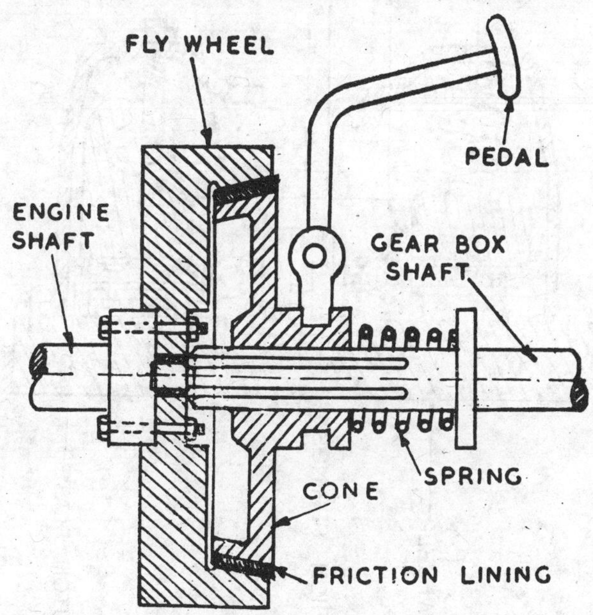

Fig. 13·2. Cone Clutch.

(*b*) **Disc clutch.** This type of clutch employs a disc having linings on its both facings. This disc is known as clutch plate and is held between the two members which are attached to the driving shaft. The clutch disc acts as a driven member and is splined over the driven shaft. The clutch is disengaged when spring pressure acting on the clutch plate is released through clutch operating mechanism by pulling back pressure plate.

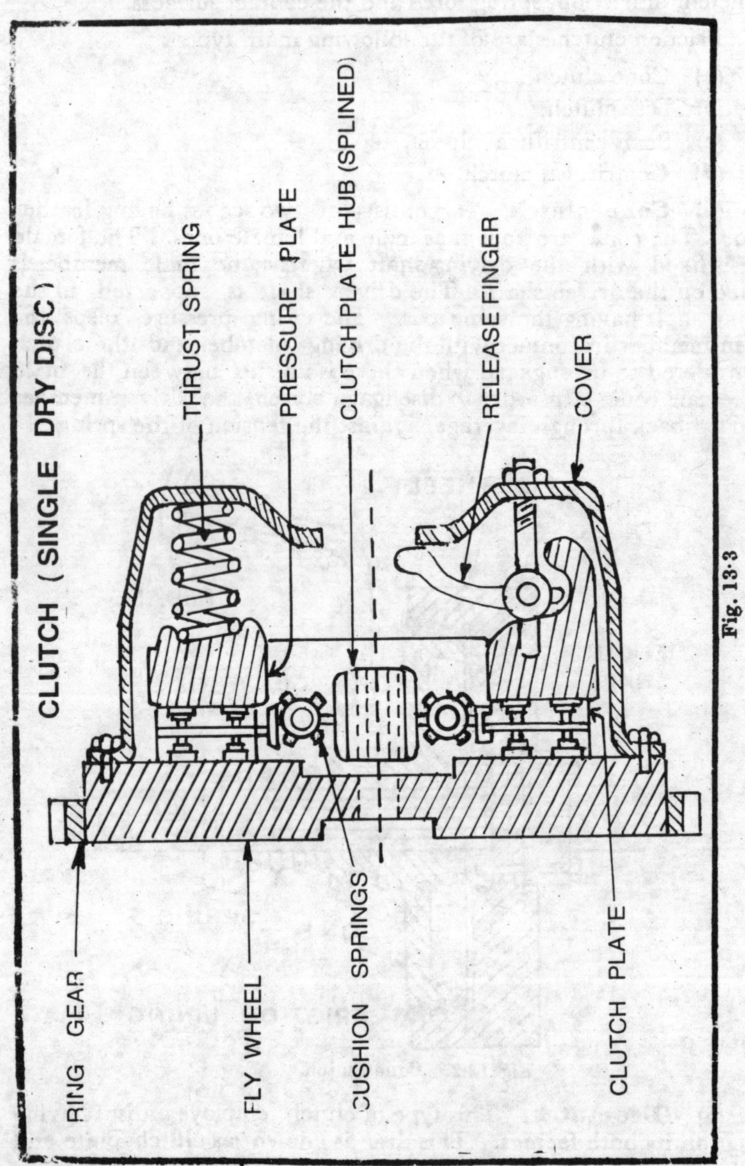

Fig. 13·3

Disc clutches are of the following types :

(*i*) Single disc clutch,

(*ii*) Multiple disc clutch.

(*i*) **Single disc clutch**. It is the most common type of clutch used with majority of the vehicles. It contains only one clutch plate which runs dry. It consists of the following main parts :—

(*a*) Driving plate,

(*b*) Driven plate,

(*c*) Pressure plate.

(*a*) **Driving plate**. Engine flywheel which is attached at the tail of crankshaft, acts as a driving plate for the operation of disc plate clutch in cars, buses, trucks, etc.

(*b*) **Driven plate**. It is the clutch plate which is held by the pressure plate with the driving member, *i.e.*, flywheel. It is a steel

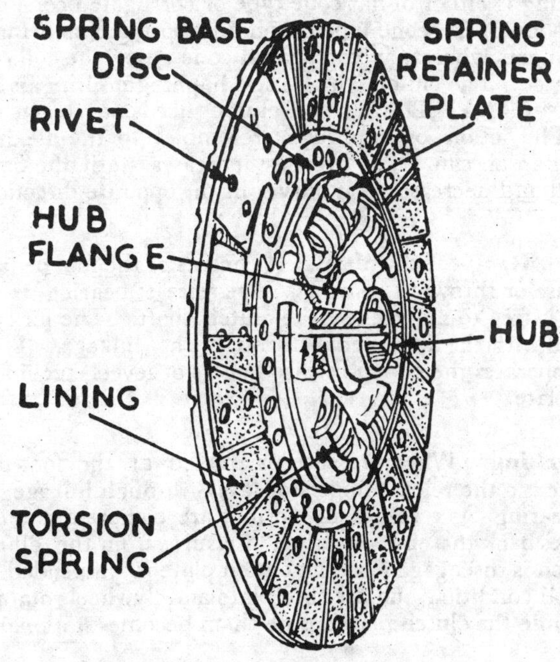

Fig. 13·4. Clutch plate.

disc which is connected to the splined hub through springs. At its outer facings, the disc contains lining of heat resisting material having a high coefficient of friction. The linings are riveted to the clutch plate and can be replaced after they wear out. The springs are mounted so that when torque is applied to the clutch plate, the springs compress, absorbing any shock before the hub turns. These springs are known as torsion damper springs because they cushion the vibrations in the clutch plate produced during engagement and disengagement.

(c) **Pressure plate.** The pressure plate assembly is connected with the flywheel and holds the clutch plate with the flywheel due to its pressure. It consists of pressure disc, springs, operating fingers and holding down cover. The pressure plate is held with the cover through the bolts. The coil springs are held between the cover and the plate. The fingers act as levers and when pressed down, help in lifting off pressure plate as the springs are pressed and pressure is released.

In some cases, crown or diaphragm type conical spring is employed in the pressure plate in place of coil springs. The diaphragm spring is either of flat cone type or corrugated cone type. The centre section of flat cone type diaphragm spring is split into distinct tapered fingers whereas in corrugated cone type, the conical spring is corrugated. The clutch which uses diaphragm spring, is known as *Diaphragm Clutch.* This type of clutch needs no fingers or release levers. The action of diaphragm resembles to the flexings of the bottom of an oil can. The pressure increases until the flat position is reached and decreases as it curves in the opposite direction.

Operating mechanism. It consists of the foot pedal, linkage and release or throwout bearing. The release bearing is held in a fork and is free to move over the clutch shaft. The fork in turn is connected with the foot pedal through the linkage. The release bearing operates the clutch release fingers or levers provided in the pressure plate.

Working. When pressure is applied at the foot pedal, it is transmitted to the release fingers or levers through linkage, fork and release bearing. As a result, the springs are compressed and pressure disc moves back thereby releasing pressure from the clutch plate. The clutch is disengaged as the clutch plate is suspended. During disengaged condition, the pressure plate and flywheel rotate with the engine while the clutch plate inside them becomes stationary.

When pressure at the release levers is released, the full pressure of the pressure plate springs is exerted through the pressure disc at the clutch plate. The clutch plate is thus held between the flywheel and the pressure plate, rotating as one unit. This is the engaged position.

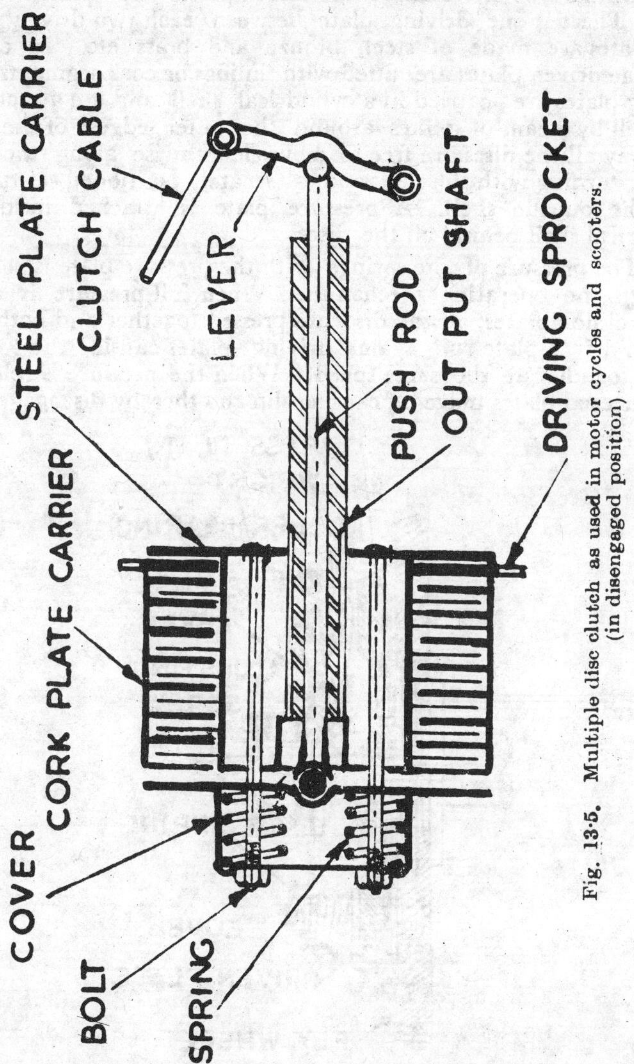

Fig. 13·5. Multiple disc clutch as used in motor cycles and scooters. (in disengaged position)

(ii) **Multiple disc clutch**. As is apparent from its name, multiple disc clutch contains more than one disc to act as clutch plate or driven member. This clutch is either of wet type or dry type. ***Wet clutch*** is that which runs in oil whereas *dry clutch* is that which does not run in oil. Multiple disc clutches are used at several places in an automobile. These are used between engine and gearbox, in planetary transmission, in steering or differential clutches etc.

In this type of clutch, all the driven plates are splined to the shaft. There is one driving plate between each two driven plates. The plates are made of steel, bronze and brass etc. In certain cases the driven plates are fitted with linings or cork segments. The driving plates are carried in a cylindrical shell and are attached to the shell by means of splines around the outer edges of the discs. This way all the discs are free to move lengthwise along the shaft, one set turning with the shaft and alternately positioned set turning with the outside shell. A pressure plate is attached inside the cylindrical shell behind all the discs.

The pressure of the springs with the pressure plate is released through the operating mechanism. When full pressure is acting on the clutch plates, all the discs are pressed together and both sides of each driven plate rub against driving plates causing the two sets to run together at the same speed. When the pressure is released, the squeezed plates move off causing slip and thereby disengagement.

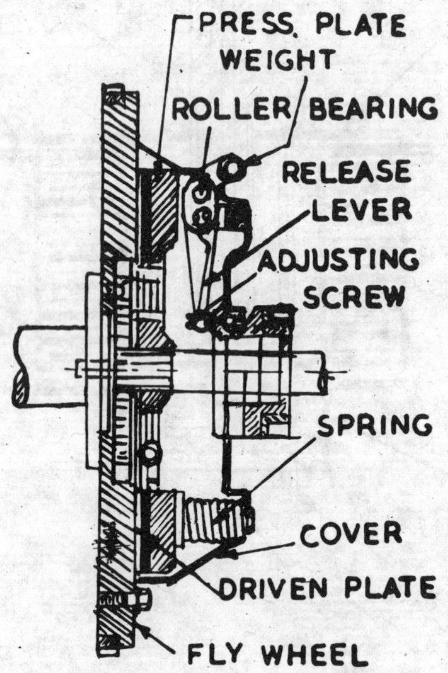

Fig. 13·6. Semi-centrifugal clutch.

(*ii*) **Semi-centrifugal clutch.** The only difference between single dry disc clutch and semi-centrifugal clutch being that the construction of operating fingers or levers is different. In semi-centrifugal clutch, extra weight is provided at the weight arm of the levers which acts on the pressure plate. When pressure plate moves, centrifugal forces through the weight arms of the operating levers, exert an extra pressure in addition to the pressure of the springs on the clutch plate through the pressure plate. It is owing to this fact, that this clutch is known as semi-centrifugal clutch.

(*iii*) **Centrifugal clutch.** It is an automatic clutch which is controlled by the engine speed through the accelerator. When the engine speed falls down, the clutch is automatically disengaged and when the speed rises above the predetermined value, the clutch is engaged. Greater the centrifugal force due to higher engine speed, more powerful the contact between the driving and driven members and better the engagement.

The simplest form of centrifugal clutch consists of two members, one fitted on the driving shaft and the other attached to driven shaft. The driven member is just a drum which encloses the driving member. The driving member contains two curved shoes having frictional linings at their backs. The shoes are anchored at one end to the back plate and are kept in position by means of coil springs which hold the free end.

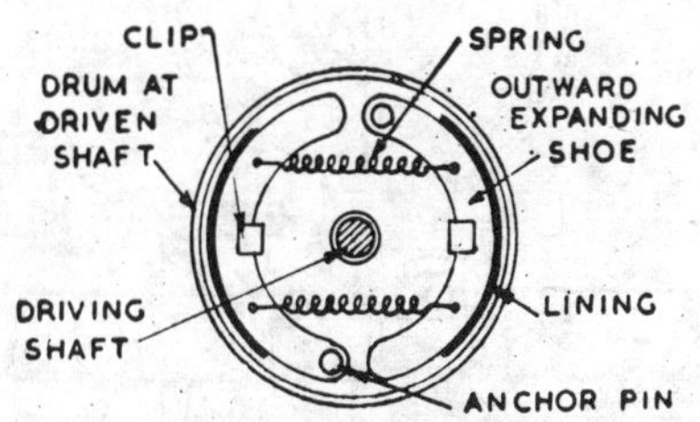

Fig. 13·7. Centrifugal Clutch.

The driving member rotates with the engine shaft. As the engine speed increases, the shoes inside the driving member drum fly outwards due to centrifugal force and come into contact with the inner surface of the driving drum. The increasing centrifugal force due to higher engine speed, binds the driving member with the driven member resulting in the movement of both the members and shafts at the same speed.

As the engine speed decreases, the centrifugal force due to which the driving and driven members are connected with each other, also decreases which results in disengagement. This type of clutch is employed in small units such as lawn mowers.

The centrifugal clutch, as employed in automobiles, consists of a clutch plate held between the flywheel and pressure plate. The actuating pressure is applied to the pressure plate through the medium of back plate and circumferentially spaced inner springs by means of centrifugal action of weights. The resistance of these springs transfers the centrifugal load gradually to the clutch plate, thus giving a smooth take up.

When the clutch is disengaged due to low engine speed, the back plate is held back by the outer springs. When the engine speed increases, the balance weights, usually three in number, tend to fly outwards until they contact the flywheel rim, while the other ends of the levers press against the back plate. The pressure in turn is transmitted through the inner springs to the pressure plate, thus overcoming the strength of the outer springs. This results in the gripping of clutch plate between the flywheel and pressure plate and thereby engagement. Thus this type of clutch acts entirely as an automatic self-contained unit.

(*ii*) **Fluid Clutch**. It is known as flywheel or coupling which couples the driving member with the driven member through a media of fluid. It consists of two members, driving and driven

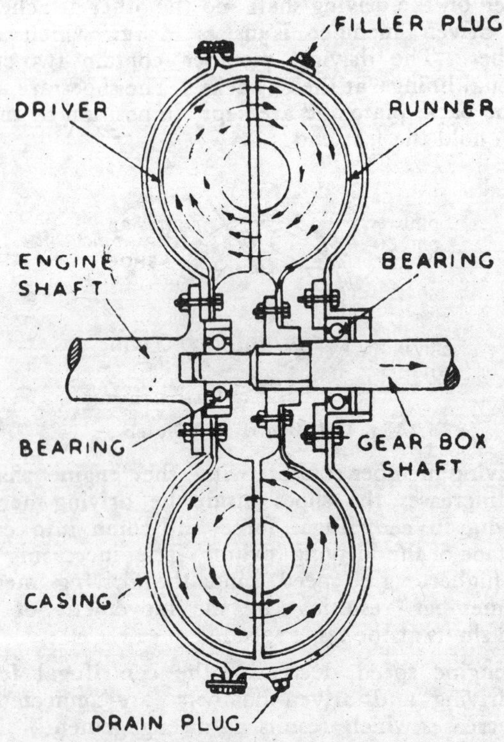

Fig. 13·8. Fluid flywheel.

(Impeller and runner). Both members are having pockets designed in their facings which face towards each other. Both the members are fitted on the respective shafts and there is small clearance between them. The members are kept in an oil tight housing which keeps oil up to a certain level.

As engine starts, the driving member or impeller begins to move inside the housing containing oil. Pockets of the moving driving member are filled with oil and the centrifugal force causes the oil to be forced outward radially. The pockets are designed in such a way that the splashed oil strikes the pockets or vanes of the driven member and it is forced to move along in the same direction. As the engine speed increases, the thrown out oil from the pockets of the driving member strikes the pockets of the driven member or runner with greater force and tends the driven member to rotate at the same speed, becoming one unit by means of oil film which combines both the members.

As the engine speed falls down, the oil film between the driving and driven members is broken away and the members are disengaged.

Advantages of fluid flywheel.

(*i*) When the engine is accelerated, it gives a smoother power take up than the friction clutch of centrifugal type.

(*ii*) The fluid in the coupling acts as a cushioning agent between the engine and gearbox to absorb shocks during braking or coasting down an incline.

(*iii*) It needs no separate pedal or lever to operate it.

(*iv*) The driving member acts as a flywheel on the crankshaft and thus smoothes out the torque variation effects.

3. Characteristics of an automobile clutch. An automobile clutch should possess the following characteristics :—

(*i*) It should be easily operated.

(*ii*) It should require minimum physical effort to disengage.

(*iii*) It should not slip when engaged.

(*iv*) Its wearing surfaces should have long life.

(*v*) Its operation should be noiseless.

(*vi*) It should possess maximum torque carrying capacity.

(*vii*) It should have proper lubrication means for the metal working parts.

(*viii*) It should have simple means of its adjustment.

(*ix*) It should be easily accessible.

4. Clutch linings. The clutch linings are made of frictional material having higher coefficient of friction. The linings are riveted to the clutch plate and are replaced by new ones when the original linings wear out. Clutch linings are of the following two types :—

(*a*) Woven type,

(*b*) Moulded or composite type

(*a*) **Woven type.** This type of lining is made by weaving asbestos fibre threads like the cloth and then impregnating with a bonding material. Sometimes weaving is done over the brass wires to increase its strength. Woven type lining is of the following classes :—

(*i*) Laminated,

(*ii*) Solid woven.

Laminated variety consists of layers of woven asbestos fibre cloth placed one above the other and held together by bonding material. Sometimes bonding is aided by stitching. In the case of solid woven variety, the required thickness is woven in one operation forming an interlocked structure. It has a much greater strength than the laminated variety which is a layered one. Both these varieties may use metallic wires, usually of brass, to strengthen them.

(*b*) **Moulded or composite type**. This type of lining is composed of asbestos fibres in their natural state, mixed with a bonding material. It is moulded in dies under pressure and at constant temperature to the required shape. Sometimes, wire netting is included in moulding to increase strength.

Cotton is sometimes mixed with asbestos fibres in the formation of lining. Cork is also used in clutch plates. Buttons made from cork sheet are usually employed in motor cycle clutch plates.

Bonding Materials. The following types of bonding materials are used in the preparation of clutch linings :—

(*i*) Vegetable gums,

(*ii*) Rubber,

(*iii*) Asphaltic bases with natural gums and oils,

(*iv*) Synthetic resins—alcohol soluble and oil soluble.

5. **Vacuum operated clutch**. Vacuum boosters or cylinders are used in some vehicles to operate the clutch. Engine manifold vacuum is utilised to supply power to the cylinder.

When throttle valve in the carburettor is closed, engine vacuum is at the highest peak. The vacuum force obtained during this condition, is used to operate a piston in the booster cylinder which disengages the clutch.

When the throttle is open, the vacuum falls off and the clutch assumes its engaged position. The clutch is thus disengaged and re-engaged by the application of accelerator pedal which controls the opening and closing of throttle valve. This sort of clutch control is known as automatic clutch control or *vacuum actuated automatic clutch control*.

Operation of vacuum booster. When throttle is in closed position, one side of the piston in the operating cylinder is subjected to vacuum. As the other side of the piston is open to atmospheric pressure, the piston moves forward, thereby disengaging the clutch. When the accelerator is relased, the control valve closes the vacuum line and opens the atmospheric line. This results in the atmospheric pressure to act at the piston in place of vacuum due to which the piston moves backward and the clutch is engaged.

QUESTIONS

1. What is clutch ? Which characteristics an automobile clutch should possess ?
2. Which are the different types of clutches ? Explain any one type of clutch.
3. Describe a positive clutch.
4. What is fluid flywheel and how it works ? What are its advantages ?
5. Illustrate the construction and working of a single plate, dry disc clutch.
6. Which are the gradual engagement clutches ? Explain any one of them.
7. Explain the construction and working of a multiple disc clutch.
8. Describe the following :
   (*a*) Cone clutch.
   (*b*) Semi centrifugal clutch.
9. What do you know about centrifugal clutch ? How a fluid flywheel differs from it in working ?
10. What is clutch lining ? What is it made of ? Which are the different varieties of linings and how they are prepared ?
11. What is vacuum operated clutch and how it works ? What is the difference between vacuum operated clutch and fluid coupling ? Discuss.
12. Illustrate the operation of vacuum operated clutch.

14

Transmission

1. Transmission (gearbox). It is a speed and torque changing device between the engine and the driving wheels. It serves the following purposes in the transmission system of an automobile :

(*a*) It exchanges engine power for greater torque and thus provides a mechanical advantage to drive the vehicle under different conditions.

(*b*) It exchanges forward motion for reverse motion.

(*c*) It provides a neutral position to disallow power flow to the rest of power train.

2. Necessity of transmission (gearbox). When a vehicle is running, various resistances oppose it. In order to keep the vehicle moving at a uniform speed, a driving force or tractive effort equal to the sum of all the opposing forces has to be applied to it. If the tractive effort increases the total resistance affecting the movement of the vehicle, the excess tractive effort will accelerate the vehicle. If the tractive effort is less than the total resistances, the excess of the resistances will lower down the speed of the vehicle.

Vehicle acceleration = Tractive effort—Total resistances affecting the movement of vehicle.

The forces which oppose the movement of the vehicle are as under :

(*a*) Air or wind resistance.

(*b*) Gradient resistance.

(*c*) Rolling resistance.

(*a*) **Air or wind resistance.** It is the resistance which the air offers to the movement of bodies through it. This resistance depends upon the shape and size of the body and its speed through the air.

A streamlined vehicle shall be subjected to less air resistance than a flat shaped vehicle moving at the same speed and accordingly more tractive effort shall be required to keep the flat shaped vehicle moving at the same speed at which the streamlined vehicle is moving. If the tractive effort affecting at both the vehicles is the same, the streamlined vehicle shall run at higher speed than the other.

At lower speed, air resistance shall be low and at high speed, the air resistance shall multiply. In practice. the air resistance is taken to vary as the square of the speed. Hence, if the speed is doubled, the air resistance increases four times.

(*b*) **Gradient resistance.** It is the resistance due to road gradient. It depends upon the steepness of the gradient and weight of the vehicle and is not affected by the speed of the vehicle up the gradient.

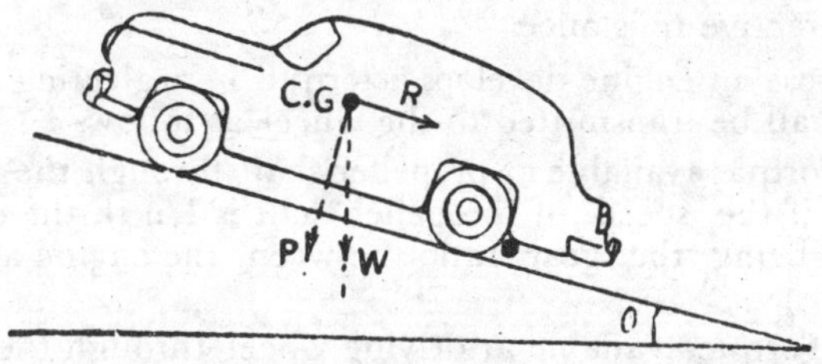

Fig. 14·1.

As is apparent from the above diagram, the weight (W) of the car acts vertically downwards. A force (R) acting on the car due to gradient, is tending to push it backwards. In order to prevent the car from rolling down the gradient, a force equal and opposite to R must be applied to it. When the car is going up the gradient, a part of the driving force or tractive effort is consumed in overcoming or neutralizing the force R. This force R is an additional resistance to the movement of the car and is known as gradient resistance.

(*c*) **Rolling resistance.** This includes all the remaining resistances to the movement of vehicle. Mainly, it includes internal frictional resistances of the transmission system and resistance due to deformation of road and tyre.

All the power developed by the engine does not reach the driving wheels to act as driving force but some of the power is lost in driving the transmission components. There is dissipation of energy through the impact of tyre with the road surface. The power losses in the transmission system depend upon its efficiency. The loss of power through the impact of tyre with the road surface depends upon the following factors:

(*i*) Nature of road surface.

(*ii*) Nature of tyres fitted with the vehicle.

(*iii*) Tyre inflation pressure.

(*iv*) Total weight of the vehicle.

(*v*) Load on the vehicle.

Total or tractive resistance. The total or tractive resistance to the movement of a vehicle is the sum of above three resistances.

Tractive resistance = Gradient and rolling resistances (Independent of vehicle speed).
+ Air resistance (Dependent on vehicle speed)

Tractive effort. It is the driving force which acts at the driving wheel to propel the vehicle. The source of power in an automobile is engine. Torque developed by the engine is transmitted to the wheels by means of transmission system. For a uniform speed, the tractive resistance should be balanced by tractive effort produced at the point of contact of wheel on the road by the driving axle torque. For acceleration, the tractive effort must be greater than the tractive resistance.

Suppose an engine develops a torque T, neglecting transmission losses, it shall be transmitted to the wheel as follows :

(*i*) Torque available at propeller shaft through the transmission (gearbox), if the speed of propeller shaft is 1/n th the engine speed $= n \times T$ (n being the gear ratio between the engine and propeller shaft speeds).

(*ii*) Torque available at driving wheel through the final drive, if the driving wheel speed is 1/m th of propeller shaft speed $= n \times m \times T$ (m being final drive ratio between the propeller shaft and axle shaft speeds).

When the driving wheel is to be driven, the torque at it shall be acting from its centre (where axle shaft is located) to the contact point at road. Tractive effort (neglecting transmission losses) thus would be as below :

$$T.E. = n \times m \times r \times T$$

where n = Gearbox gear ratio
m = Final drive ratio
r = Wheel radius
T = Engine Torque.

The tractive effort is proportional to the engine torque, since the engine is geared to the driving wheel, so the engine speed is related to the vehicle speed. The variation of the tractive effort with the variation of engine torque depends upon the variation of engine torque with the variation of engine speed. For a given power, the torque is inversely proportional to the speed and if the speed is reduced, the torque will be increased in the same ratio.

The tractive effort required at the driving wheels varies according to the conditions of use. A high torque is required at the wheels to move off from the standstill or climbing a hill which require high engine speed. Since the engine speed is kept constant, so the gearbox helps in varying the speed of driving wheels in relation to engine speed.

The gearbox (transmission) is thus necessary in the transmission system to maintain engine speed at the most economical value under all conditions of vehicle movement. An ideal gearbox would provide an infinite range of gear ratios so that engine speed could be kept at or near that at which maximum power was developed whatever the speed of the vehicle.

3. Types of Transmission (gearbox).

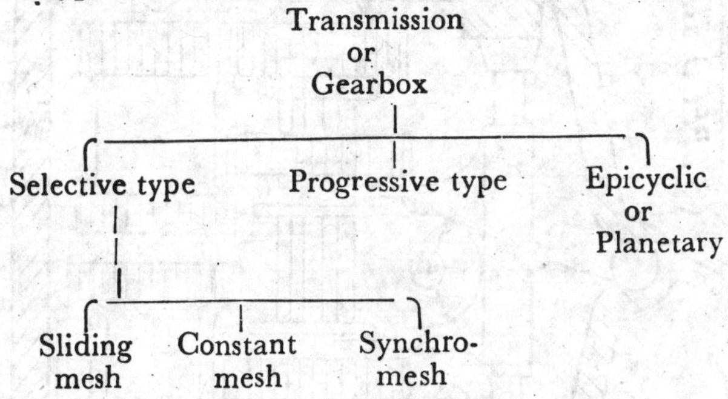

These are the following types of transmission :

(*a*) Selective type, (*b*) Progressive type,

(*c*) Epicyclic or planetary.

(*a*) **Selective type transmission.** It is that transmission in which any speed may be selected from the netural position. In this type of transmission, neutral position has to be obtained before selecting any forward or reverse position. Selective type transmissions are of the following types :

(*i*) Constant-mesh,

(*ii*) Sliding-mesh,

(*iii*) Synchromesh.

(*i*) **Constant mesh gearbox.** It is that gearbox in which all the gears are in constant mesh with each other. They are free to move on the shafts which are splined. In order to obtain different speeds, the requisite train of gears is made to rotate the shaft by means of sliding dog-clutch which is operated with the help of gear-shift lever.

The constant-mesh gearbox obtains its name from the fact that all the counter-shaft gears and main shaft gears are in constant mesh.

(*ii*) **Sliding mesh gearbox.** It is that gearbox in which the gears on the splined main shaft are moved right or left for meshing them with appropriate gears on the layshaft for obtaining different speeds. This type of gearbox derives its name from the fact that

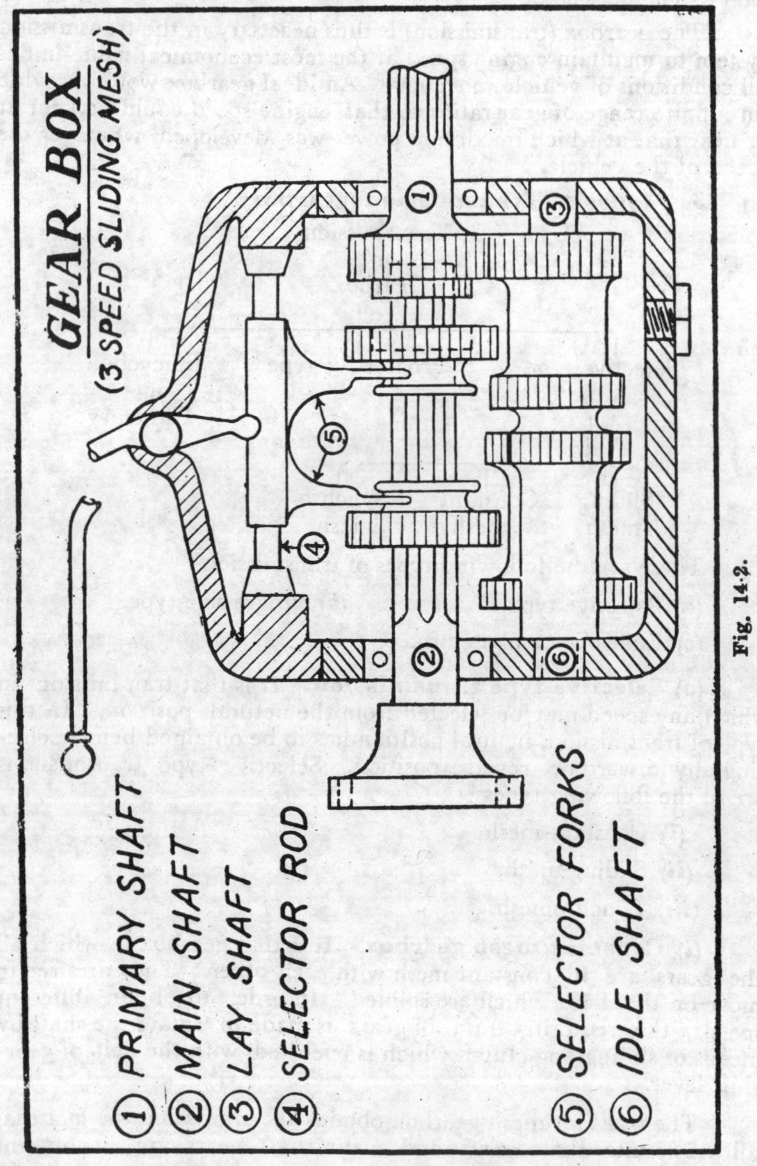

Fig. 14·2.

the gears are meshed by sliding or crashing one on to the other. This gearbox is also known as *crash-type gearbox*.

(iii) **Synchromesh gearbox.** It is that gearbox in which sliding synchronizing units are provided in place of sliding dog clutches

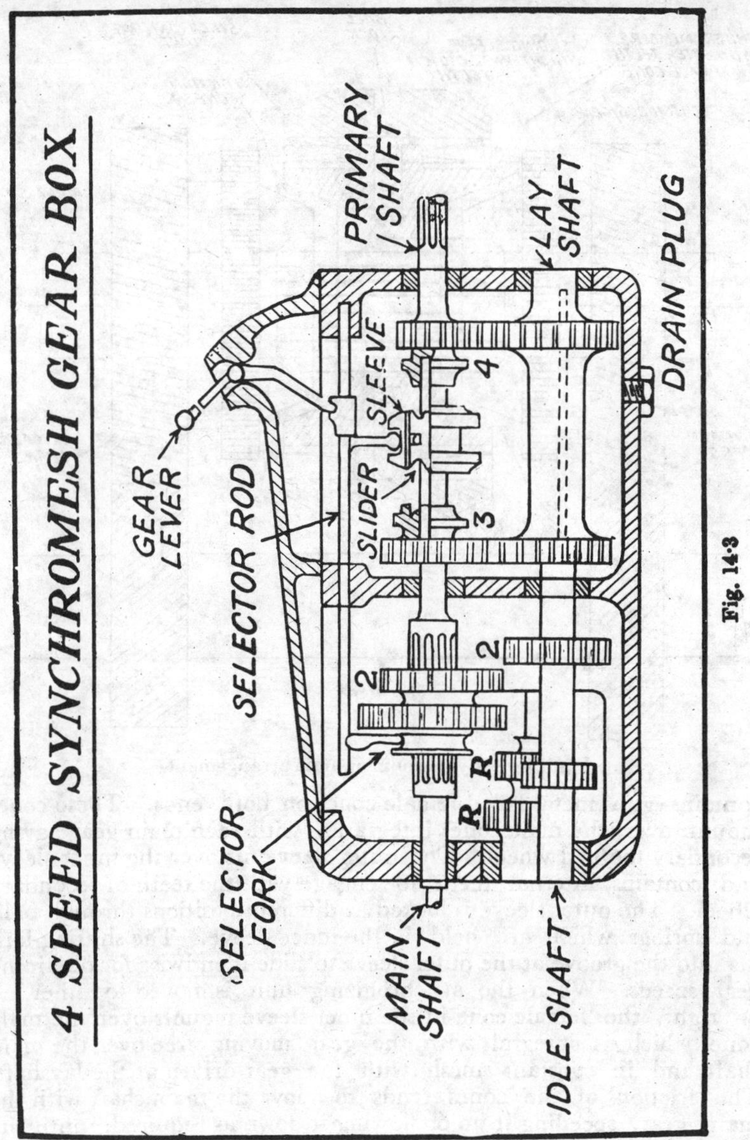

Fig. 14·3

as in case of constant-mesh gearbox. With the help of synchronizing unit, the speed of both the driving and driven shafts is synchronized before they are clutched together through the train of gears. The synchronizing unit makes gear changing easy and minimizes noise.

The synchronizing unit consists of a set of sleeves which slide endwise. The inner sleeve is splined on the mainshaft and

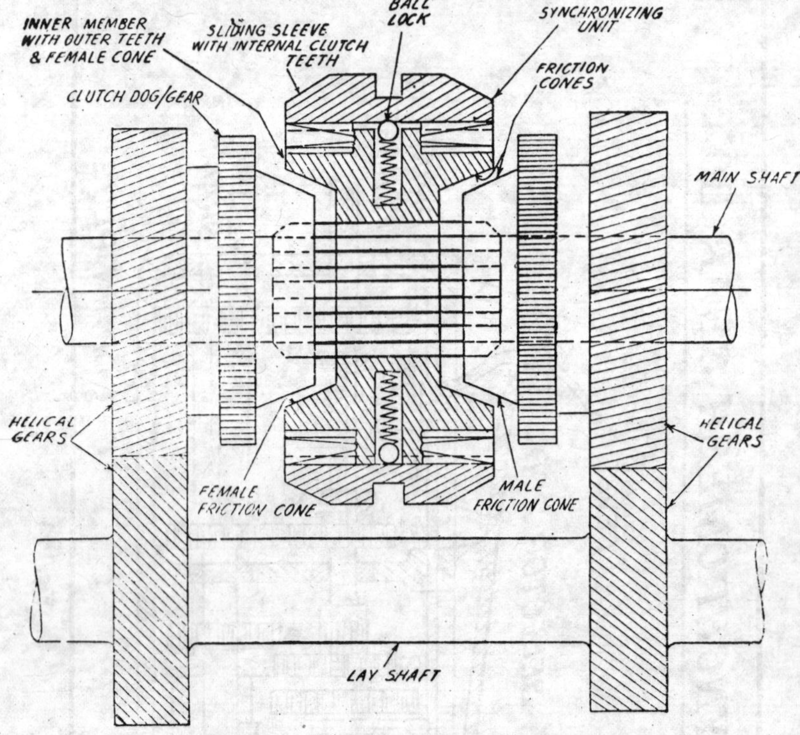

Fig. 14·4. Synchronizing unit arrangement.

contains gun metal faced female cones on both ends. These cones mount over the male cones integrated with each main gear having secondary toothed wheel. The outer sleeve fits over the inner sleeve and contains internal teeth to engage with the teeth of secondary wheel. The outer sleeve is locked at different positions through balls and springs which are held in the inner sleeve. The shifting fork fits into the groove at the outer sleeve to slide it endwise for obtaining gear speeds. When the synchronizing unit is moved to either left or right, the female cone in the inner sleeve mounts over the male cone which is integral with the gear moving free over the main shaft and in constant mesh with the gear driven at the layshaft. The friction of the cones tends to move the main shaft with the main gear, speeding it up or slowing it down as required, until the speeds of the mainshaft and gear are the same. When the speed is synchronized, it is easier to engage or clutch the gear with the main shaft through the synchronizing unit. The synchronization of speed occurs with a partial movement of the gear shift lever. Further movement of the gear lever causes the outer toothed sleeve to slide relative to the cones and to engage with the secondary wheel, in the form of a dog clutch. This results in the driving of mainshaft with the main gear. In a nutshell, the synchronizing unit slides on the splines of the mainshaft to engage the cones and then the outer sleeve slides over the inner sleeve or hub to engage the gears through dog clutch.

(*b*) **Progressive type gearbox.** Usually these gear boxes are used in motor cycles. In these gearboxes, the gears pass through the intervening speeds while shifting from one speed to another. There is a neutral position between two positions as shown below in the case of a four speed gearbox :—

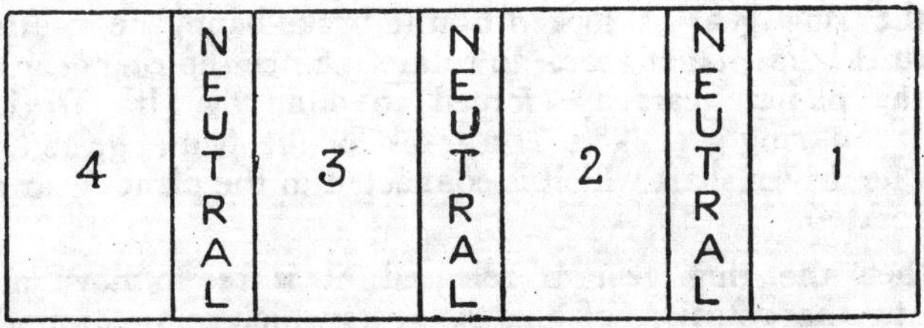

4	NEUTRAL	3	NEUTRAL	2	NEUTRAL	1

Fig. 14·5. Gear change positions in a 4-speed gearbox of progressive type.

These gearboxes are a combination of sliding and constant mesh gearboxes. The various gear speeds are obtained by sliding the dog clutch or gear to the required position.

(*c*) **Epicyclic or Planetary type transmission.** The epicyclic or sun and planet type transmission uses no sliding dogs or gears

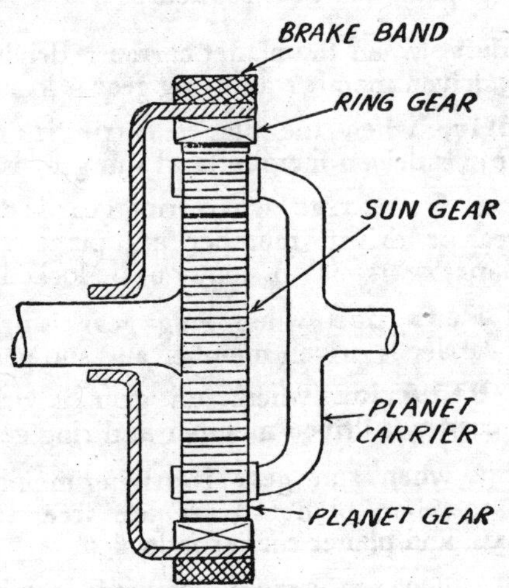

Fig. 14·6. Epicyclic gear set.

to engage but different gear speeds are obtained by merely tightening brake-bands on the gear drums, which simplify gear changing.

A planetary gear set consists of a ring gear or annular wheel, sun gear and planet gears with carrier. In order to obtain different speeds any one of these three units can be held from rotation by means of brake bands.

The ring gear contains teeth on its inner circumference and is surrounded by a brake band. The brake band is operated by a gear stick or lever to grip the ring gear and hold its movement. The sun gear is rotated by the driving shaft from the engine and thus moves along with the movement of engine crank shaft. The planet gears are in constant mesh with both the sun gear and ring gear or annular wheel and are free to rotate on their axes carried by the carrier frame which in turn is connected to the driven shaft. When the ring gear is locked by the brake-band, the rotating sun gear causes the planet gears to rotate. Since the ring gear cannot move, the planet gears are forced to climb over it. During this position, the ring gear acts as a track for the planet gears to move over. The driven shaft which is connected to the planet gear carrier is thus rotated.

When the ring gear is released, it is free to move in consequence to the rotation of planet gears which rotate around their axis. During this position, there is no movement of planet carrier and hence the driven shaft remains stationary.

A planetary gearbox contains a number of such units in different sizes to obtain various speed reductions.

Operating characteristics of the planetary gear depend entirely on which members are locked and which are free to revolve about their own axes. A planetary gear set acts as :—

(*a*) **Overdrive** when the planet carrier is driving member, sun gear is driven member and ring gear is locked.

(*b*) **Overdrive** when the planet carrier is driving member ring gear is driven member and sun gear is locked.

(*c*) **Reverse Overdrive** when ring gear is driving member, sun gear is driven member and planet carrier is locked. The planet gears may or may not be locked.

(*d*) **Speed Reduction** when ring gear is driving member, planet carrier is driven member and sun gear is locked.

(*e*) **Speed Reduction** when sun gear is driving member, planet carrier is driven member and ring gear is locked.

(*f*) **Reverse** when sun gear is driving member, ring gear is driven member, planet gears are free to rotate around their axis, and planet carrier is locked.

(*g*) **Clutch** when sun gear is driving member, ring gear is driven member and planet carrier is locked. The planet gears may or may not be locked.

4. Construction of Conventional Transmission. A transmission consists of the following two units :—

(*i*) Housing containing shafts, bearings and gears etc.

(*ii*) Top cover containing selective mechanism.

A selective or progressive type gearbox usually contains the following shafts which carry the gears :—

(*i*) Primary shaft,

(*ii*) Main shaft,

(*iii*) Lay or counter shaft,

(*iv*) Idler shaft.

The primary and main shafts are in one line and the lay shaft runs parallel below them. In progressive type gearbox, the primary shaft usually passes through the main shaft. The shafts are supported in the housing through plain, ball and roller bearings. The primary and main shafts are projecting outside the gearbox. The primary shaft is known as clutch shaft and carries clutch plate when the transmission is installed with the engine. The main shaft contains a coupling at its outer end for connection with the propeller shaft.

The idler shaft is placed to one side of layshaft and carries idler gears to obtain reverse speed. The main shaft is splined over which the gears, dog clutch or synchronizing unit slide endwise. In sliding mesh gearboxes, the layshaft is a simple shaft which carries a fixed train of gears having different number of teeth. In progressive type gearbox, the layshaft is usually pegged or splined over which gears or dog clutches move endwise.

The gears used in the transmission are spur, helical or herringbone. The spur gears have straight teeth whereas helical gears contain inclined or helical teeth. The teeth on herringbone gears form the shape of V. Spur gears are employed in sliding mesh gearboxes whereas helical and herringbone gears are employed where the gears are to remain in constant mesh. The latter two types of gears are usually used in constant and synchromesh gearboxes.

The selective mechanism is employed to obtain various gear speeds. It consists of shafts, forks and balls and springs. The

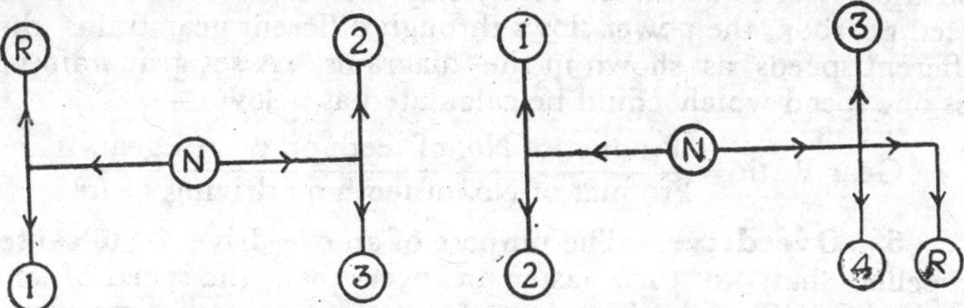

Gear change positions on a 3 speed gearbox.

Gear change position on a 4 speed gearbox.

Fig. 14·7

shafts carry the forks to operate or slide gears, dog clutches or synchronizing units. Balls and springs lock the position of forks or shafts as the case may be. In some cases the forks are fixed with the shafts and move as one assembly. In other cases, the shafts are fixed and the forks slide over them.

The selective mechanism is operated through the gear change lever which is either directly mounted at the gearbox or is a remote control by the side of steering column. The path of gear lever travel is usually like the letter H in selective type gearboxes. Reverse speed is obtained by shifting the lever to the right or left on neutral line beyond the legs of H, and then moving it up or down.

The power flows from the primary shaft to the layshaft and then through the set gear train to the output coupling of main shaft.

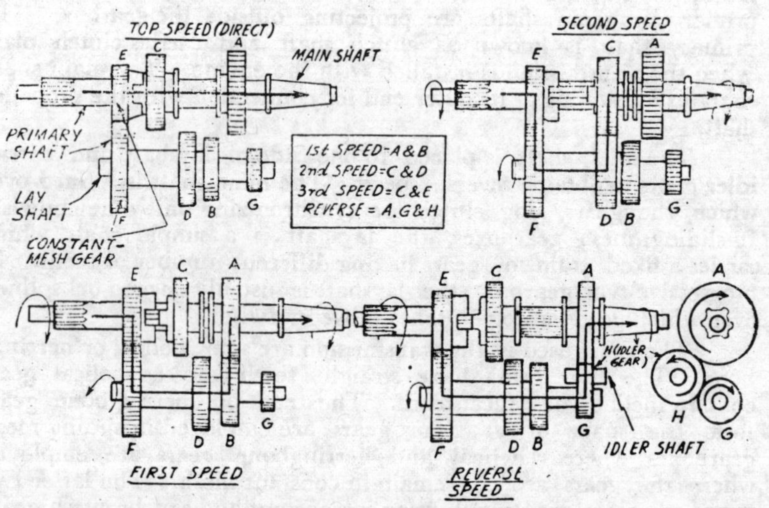

Fig. 14·8. Power flow in various gear speeds (3 speed gearbox).

In direct drive, the main shaft is clutched with the primary shaft and the power flows direct from primary to main shaft. In a three speed gearbox, the power flows through different gear trains during different speeds as shown in the diagrams. A set gear train provides one speed which could be calculated as below :—

$$\text{Gear Ratio} = \frac{\text{Product of No. of teeth on driven gears}}{\text{Product of No. of teeth on driving gears}}$$

5. Overdrive. The purpose of an overdrive is to cause the propeller shaft to turn faster or over than the speed of engine crankshaft. The overdrive in a planetary or epicyclic gearbox has already been explained under epicyclic transmission.

6. Free-wheeling Device or over-running clutch. Free-wheeling device enables the vehicle to free-wheel or overrun while going down a gradient or when the accelerator is released. It does not prevent the vehicle to be driven positively in a forward direction. It is fitted at the back of gearbox and in case an overdrive is attached to the gearbox, it is either incorporated in over-drive or attached at its back. In every case, it is fitted between the propeller shaft and transmission.

Free-wheeling device works on the same principle as the free-wheel in a bicycle. It consists of a central driving member which is attached to the splined shaft forming rear end of the gearbox output shaft. An annular outer driven member is mounted over the driving member. The driven member is connected to the propeller shaft.

The driving member contains wedge shaped recesses at its outer edge, equally spaced and usually three or four in number in a free wheel. In these recesses are contained a series of hardened steel rollers of different sizes as shown in the diagram.

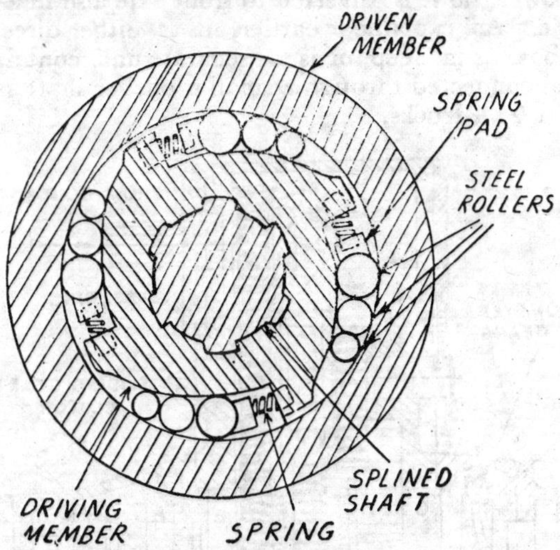

Fig. 14·9. Free-wheeling device.

In the positive drive, the rollers are forced to the right by the spring pad. They become wedged between the driving and driven members as a result of spring pressure and driving effect of driving member.

When the driven member over-runs the driving member, the rollers move to the left and the drive between them is disconnected.

There is dog-clutch type arrangement for locking the free-wheeling unit. In order to lock the free-wheel, the toothed dog on the gearbox splined shaft is moved to the right so as to engage with the internal teeth of the outer member of the free-wheel. The sliding dog is controlled by means of a fork and flexible cable having hand operated control at the instrument panel.

7. Advantages of free-wheeling device. (*i*) Gear shifting is simplified.

It is usually not necessary to employ clutch once the car is in motion.

(*ii*) The free-wheeling device enables the car to free wheel while going down a gradient or when accelerator is released.

(*iii*) It saves transmission from wear whenever the vehicle free wheels.

(*iv*) There is less frictional resistance to the movement of the vehicle since the engine and transmission are disconnected.

(*v*) Fuel consumption is reduced.

8. Transfer case. It is an auxiliary transmission by means of which power flow is diverted to front axle also in a four wheel drive vehicle. As explained earlier, it is either directly attached to the gearbox as in Jeep or is a separate unit contained next to gearbox and connected through a small propeller shaft as in Chevrolet and Ford 4×4 trucks.

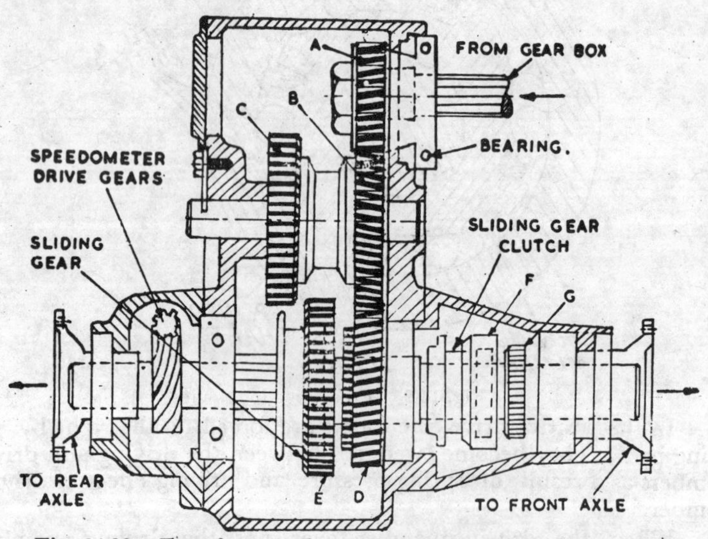

Fig. 14·10. Transfer case (Low speed through A×B and C×E ; High speed through A×B and D×E (clutching) ; Power flow to front wheels by clutching (F×G).

It is not only the power transferring device but is a speed and torque changing device too.

The drive to the propeller shaft goes through separate gear trains. It usually provides two speeds : one low and other high. There are in all five positions of the gear levers as shown in the diagram :—

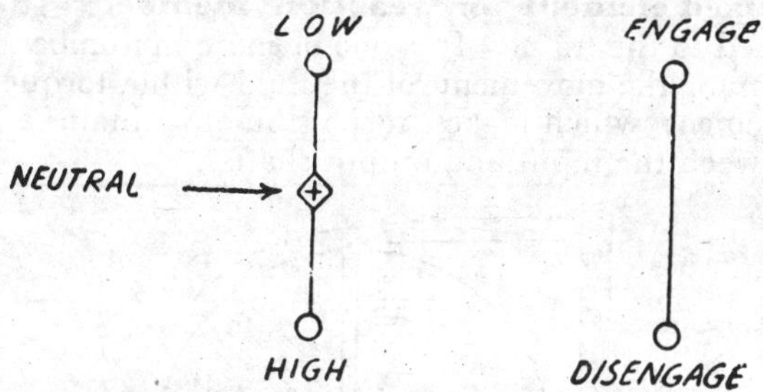

Fig. 14·11. Gear lever positions.

The front wheel drive could be engaged or disengaged by one lever whereas the other lever selects low or high speeds.

When the rear drive only is used, the transfer case gears simply act as idlers to transmit power from the main transmission to the rear propeller shaft. When both rear and front axles are to be powered, the change lever is put in the engage position.

9. Torque converter. It is a kind of hydraulic transmission which increases torque while reducing speed. Its function is similar to that of a gearbox A gearbox provides only a small number or fixed gear ratios but the torque converter provides a continuous variation of ratio from the lowest to the highest.

Torque converter is somewhat similar to fluid coupling. It differs from fluid coupling in one respect that it consists of three principle components instead of only two. Like fluid coupling, it transmits power through the media of fluid. The blades of the torque convert are designed to impart movement to the fluid. This is the reason why the torque converter never locks up as the fluid coupling does when centrifugal force on the driving and driven members is nearly the same and the movement of fluid stops.

A torque converter consists of the following three principal components ;—

(*a*) Driving element or impeller,

(*b*) Driven element or rotor,

(*c*) Fixed element or reaction member.

(*a*) **Driving element or impeller.** It is connected to the engine flywheel and acts as a pump. It is the driving member and is also known as driving torus.

(*b*) **Driven element or rotor**. It is a sort of turbine which is connected to propeller shaft. This is one or more in number. Its blades are inclined having pitch. Energy from the impeller is imparted to it.

(*c*) **Fixed element or reaction member**. It is a stator which is fixed to the frame. It is one or more in number. It changes the direction of the movement of the fluid within torque converter. It is the element which makes it possible to obtain a change of torque between the input and output shafts.

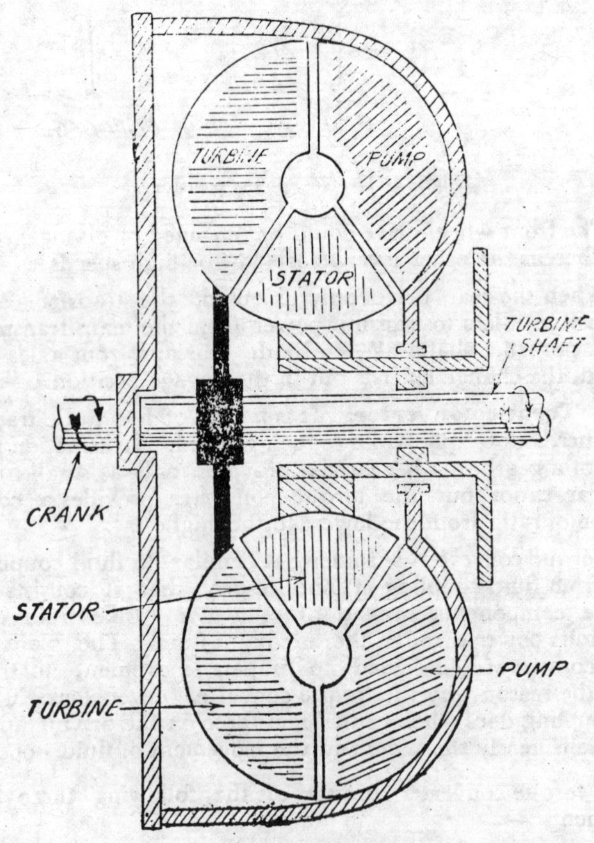

Fig. 14·12. Torque converter.

When the impeller starts moving, the fluid from the driving torque or pump strikes the rotor or turbine blades. The thrust of the fluid on the blades applies torque to the turbine and the rotor starts moving.

Fluid leaving the turbine is then directed against the stator or reactor. Torque multiplication occurs due to change in fluid direction. In case in the converter contains more than one turbines, the fluid entering the next or second turbine is capable of exerting more push or thrust than it did on the first turbine. The process is repeated for as many combinations of reactor and turbine stages as are built into the torque converter. Torque multiplication takes place during each redirection of fluid from a reactor to turbine.

10. Comparison between fluid coupling and torque converter.

Fluid coupling	*Torque converter*
1. Transmits power through fluid.	1. Also transmits power through fluid.
2. Automatic device.	2. Also automatic device.
3. It has two main components : impeller and runner.	3. It has three main components : pump, stator and turbine.
4. It is simply a means to connect driving and driven members.	4. It is a torque multiplication unit.
5. Impeller and runner are locked up and movement of oil stops during engagement when centrifugal force is approximately the same on both members.	5. It never locks up and flow of oil never stops but continues.
6. Its blades are merely fins.	6. The turbine blades are inclined having pitch.
7. It serves the purpose of an automatic clutch.	7. It acts as an automatic clutch and serves the purpose of automatic gearbox to increase torque.
8. It is efficient at highway speeds.	8. It is not as efficient as fluid coupling at highway speeds but is slightly more efficient under load.
9. It is not assisted by friction clutch.	9. It is usually used in conjunction with automatic clutch (mostly fluid flywheel) to eliminate the slight loss of efficiency at highway speeds.

11. Torque converter oil. The oil or fluid used in torque converter is mineral oil having anti-oxidant additive. The converter fluid should possess the following properties :—

(*i*) It must have the correct viscosity at both higher and lower working temperatures.

(*ii*) It should possess good anti-corrosive properties.

(*iii*) It should not be injurious to the oil seals and shafts.

(*iv*) It should offer maximum resistance to foaming.

(*v*) It must have good lubricating properties.

12. Pre-selective gearboxes. In these gearboxes a pre-selector lever mounted on the steering wheel takes the place of ordinary gear change lever. The lever is moved in an arc to the marked positions relating to 1st, 2nd, 3rd, 4th and reverse, for selecting different gear speeds. In order to change gear, the desired speed is selected by moving the lever to the required gear position marked on the steering wheel sector. After selecting the required gear speed, the gear

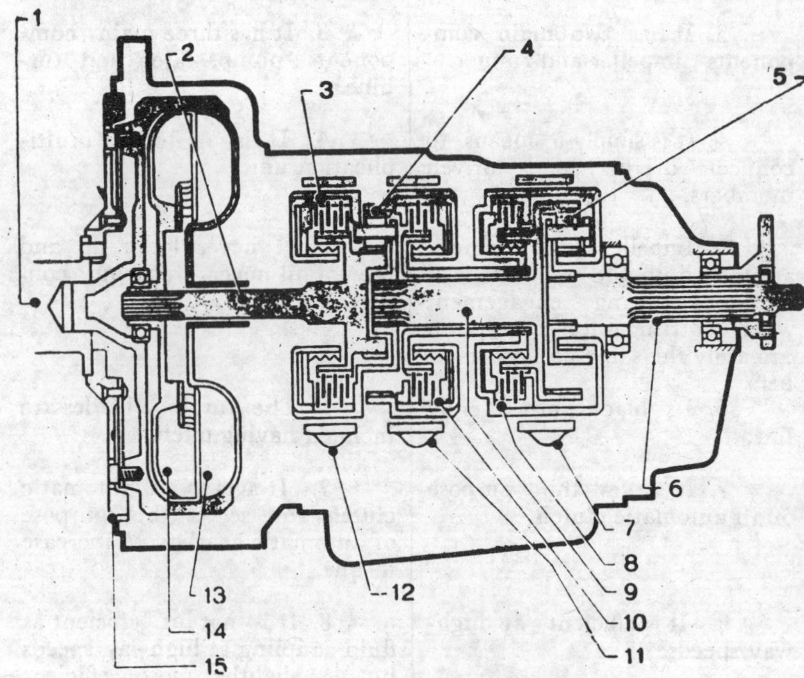

Fig. 14·13. Automatic Transmission.

1.	Crank shaft	2.	Front shaft
3.	Clutch	4.	Front planetary unit
5.	Rear planetary unit	6.	Rear shaft
7.	Brake band 2	8.	Clutch 2
9.	Intermediate shaft	10.	Clutch 3
11.	Brake Band 1	12.	Brake Band 3
13.	Pump	14.	Turbine
15.	Flywheel.		

change pedal is depressed and then released when the selected gear is automatically engaged. The next gear position thus can be preselected at any convenient time before actually changing to that gear speed.

The preselective arrangement is usually used with planetary transmissions in which gear changing is done by loosening and tightening brake bands.

13. Automatic Transmission. Transmissions in which various speeds are obtained automatically are known as automatic transmission. The driver merely selects the general car condition such as neutral, forward or reverse. The selection, timing and engagement of gear for the required gear speed are accomplished automatically when the accelerator is pressed or depressed. Automatic transmission need no gear change lever and clutch pedal since clutch and transmission is a combined unit and works automatically. The transmission is operated hydraulically and is often known as *hydramatic transmission*. This transmission employs fluid coupling, torque converter and epicyclic gear arrangement. When all the different devices are combined into one unit, they perform their duties collaborately.

Automatic transmissions are popular under different names as prescribed by the manufacturers. They differ in construction a little bit. Some employ only fluid coupling with the planetary transmission whereas the others include torque converter with fluid coupling and planetary transmission. The important transmissions as applicable to American cars are well known under the following names :

(*i*) Buick Dynaflow.
(*ii*) Chrysler Powerflite.
(*iii*) Chevrolet Powerglide or Turboglide.
(*iv*) Studebaker Flightomatic.
(*v*) Ford Fordomatic and Mecomatic.
(*vi*) Packard Ultramatic.
(*vii*) Lincoln Turbodrive.

In hydramatic gearbox, the planetary gear sets are placed in series so that power may be transmitted through them. A centrifugal governor incorporated in the transmission, selects the proper gear for each speed and throttle position. The change from one gear to another is accomplished through hydraulically operated pistons assisted by springs which control brake bands on the planetary gear sets and clutches within the planetary unit. The speed at which the various shifts occur, is governed by the throttle and centrifugal governor.

Driver control is provided by a lever and sector segment which are usually located on the steering column, below the steering wheel. The control could be adjusted to any of the following positions :

Neutral—Drive--Low—Reverse

In the drive position, the transmission shifts automatically for any one of the forward speeds. The low position is for hill climbing or muddy road or under heavy load, and only first and second gear speeds are provided. Reverse position is for driving back. In neutral position, there is no power transmission.

14. Daimler Benz (DB) automatic-transmission. This transmission consists of two major units; the fluid coupling and a four speed planetary transmission. For the four forward speeds and reverse, this transmission has only two simple planetary gear sets. For changing of gears, three multidisc clutches and three brake bands have been provided. Shifting occurs automatically depending on engine torque as well as on vehicle speed.

The hydraulic coupling is of novel design to improve efficiency at low speeds. Its wall has a sharp edge at the outer diameter. It

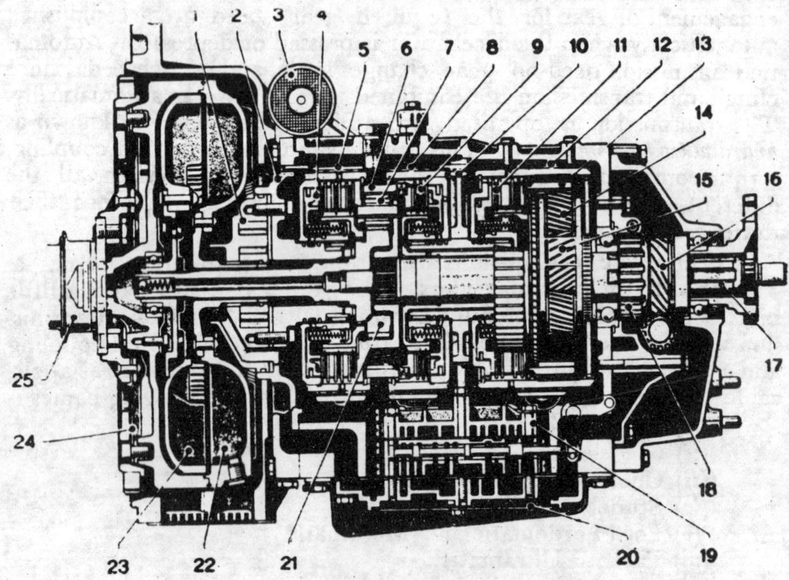

Fig. 14·14. Benz Automatic Transmission.

1. Front oil pump.
2. Front cover.
3. Solenoid.
4. Clutch 1.
5. Brake Band 3.
6. Outer gear, front planetary unit.
7. Front planetary gears.
8. Brake band.
9. Clutch 3.
10. Central partition.
11. Clutch 2.
12. Brake band 2.
13. Outer gear, rear planetary unit.
14. Rear planetary gears.
15. Sun gear, rear planetary unit.
16. Driving gear for governor and secondary pump.
17. Output shaft.
18. Parking Block.
19. Shift plate.
20. Oil strainer.
21. Sun gear, front planetary unit.
22. Impeller.
23. Turbine member.
24. Driving disc.
25. Crankshaft.

helps in abruptly changing the fluid flow from radial to axial direction at the pump and *vice versa* at the turbine. This design has at low slip. the high torque capacity of a large coupling and at stall, the low capacity of a small coupling.

The hydraulic coupling is mounted at the crankshaft by a flexible disc. The impeller transforms mechanical power of the engine into hydro-kinetic power of the oil flow which in turn is reconverted into mechanical power by the turbine. In order to keep the oil in the coupling at operating temperature, it is constantly circulated through a heat exchanger.

The front cover of the transmission is housing the primary or front oil pump and also its suction and pressure oil channels. The pump is of internal gear type. Contrary to the torque converter, the hydraulic coupling does not have a reaction member, thus keeping the pump diameter very small which also reduces the power losses of the pump.

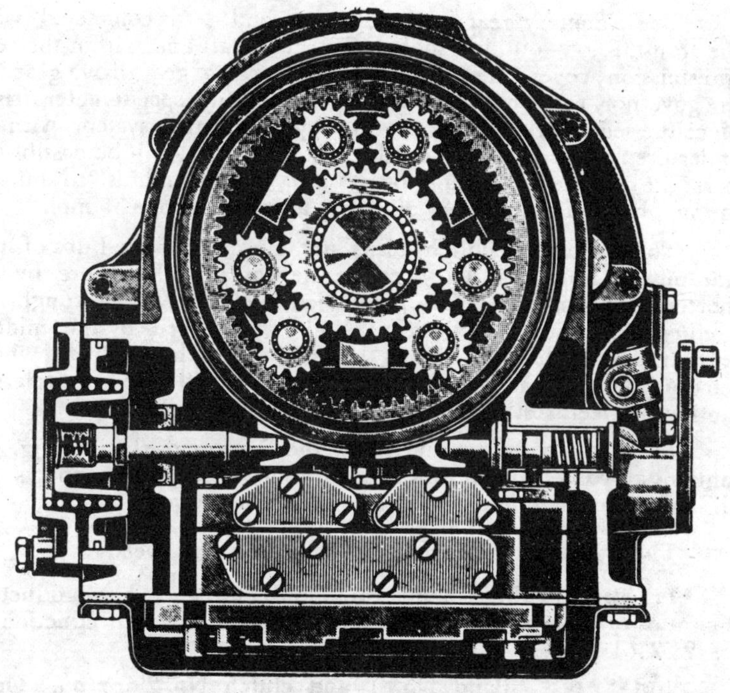

Fig. 14·15. Cross-section of Daimler Benz automatic transmission with 6 planetary gears.

Each planetary gear set consists of three identical planetary gears; one sun gear and one internally toothed ring gear. The power flow goes from the turbine to the sun gear of the first planetary gear set. When the ring gear of the first gear set is held stationary, the speed ratio of planetary carrier to sun gear is 1 : 2·52. The planetary carrier can be locked up with the sun gear by clutch, giving a ratio 1 : 1. The planetary carrier however, can also be held stationary by brake band. This condition reverses the rotation between ring and sun gears in the ratio of 1 : 1·52.

To sum up, the first planetary gear set results in three different ratios. The carrier of the first planetary gear set is connected with the ring gear of the second planetary gear set. When the sun gear of the second set will be held stationary by brake band, the ratio between the carrier and ring gear of second planetary set is 1 : 1·58. When the clutch locks up ring and sun gears of second planetary set, the entire secondary set will turn as one unit (ratio 1 : 1). When, however, the ring gear is held stationary by third brake band, the sun gear is driven by third clutch, resulting in a ratio of 1 : 2·72 between the planetary carrier and sun gear. This way the second planetary gear set can also produce three different ratios.

The planetary gear carrier of the second set is connected with the transmission output shaft. On this shaft and within the rear transmission cover, there are also the parking gear, drive gear for the governor, rear or secondary oil pump and speedometer drive. Since the secondary oil pump will supply the entire system with oil under pressure as long as the vehicle is moving, it will be possible to start the engine by towing or pushing the car and also to shut-off the supply from the primary oil pump at speeds above 35 mph.

The revolving clutch units are running on the hubs of the transmission housing. They receive the working oil pressure by the shortest possible way. Oil is supplied to clutch No. 1 through the front cover whereas clutches No. 2 and 3 are served by the middle portion. Each clutch unit is at the same time a brake drum which remains stationary as long as its band is applied. There is no relative motion between resting drum and its supporting hub.

The valve body is mounted to the bottom of the housing. It contains all valves and springs necessary for shifting and connects the oil lines from the pump, the governor and the servos.

The transmission provides the following gear speeds:

1st Speed. Brake bands No. 1 and 2 are on. The product of the ratios of both planetary gear sets will result in total reduction of 1 : (2·52 × 1·58) = 1 : 398.

2nd Speed. Band No. 1 and clutch No. 2 are on. Only planetary gear set No. 1 is used for ratio of 1 : 2·52.

3rd Speed. Clutch No. 1 and brake band No. 2 are on. Only second planetary gear set is working for the ratio of 1 : 1·58.

4th Speed. Both clutches are on. Direct drive results and the ratio is 1 : 1

Reverse. Brake Band No. 3 and clutch No. 3 are on. In the first planetary gear set, the reverse speed is obtained and the second gear set is employed for necessary reduction. The result of the combination is $1:(-1\cdot52\times2\cdot72)=1:4\cdot13$.

The arrangement of planetary gear sets has been done in such a way which permits both sets to operate only in first or reverse speeds. In second or third speeds, only one planetary gear set is active and in the fourth speed, the drive is direct and none of the planetary sets is used.

The selector lever has the following ranges :

(*i*) "P" = Parking when a pawl blocks the movement of the parking gear in the transmission.
(*ii*) "R" = Reverse.
(*iii*) "O" = Neutral.
(*iv*) "4" = Drive—Normal driving position under all traffic conditions.
(*v*) "3" = Mountain—mountain range.
(*vi*) "2" = Low—up to second speed only.

For driving under special conditions such as trailing a caravan over mountain passes, on steep garage ramps and whenever severe engine braking is required.

15. Advantages of Automatic Transmission.

(*i*) Simplified driving control.
(*ii*) Less physical and mental fatigue to the driver.
(*iii*) No clutch pedal and gear lever and hence simplification of driver's compartment.
(*iv*) Smoother running under all conditions due to hydraulic coupling and automatic gear change.
(*v*) No shocks or jerky driving.
(*vi*) Improved acceleration and hill climbing.
(*vii*) Reduced fuel consumption.
(*viii*) Less wear and tear due to planetary gearing.
(*ix*) Noisefree gear shifting.
(*x*) Longer life.

QUESTIONS

1. What is transmission and which objects are fulfilled by it ?
2. What is the necessity of transmission ? Explain.
3. Which resistances affect the movement of an automobile ? Describe.
4. Define tractive effort and tractive resistance. What is the relationship between the two ?
5. If an engine produces a torque T, what torque shall act the driving wheels if gearbox gear ratio is 1 : n and final drive reduction is in the ratio of 1 : m ?
6. Which are the different types of transmission ? Define them.

7. How selective type transmission differs from progressive type transmission ?

8. Which are the different types of selective type transmission ? Describe any one of them.

9. What is the difference between constant-mesh and synchromesh transmissions ? Describe any one of them.

10. How a planetary transmission acts as (*a*) Clutch (*b*) Over Drive (*c*) Speed reduction (*d*) Reverse drive ?

11. Describe the following :

(*a*) Synchronizing Unit.

(*b*) Planetary Transmission.

12. Describe the construction of a sliding mesh gearbox. Show how the power flows in various speeds.

13. What is free-wheeling device ? How it works and what are its advantages ?

14. What is torque converter ? How it differs from gearbox ?

15. Illustrate the construction and working of a torque converter.

16. Compare hydraulic coupling and torque converter.

17. What do you know about the following ?

(*a*) Transfer case.

(*b*) Overdrive.

(*c*) Automatic transmission.

(*d*) Pre-selective gearbox.

18. Illustrate the construction and working of Transfer case.

19. Illustrate an automatic transmission.

20. What are the advantages of automatic transmission ? Which are the usual selector positions in an automatic drive ?

15

Drive Shaft and Drive Axle

1. Drive shaft. It is the shaft by means of which drive is given to the driving axle. Drive shaft is known as propeller shaft by means of which power from transmission is transmitted to the driving axle at varied lengths and varied angles. This shaft connects the transmission with the driving axle. In vehicles having long wheel base, the drive shaft is in two pieces and is supported at the centre to avoid whipping action of a long shaft.

The shaft contains a *slip joint* which enables the shaft to squeeze and increase in length. The splined end of the shaft slides inside the other end to increase and decrease in length.

The drive shaft is connected with the transmission and driving axle by means of universal joints. In Hotchkiss Drive, there is a universal joint at each end of the drive shaft whereas in Torque Tube Drive there is only one universal joint on the transmission side. In Hotchkiss Drive, the drive shaft is open whereas in Torque Tube Drive, the shaft is enclosed by a torque tube.

2. Universal Joint. It is the joint which enables the drive shaft to transmit power at varied angles.

The transmission which is connected with the driving axle by means of drive axle is at higher level than the drive axle. The propeller shaft is never in line with the engine crankshaft. So, the drive shaft is subject to transmit power at an angle. Moreover, this angle is never uniform. It is varied when the road wheels move up and down due to road irregularities and the frame moves up and down in relation to wheels, depending on the amount of

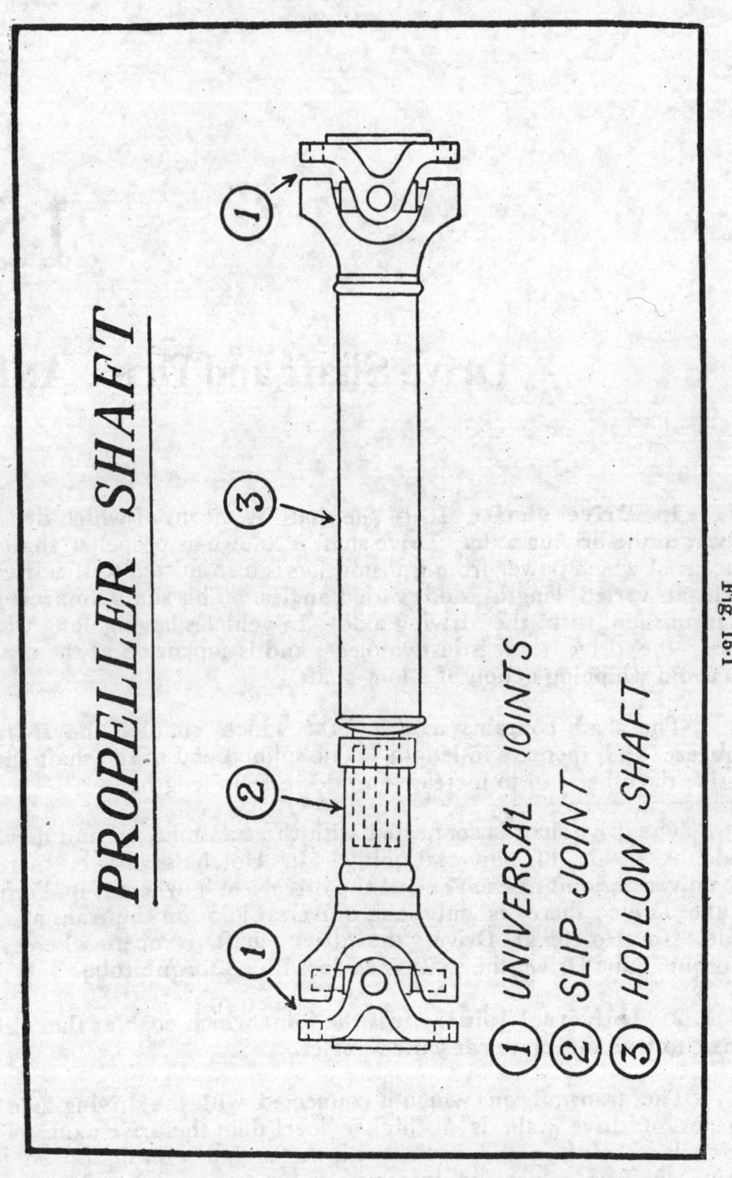

Fig. 15·1

weight in vehicle body. The universal joints provide the ability to the shaft to work through varied angles.

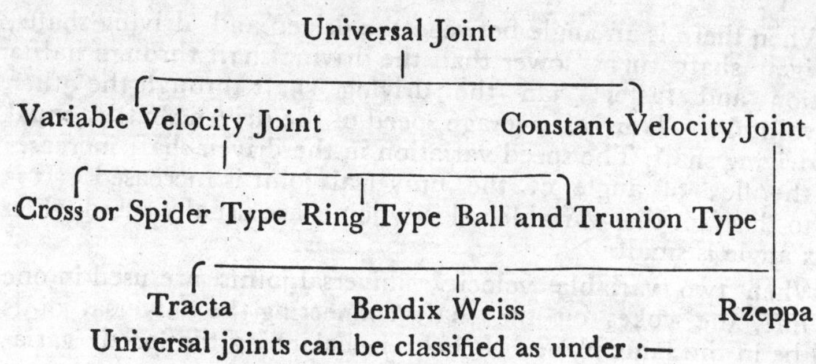

Universal joints can be classified as under :—

(*a*) Variable Velocity Joints,

(*b*) Constant Velocity (C.V.) Joints.

(*a*) **Variable Velocity Joints**. In variable velocity joints, the driven and driving shafts do not turn at the same speed through each part of a revolution although they turn at the same r.p.m. The driven and driving shafts should therefore be in a straight line so that they may turn at the same speed through each part of a revolution. But in an automobile, it is not feasible as the drive shaft is inclined.

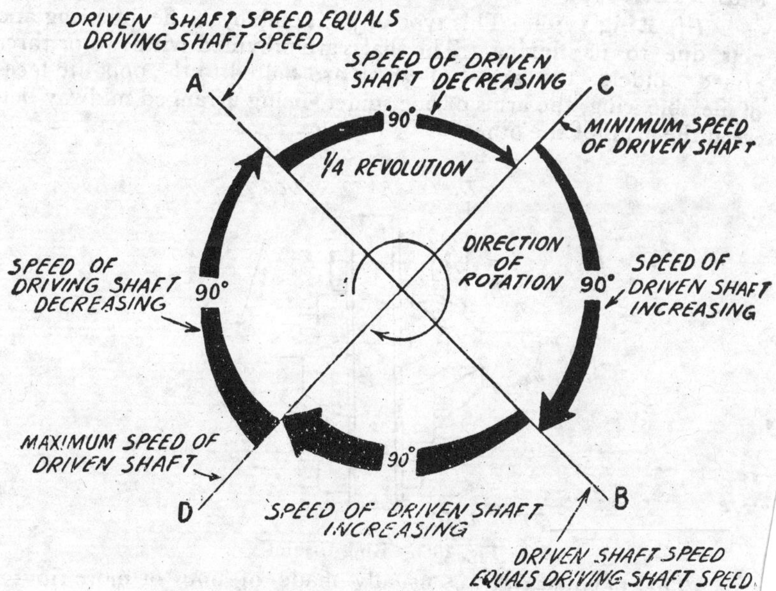

Fig. 15·2. Working of variable velocity universal joint in one revolution (Thickness of arrows indicates increase and decrease of speed)

At A and B driven shaft speed equals driving shaft speed. At C, speed of driven shaft is minimum and at D, the speed of driven shaft is maximum.)

When there is an angle between the driven and driving shafts, the driven shaft turns lower than the driving shaft through half a revolution and faster than the driving shaft through the other half revolution. Thus the average speed of the driven shaft is equal to the driving shaft. The speed variation in the driven shaft increases when the flex of angle of the universal joint is increased. It is owing to this fact that variable velocity joints are usually used when the flex angle is small.

When two variable velocity universal joints are used in one drive line, the yokes on the shafts connecting the universal joints should be in the same plane. It helps in balancing the speed variations.

Variable velocity joints are of the following types :—

(*i*) Cross or spider type,

(*ii*) Ring type,

(*iii*) Ball and trunion type.

(*i*) **Cross or spider type.** This type of universal joint consists of two yokes connected respectively to the driving and driven shafts and at right angles to each other by means of a cross or spider. Needle type bearings are employed between the yokes and cross ends. These types of joints are mostly used with the drive shafts.

(*ii*) **Ring type.** This type of joint employs a flexible ring and acts due to its flexing. The shafts are provided with two or three armed spiders, the arms of which are bolted to the opposite faces of flexible ring, the arms of one spider being arranged midway between the arms of the other.

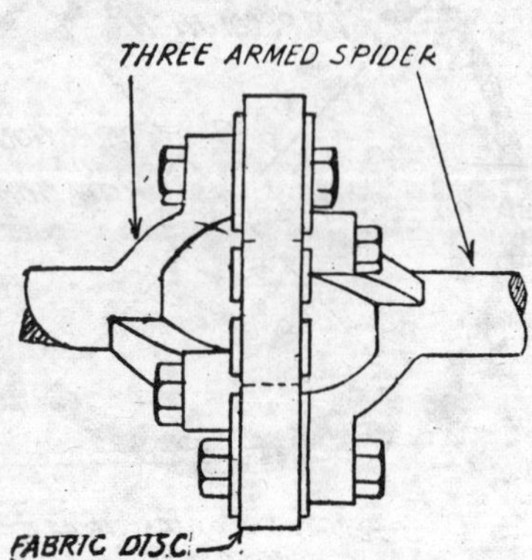

Fig. 15·3. Ring-U-joint.

The flexible ring is usually made of one or more rings of rubberized fabric made in a special way for providing necessary strength. Sometimes a number of thin steel discs are used instead of fabric rings. When the shafts are revolving about their axes, there is a continuous flexing of the ring to enable drive through varied angles.

This joint can also accommodate a considerable amount of axial movement of the shaft. It also helps in smoothing out torque fluctuations and needs no lubrication. Its main drawback being that the ring does not withstand for a long period.

(*ii*) **Ball and trunion type.** This type of joint combines both the universal and slip joint in one assembly. In this type, a pin or cross shaft is mounted crosswise in T fashion in the end of universal joint shaft. Each end of the cross shaft carries a ball mounted on needle bearings. The balls fit into grooves in a trunion which is attached to the other shaft. The complete assembly is free to slide in grooves machined in the outer body of the joint. A heavy spring tends to resist excessive longitudinal movement of the shaft. Power is transmitted through the trunion, balls and cross shaft. The bending moment is accomplished in one direction by the rolling action of balls and in the other direction by the balls moving lengthwise in the trunion grooves. A leather or rubber boot covers the open end of the shaft.

(*b*) **Constant Velocity Joints.** In these types of joints, the driven shaft turns at the same speed as the driving shaft, through each part of its revolution and at any degree of flex. Mostly these joints are used in front drive axles where the joint must transmit power through a large angle and at the same time not introduce a speed variation or vibration which would cause steering difficulty and excessive tyre wear. Cadillac cars employ ball and socket type constant velocity joints in their propeller shafts.

These types of joints are of the following main types :—

(*i*) Rzeppa,

(*ii*) Bendix Weiss,

(*iii*) Tracta.

(*i*) **Rzeppa.** It consists of spherical inner and outer ball races which have grooves cut parallel to the shafts. Steel balls are held in the grooves on the spherical recess. The torque is transmitted from one race to another by the balls. The circular pattern of balls results in both shafts to turn at the same velocity.

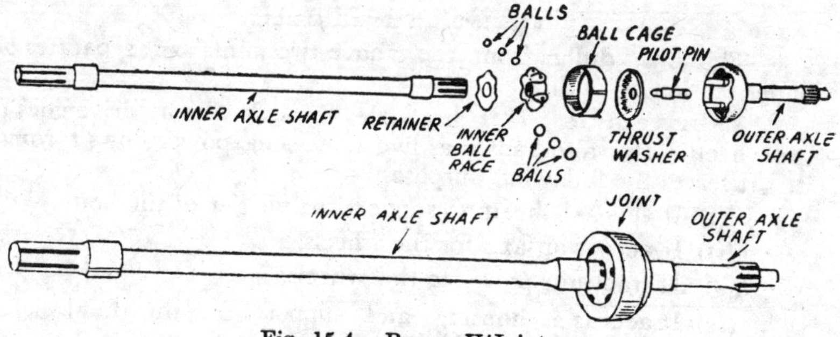

Fig. 15·4. RzeppaU/Joint.

(*ii*) **Bendix Weiss.** This type of joint also uses the principle of driving through balls held in a circle around a sphere. Driving balls four in number, are placed into races which are machined in close fitting yokes. There is a fifth or centre ball placed between the two yokes, as an inner race. The driving balls arrange themselves in a circle in a manner similar to Rzeppa joint. The aligning action of the balls results in a constant velocity joint.

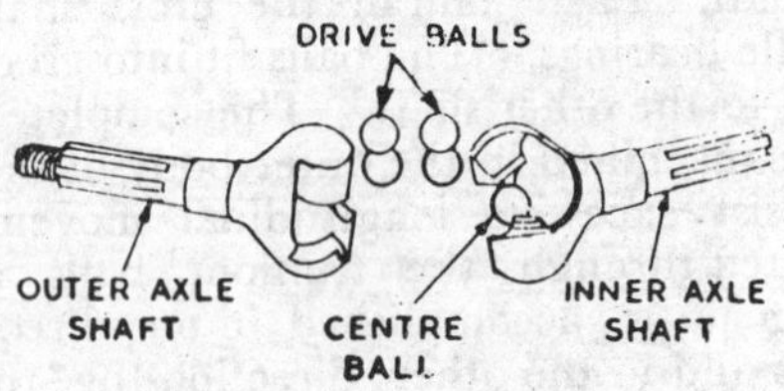

Fig. 15·5. Bendix weiss.

(*iii*) **Tracta.** There are four yokes in this joint ; two of them fastened to the shafts and the other two floating in the centre of the joint. The mating parts of the yokes are segments of a circle. The circular segments plus the floating action of the two yokes result in a constant velocity joint.

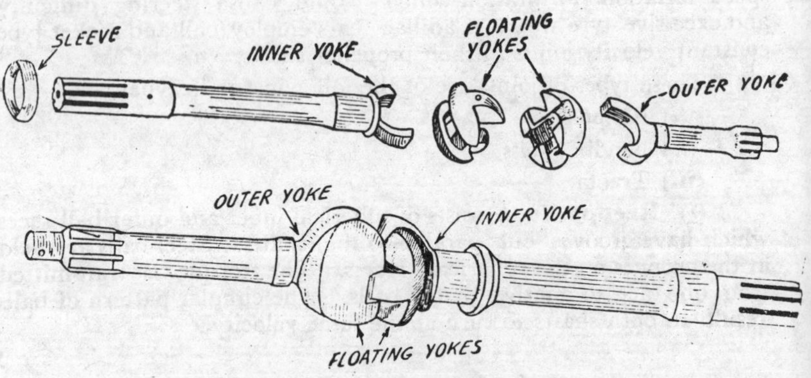

Fig. 15·6. Tracta U/Joint.

This joint differs from the above two joints as it separates off under tension or pull.

3. Drive Axle. It is that axle through which drive goes to the wheels. It is also known as live axle since power flows through it. It serves the following purposes :—

(*i*) It acts as a beam to support the weight of the body.

(*ii*) It acts as an axis for the wheels.

(*iii*) It transmits power to the wheels.

(*iv*) It acts as a housing and support for the final drive, differential and half shafts.

The drive axles are of the following two categories :—

(*a*) Front drive axle,

(*b*) Rear drive axle.

(*a*) **Front drive axle.** It is front axle of the vehicle through which power flows to the front wheels. The front drive axle has to

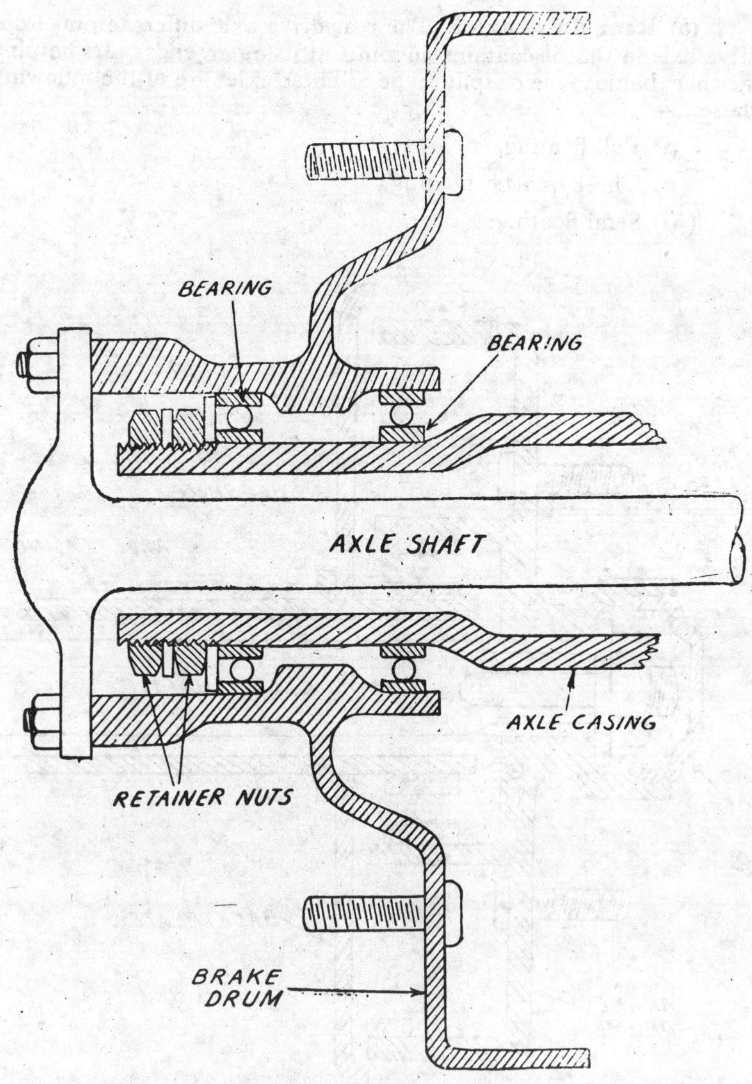

Fig. 15·7. Full floating,

do the task of steering the vehicle too in addition to its other functions. It contains joints at its outer ends to enable the wheels to swivel around for taking a turn. The main axle casing or housing may be one piece of banjo type or split type in two halves. The axle houses final drive, differential and half shafts. The half shafts or axle shafts contain constant velocity joints to enable them to transmit power through varied angles.

(*b*) **Rear drive axle.** The rear drive axle differs from front drive axle in that it contains no joints at its outer ends. Its housing is either banjo type or split type. These axles are of the following classes :—

(*i*) Full floating,

(*ii*) Three quarter floating,

(*iii*) Semi floating.

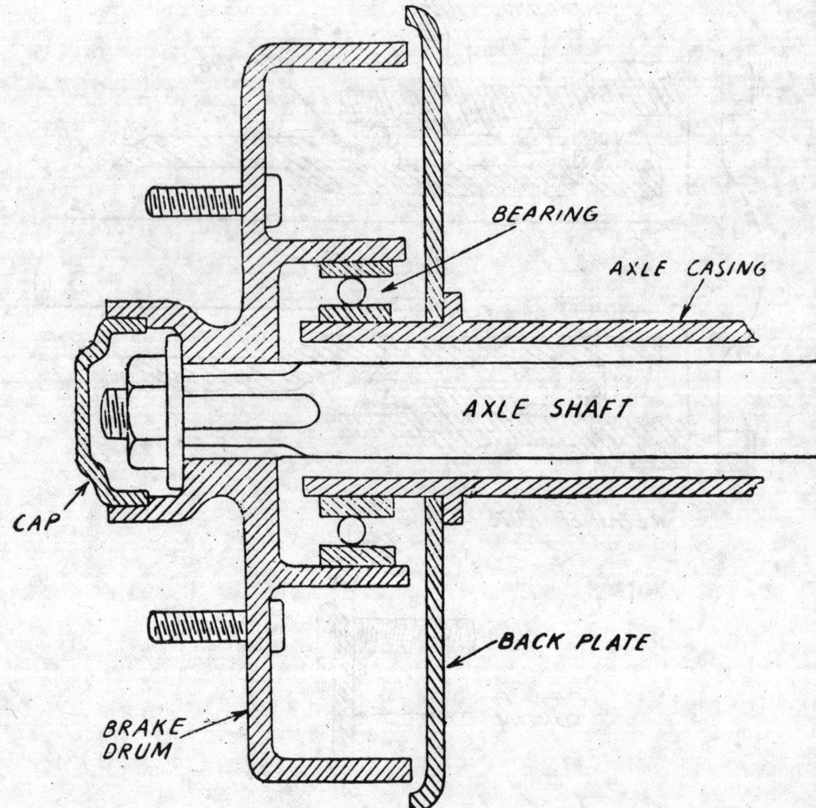

Fig. 15·8. Three quarter floating.

(*i*) **Full floating**. It is that axle in which all the load of rear portion of the vehicle is borne by the axle housing or tube only. In this type of axle, both the brake drum hub bearings are fitted on the axle tube and there is no load on the axle shaft which can be removed without jacking up the wheels.

(*ii*) **Three quarter floating**. It is that axle in which part load of the rear portion of the vehicle is borne by the axle tube and part by the axle itself. In this type of axle, one bearing of brake drum hub is fitted on the axle whereas the other on the axle tube.

(*iii*) **Semi-floating.** It is that type of axle in which all the load of rear portion of the vehicle is carried by the axle shafts. In this type of axle, the bearings of the brake drum hub are fitted on the half shafts.

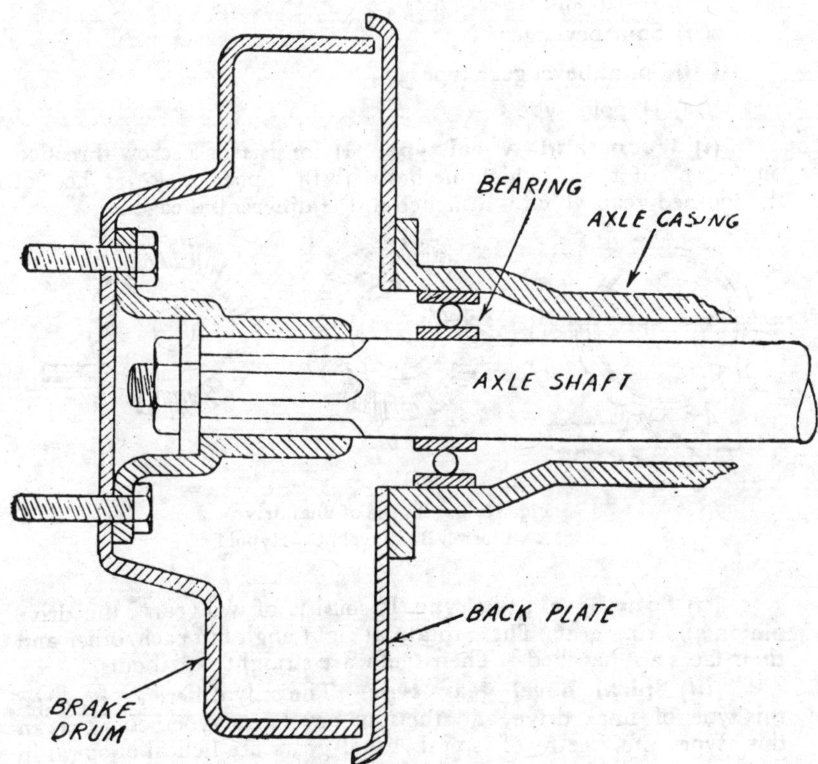

Fig. 15·9. Semi-floating.

4. Final Drive. Upto the drive axle, the power flows is one straight line. At the drive axle, it is diverted at right angles to flow towards the wheels. The change in the direction of power flow is obtained by means of final drive. The final drive also provides a fixed speed reduction between the drive shaft and the driving axles. Final drive is of following two main classes : (*a*) chain type (*b*) gear type.

(*a*) **Chain type**. This type of final drive is obsolete in motor cars and trucks now-a-days. In this type, the drive wheel is connected with the gearbox by means of chains and sprockets. Motor cycles employ this type of drive.

(*b*) **Gear type**. It consists of a ring gear and a drive pinion. The ring gear is riveted with the differential cage and the drive pinion is connected with the propeller shaft. Power from the propeller shaft flows towards the axle shafts through the drive pinion and ring gear.

Gear type final drive is of the following main kinds :

(*i*) Worm and wheel type,

(*ii*) Spur bevel gear type,

(*iii*) Spiral bevel gear type,

(*iv*) Hypoid type.

(*i*) **Worm and wheel type**. It consists of a screw threaded on the end of a shaft which meshes with the upper or lower face of the toothed gear which is attached to the differential cage.

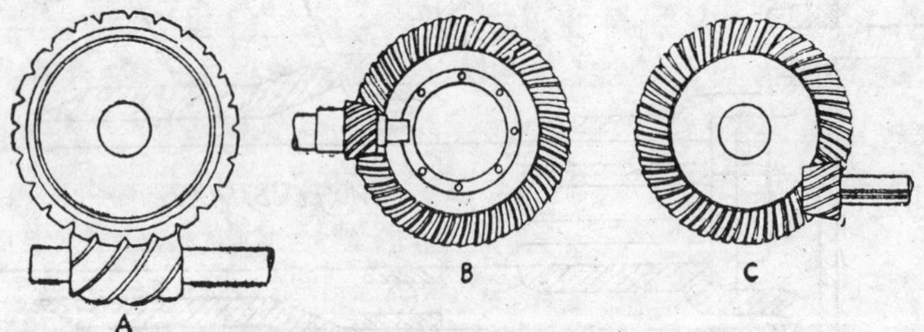

Fig. 15·10. Types of final drive.
A—Worm B—Bevel C—Hypoid.

(*ii*) **Spur bevel gear type.** It consists of two gears : the drive pinion and ring gear. These rotate at right angles to each other and their faces are bevelled. Their teeth are straight radial cuts.

(*iii*) **Spiral bevel gear type**. The only difference between this type of final drive and the spur bevel gear type being that in this type the teeth of spiral bevel gears are helical or spiral in shape.

(*iv*) **Hypoid type**. It resembles the spiral bevel type in some respects. The main differences are as under ;

(*a*) The gear teeth are cut in a hyperbola curve.

(*b*) The pinion meshes with the ring gear below the axle shaft level.

5. **Differential.** It is the mechanism by means of which outer wheel runs faster than the inner wheel while taking a turn or moving over upheaval road. It is of the following types :

(*a*) Conventional,

(*b*) Power-lock or Non-slip,

(*c*) Double reduction type.

(*a*) **Conventional.** It consists of a cage which contains differential gears. The differential gears consist of two sun gears and two or four star pinions, all of bevel gear type. The star pinions are fitted on a pin if these are two in number and on a spider, if these are four in number. The pinions are free to move around their axes. The pin or spider is held in between the two parts of the cage which encloses the differential gears. The sun gears and star pinions are always in mesh with each other. The sun gears are free to move inside the cage. The differential assembly is supported on taper roller bearings provided on both sides of the cage. When installed in the drive axle, the whole assembly moves around the bearings.

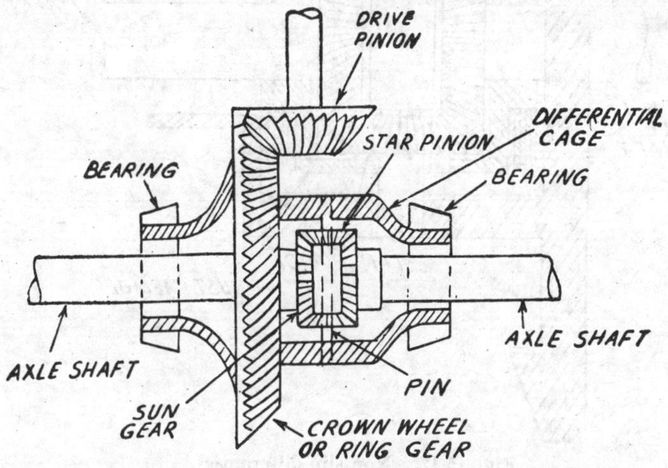

Fig. 15·11. Differential.

At the differential cage is attached the ring gear or crown wheel which forms part of the final drive. Drive is given to the ring gear by means of drive pinion to which propeller shaft is attached.

The sun gears are located parallel to ring gear inside the differential cage and face towards each other. Shaft to each wheel is splined into the sun gear of that side.

When the vehicle is moving on straight level road and the resistance affecting both the driving wheels is the same, there is no relative movement amongst the differential gears. The whole arrangement meshed together moves as one unit and both the half shafts in the driving wheels rotate at the same speed. But this actually seldom happens as the road is mostly never level.

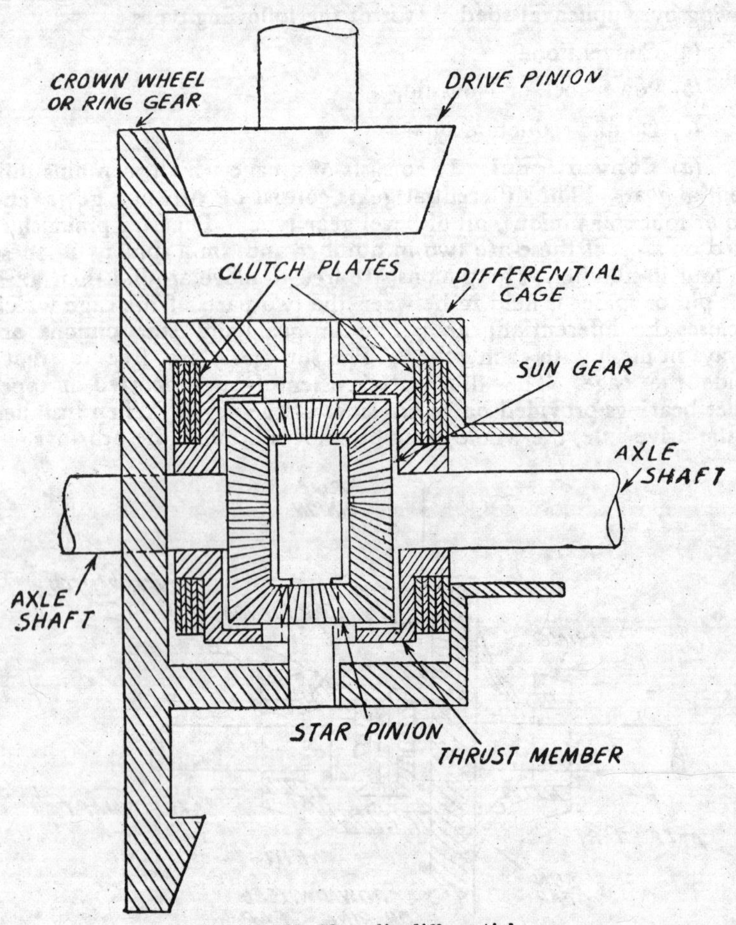

Fig. 15·12. Non-slip differential.

When the front wheels are turned to any direction to take a turn, a binding force acts on the inner wheel being nearer to the point around which the wheels are to move in a circle. The sun gear of that side is held slow in relation to the movement of the complete cage or crown wheel. When the vehicle is going straight on a level road, the power is divided equally at the differential, one half flowing to one side wheel and the other half flowing to other side wheel. While taking a turn when the binding force acts on the inner side sun gear and its speed is slowed down, the star pinions rotate the other side sun gear at a speed, as a result of loss on the inner side and gain on the outer side plus the speed at which the complete differential assembly is rotating. This results in a faster movement of the outer wheel than the inner one.

(*b*) **Power-lock or Non-slip**. The conventional differential delivers the same amount of torque to each wheel on the drive axle. If one wheel slips on a slippery road, mud or ice, the wheel becomes stationary, all the power being flowing to the slipping wheel. This results in no movement to the vehicle. In order to overcome this drawback, the vehicles are now provided with a non-slip or self-locking differential.

Non-slip differential is very much similar in construction to conventional type except that it has two sets of clutch plates in addition. In non-slip differential, the ends of pinion shafts lay loosely in notches, in the two halves of the differential cage.

Rounding a Corner: When taking a turn, the non-slip differential functions in the conventional way to tend the outer wheel to run faster than the inner wheel and thus to cover more distance within the same time period. This action is permitted by the slipping of clutch plates.

Straight ahead drive. The rotating differential cage carries the sun gear shafts around with it. Since there is considerable side thrust, the axle shafts in sun gears tend to slide up the sides of the notches in the two halves of the differential cage. As they slide up, they are forced outward and this force is transmitted to the clutch plates. Clutch plates thus lock the sun gears and the axle shafts to the differential cage. In this case, if one wheel encounters a muddy or slippery surface, that causes it to lose traction temporarily avoiding it from slipping or spinning since it can't turn faster than the other wheel as the half shaft is locked with differential cage by means of clutch plates.

(*c*) **Double reduction.** This type of differential provides one additional gear reduction between the engine and the driving wheels. This is employed in heavy duty vehicles which require higher gear ratio between the engine and the driving wheels.

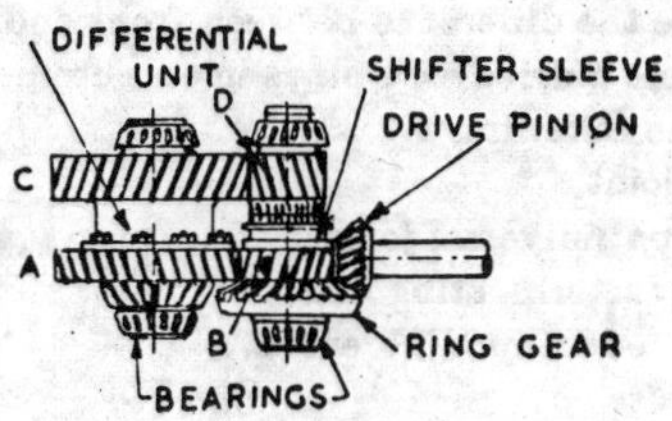

Fig. 15·13. Double reduction differential ;
(Low speed through C and D gears)
(High speed through A and B gears)

In this type of differential, the drive pinion meshes with a ring gear mounted on a straight shaft which contains a reduction-drive gear set. The reduction-gear drive set drives a driven gear set attached to the differential cage. The driven gear set contains more number of gear teeth as compared to reduction-drive gear set. Gear reduction is thus obtained between the drive pinion and ring gear and also between the two-reduction gear sets.

6. **Axle shafts or half shafts.** These are the jack shafts through which power flows from the differential to the driving wheels. There is a separate shaft for each half of the drive axle. A half shaft connects sun gear of the differential with the brake drum over which wheel is mounted.

Half shafts for the rear drive axle are in one piece and splined at the front end to fit into the differential sun gear. The rear end of the shaft usually contains a flange by means of which connection is made with the brake drum.

Axle shafts for the front drive axle contain constant velocity joint to accommodate steering action. In case of independent suspension; sun gear shaft is connected with the stub axle through a carden shaft which contains universal joints.

QUESTIONS

1. **Illustrate the drive shaft. Explain the necessity of universal joint.**
2. **Explain the type of universal joint which eliminates the necessity of slip joint in a drive shaft.**
3. **Which are the two main types of universal joints and how they differ from each other ?**
4. **Explain the different types of constant velocity universal joints.**
5. **Describe variable velocity universal joints. Why this type of joint is not suggested for front axle half shafts ?**
6. **Explain final drive.**
7. **What is the necessity of differential in an automobile ? Explain one type of differential.**
8. **Which are the different types of differentials ? Explain non-slip differential.**
9. **Illustrate the working of conventional differential.**
10. **Explain the various types of rear drive axles.**
11. **(*a*) Describe the difference between front and rear drive axles.**
    (*b*) What role a drive axle plays in an automobile ?
12. **Describe the following :**
    (*a*) Tracta joint.
    (*b*) Ring type universal joint.
    (*c*) Three quarter floating axle.
    (*d*) Double reduction differential.
    (*e*) Half shafts.

16

Steering System

1. **Steering system.** It is the system which provides directional change in the performance of an automobile. This system converts rotary movement of the steering wheel into angular turn of the front wheels. It multiplies driver's effort by mechanical advantage, enabling him to turn the wheels easily.

The steering system should possess the following qualities :—

(*i*) It must be capable to keep the wheels, at all times, in true rolling motion without rubbing or scuffing of tyres on road.

(*ii*) Steering in the system must be easily operateable.

(*iii*) It should have a certain degree of self-centring action to keep the vehicle on a straight course.

2. **Types of steering system.** Steering system is of the following types :—

(*a*) Fifth wheel steering system,

(*b*) Ackerman steering system.

(*a*) **Fifth wheel steering system.** It is single pivot steering system in which the front axle along with the wheels, moves to right or left. The movement to the whole axle and wheel assembly is affected by means of a steering and a wheel which is placed between the chassis frame and axle. The fifth wheel acts as a turntable. The axle assembly is connected with the frame by means of a pin which serves as a pivot around which the axle assembly moves. The fifth wheel contains a ring gear mounted at its rim and is moved by means of a steering. Movement of the steering wheel tends the front axle and wheel assembly to move away.

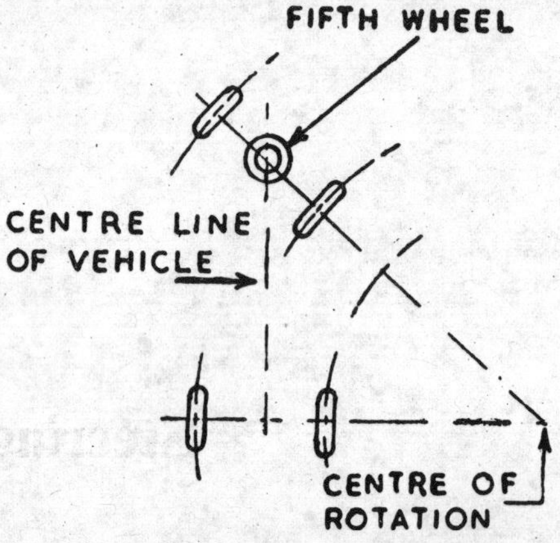

Fig. 16·1. Fifth wheel steering.

(*b*) **Ackerman steering system.** It is double pivot steering system which was invented by Lankensperger, a Munich carriage builder in 1817. The English patent on it was taken out by Lankensperger's agent Rudolph Ackerman against whose name this system became popular.

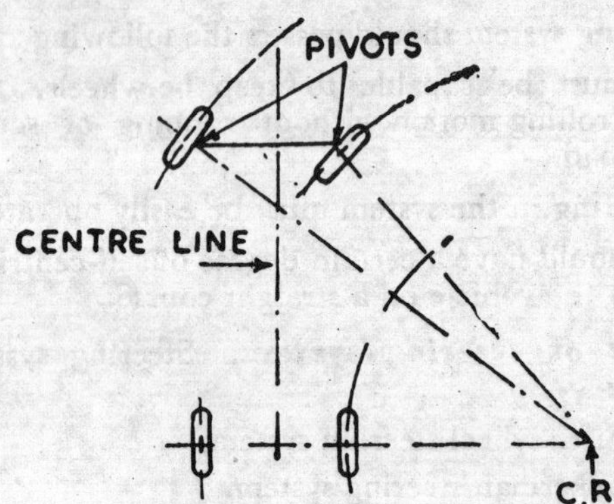

Fig. 16·2. Ackerman Steering.

This divided axle steering system works upon the Ackerman principle in which a line intersecting each steering king pin and tie rod end would intersect at or near the differential.

In this type of steering system, the front axle is fixed with the frame through the springs in conventional suspension system. The stub axles are pivoted with the axle beam by means of king pins. The steering knuckle arm is attached to the stub axle by means of which the wheel is turned through the steering. Steering knuckle arms on both stub axles are connected through tie rod.

In case of independent suspension, there is no axle beam. The steering knuckles are supported in the individual suspension and are connected through linkages.

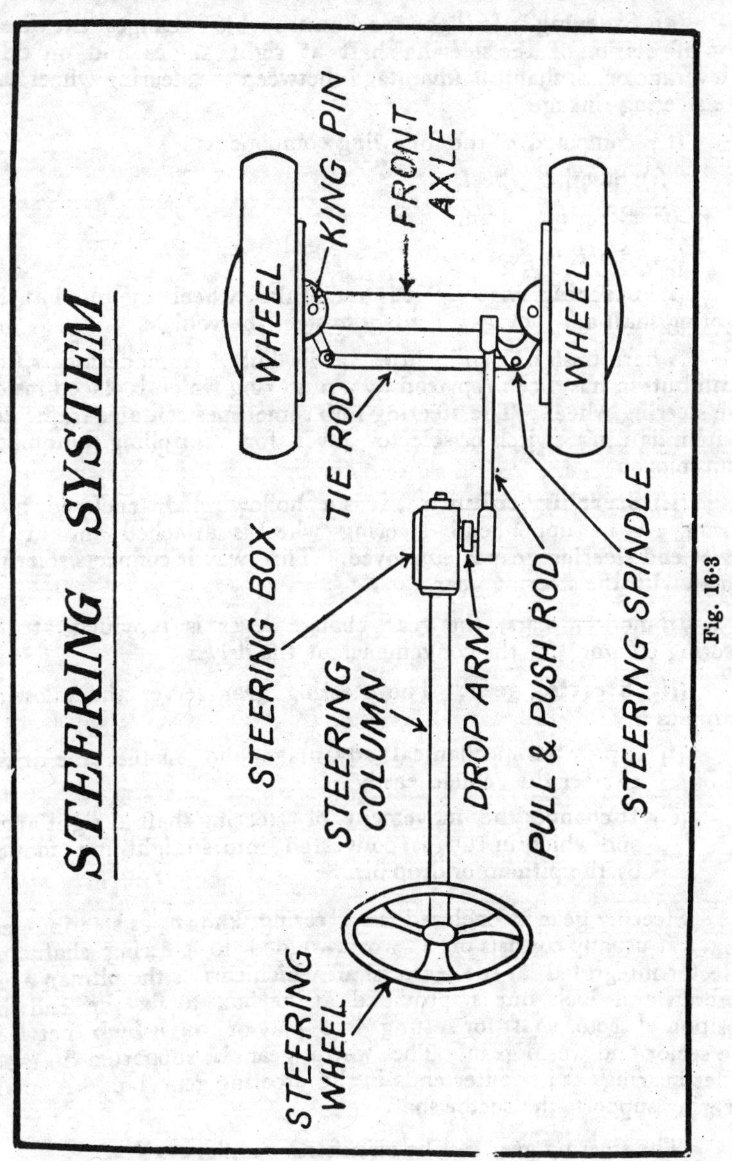

Fig. 16·3

3. Components of steering system. The steering system is composed of the following elements :—

(*a*) Steering,

(*b*) Steering linkages,

(*c*) Steering knuckle.

(*a*) **Steering.** It is the mechanism which changes the direction of rotation of the steering shaft at right angles and provides a leverage or mechanical advantage between the steering wheel and the steering linkage.

It is composed of the following components :—

(*i*) Steering wheel,

(*ii*) Steering column,

(*iii*) Steering gear.

(*i*) **Steering wheel.** It is a circular wheel mounted at the steering shaft and acts as a control to steer the vehicle.

A horn push button is fitted at its hub. In modern cars, the push button has been replaced by a push ring which is placed inside the steering wheel. The steering hub sometimes contains trafficator switch, lighting switch or selector lever for controlling automatic transmission.

(*ii*) **Steering column.** It is a hollow shaft enclosed by a casing. At its upper end steering wheel is attached and at the lower end steering gear is employed. This way it connects steering wheel with the steering gear.

In modern cars, the gear change lever is provided at the steering column for the convenience of the driver.

(*iii*) **Steering gear.** The steering gear serves the following purposes :

(*i*) It provides mechanical advantage and enables the driver to steer the vehicle easily.

(*ii*) It changes the movement of steering shaft at right angle and which in turn is converted into straight-line motion by the pitman or drop arm.

Steering gear is enclosed in a casing known as steering-gear box. It usually consists of a worm welded to steering shaft and a sector integrated at the sector shaft which carries the pitman arm. A screw and lock nut is provided at the box to fix the endwise position of sector shaft for setting free play or **back lash** between the sector and worm gear. The worm gear is supported in taper roller bearings at its outer ends in the steering-gear box. A plain bearing supports the sector shaft.

The steering gear can be classified as under :—

(*i*) Worm and wheel.

(*ii*) Worm and sector.

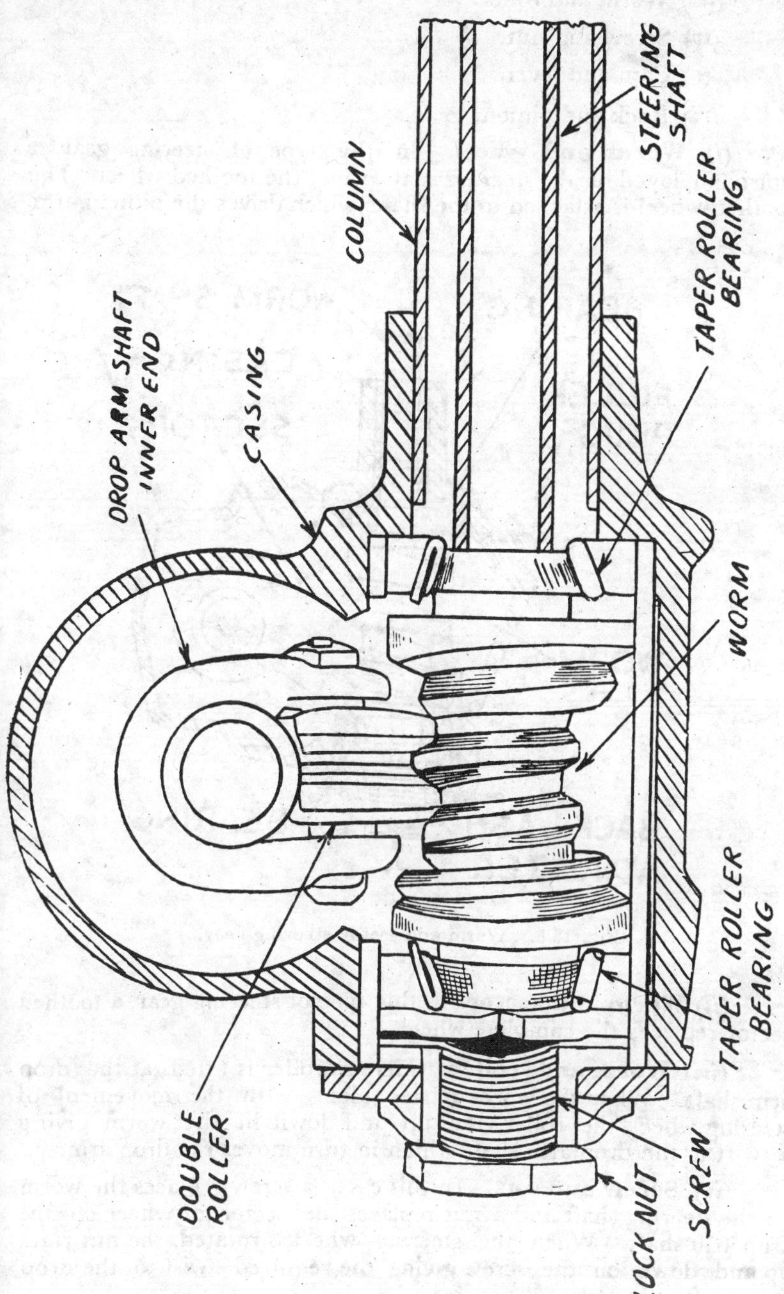

Fig. 16·4. Worm and roller type steering gear.

(*iii*) Worm and roller.

(*iv*) Screw and nut.

(*v*) Cam and lever.

(*vi*) Rack and pinion.

(*i*) **Worm and wheel**. In this type of steering gear, a worm employed on the steering shaft drives the toothed wheel. The toothed wheel is attached to the shaft which drives the pitman arm.

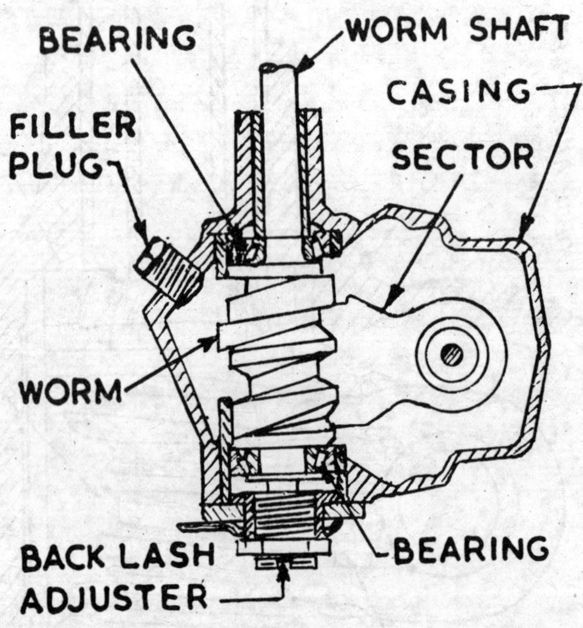

Fig. 16·5. Worm and sector steering gear.

(*ii*) **Worm and sector**. In this type of steering gear, a toothed sector replaces the complete wheel.

(*iii*) **Worm and roller**. Here a roller is fitted at the drop arm shaft. When the worm moves along with the movement of steering wheel, the roller rides up and down in the worm giving a twist to the drop arm shaft which in turn moves the drop arm.

(*iv*) **Screw and nut**. In this case, a screw replaces the worm on the steering shaft and a nut replaces the sector or wheel on the drop arm shaft. When the steering wheel is rotated, the nut rides up and down on the screw giving the required twist to the drop arm shaft.

In order to reduce friction between the nut and screw, a recirculating ball race is usually provided between them. The steering gear is then known as re-circulating ball nut type gear. In this case, the balls roll between the screw and nut and as they reach the upper end of the nut, they enter the return guide and then roll back to a lower point where they re-enter the groove between the screw and nut.

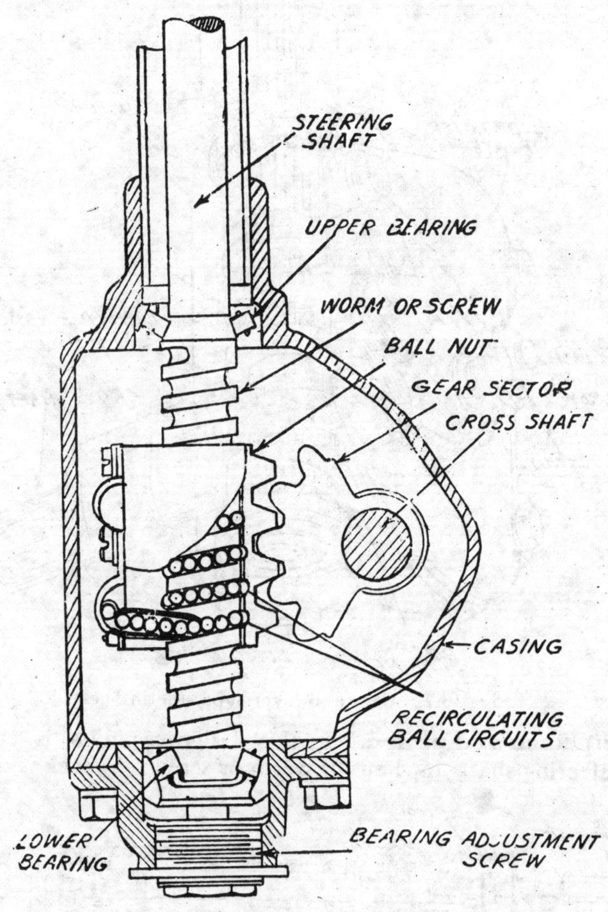

Fig. 16·6. Recirculating Ball, screw and nut steering gear.

(*v*) **Cam and lever**. In this type of steering gear, a helical groove around the steering shaft lower end, in lieu of worm or screw, acts as a cam. The drop arm shaft contains one or two pegs which act as levers. As the steering shaft moves, the pegs at the drop arm shaft ride up and down the helical groove causing movement to the drop arm shaft.

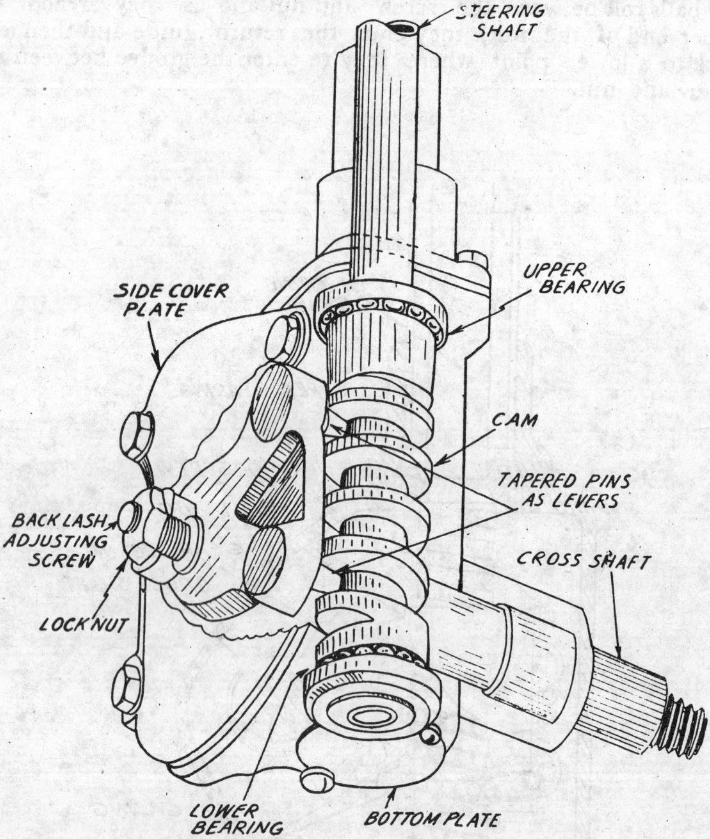

Fig. 16·7. Cam and lever type steering gear.

(ii) **Rack and pinion**. In this design, a pinion is employed at the steering shaft in lieu of worm or screw. A rack is operated

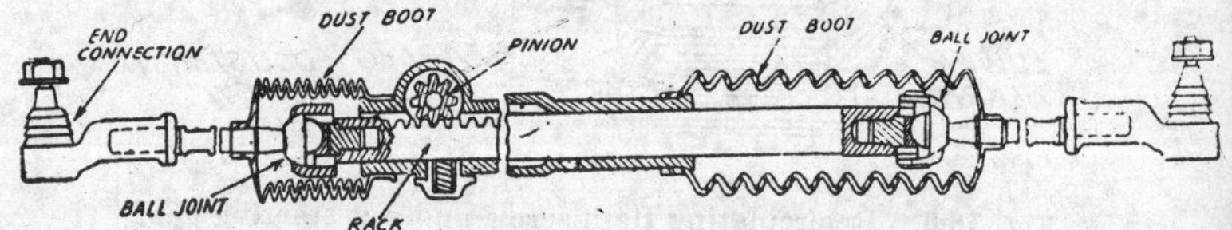

Fig. 16·8. Rack and pinion steering gear.

by the pinion resulting in change in the rotary motion of steering shaft to straight-line motion. In the modern design, the rack is connected in the centre by two halves of the tie rod.

Pitman or drop arm. It is a lever which is attached to the steering sector shaft. It converts rotary motion of steering shaft into straight line motion through the steering gear and sector shaft. It swings like the pendulum of a clock when the steering wheel is rotated. It creates a pulling and pushing effect at the steering knuckle through drag link or pull and push rod resulting in turning of the road wheels.

(*b*) **Steering linkages**. These linkages connect steering gear knuckle arms of the front wheels and drop arm at the steering gear. They carry backward and forward swinging effect of drop arm to the steering knuckles.

In the conventional system, steering knuckle arms of both the wheels are connected by a rod known as tie rod or track rod which runs parallel to the axle beam. The steering side knuckle arm is projecting out and is connected with the pitman arm by means of a drag link or pull and push rod.

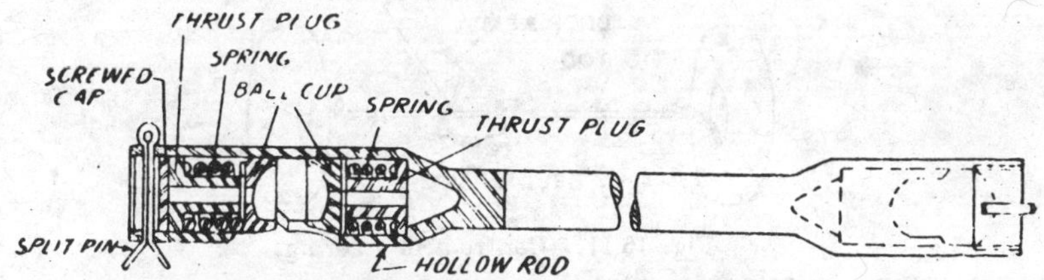

Fig. 16·9. Pull and push rod.

The drag link is usually of tubular construction and contains a pair of divided ball sockets at each end. One half-socket is fixed and the other is held against the ball end of the drop arm or steering knuckle arms by means of a strong compression spring which helps in absorbing shocks at the ball and socket joint. The compression

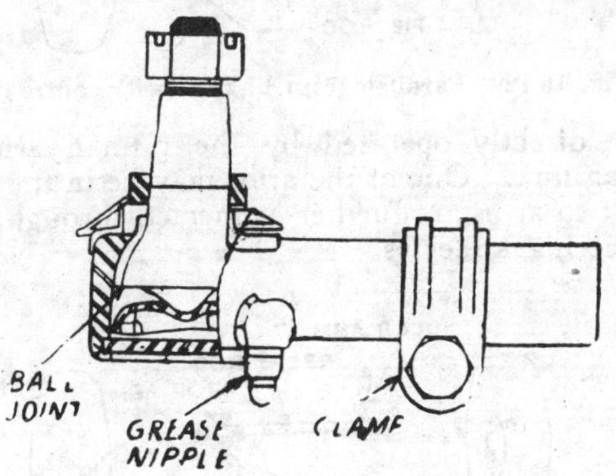

Fig. 16·10. Tie rod end.

of the spring is varied through a screwed end plug which is locked with a split pin.

The tie rod is also a hollow tube which contains ball and socket joint type ends. The length of the rod could be adjusted to suit the toe in.

The steering linkages as employed in different vehicles are of the following types : –

(*a*) Conventional.

(*b*) Centre arm steering type.

(*c*) Parallelogram linkage.

Conventional linkage has been described above. The centre arm steering type is applicable in Jeep. It contains a bell crank in the centre of two separate tie rods which connect the individual steering knuckle arms at the wheels. The pull and push rod connects the centre arm of the bell crank and pitman arm.

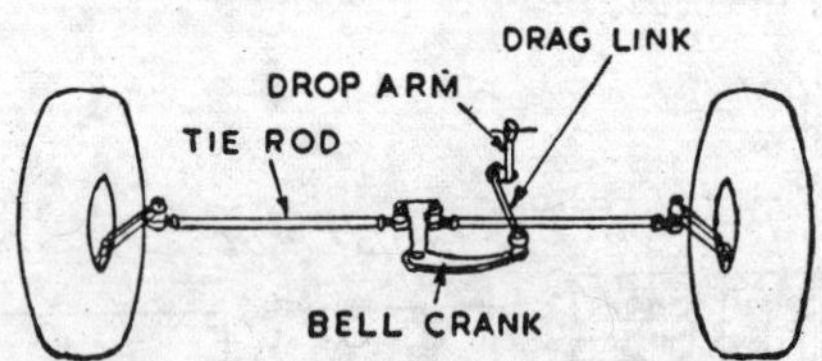

Fig. 16·11. Centre arm steering.

The parallelogram linkage is of different forms. In centre link arrangement, there are two parallel arms, one on each side, connected through a centre link or connecting rod. One of these

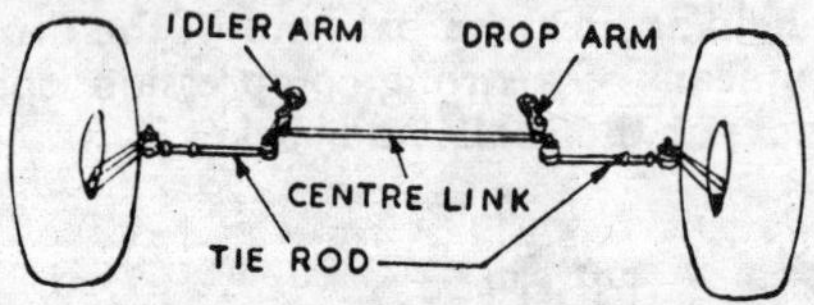

Fig. 16·12. Parallelogram linkage with centre link.

arms is either directly operated by the pitman arm or operated through a drag link. One of the arms may be in the shape of bell crank. The two arms are further connected through tie rods to the individual steering knuckles.

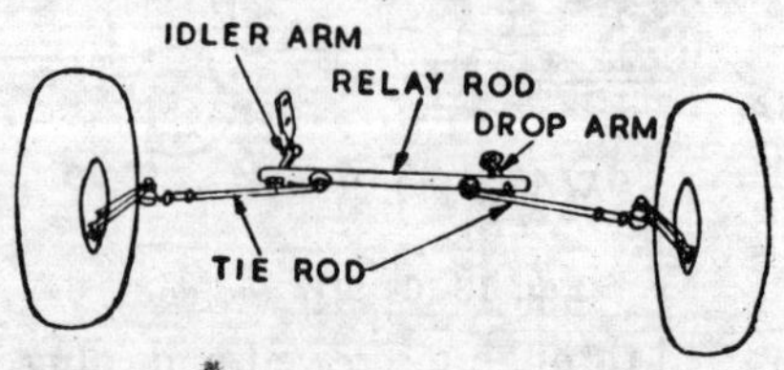

Fig. 16·13. Parallelogram linkage with relay rod.

In another form known as ***relay linkage***, a relay rod connects the pitman arm and an idler arm is placed parallel to it on the other side of the frame. The relay rod is further connected throught the tie rods placed parallel to relay rod, with the individual steering knuckles.

The parallelogram linkage is usually used in vehicles having independent suspension at the front where one wheel works independent of the other. All the connections are made in the linkages through bali and socket joints as they transmit angularity movement.

(*c*) **Steering knuckle.** It is the arm, sometimes known as steering spindle, which is integrated with the stub axle of its movement around the pivot which is known as king pin. The steering efforc is conveyed to the steering knuckle by means of steering linkages for turning the road wheels to the right or left direction. Each stub axle has its own steering knuckle arm which is connected with each other through tie rods.

Stub axle. It is the spindle over which the road wheel is mounted through the brake drum. It acts as an axis for the wheel. It takes load of the wheel and enables it to spin around. Steering knuckle arm is connected to it. In conventional arrangement, the stub axle is connected to the axle beam though king pin and in independent suspension, it is connected to the suspension cradle through the support.

The connection of stub axle with the axle beam or steering knuckle support is through a yoke and king pin. The yoke may be on the stub axle or on the axle beam. There are the following ways of connecting stub axle with the axle beam :

(*a*) Yoke for hinging stub axle with the axle beam, at the end of axle beam, known as *elliot* type arrangement or front axle.

(*b*) Yoke for the hinge at the stub axle, known as *reverse elliot* type arrangement or front axle.

(*c*) *L*-shaped spindle in lieu of yoke, being connected with axle beam, known as *lemoine* type arrangement or front axle.

4. **Factors affecting steering.** The following factors affect the steering in an automobile :—

(*i*) Steering gear ratio and back lash,

(*ii*) Steering linkage connections,

(*iii*) Tyre pressure,

(*iv*) Play in the wheel bearings,

(*v*) Condition of king pins and bushings,

(*vi*) Wheel alignment.

(*i*) **Steering gear ratio.** The steering gear provides mechanical advantage for the driver to guide the vehicle by exerting only

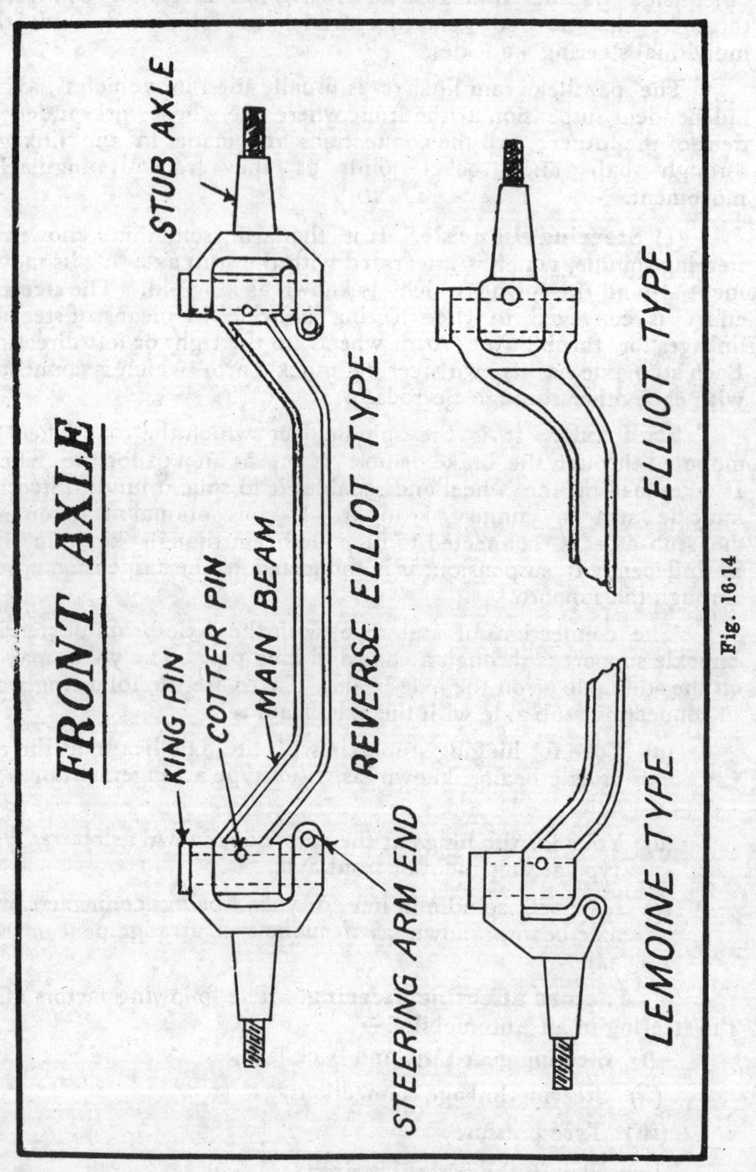

Fig. 16·14

a small amount of physical effort at the steering wheel. Steering gears usually have a mechanical advantage of 12 to 1 and higher. The ratio differs from vehicle to vehicle. The steering gear ratio affects the ease with which the steering wheel is moved. The free play or back lash should be neither less nor more otherwise hard steering or loose steering shall result in.

(*ii*) **Steering linkage connections**. They must provide free movement of the linkages to transmit driver's effort at the steering wheel to the knuckles. If there is more play in the joints, most of the effort shall be going waste and if there is less play, hard steering would result in. More play in the joints shall lead to car wander.

(*iii*) **Tyre pressure**. It should be as per specifications. Less or uneven pressure of air in the wheels shall result in hard steering and more tyre pressure shall lead to wobbling.

(*iv*) **Play in wheel bearings**. The wheel bearing should be properly adjusted Less play shall result in hard steering and more play shall affect wheel alignment and lead to wheel wander.

(*v*) **Condition of king pins and bushings**. The front portion vehicle load acts through the pins. If the pins are tight in the bushes, hard steering shall result in and if there is more play, wheel alignment shall be affected, which has great influence on the steering operation.

(*vi*) **Wheel alignment** This relates to the relative position of the wheels for obtaining a true and free rolling movement over the road. The smooth operation of steering depends much upon the wheel alignment.

Wheel alignment is the mechanics of adjusting the interrelated factors which influence steering. The important alignment factors are as under :—

(*a*) Caster,

(*b*) Camber,

(*c*) Toe-in,

(*d*) King pin inclination,

(*e*) Toe out or steering geometry,

(*f*) Tracking.

(*a*) **Caster**. It is backward tilt of the king pin and axle at the top. The purpose of caster is to give the vehicle directional-stability or ability to travel straight ahead with a minimum of actual steering by the driver. This is accomplished by the fact that the projected centre line of king pin strikes the road ahead of the contact point of the tyre. This has a tendency to lead or drag the wheel behind it, giving the vehicle directional stability.

If the axle is set horizontally and the king pin vertically, the weight of the vehicle would be directly above the point of contact. As a result, the wheels would wander and the car would lack steering stability. The caster or backward tilt of the axle prevents wheel wander and makes the vertical load precede the point of contact which causes the wheels to run straight normally and to straighten up after turn.

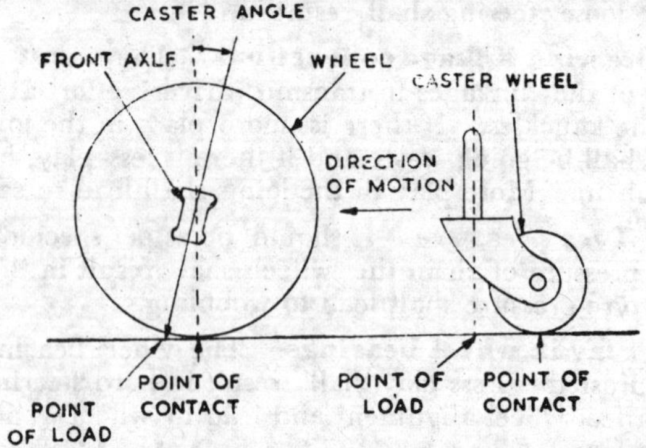

Fig. 16·15. Caster. (Point of load falling ahead of point of contact as in chair caster wheel).

(*b*) **Camber.** It is outward tilt of the wheels at the top. It is to allow the load to bear more directly down through the king pin thereby relieving any binding action of pivots in their bearings.

If the front wheels of an automobile are perpendicular to the outer, the load and thrust of the wheels would be placed on the outer end of the spindle and hard steering would result. In order to relieve this strain and place it on the large inner wheel bearing, the wheels are tilted outward at the top.

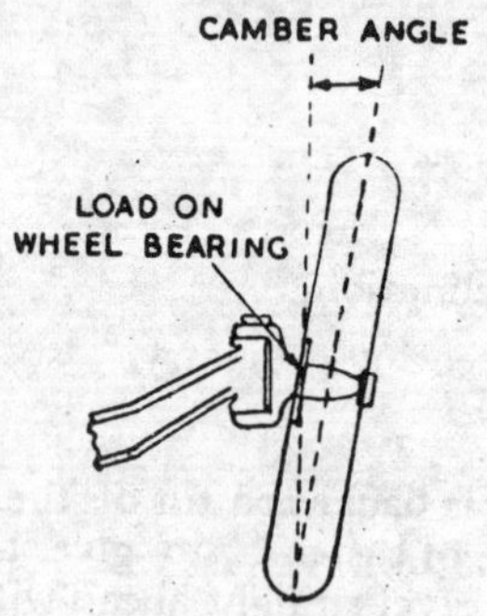

Fig. 16·16. Camber.

When the vehicle is loaded and rolling along on the road, the load will tend to bring the wheels to near about vertical position if the wheels are cambered.

(*c*) **Toe-in.** The adjustment for gathering the wheels inward at the front is known as toe-in. It means that the distance between the front wheels at the front is less than the distance at the back.

Toed-in wheels have the tendency to run inward. Positive cambered wheels have a tendency to roll outward. When toe-in and camber are properly combined, it results in rolling the wheel in a straight line.

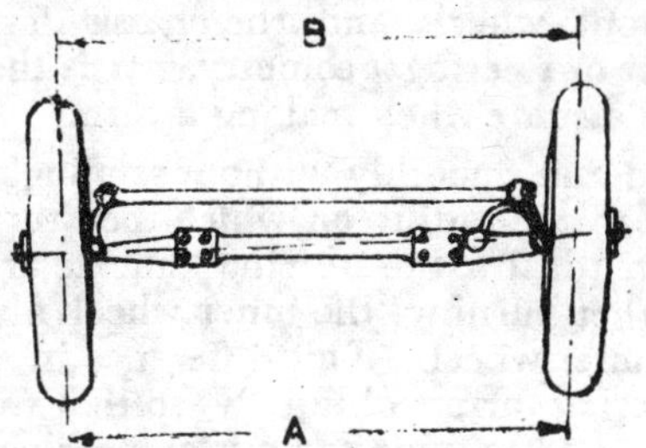

Fig. 16·17. Toe-in (B—A).

The purpose of toe-in is to counteract or neutralize camber. The greater the camber angle, the more toe-in that will be needed and *vice-versa*. Without toe-in the cambered wheels would want to run away from the vehicle and would cause excessive tyre wear.

(*d*) **King-pin inclination**. It is inward tilt of king-pin at the top. It tends to keep the stub axle pointed outward in line with the axle just as caster tends to keep the wheels pointed ahead.

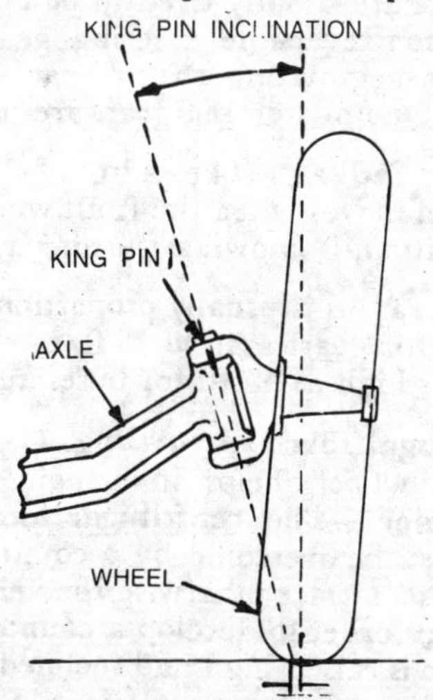

Fig. 16·18. Kingpin inclination.

If the kingpin is allowed to remain vertical, the cambered wheels would exert a steering action on kingpin and hard steering would result. Kingpin inclination tends to keep the wheels straight ahead.. It helps the return of wheels to the straight ahead position after a turn has been made. When the wheel is swung away from straight ahead, the kingpin and supporting parts are moved upward. This means that the vehicle body is actually raised up. The kingpin inclination causes the vehicle to be raised every time the front wheels are swung away from straight ahead. Then the vehicle weight brings the wheels back to straight ahead after the turn is completed and the steering wheel is released. This is known as *self-centering action.*

(*e*) **Toe-out or steering geometry**. The difference in angles between the front wheels and the chassis frame during turns is known as toe-out or steering geometry. It is the toed out position the front wheels assume when making a turn.

In order to run smoothly without scuffing, the wheels must be at right angles to the radius on which they turn. In order to get this action automatically, the steering knuckle arms are bent inward at an angle. When turning, the inner wheel always turns a sharper angle than the outer wheel. This difference in angles is obtained by setting the steering arms obliquely so that they point inward the centre line of car and provide toe-out or steering geometry.

(*f*) **Tracking**. Rear wheels should follow in the tracks of front wheels in a parallel position. Tyre scuffing and hard steering are caused by improper tracking which is due to bent chassis frame.

5. **Reversible steering**. A steering gear is known as reversible when it transfers road shock back through the steering wheel to the driver.

Steering gears are usually tried to be made nearly non-reversible. A strictly non-reversible steering gear would not tend to straighten out after negotiating a turn, that is why the intermediate or semi-reversible type of steering gears are used on most vehicles.

6. **Turning radius**. The radius of the circle on which the outside front wheel moves when the front wheels are turned to their extreme outer position, is known as turning radius.

The turning radius is usually proportional to the wheel base of the car. This radius varies from 15 ft. to over 23 ft. for passenger cars and may be as high as 45 ft. for buses and trucks.

7. **Centrifugal force in turning**. Centrifugal force developed by the rotating wheel, helps in changing the direction of wheel travel during turning. The centrifugal force which depends upon wheel speed, must be overcome by a counter force acting through the tyres at the road surface otherwise the vehicle will continue in linear motion. In order to develop a counter force, the outer side of the road surface is raised high and inclined towards the centre of curved path.

8. **Cornering force and its effect on steering**. The frictional force between the tyre and road surface required to counteract the centrifugal force is known as *cornering force.* When a vehicle is steered, the tyres are distorted upto some extent in the direction of tyre axis which gives rise to the cornering force. The distortion is greater at the front part of contact area of the tyre with

the road surface than at the back. The amount of cornering force developed during turning, depends upon slip angle, tyre pressure, load on tyre and type and condition of road surface.

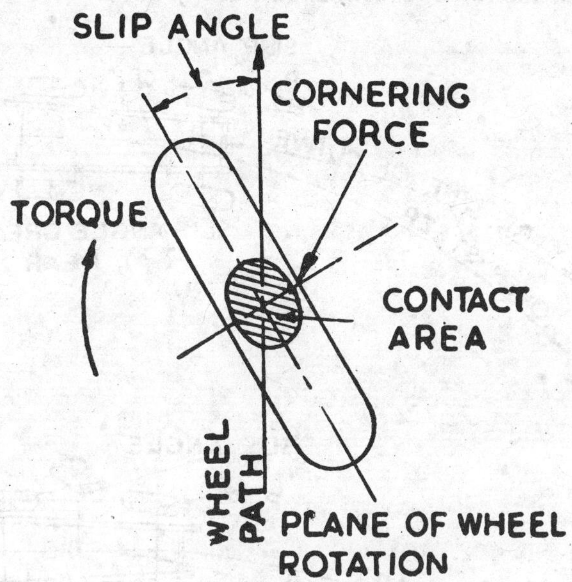

Fig. 16·19. Slip angle.

There is maximum limit to the total cornering force which may be obtained by increasing the *slip angle* under any given condition of road surface and tyre. If the cornering force required to balance the external forces becomes greater than this maximum, the wheel will begin to skid and the vehicle will run out of control.

Slip angle. The angle between the central plane of tyre and the direction of motion is known as slip angle. The torque at the wheel tends to restore the direction of tyre or reduce the slip angle. This angle ranges from 8° to 10° for dry and slippery pavements respectively.

9. Under-steer and over-steer. A vehicle is said to be under-steer when it has a tendency to change direction from the exciting force. When a vehicle is being driven along a straight level road and subjected to a side force by a cross wind, the vehicle tends to drift away. When a vehicle is being steered into a corner, it has a tendency to run away from centrifugal force. An over-steering vehicle tends to steer into the exciting force.

Owing to side wind pressure or cornering effect, the tyre tread tends to remain stationary on the road but the wheel axles tend to move side-ways or laterally in the direction of side or centrifugal force. This results in causing distortion because the central plane of the tyre moves away from the centre of tread. Difference between the direction of vehicle motion and tyre direction makes an angle between the two directions known as ***slip angle***.

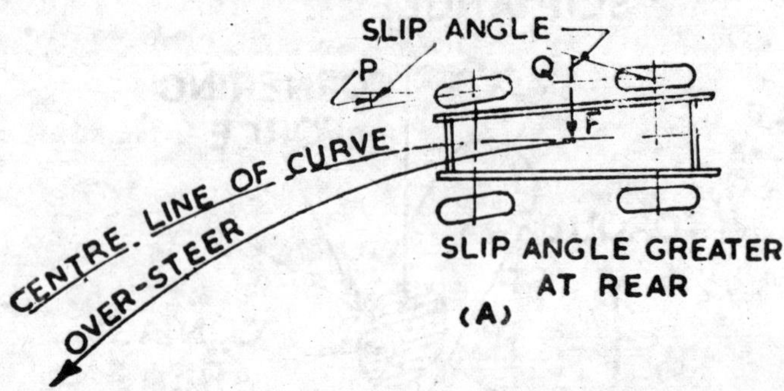

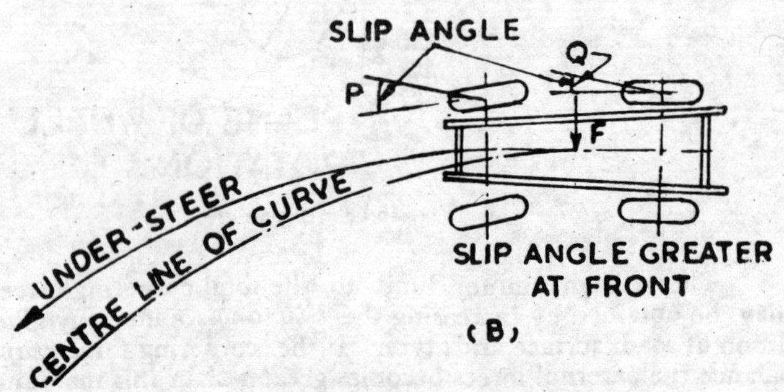

Fig. 16·20. (A) **Over-steer.** (B) **Under-steer.**

All the wheels of the vehicle are subjected to tyre creep or slip angle effects. When the slip angle at front and rear wheels is same, there is no tendency of the vehicle to move away from the path of neutral straight line. This is called *neutral steer*. When the rear slip angle is greater than front slip angle, ***the vehicle tends to adopt a course away from the direction of side force F*** as shown in the diagram. This is known as ***over-steer***. It reduces driver effort when cornering and has an advantage while driving along roads having many twists and bends.

When the front slip angle is greater than rear one, ***the vehicle tends to steer in the direction of side force F***. This is called ***under-steer***. It gives greater driving stability especially when there are strong side winds. It is better for fast driving on long straight roads.

10. Power steering. It is that type of steering in which external power is utilized to assist in operating the steering. It uses compressed air, electrical mechanism and hydraulic pressure. A vast majority of the power steering mechanisms use hydraulic pressure to operate them which is supplied by an engine driven pump.

Power steering is actually power-assisted steering in which a booster arrangement is set in operation when the steering wheel is turned. The booster then takes over and does most of the steering work. The construction of power steering is in such a way so that vehicle may be steered in the usual way when the engine is not working or any breakdown occurs in the power source.

Power steering is of the two main classes which are as follows :

(*a*) Integral type,

(*b*) Linkage type.

(*a*) **Integral type.** It is that type of power steering in which power operating assembly or booster assembly is built into the steering gear unit.

In one form, the booster assembly is integrated with recirculating ball steering gear and consists of two units : power cylinder unit and valve unit. The power cylinder contains a piston which is connected to a rack that operates a sector on the drop arm shaft.

The valve unit contains a spool in the housing which slides up and down due to turning effect of steering wheel, admitting fluid

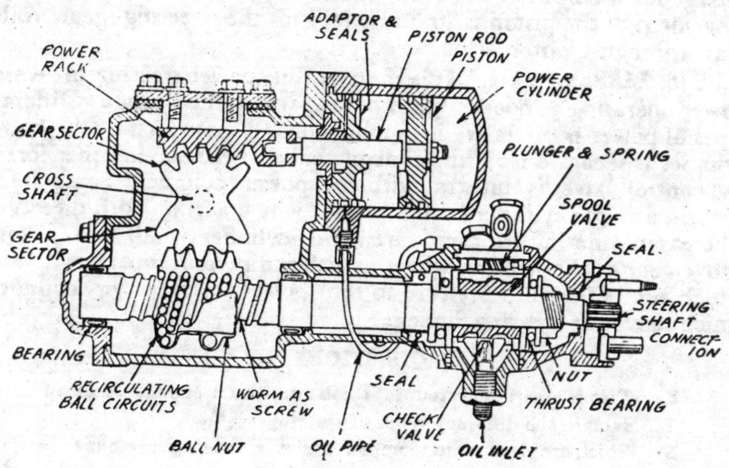

Fig. 16·21. Power steering.

pressure to act on one or the other side of the piston. The pressure in the cylinder causes the piston to move, resulting in the movement of sector shaft which turns the drop arm for turning the wheels.

In *straight ahead position*, the oil is circulating at low pressure throughout the steering gear unit. The whole unit is full of hydraulic fluid at all times and oil pressure is the same on each side of the piston. The piston is therefore balanced and no movement occurs.

When the steering wheel is turned to left, the worm or screw on the steering shaft turns in the ball nut causing the worm to slide up and out of nut due to screw and nut action. There is resistance to the movement of ball nut from the ball nut to the front wheels. If the ball nut is held stationary, the worm and steering shaft will move upward. This upward movement carries the valve spool upward. This results in the flow of fluid pressure to the left turn chamber of the power cylinder, and turning of the wheels to the left. When the steering wheel is turned to the right, the valve spool moves down, directing fluid pressure to the right turn chamber and cutting of oil supply to the left side chamber. This results in the movement of front wheels to the right.

The above form of power steering is known as *saginaw* power steering and is applicable in Buick and Chevrolet cars. The another form of integral power steering is known as *Gemmer* power steering and is applicable in Chrysler and Desoto cars.

The gemmer power steering contains two opposed power pistons and the valve body assembly having four valves. One or the other of the two distribution valves directs the flow of fluid under pressure to one piston while one of the reaction valves opens a passage for the fluid at the other piston to return to the reservoir. Movement of the piston is transmitted to the steering gear roller shaft through a piston arm.

(*b*) **Linkage type**. It is that type of power steering in which power operating or booster assembly is part of the linkage. Different types of power boosters are in application. In some forms, the power cylinder is separate from the control valve whereas in other forms, the control valve is integral with the power cylinder. Some forms contain a single cylinder with a piston which acts on both directions whereas in other forms, there are two cylinders and two pistons. Movement of the steering wheel in either direction operates a valve which allows the fluid pressure to the power cylinder or cylinders which move the steering linkage.

QUESTIONS

1. What is steering system ? Which qualities it should possess ?
2. Explain the different types of steering systems.
3. Which are the main components of a steering system ? Describe.
4. Which are the different types of steering gears ? Explain them.
5. Illustrate recirculating ball nut type steering gear.
6. Write notes on the following :
   (*a*) Pitman arm.
   (*b*) Drag link.

(*c*) Tie rod.

(*d*) Parallelogram linkage.

(*e*) Rack and pinion steering gear.

(*f*) Ackerman principle.

7. Which are the different types of steering linkages ? Explain.

8. Which are the different forms of connecting stub axle with the axle beam ? Illustrate.

9. Which factors influence steering ? Explain.

10. What is wheel alignment and which are its inter-related factors ? Explain.

11. Explain the following :

(*i*) Reversible steering.

(*ii*) Turning radius.

(*iii*) Self centering steering action.

(*iv*) Toe-out.

(*v*) King pin inclination.

(*vi*) Directional stability.

(*vii*) Saginaw power steering.

12. Which are the different types of power steering systems ? Explain them.

13. Explain the following :

(*i*) Slip angle.

(*ii*) Cornering force.

(*iii*) Over-steer and under-steer.

17

Brakes

1. Brakes. These are mechanical devices which use the force of friction to overcome friction. The brakes in an automobile, stop it while moving and also hold it from rolling off when not in motion.

When the brakes are applied on a moving vehicle, the kinetic energy or energy of motion of the vehicle is transformed into heat generated by the friction between the brake lining and drums. The heat generated is dissipated into the surrounding air.

The force of friction or retarding force between the linings and the drums depends upon the coefficient of friction of the two materials and the force exerted on the shoes by the retarding mechanism.

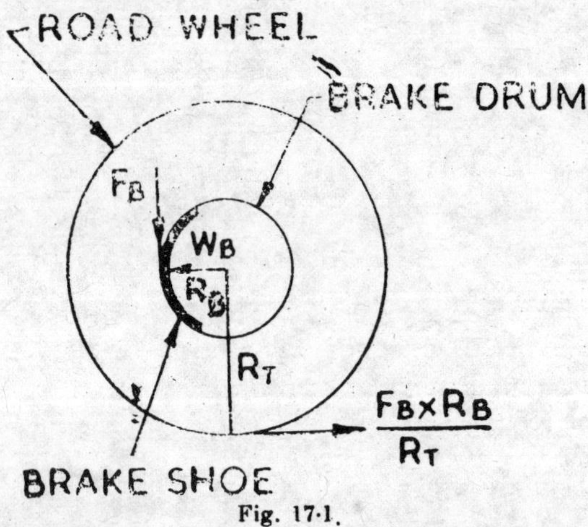

Fig. 17·1.

2. Braking principles. When brakes are applied on a running vehicle, the energy of motion or kinetic energy (K.E.) is converted into heat energy generated by the friction between the brake linings and drums.

The retarding force or force of friction between the linings and drum depends upon the coefficient of friction for the two materials and the force exerted on the shoes by the brake mechanism.

Retarding force on the brake drum,

$$F_B = \mu_B \times W_B \qquad \ldots(i)$$

where μ_B = Coefficient of friction between linings and drum.

W_B = Normal force on brake shoes.

Retarding torque

$$= F_B \times R_B \qquad \ldots(ii)$$

where F_B = Retarding force on brake drum.

R_B = Radius of drum.

Retarding force on ground

$$= \frac{F_B \times R_B}{R_T} \qquad \ldots(iii)$$

where R_T = Radius of tyre

Retarding force on road wheel or force of friction between tyre and road,

$$F_T = \mu_T \times W_T \qquad \ldots(iv)$$

where μ_T = Coefficient of friction between tyre and road.

W_T = Normal force on tyre.

3. **Braking force.** The braking force is equal and opposite to the linear inertia effects on the vehicle. Neglecting retarding forces caused by rolling, air and gradient resistances, the force required to stop a vehicle is dependent on the weight of vehicle and the deceleration rate or rate at which it is stopped.

Braking force, $$B_F = \frac{wd}{g}$$

where w = weight of vehicle

d = Rate of deceleration

g = 9·81 m/sec² (in C.G.S.)

or 32·2 ft./sec² (in F.P.S.)

If the sum of retarding forces at all the wheels is equal to the weight of vehicle, the deceleration is equal to g which is the rate of change of velocity of a body falling under the influence of its own weight.

The limiting value of braking force depends upon the grip of wheel on the road surface. The ratio of grip to the weight of vehicle is known coefficient of friction or μ.

$$\mu=\frac{BF}{W}$$

and rate of deceleration,

$$d=\frac{g\times B_F}{W}$$

4. Weight transfer during braking. If maximum deceleration is to be achieved, the braking effort of each wheel must have the best possible relationship to the weight on the wheel. This is required due to the fact that during deceleration, there is a change in weight distribution which is proportional to retardation. For a given deceleration, the ratio of height of C.G. to the length of wheel base determines the extent of weight transfer.

Amount of weight transfer

$$=\frac{B\times h}{L}$$

where,
$$B=\frac{\%g\times W}{100}$$

h=Height of C.G.

L=Length of wheel base.

If the brakes are applied on level road, the front wheel weight shall become $W_F+\frac{B\times h}{L}$ and rear wheel weight shall amount to $W_R-\frac{B\times h}{L}$ (the sum of weight on the wheel remaining constant),

where W_F=weight on front wheels (static)

W_R=weight on rear wheels (static)

During weight transfer on braking, the nose of the vehicle dips down. This tendency of the vehicle is known as *brake dip*.

The height of centre of gravity of the vehicle above the ground, in relation to its wheel-base and the suspension characteristics, determine the nose dip.

The retarding force acts at ground level during deceleration, whilst the inertia of vehicle produced by its weight and speed acts through C.G. in the opposite direction.

These two equal and opposite forces produce an overturning couple tending to lift the rear portion of vehicle. As a result, perpendicular force between the wheels and the ground is increased at the front and decreased at the rear to form a resisting couple during deceleration. This weight transfer to the front wheels during braking, enables a greater braking torque to be applied to the front wheels than to the rear without skidding.

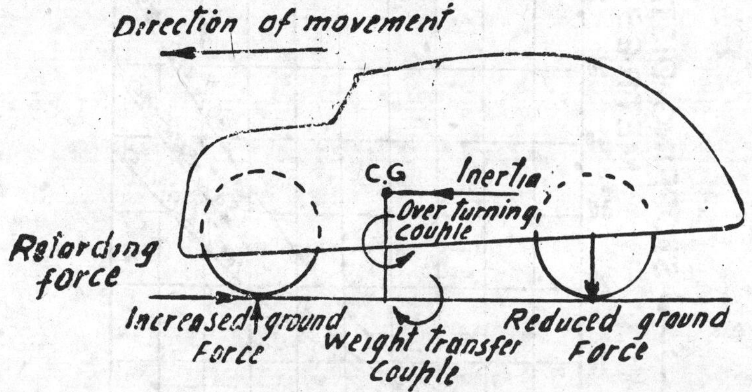

Fig. 17·2. Weight transfer.

5. Stopping distance and time. The distance required to stop a vehicle is proportional to the square of speed at which the brakes are applied. It is inversely proportional to brake efficiency and is measured in feet or metres from the point the brakes are applied and to the point the vehicle comes to stop. Different types of *decelerometers* are in use to check up stopping distance and thus to find out brake efficiency.

$$\text{Stopping distance} = \frac{V^2}{2a} \text{ metres}$$

where, $V =$ initial velocity in m/sec

and $a =$ deceleration in m/sec^2.

The *stopping time* is the duration of brake application to a complete stop. It is inversely proportional to brake efficiency and is proportional to speed.

$$\text{Stopping time} = \frac{V}{a}$$

6. Brake efficiency. It is the ratio between the *retarding force* and the *weight* of *vehicle*. It is expressed as a percentage.

If the sum of retarding forces on all the wheels of a vehicle is equal to the total weight of the vehicle, the retardation is 9·81 m/sec^2 or 32·2 ft./sec^2. (g). The brake efficiency is then considered as 100%. Thus the retardation or *deceleration*

$$= \frac{g \times B_F}{W}$$

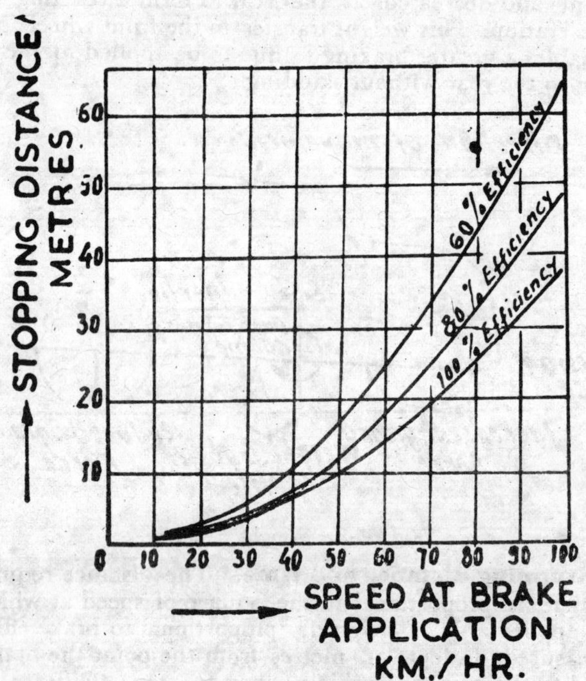

Fig. 17·3. Brake efficiency.

where,

$g = 9{\cdot}81$ m/sec²

B_F = braking force

W = weight of vehicle

In terms of brake efficiency, it is as under :

$$\frac{100 \times B_F}{W}$$

The above relation between the retarding or braking forces and deceleration applies only to level road conditions. On gradient, there will be either increase or decrease in deceleration for a given force depending on whether the vehicle is going up or down. For example, if the gradient is 1 in 4, there will be a force of 1/4 of the vehicle weight either to retard the motion or to accelerate it. Thus for ascending a gradient of 1 in 4, the expression for percentage braking efficiency would be as follows :—

$$\eta_B = \frac{100(B_F + W/4)}{W}$$

If the vehicle is descending,

$$\eta_B = \frac{100\,(B_F - W/4)}{W}$$

Drag or braking force shall be lower under gradient than on a level road due to weight transfer.

7. **Types of brakes.** Automobile brakes belong to the following two classes :

(*a*) Foot brake or service brake.

(*b*) Hand brake or parking brake.

(*a*) **Foot brake or service brake.** Service brake is operated by the foot pedal that is why it is known as foot brake. This brake is used to stop the vehicle which it is in motion.

(*b*) **Hand brake or parking brake.** This brake is operated by a hand lever and is used to hold the vehicle while it is stationary. It is used during parking so that the vehicle may not roll off due to road gradient or fast blowing wind. It is also used in emergency when the service brake fails or proves ineffective. Owing to its use during emergency, it is also known as emergency brake.

According to the method of braking, the automobile brakes can be classified as below :

(*a*) External contracting brakes } (Drum Brakes)
(*b*) Internal expanding brakes }

(*c*) Calliper or disc brakes.

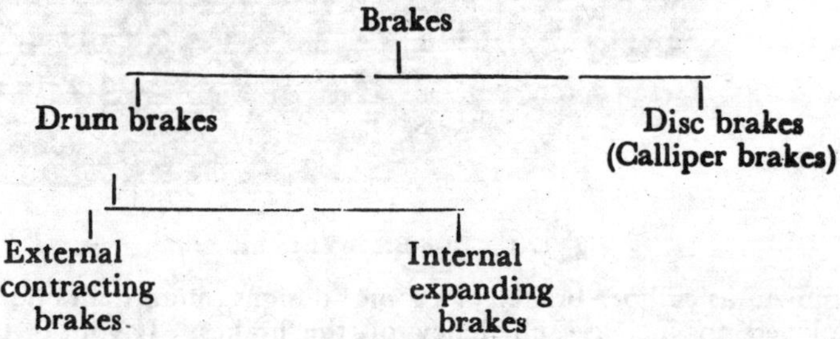

(*a*) **External contracting brake.** This type of brake consists of a drum, an external contracting brake band, linkage and operating lever. It is usually the hand brake of the automobile. The drum is fitted with the transmission output shaft and rotates with it. The brake band having lining of frictional material encircles the brake drum. In order to hold the rotating brake drum, the brake band is contracted about the drum by means of a lever and linkage.

(*b*) **Internal expanding brake.** This type of brake consists of a brake drum inside which there are two brake shoes anchored at the lower ends to the back plate and connected with each other at the top through a spring. The upper ends of the shoes rest at an operating cam or wheel cylinder. The shoes are expanded outward to hold the rotating brake drum through the cam or wheel cylinder.

(*c*) **Disc brake.** It consists of a disc held between two pads The disc is attached to the axle in lieu of brake drum. The rotating disc is held up through the frictional force exerted by the brake pads. The action of brake pads is similar to the action of calliper brakes in an ordinary bicycle. It is due to this fact that this brake

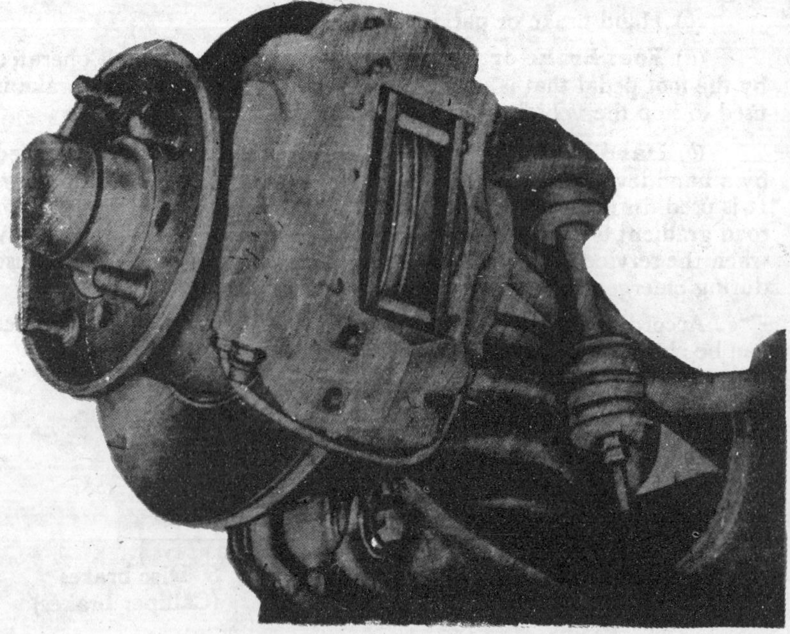

Fig. 17·4. Disc Brake at front.

is known as calliper brake. In some designs, more than one disc is employed to increase efficiency of the brakes. Layout of typical disc brakes is shown in Fig. 17·2.

According to the mode of operation, the automobile brakes can be divided into the following categories :

(*a*) Mechanical brakes.

(*b*) Hydraulic brakes.

(*c*) Electric brakes.

(*d*) Vacuum brakes.

(*e*) Hydrovac brakes.

(*f*) Air brakes.

(*a*) **Mechanical brakes.** The brakes which are operated mechanically by means of levers, linkages, pedals, cams, bell cranks, etc., are known as mechanical brakes. The external contracting

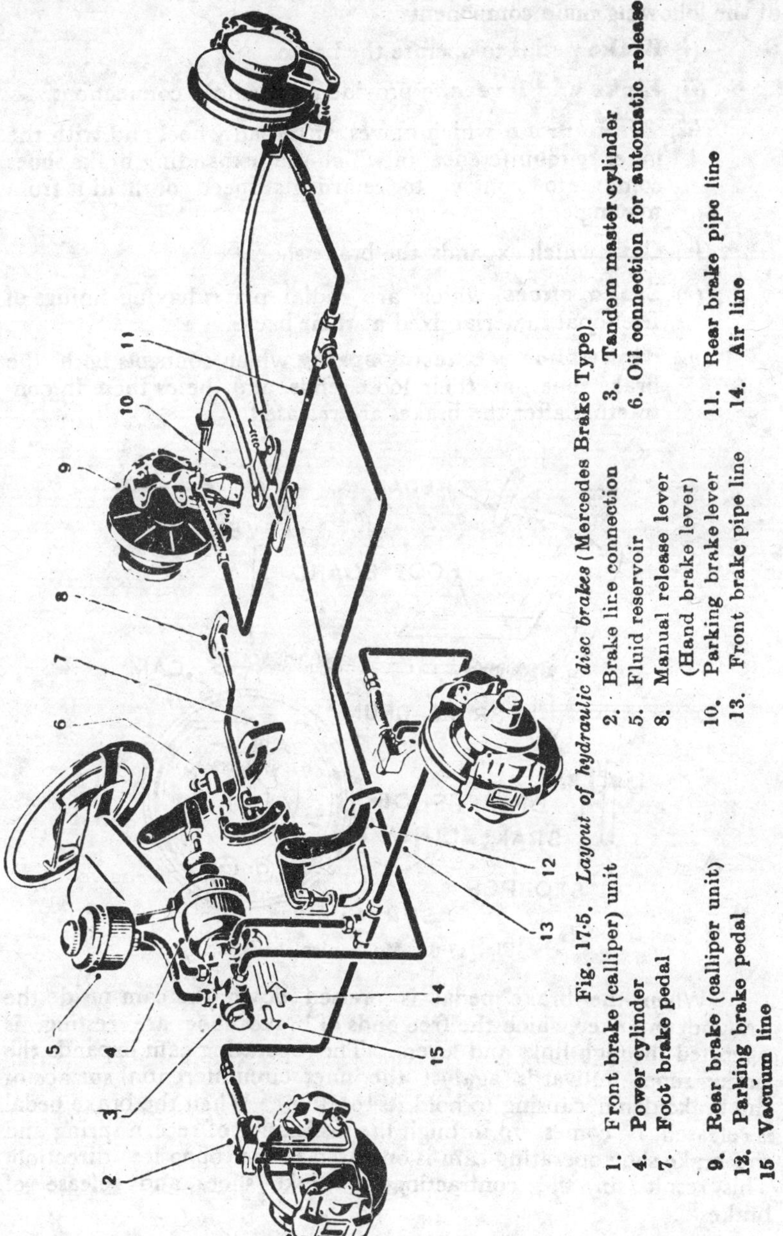

Fig. 17·5. *Layout of hydraulic disc brakes* (Mercedes Brake Type).

1. Front brake (calliper) unit
2. Brake line connection
3. Tandem master cylinder
4. Power cylinder
5. Fluid reservoir
6. Oil connection for automatic release
7. Foot brake pedal
8. Manual release lever (Hand brake lever)
9. Rear brake (calliper unit)
10. Parking brake lever
11. Rear brake pipe line
12. Parking brake pedal
13. Front brake pipe line
14. Air line
15. Vacuum line

brake which is usually hand brake in automobiles, is mechanical brake. Automobiles contain service brakes operated mechanically.

An internal expanding mechanically operated brake usually consists of the following main components :

(*i*) **Brake pedal** to operate the brakes.

(*ii*) **Links and levers** to provide mechanical connections.

(*iii*) **Brake drum** which moves with road wheel and with the inner circumference of which the expanding brake shoes come into contact to retard its speed or hold it from moving.

(*iv*) **Cam** which expands the brake shoes.

(*v*) **Brake shoes** which are radial plates having linings of frictional material fixed at their backs.

(*vi*) **Brake shoe retracting spring** which connects both the brake shoes at their loose ends and helps them in contracting after the brakes are released.

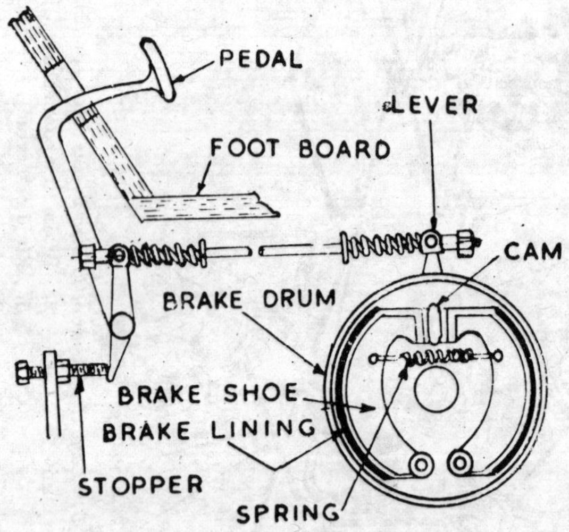

Fig. 17·6. Mechanical brakes.

When the brake pedal is pressed down, the cam inside the brake drum, over which the free ends of brake shoes are resting, is operated through links and levers. The operating cam expands the brake shoes outwards against the inner circumferential surface of the brake drum causing to hold its rotation. When the brake pedal is released, it comes up through the assistance of return spring and the brake shoe operating cam is operated in the opposite direction. This results in the contracting of brake shoes and release of brakes.

(*b*) **Electric brakes.** These brakes are operated electrically. The electric current is supplied by the battery and controlled by a rheostat in the brake pedal. Each brake drum encloses an armature which is operated by an electromagnet. The amount of electric current passing through the electromagnetic circuit is subject to the armature. The armature operates the cam to expand the brake shoes in the same way as in mechanical brakes.

(*c*) **Vacuum brakes.** These brakes are operated with the vacuum of engine manifold. The brakes include a vacuum booster to operate the cam inside the brake drum. The vacuum pump or booster operates upon the engine manifold vacuum and is put into action by the brake pedal.

The vacuum booster consists of a cylinder inside which moves a piston which is connected to the brake shoe operating cam through the links and levers. The cylinder chamber contains two valves ; atmospheric valve and vacuum valve which are operated by the brake pedal.

As the brake pedal is pressed down, it closes the air or atmospheric valve and opens the vacuum valve which connects the booster cylinder to the induction manifold vacuum. A connection is made by means of pipe line from booster to induction manifold. The manifold vacuum affects the piston inside the cylinder and the piston is pulled or pushed up. This causes the brake shoe operated cam to move and expand the brake shoes.

When the pressure is released from the brake pedal, the moving up pedal closes the vacuum valve and opens the atmospheric valve. The atmospheric pressure acting on the piston inside the cylinder pushes it back resulting in the reverse movement of cam and contracting of brake shoes. The brakes are thereby released.

(*d*) **Air brakes.** In these brakes, the brake shoe operating cam is operated by means of air pressure which is developed by an air compressor driven by the engine. There are separate brake chambers for the separate brake shoe operating cams. The brake chambers are connected with the air reservoir by means of pipe line. A brake valve operated by the foot pedal, controls the pressure of air which affects the brake chambers.

As the foot pedal is pressed down, air pressure acts on the diaphragm of brake chamber. The diaphragm is linked with the brake shoe aperating cam-shaft. The diaphragm is pushed outward in the brake chamber causing the movement of brake shoe operating cam. The brake shoes expand outwards and hold the moving brake drum as they come into its contact.

When the pressure is released from the brake pedal, it comes back with the help of return spring. This results in the closing of brake valve and release of pressure inside the brake chamber. The brake shoe operating cam moves in the reverse direction as a result of pressure release on the brake chamber. The brake shoes contract inward with the help of retracting spring, releasing the brake drum of the binding effect. The brakes are thus released.

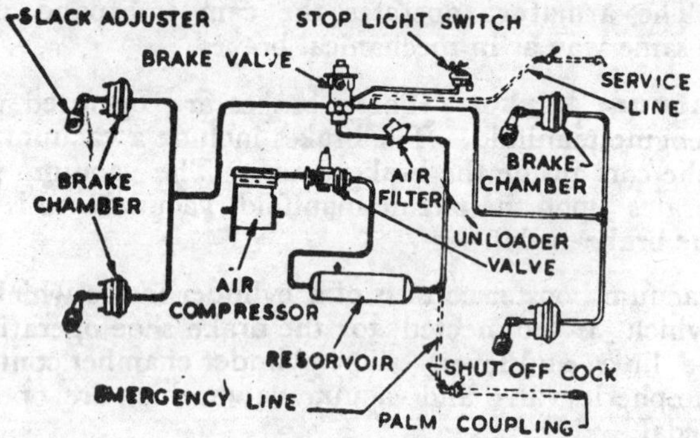

Fig. 17·7. Typical air brake system.

An air brake consists of the following main parts :—

(*i*) **Air compressor**. It is a machine which builds up air pressure in reservoir. A piston type air compessor is commonly employed in the brake system. Air is drawn into the cylinder through intake valve when piston moves downwards. When the piston moves upward, the intake valve is closed and outlet valve is opened by the air pressure and the air is forced out into the reservoir.

(*ii*) **Governor or unloader valve**. A governor in the form of safety valve prevents the building up of excessive and dangerous pressures in the reservoir. The valve opens and relieves excessive pressure when it reaches the predetermined limit.

(*iii*) **Reservoir.** It is a steel tank designed to withstand pressures greatly in excess of the pressures required to operate the brake system.

(*iv*) **Brake valve**. It is the control valve which is operated by the brake pedal. It is located between the reservoir and air lines leading to individual brake chambers.

(*v*) **Brake chamber**. It consists of a housing which encloses a moveable diaphragm connected by a rod linked to the brake shoe operating cam-shaft. The chamber is divided into two parts by the diaphragm, the side opposite to rod being air tight. Air pressure acts in the air tight portion of the chamber which causes deflection of diaphragm and application of brakes.

(*vi*) **Quick release valve.** This valve is used in the front brake lines to accelerate the release of air from the brake chambers. It directly releases pressure to the atmosphere rather than through the brake valve.

(*vii*) **Relay valve.** It speeds up the application and release of air from the brake chambers. It supplies air to the brake

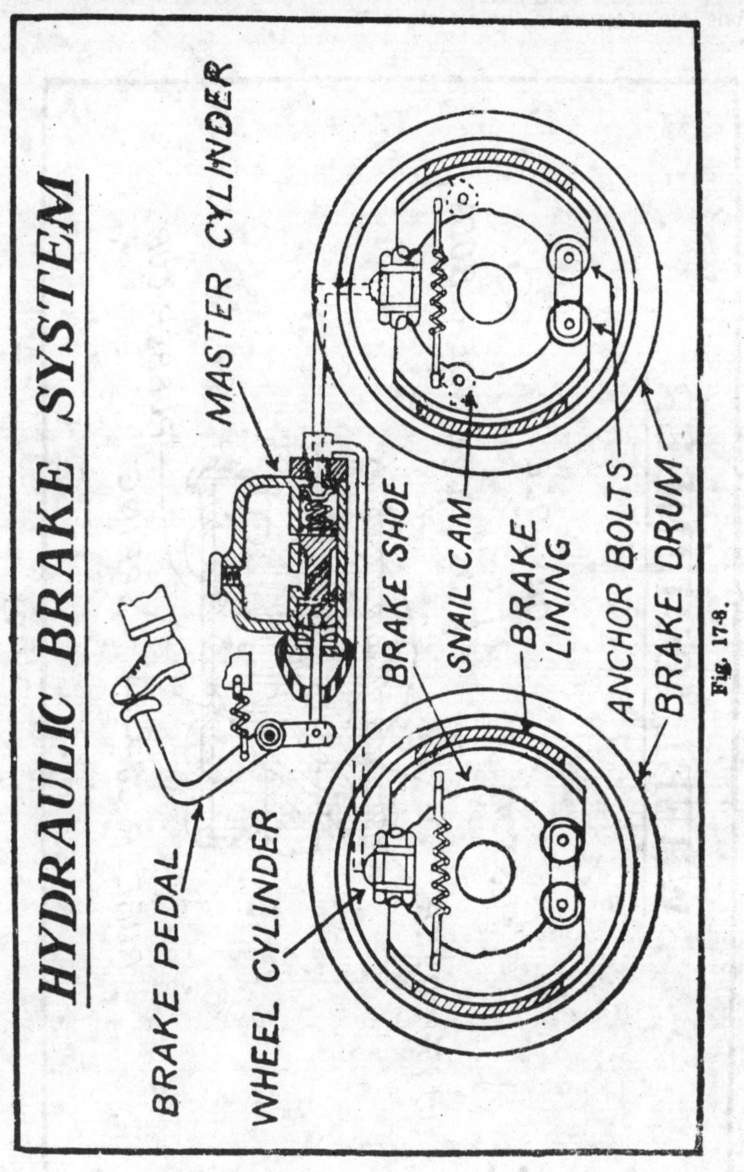

Fig. 17-8.

chambers directly from the reservoir for quick application of the brakes. It also exhausts compressed air from the rear brake chambers directly to the atmosphere rather than through the brake valve.

(*viii*) **Warning signal.** It is a warning light or buzzer which warns low pressure in the reservoir.

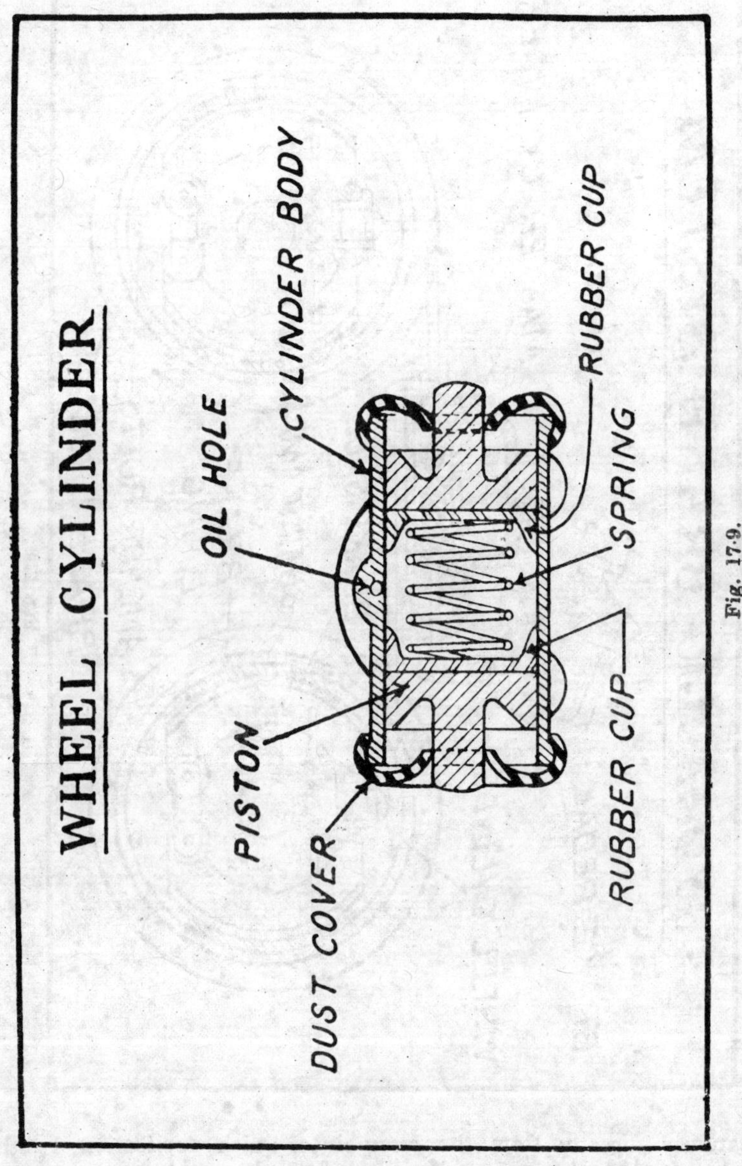

Fig. 17·9.

(*e*) **Hydraulic brakes.** Brakes which are operated by means of hydraulic pressure are known as hydraulic brakes. They function on the principle of Pascal's law which reads as below :—

"Pressure applied to a liquid is transmitted equally in all directions."

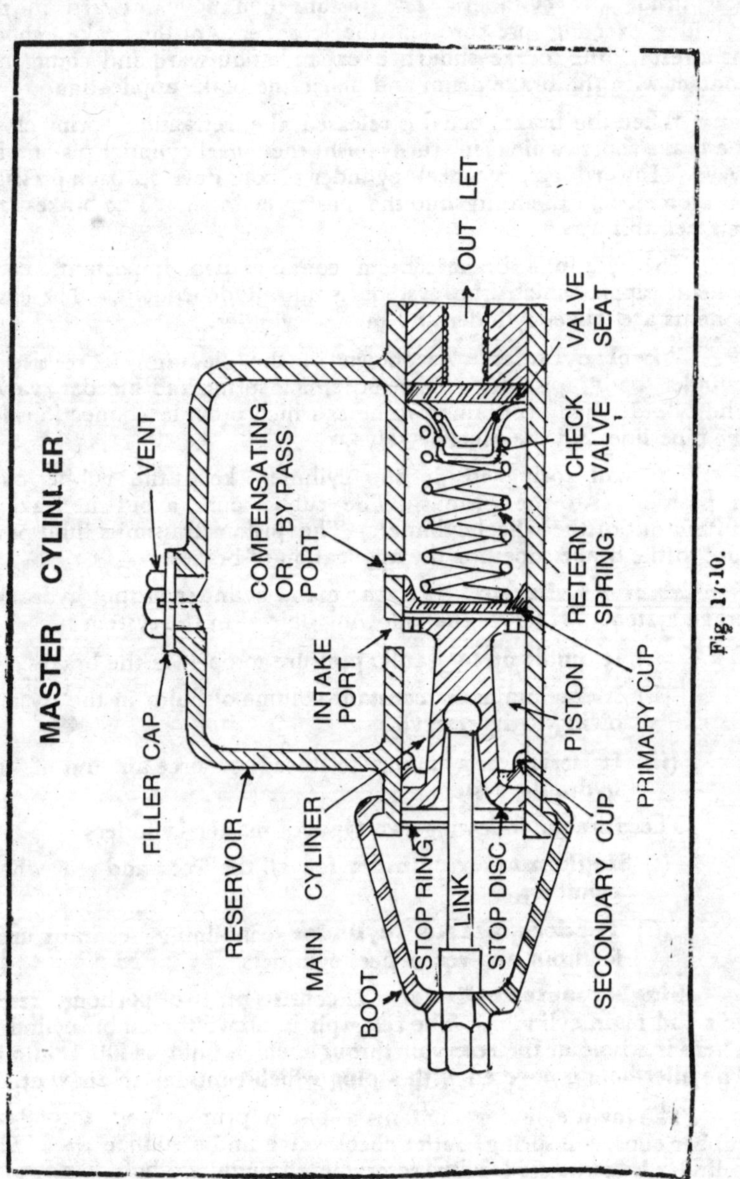

Fig. 17·10.

In this type of brake, wheel cylinders take the place of brake shoe operating cam. The wheel cylinder contains two pistons, at the ends of which the loose ends of brake shoes rest. The pistons inside the wheel cylinder move out when greater hydraulic pressure acts inside the cylinder. The pistons then move outward in the cylinder exerting pressure on the loose ends of the brake shoes. As a result, the brake shoes are expanded outward and come into contact with the brake drum and hence the brake application.

When the brake pedal is released, the retracting spring closes the brake shoes which in turn push the wheel cylinder pistons inward. Inward going wheel cylinder pistons develop back pressure and force back the fluid into the master cylinder. The brakes are released this way.

The hydraulic brake system contains two important components upon which the system is mostly dependent. The components are wheel cylinder and master cylinder.

Wheel cylinder. As shown in the diagram, it consists of cylinder body, pistons, rubber cups, coil spring and bleeder valve. The cylinder body contains two holes which provide connections for the pipe line and the bleeder valve.

The coil spring inside the cylinder, keeps the rubber cups in position with the pistons. The rubber cups avoid the leakage of fluid out of the wheel cylinder. The piston transmits fluid pressure to the brake shoes for the application of brakes.

Master Cylinder. It is the main cylinder in the hydraulic brake system. It serves the following objects in the system :—

(*i*) It builds up hydraulic pressure to operate the brakes.

(*ii*) It maintains a constant volume of fluid in the system owing to its reservoir.

(*iii*) It serves as a pump to bleed or force air out of the hydraulic system.

There are the following two types of master cylinders :—

(*i*) **Single master cylinder** for all the front and rear wheel cylinders.

(*ii*) **Tandem master cylinder** containing separate units for front and rear wheel cylinders.

Single master cylinder. It consists of two portions, reservoir and main cylinder. The reservoir is above the main cylinder. There is a hole at the reservoir through which fluid is filled into it. The filler hole is covered with a plug which contains an air vent.

The main cylinder contains a piston, primary and secondary rubber cups, coil spring, outlet check valve and a rubber seat. The cylinder is connected with reservoir through two holes known as main port and compensating port.

The piston works inside the cylinder and is operated by the brake pedal through linkage. As the brake pedal is pressed down, the piston inside the cylinder pumps out fluid into the brake lines through the check valve. As a result, a fluid pressure is built up in the system. The fluid pressure acts at the pistons in the wheel cylinders. The moving out wheel cylinder pistons expand the brake shoes and the brakes are applied.

The fluid enters back into the master cylinder due to the following facts after the brakes are released :—

(*i*) The tension of the retracting spring contracts the brake shoes which force the pistons inside the wheel cylinders. The inward moving pistons develop back pressure and the fluid is pushed back into the main cylinder when outlet check valve is lifted off the seat.

(*ii*) The sudden release of brake pedal pressure tends the piston to move back quickly due to the coil spring contained between the outlet check valve and the piston. Backward moving piston causes vacuum in the master cylinder which assists in lifting the outlet check valve off the seat allowing the fluid enter back into the master cylinder.

Action in the master cylinder while applying brake. The push rod forces the piston and rubber cup towards the outlet.

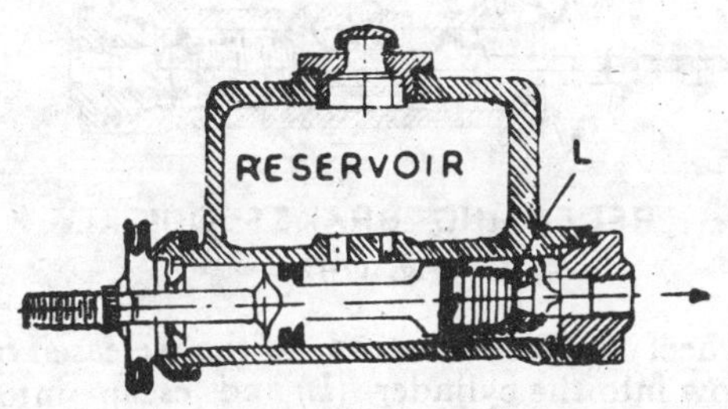

APPLYING BRAKES

Fig. 17·11.

As soon as the primary rubber cup covers the compensating port, pressure is built up in the cylinder (L) in front of primary cup. This forces the fluid to pass through the outlet check valve into the system, resulting in the pressure in each wheel cylinder being the same as the pressure inside the cylinder (L) before primary cup. The cylinder pressure is proportional to the pressure exerted on brake pedal.

Action in master cylinder while releasing brake.

(*a*) **When brake pedal is released slowly**, primary rubber cup follows the piston in release direction. Cylinder (L) in front of primary cup fills with fluid returning from wheel cylinders. The outlet check valve is pushed back of its seat due to the pressure of returning fluid as a result of which the fluid passes into the cylinder through the valve seat and hexagonal sides of check valve cage.

The spring inside the main cylinder (L) always maintains enough tension on the check valve to provide about 6 to 8 lbs. pressure in the brake lines and wheel cylinders after the brakes are fully released.

(*b*) **When brake pedal is released quickly**, the brake pedal return spring returns piston and primary cup to the released position before the cylinder (L) is filled by fluid flow from the wheel cylinders. During this quick release action, additional fluid flows through the piston holes and past the edges of primary rubber cup. This results in the supply of sufficient fluid to cylinder (L) to fill it even though wheel cylinders have not fully returned.

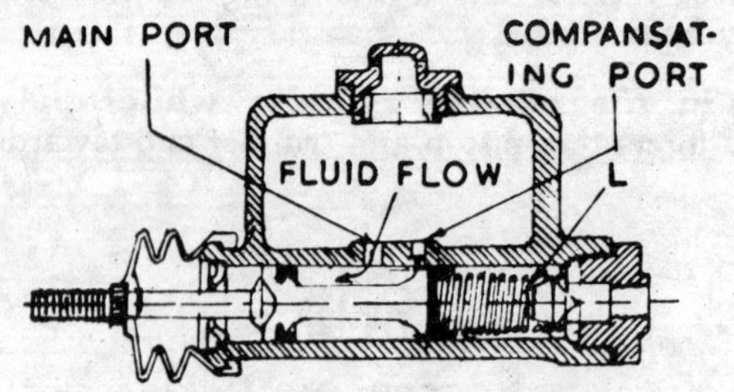

RELEASING BRAKES QUICKLY

Fig. 17·12.

As the wheel cylinders return to the released position, fluid continues to flow into the cylinder (L) and escape into the reservoir through the compensating port. The piston is always surrounded by fluid from reservoir because the main port connects the two.

If compensating port becomes plugged or is covered partially by the primary cup, there will be no escape for additional fluid accumulated in cylinder (L) at the time of quick release action. This will result in dragging of brakes and continuous glowing of stop light.

Master cylinder position when brakes are released : When brake pedal is fully released, spring in the cylinder (*L*) holds the check valve against the rubber seat with sufficient pressure to maintain 6 to 8 lbs. pressure in brake lines and wheel cylinders.

Cylinder (*L*) is connected with reservoir through compensating port. The piston is surrounded by fluid from reservoir because of connecting main port.

(*ii*) **Tandem Master Cylinder.** Master cylinder with two separate cylinders and reservoirs in the same master cylinder

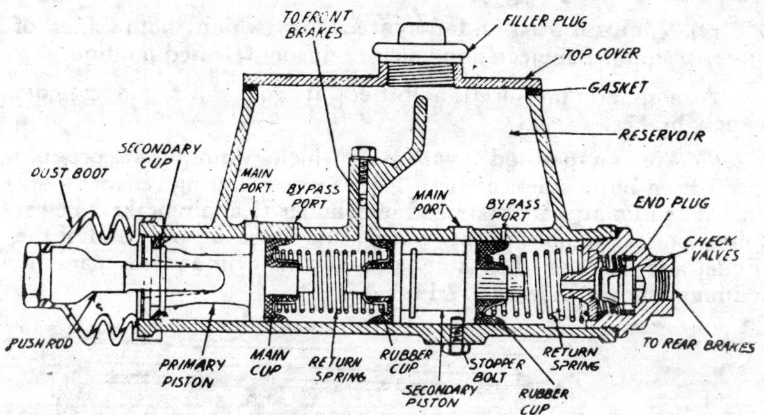

Fig. 17·13. Tandem Master Cylinder.

assembly, one operating front brakes and the other cylinder operating rear brakes, is known as Tandem Master Cylinder. This master cylinder avoids the possibility of all the brakes of a vehicle being put out of order by a leak or fracture in the pipe line leading to one wheel cylinder.

As shown in the diagram, the master cylinder contains two pistons, the rear being operated directly by the brake pedal. The space between the pistons connected to the front brakes while the connection to rear brakes is made at the front end opposite to operating link.

When the brake pedal is operated, rear piston moves inward developing pressure in the operating cylinder for front brakes. Since the front piston is free to move along the cylinder, so it also moves ahead developing an equal pressure in the operating cylider for rear brakes.

The return motion of the front piston is limited by the stopper screw.

(*f*) **Hydrovac Brakes.** It is a combination of hydraulic and vacuum brakes. In this type, vacuum system assists in the operation of hydraulic brakes. Instead that the piston in the vacuum booster directly operates the brake shoe operating cams, it merely helps in operating the master cylinder which develops hydraulic pressure in the hydraulic brake system for the effective application of brakes.

The vacuum booster consists of a cylinder inside which a piston operates and a control valve to admit and stop engine vacuum and atmospheric pressure.

This type of brake is also known as vacuum or power assisted hydraulic brake. The vacuum system as employed in this arrangement, is of the following types :

(*a*) **Vacuum suspended system** in which both sides of booster piston are subjected to vacuum during released position.

Atmospheric pressure is admitted at one end of the cylinder to apply brakes.

(*b*) **Air suspended system** in which atmospheric pressure is present on both sides of the booster piston during released position. Vacuum affects at one end of cylinder to apply brakes. Power application is accomplished by closing air inlet at one end of the cylinder and opening the same end of the cylinder to manifold vacuum as shown in figure 17·14 A & B.

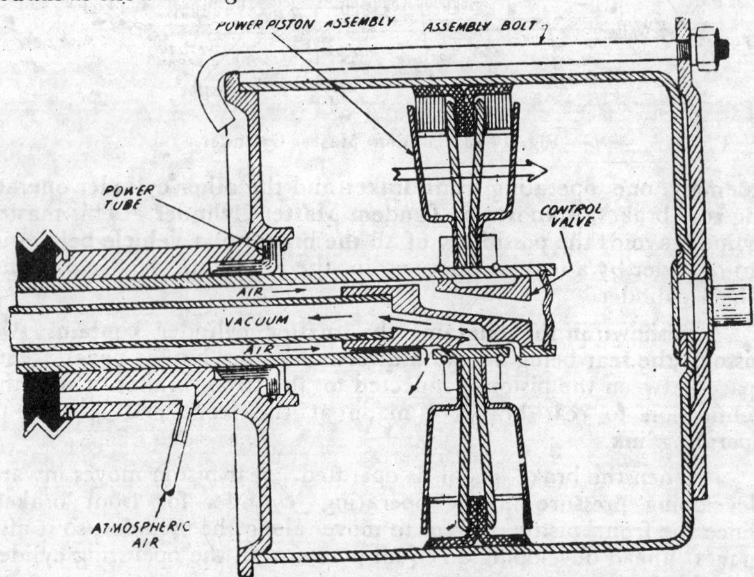

Fig. 17·14. (A). **Vacuum Booster.** Piston moving backward. Vacuum acting at the back side and front side opened to atmospheric pressure due to inward movement of inner sleeve valve.

The linkage of the power cylinder can be designed to either pull or push as desired for operating the master cylinder, by changing the operating end of the power cylinder. Vacuum booster or power cylinder is connected with the master cylinder by two different ways which are as under :—

(*i*) Vacuum booster or power cylinder mechanically connected to foot brake pedal and to master cylinder.

(ii) Vacuum booster connected by hydraulic lines to master cylinder and to wheel cylinder.

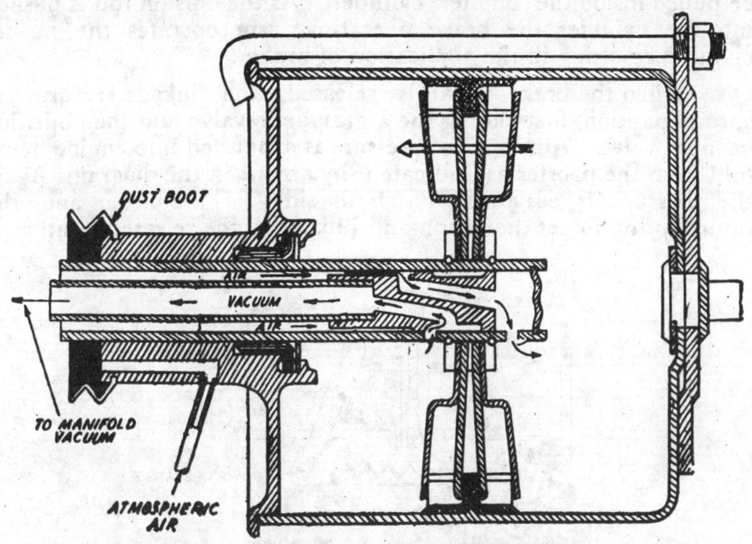

Fig. 17·14. B. Vacuum Booster. Piston moving forward. Vacuum acting at forward side and atmospheric pressure acting at the back of piston due to forward movement of inner sleeve valve.

(*i*) **Mechanical connection through linkage.** In this type, the booster is operated by means of linkage by the foot brake pedal. In vacuum suspended system, when the brake pedal is operated, the reactionary linkage as shown in the diagram, moves inward until

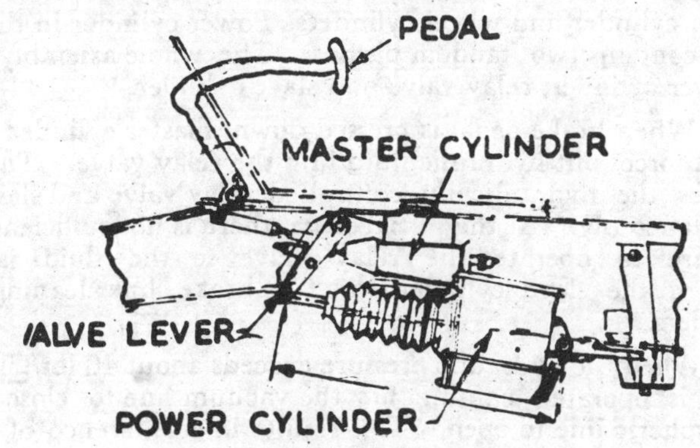

Fig. 17·15. Booster layout.

the adjusting screw contacts the brake operating arm. The movement of the reactionary linkage forces the control rod inward, closing the vacuum valve and opening atmospheric valve. This seals the cylinder on back side of piston and admits atmospheric pressure into chamber in front of piston. A difference of pressure is created between both the chambers which causes the piston rod to be pulled inside the booster cylinder. As the piston rod is pushed into the cylinder, the brake operating arm operates the master cylinder resulting in the application of brakes.

When the brake pedal is released, the linkage returns to normal position, first closing the atmospheric valve and then opening vacuum valve. Atmospheric pressure is exhausted into engine manifold from the booster as indicated by arrows in the diagram. Again the pressure is equalized on both sides of the piston and the return spring forces the piston and linkage to the released position.

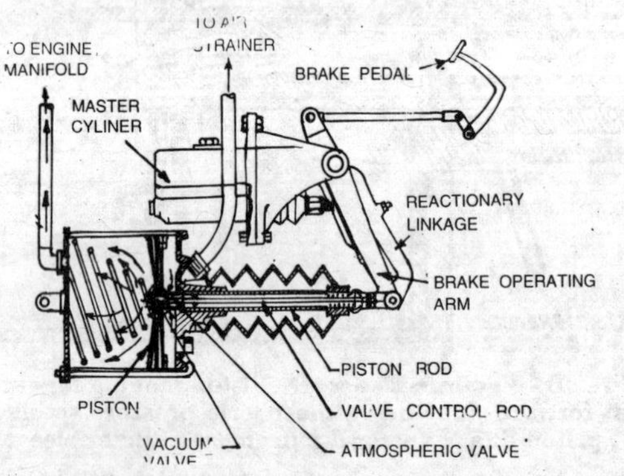

Fig. 17·16. Released position.

(*ii*) **Hydraulically connected vacuum booster.** This type of arrangement needs no mechanical linkages and hence the power cylinder could be placed anywhere in the chassis frame between the master cylinder and wheel cylinders. Power cylinder in this arrangement contains two tandem pistons. The whole assembly consists of a power cylinder, relay valve and slave cylinder.

When brake pedal is pressed down, master cylinder is operated which forces out hydraulic fluid into the relay valve. The pressure reaches the hydraulically controlled relay valve and slave cylinder simultaneously. At the initial stage, there is not sufficient hydraulic pressure to operate the relay valve so the fluid is by-passed through the slave cylinder into the brake lines leading to wheel cylinders.

When the hydraulic pressure exceeds about 40 lbs/□″, the relay valve is operated causing first the vacuum line to close and then atmospheric line to open. This results in a difference of pressure in the power cylinder containing pistons due to which the power cylinder pistons begin to move forward first, closing the bypass in the slave cylinder piston and then pushing the slave cylinder piston forward, exerting additional pressure on the fluid in the brake lines resulting in application of brakes.

When brake pedal is released, master cylinder pressure falls down. This results in the relay valve, power cylinder and slave cylinder in the released position. The atmospheric line is closed and vacuum line is opened in vacuum suspended system.

8. **Arrangement of brakes in different vehicles**. As explained earlier, there are two types of brakes in an automobile, *i.e.*, service brake and hand brake. The arrangement of both types of brakes in different automobiles is usually as under :

(*i*) Service or foot brake applicable to all the front and rear wheels and hand brake applicable to rear wheels alone.

(*ii*) Service or foot brake on all wheels and hand brake on drum at transmission shaft.

9. **Brake linings**. The following two types of linings are employed with the brake shoes :

(*a*) Organic,

(*b*) Metallic.

(*a*) **Organic lining**. It is of two types ; moulded and woven. The moulded type is made from a thoroughly mixed compound of asbestos, filler materials and powdered resins. The compound is moulded in dies to form into shape and is placed under heat and pressure until a hard slate like board is formed. It is then cut and bent into individual segments for attachment to the brake shoes.

The woven type organic lining is woven from strands of asbestos and threads of other materials and impregnated with a rubber compound.

In some cases the organic brake lining is reinforced with fine metal wires or wire nettings.

The organic type brake lining is used almost exclusively for ordinary brake service.

(*b*) **Metallic lining**. It is made of sintered metal and is composed of finely powdered copper or iron, graphite and some amount of inorganic fillers and friction modifiers. After thoroughly mixing the constituents, a lubricating oil is added to avoid segregation of different materials. It is then made into the required form by means of a special process.

Metallic linings are used for extreme braking conditions as encountered in police cars, fire brigade vehicles and sports cars. Under such extreme service, the frictional qualities of the metallic lining are more constant than that of organic lining.

10. **Brake drum**. The brake drum rotates with the road wheel and provides a contacting surface for the brake shoes to come into contact for braking action. Brake drums are usually made of

cast iron. In some cases brake drums are made of steel or aluminium having inner liner of cast iron.

In most cases, fins are provided at the outer surface of the brake drum for the rapid dissipation of heat and thus help in its cooling.

11. Brake effectiveness. The following factors contribute to the effectiveness of the brakes :

(*i*) Area of brake linings.

(*ii*) Amount of pressure applied to brake shoes.

(*iii*) Radius of brake drum.

(*iv*) Radius of car wheel.

(*v*) Coefficient of friction of braking surfaces.

(*vi*) Coefficient of friction between tyre and road surface.

12. Factors controlling the stop of an automobile. The following factors control the stop of an automobile ;

(*i*) **Speed and load.** Lesser the speed and lesser the load, the less energy shall be absorbed to stop the vehicle as the brakes convert energy of motion into heat. The higher the speed and more the load, the more energy shall be absorbed for stopping a vehicle.

(*ii*) **Road surface.** The nature of road affects the coefficient of friction between the road and tyre. If the road is slippery, the vehicle shall stop at a greater distance than if the road is dry.

(*iii*) **Tyre tread.** A good tyre tread shall have good holding ability. It provides a greater coefficient of friction than a smooth tyre.

(*iv*) **Gradient.** If the vehicle is going up a gradient or hill, the gravitational force assists in stopping the movement of the vehicle. When a vehicle is going down a gradient or hill, the force of gravity tends to keep the vehicle moving and hence a greater braking force is required to overcome its movement.

(*v*) **Number of wheels braking.** For maximum braking effort, all the wheels should be equipped with brakes. If any of the wheels are not provided with brakes, the time and distance for stopping the vehicle shall increase.

(*vi*) **Coefficient of friction of braking surfaces.** The effectiveness of the brakes much depends upon the braking surfaces apart from road surface and tyre tread. The ability to hold a moving wheel is controlled by the coefficient of friction between the brake lining and brake drum. If the linings are glazed with oil or grease, they will tend the brake drums to slip as the coefficient of friction is lowered down.

(*vii*) **Pressure applied by leverage.** The s multiply the applied force. The increase over applied physical force depends upon the ratio of levers between the brake pedal and brake shoes.

(*viii*) **Pressure applied through energization**. The natural tendency of the brake shoe is to follow the direction of rotation of brake drum, when brakes are applied. Through design, the tendency to wedge the brake shoe more tightly against the brake drum is used to increase the applied pressure. More the self-energizing ability of the brake shoes, better the braking effect.

(*ix*) **Weight transfer**. When brakes are applied, there is a natural transfer of weight from the rear wheels to the front caused by the tendency of the mass to continue in its forward motion. The greater the deceleration, the greater is the load or weight transfer from rear to front wheels. During the stop, the weight holding the rear wheels to the road is lessened while the weight on the front wheels is increased by exactly the same amount. It is owing to this fact that the brakes are designed so that applied effort is greater on front wheels than on the rear.

Trucks are often designed so that the major portion of the load is carried by the rear axle. In most cases, even with the transfer of weight during stop, the rear wheels still carry most of the load. Thus larger brakes and greater applied effort are used on the rear wheels.

Vehicles in which there is equal distribution of load at front and rear axles, usually require larger brakes and greater applied effort at the front wheels to combat the load transfer. This is accomplished by increasing the size of brake drums, brake shoes, wheel cylinder etc.

(*x*) **Braking force of engine.** The engine is also used as a brake when going down a hill. The braking effect of the engine is more in lower gear speed than in direct drive.

8. Self-energization and servo action of brake. When the brake is applied and the brake shoes come in contact with the brake drum, the shoes have a tendency to follow the drum and to rotate with it. After the initial movement, the anchor prevents the further movement at the anchored end of the shoe. The anchored end, however, still tries to rotate with the drum, wedging the shoes more tightly against the drum. This action known as self-energization, assists in the ability to brake and is utilised to lessen the force required to apply the brakes.

In some brake shoe arrangement, the secondary brake shoe assists in the self-energization of the brake. The primary shoe pushes the secondary shoe during wedging action due to which self-energization is increased or amplified. When such is the case, the amplification of forces is known as *servo action*. In order to accomplish servo action, the brake shoes are linked together at the bottom and clearance is provided between the ends of the shoes and the anchor pin at the top, so that the brake shoes can be rotated slightly in relation to the axle shaft or wheel spindle.

In this case when the brake is applied, both the shoes will move with the rotation of brake drum until the secondary shoe is stopped by the anchor pin and the primary or leading shoe is also stopped as it is linked to the secondary or trailing shoe through the connecting link or adjusting screw. The primary shoe starts the servo action. The nearer the centre of rotation from the anchor pin, the more powerful will be the wedging action. The wedging action starts at the toe of the brake shoe and leads to its heel. The self-energization and servo action greatly multiply the force pressing the shoes against the brake drum.

9. **Engine as a brake.** When accelerator is released during the movement of a vehicle, a retarding effect results in because the engine wants to idle but the drive from the drive axle forces it to rotate at a considerably higher speed, the engine thus exerts a braking effect through the driving wheels. The braking effect of the engine can be increased by engaging a lower gear for raising r.p.m. of the engine.

10. **Power brakes.** Brakes operated by power other than physical effort applied at brake pedal, are known as power brakes. The power may be exerted by air pressure, engine vacuum or electrical energy. The following types of brakes described already, are power brakes :

(*i*) **Air brakes** employing compressed air to operate the brake mechanism.

(*ii*) **Vacuum brakes** utilizing engine vacuum to actuate brakes.

(*iii*) **Electric brakes** using electrical energy from the battery to operate the brakes.

There are semi-power or power assisted brakes too, in which the power assists or supplements the power in actuating the brakes. Hydrovac brakes are semi-power brakes. There the engine vacuum assists in the actuation of hydraulically operated brakes.

The main features of the power brake systems are as below :

(*i*) "Time-lag" or the time interval between the moment when brake pedal is pressed down and the moment when the brakes are put on, must be very small.

(*ii*) The force applying the brakes should be closely proportional to the force exerted by the driver on the brake pedal.

The brake system should be such so that driver may judge the intensity of the application of brakes fairly accurately.

11. Girling brakes. Girling brakes as famous by their trade name, contain brake expander in lieu of brake shoe operating cam.

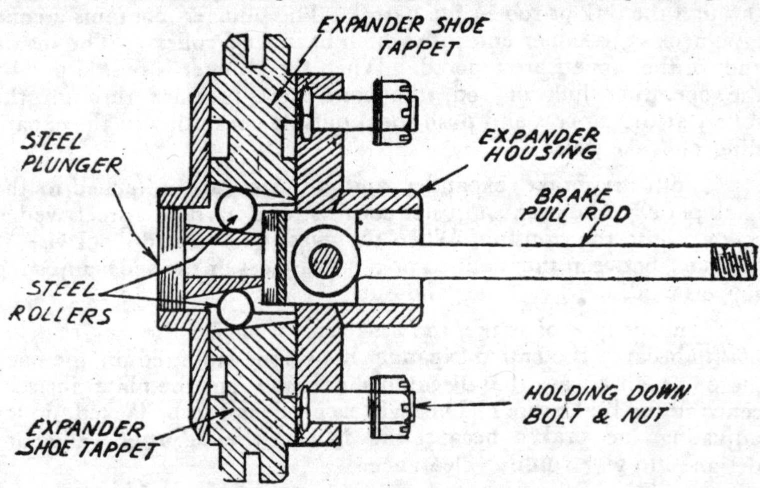

Fig. 17·17. Girling brake expander.

The brake shoes are not pivoted at their lower ends but instead, are resting in the slots of individual tappets of the expander mechanism. The upper ends of the brake shoes rest over the tappets of brake adjuster. The shoes are connected with each other through two springs, one placed in the lower part and the other at the upper portion. The springs keep the brake shoes contracted inward and

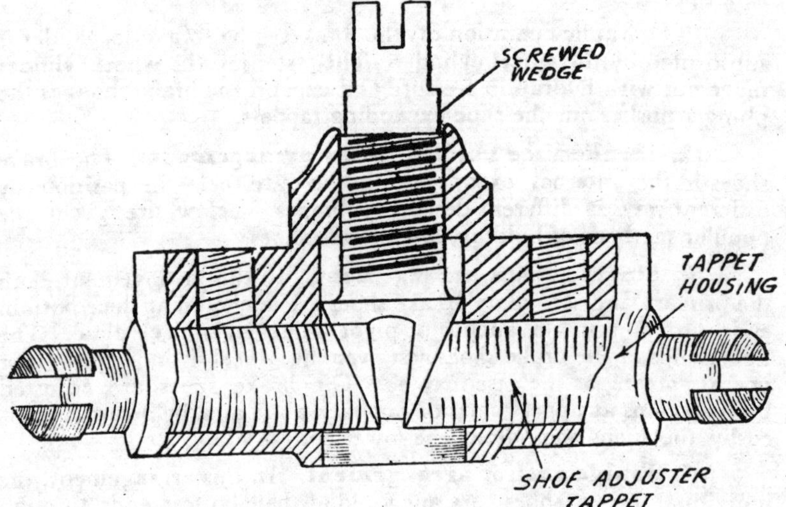

Fig. 17·18. Girling brake adjuster.

held in place over the tappets of brake expander at the lower end and brake adjuster at the upper end.

The brake expander consists of a housing, tappets, steel rollers, plunger and operating link. The plunger is operated through the link or rod in the barrel. The plunger contains a cone type neck at its upper end into which fit in steel rollers. The inside ends of the tappets are tapered. When the plunger is pulled up by the operating link or rod, the cone exerts pressure through the rollers at the tappets and push them outside, resulting in the expanding of brake shoes.

Both the brake expander and adjuster are attached to the back plate. The brake adjuster consists of a hardened-steel wedge screwed into the housing. When the wedge is turned clockwise, it is forced between the inclined or tapered faces of the shoe adjusting tappets tending them to expand out.

In this type of brake, the brake shoes possess the *self-centring ability* because the entire expander housing is mounted on the back plate in such a way that it could slide away on the plate for self-centring the brake shoes. The wheels need not to be jacked up for adjusting the brakes because the flats on the adjuster wedge are designed to give running clearance.

For adjusting the brakes, the adjuster is rotated clockwise until the resistance of the shoes contacting the brake drum is felt and then it is slacked back to the nearest notch. After all the wheels are adjusted, the brake pedal is pressed down to centralize the brake shoes in the drums. Once this is done, the shoes remain centralized. Before making adjustment, it should be ensured that the hand brake is in fully released position and foot brake pedal is not pressed.

In hydraulic operation of the brakes, the brake expander is substituted by the wheel cylinder. The pistons in the wheel cylinder move out with hydraulic pressure and expand the brake shoes as the plunger pushes out the shoe expanding tappets.

12. Brake-shoe holding down arrangements. The brake shoes in the internal expanding brakes are held in position by different ways in different designs of brakes. Below are given the popular methods of holding down brake shoes.

(*i*) **Single anchor arrangement.** In this arrangement, both the primary and secondary br ake shoes are anchored at their bottom ends through a single anchor or pivot attached to back plate. The upper ends of the brake shoes rest over the wheel cylinder pistons or operating cam, as the case may be. The brake shoes are adjusted by snail cams at upper parts of the back plate and by a cam operated by the anchoring bolt at the lower end.

(*ii*) **Double anchor arrangement.** In this arrangement, the individual brake shoes are anchored at their bottom ends through separate anchor bolts. The upper ends of the brake shoes rest over thewheel cylinder pistons or operating cam. The adjustment of brake shoes is in the similar method as employed in single anchor arrangement. In certain cases, adjusment of brake shoes at their upper ends is through notched collars provided at the wheel cylinder pistons. The collars are screwed in or out to adjust the brake shoes at their upper ends.

(*iii*) **Self-centring or adjusting arrangement**. In this type of arrangement, the brake shoes are not pivoted at their lower ends. The lower ends simply rest over the brake adjuster. The brake shoes are held over the brake adjuster and wheel cylinder or brake expander by means of springs, one at the lower and the other at the upper end and holding down springs and clips. The brake adjuster may be of girling type or of bendix or screw type.

In some cases, there is a peg or anchor placed between the uppermost ends of the brake shoes In some cases, the anchor is fixed whereas in some designs, it is adjustable. The wheel cylinder is located below that peg. The peg assists in the self-energization action and servo action of the brakes.

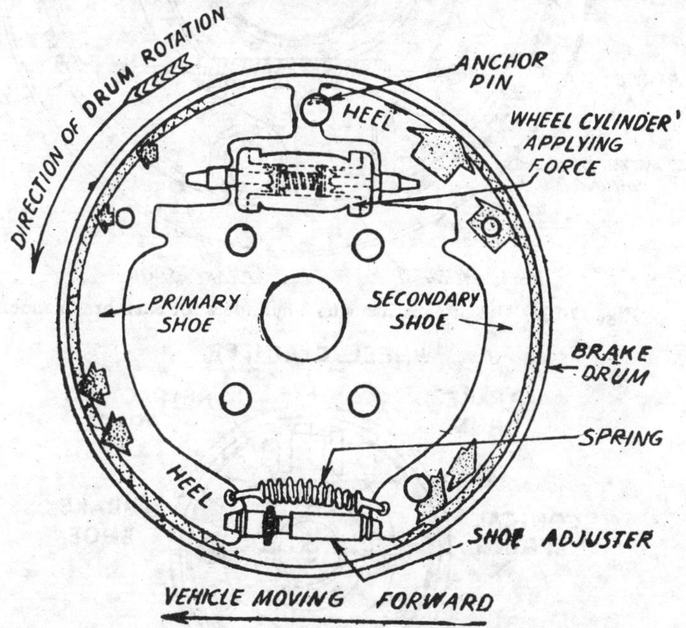

Fig. 17·19. Self-centring arrangement (servo action of brake).

(*iv*) **Separate wheel cylinders for each brake shoe.** In this arrangement there are separate wheel cylinders for the individual brake shoes. One end of each brake shoe is anchored to one wheel cylinder which operates the opposite brake shoe. There are two retracting springs, one at the lower and the other at the upper end. The adjusment of the brake shoes is through snail cams and anchor bolt cams.

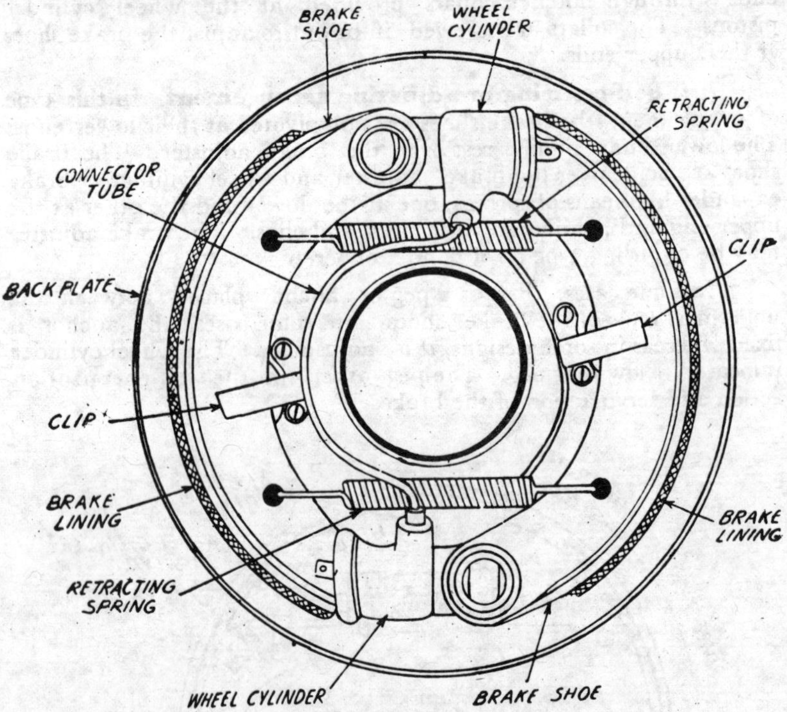

Fig. 17·20. Use of separate wheel cylinders for each brake shoe.

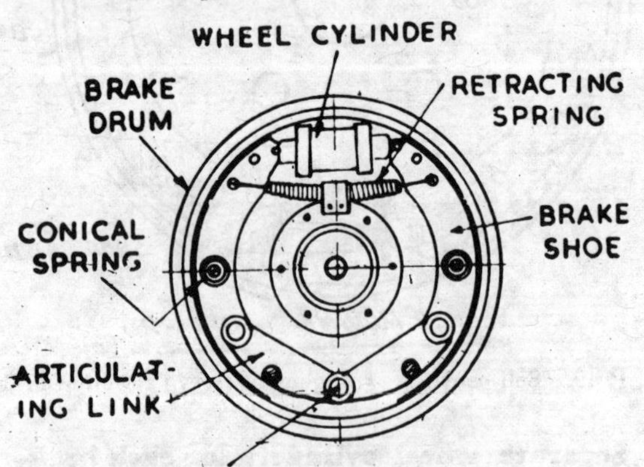

Fig. 17·21. Articulating link arrangement.

(*v*) **Articulating link arrangement.** In this arrangement the brake shoes are connected to single anchor pin through articulating links. The articulating links connect the lower ends of the brake shoes together. The pivot pin for the articulating link is anchored to the back plate. The upper ends of the brake shoes rest over the piston ends of wheel cylinder.

The articulating links provide double lever action on the brake shoes which results in self-energization.

13. Characteristics of hydraulic brake fluid. The hydraulic brake fluid should possess the following characteristics :

(*i*) It should not soften the rubber parts used in the hydraulic brake system.

(*ii*) It should not corrode or rust metallic parts in the brake system.

(*iii*) It should not vaporize at high temperatures encountered in actual service.

(*iv*) It must remain fluid at low temperature.

(*v*) It must act as a lubricant to the moving parts inside the system.

(*vi*) It must retain all its characteristics for a maximum long period.

(*vii*) It should be non-compressible.

(*viii*) It must mix satisfactorily with other makes of hydraulic brake fluids.

14. Bleeding of hydraulic brakes. When any part of hydraulic line is replaced, air traps into the system. Air being compressible, the effort of brake pedal goes waste in applying brakes. Until and unless air from the system is removed, the brakes wouldn't function properly. The process of removing air from the brake system is known as bleeding.

Bleeding is started from the farthest wheel cylinder from master cylinder. The nearest wheel cylinder is bled in the last so that air may not re-enter into the brake lines during bleeding. A bleeding tube is connected at the bleeder valve on wheel cylinder. The other end of bleeding tube is immersed in the glass jar containing some brake fluid. The brake pedal is operated 3 or 4 times till the pressure is developed and the pedal comes up. Brake pedal is kept pressed and bleeder valve is unscrewed. As soon as the bleeder valve is opened, the fluid will come out and enter into the glass jar. Air bubbles shall be seen coming along with fluid in the jar. Brake pedal shall come down to the foot board. It is released and pressed down again. This process is repeated two or three times till all the air is expelled from the wheel cylinder. Pedal kept pressed down, the bleeder valve is screwed on (closed). Fluid is transferred from glass jar to the master cylinder so that air may not enter into the system due to deficiency of fluid in the master cylinder.

One by one all the wheel cylinders are bled, starting from the farthest and ending at the nearest. A watch is kept at the master cylinder lest fluid level in the reservoir may fall too low.

QUESTIONS

1. Illustrate mechanical service brake system.
2. Describe the difference between drum and disc brakes.
3. Illustrate external contracting type brake.
4. Classify the brakes and illustrate any one type of brake.
5. Illustrate air brakes.
6. Describe any two of the following :
   (*i*) Hydraulic brakes.
   (*ii*) Electric brakes.
   (*iii*) Hydrovac brakes.
   (*iv*) Disc brakes.
   (*v*) Parking brake.
7. Describe the construction and working of master cylinder with the help of suitable sketches.
8. Illustrate the action of master cylinder when brake pedal is quickly released.
9. Describe the main parts of the air brake system.
10. Illustrate a wheel cylinder.
11. Write short notes on the following :
    (*i*) Tandem master cylinder.
    (*ii*) Brake lining.
    (*iii*) Brake drum.
    (*iv*) Engine as a brake.
    (*v*) Self-energization of brakes.
    (*vi*) Power brakes.
12. Explain the factors which effect the stop of an automobile.
13. Which are the different types of vacuum boosters ? Explain.
14. Illustrate hydrovac brake.
15. How a power cylinder is operated ? Illustrate one arrangement as used in hydrovac brake system.
16. Explain mechanical type of Girling brake.
17. Illustrate the following :
    (*i*) Girling brake expander.
    (*ii*) Girling brake adjuster.
    (*iii*) Wheel cylinder.
    (*iv*) Servo action of braka.
18. Explain the different arrangements of brakes as provided in different vehicles.

19. **Illustrate the different arrangements for holding down brake shoes in the different designs of brakes.**

20. **Explain how self-energization or servo action is obtained in articulating link arrangement.**

21. **Which characteristics hydraulic brake fluid should possess?**

22. **What is bleeding? Explain the process.**

23. **Explain the braking principles.**

24. **Explain the following :—**

(*i*) **Braking force.**

(*ii*) **Weight transfer.**

(*iii*) **Brake dip.**

(*iv*) **stopping distance and time.**

(*v*) **Brake efficiency.**

18

Air-Conditioning in Automobiles

1. Air-conditioning. The object of air-conditioning is to improve passenger comfort by controlling temperature. The job is done by the air-conditioner, abbreviated as A/C. The principles of operation of automobile A/C are basically same as those of refrigerators and home air-conditioners.

Most refrigeration systems cool through evaporation. When a liquid vaporizes, it absorbs a big amount of heat. If you put some alcohol on your hand, it will cool you by evaporation. As soon as alochol evaporates, it will carry away heat from your hand and you shall feel cool.

Although air-conditioner is designed to cool air, yet it controls humidity, cleanliness and circulation of air too by reducing moisture-content in the air.

2. Air conditioning cycle. In air-conditioning systems, air to be cooled passes through a finned heat exchanger known as evaporator. The substance known as *refrigerant* to carry heat out of automobile, boils in the evaporator, absorbing heat as it changes into a vapor.

The refrigerant is circulated in the system by the compressor driven by the engine. When the system is operating, the compressor draws low pressure refrigerant vapor from the evaporator. Then the compressor compresses this vapor, puts it under high pressure and passes it to the condenser. The condenser does the same job of removing heat.

When refrigerant vapor is compressed, it becomes very hot. the hot compressed vapor from the compressor, enters the condenser. The vapor heat is carried away by the surrounding air which passes through the condenser fins. This action is just like the radiator in the engine cooling system.

The cooling down vapor in the condenser looses heat and changes to liquid form. The liquid flows out of condenser and enters into the receiver. The receiver is a store for liquid refrigerant, from where it flows through the high pressure line to the orifice tube. In this line, a sight glass is also fitted in some cars, to see that liquid refrigerant flow is normal.

The orifice is a sort of restriction to control the flow of refrigerant. After passing through orifice tube, the refrigerant enters the evaporator. The evaporator is at low pressure which tends the refrigerant to evaporate. As it evaporates, it takes heat from the air passing through air passages in the evaporator. This air is from inside the car which is cooled by passing through the evaporator. A blower directs cooled air into the vehicle. Now-a-days a thermostatic expansion values is fitted in place of orifice.

3. Main parts of air-conditioner. the following are main parts of an automobile air conditioner. (*i*) Compressor, (*ii*) Condenser, (*iii*) Receiver/ Drier, (*iv*) Expansion valve, (*v*) Evaporator, (*vi*) Blower.

(*i*) *Compressor.* It is a pump to circulate and increase the pressure of refrigerant in the system. It is fitted in the front portion of the engine and is driven by a V-belt. It is operated through a magnetic clutch when the air-conditioner is switched on.

(*ii*) *Condenser.* It is placed in front of engine radiator. Air flowing over its fins takes away heat from the refrigerant and cooles it down. The refrigerant condenses as it passes through the condenser.

(*iii*) *Receiver/Drier.* It is a reservoir to store and dry the precise amount of refrigerant required by the system.

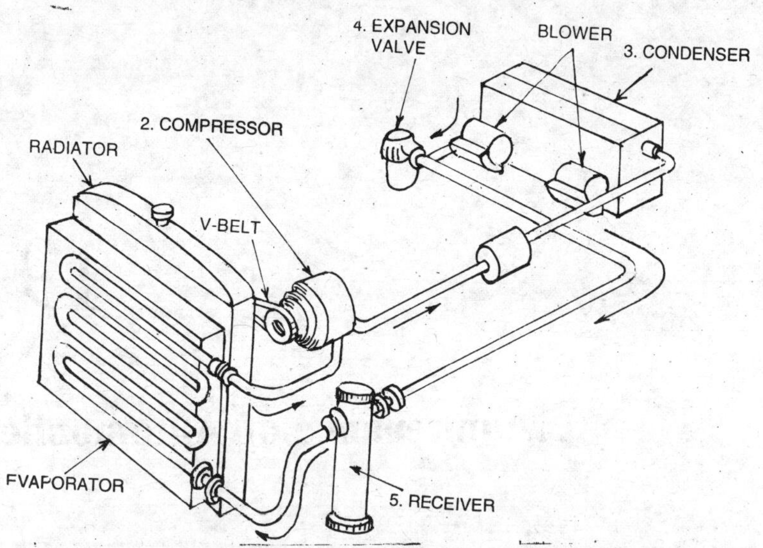

Fig. 18.1 Automobile air-conditioning system.

(*iv*) *Expansion Valve.* It is a thermostatic valve which is fitted at the inlet side of evaporator. It regulates refrigerant flow to the evaporator. Its regulating action is controlled by the temperature-sensing bulb placed on the suction (outlet) line of evaporator.

(*v*) *Evaporator.* It is an air cooler and dehumidifier placed at the back of system.

(*vi*) *Blower.* It is a small fan, driven by an electric motor, to direct cooled air into an automobile.

4. Refrigerant. The substance which takes away heat out of an automobile or a refrigerator cabinet is known as refrigerant. Any liquid which boils near the freezing point of water may be used as refrigerant. It should be non-poisonous, non-explosive, non-corrosive and capable to mix with oil. A volatile refrigerant called R-12 is used in automobile air-conditioning systems. It boils at – 30°C under atmospheric pressure.

Refrigerant R-12 can be stored in liquid form only under high pressure. If pressure is removed, it evaporates many times faster than alcohol.

5. Technical terms in air-conditioning system servicing:

(*i*) *Purging* (*Discharging*): It is the act of releasing refrigerant from the high and low sides of the system until no pressure exists.

(*ii*) *Evacuating*: It is the process by which all air and moisture is removed from the system.

(*iii*) *Charging*: It is the method of filling refrigerant into the system.

QUESTIONS

1. What is the object of air-conditioning?
2. Define a refrigerant.
3. Explain air-conditioning cycle.
4. Define the role of following components:
(*i*) Compressor (*ii*) Condenser (*iii*) Evaporator (*vi*) Expansion valve (*v*) Receiver.
5. Describe the following terms: (*i*) Purging (*ii*) Evacuatng (*iii*) Charging.

19

Maintenance of Automobile

1. **Maintenance.** An automobile has to work under severe operational conditions owing to which there is great strain on its working parts. It is neither the age of an automobile nor the mileage which determines its usefulness. The life of an automobile depends mainly on the following two points :—

(*a*) How the vehicle has been driven ?

(*b*) How the vehicle has been maintained or looked after ?

Although most men could operate one or other type of automobile yet there are right and wrong ways in driving which need serious thinking.

By acquiring good driving habits, you do only one part of the job. The other important part is maintenance. In order to preserve full working life of an automobile, its maintenance should be carried out religiously at the prescribed intervals and in the prescribed manner.

Maintenance can be divided into the following two parts :—

(*a*) Lubrication.

(*b*) Cleaning, inspection and adjustment.

(*a*) **Lubrication.** In an automobile, different parts need different type of lubricant owing to their state of operation. The parts moving at the normal to high speeds need lubricant oil. Such parts are engine bearings, cams and gears and transmission bearings and gears. The slow moving and intermittently moving parts require a heavier lubricant or grease. Such parts are steering mechanism joints, pivot pins, spring shackles etc. One must know which lubricant is to be used where and when.

The most common cause of damage or failure is lack of lubrication but the danger of over-lubrication also exists there. It is as important to use the *right amount* of lubricant as the *right lubricant* at the *right time* in the *right place*. More oil in engine shall lead to oiling-up and dark white smoke. Excessive oil in the transmission shall resist the movement of gears and shafts. Too much grease on C. B. arm heel shall run onto the breaker points and lead to fouling. Similarly low engine oil level shall lead to engine overheating in addition to damage to working parts. Lack of lubrication at steering linkage joints, pivot pins etc. shall result in hard steering. Poor lubrication also causes metallic noise in addition to increase in wear rate. Failure in oil supply leads the engine to seizure. Lack of lubrication at wheel hub bearings leads the brake drums to overheat and break away resulting in serious accident. Improper lubrication at the speedometer cable twists it to break away. Accelerator, clutch and brake cables etc., stick into their outer casings and hold or restrict the movement of important controls.

The manufacturers supply lubrication charts in the operation as well as service manual as a guidance in the maintenance of vehicle. The lubrication chart supplies the following information :—

(*i*) Lubrication points.

(*ii*) Type of lubricant required for a particular point.

(*iii*) Period or interval after which to lubricate.

(*iv*) Quantity and grade of lubricant required for different points.

The grade of lubricant differs from season to season, region to region and condition to condition. There is temperature effect upon the viscosity of oil. During winter when temperature is low, thin oils are required and during summer thick ones are needed. Similarly the grade of oil is different in cold regions than hot ones.

The lubricant should provide a perfect bed (cushion) for the moving parts under different conditions of load. speed and temperature. Lubricant film should not break under these varied conditions. If the lubricating film is subjected to greater loads, speed and temperature, the lubricant should be of higher viscosity and if it is subjected to lower loads, speed and temperature, the lubricant should be of lower viscosity.

Only high quality of engine oils having excellent lubricating properties should be used as the engine life much depends upon them. They should have good detergent and dispersing effects. They should have sufficient protection against corrosion, good resistance against deterioration, prevention of formation of foam and acid, etc.

Heavy duty (HD) oils fulfil all these requirements. Since special conditions prevail during running-in period so special quality of oils should be used in new or newly overhauled engines. Viscosity and quality should be kept in mind when topping or changing engine oil.

Hindustan motors Ltd., Calcutta, recommend the following type and grade of lubricants for their Ambassador cars :—

Component	*Climatic conditions*	*Type and grade of lubricant*
(*i*) Engine and air cleaner	(*a*) Tropical and temperature down to 32°F	Caltex RPM Motor oil SAE 30 HD
	(*b*) Extreme cold down to 10°F	Caltex RPM motor oil SAE 20 HD
	(*c*) Arctic consistently below 10°F	Caltex RPM Motor oil SAE 10/10 W
(*ii*) Gearbox, steering gear box and differential	(*a*) Tropical and temperate down to 10°F (*b*) Extreme cold below 10°F	Caltex Universal Thuban 90
(*iii*) Wheel hubs, fan bearings, chassis greasing, (nipples), cables and control points	All conditions	Caltex Marfak 2 HD
(*iv*) Oil can and carburettor	All conditions	Caltex RPM Motor oil SAE 20 HD
(*v*) Upper cylinder lubrication	All conditions	Caltex upper cylinder lubricant
(*vi*) Battery terminals and earthing points.	All conditions	Petroleum jelly.

There are also alternative lubricants supplied by Castrol Ltd., Indian Oil, Burmah Shell, Standard Eastern, INC, under the trade name of Castrol, Mobiloil, Shell and Esso respectively.

During running-in period (first 1500 kms) of "Vijai Super" scooter, SAE 30 mobiloil 4% and after running-in period of 1500 kms. SAE 30 mobiloil 2% is used in petroil mixture. This means that more quantity of lubricating oil is needed during running-in period than the routine operation. Hence special attention is needed during running-in period.

Premier Automobiles Ltd., Bombay has drawn the following lubrication schedule for their *Premier* cars.

Every 500 kms

Check engine oil level. Add oil up to the max. mark on dip stick (oil level indicator rod) if the level is low.

Every 3000 km

(*i*) Replace engine oil. If the engine is new, replace oil after first 1500—2000 km and 3000—4000 km.

(*ii*) Screw on greaser cap on distributor shank two or three turns.

(*iii*) Inject chassis grease through grease nipple on clutch pedal shaft.

(*iv*) Inject some chassis grease in the two lubricators for upper and lower swinging arm spider bushes on each wheel.

(*v*) Inject some grease in the lubricators of track rod, link rod and relay lever.

(*vi*) Inject some chassis grease in the steering box lubricator.

Every 6000 km

(*i*) Pack greaser on distributor shank with high melting point grease and smear slightly some grease on the breaker cam.

(*ii*) Check up oil in the gearbox through plug on right side of casing. Add SAE 90 EP oil up to lower edge of plug hole if the level is low.

(*iii*) Inject chassis grease in lubricator on front slip yoke of propeller shaft.

(*iv*) Inject chassis grease in lubricator at universal joint spider (axle end).

(*v*) Check oil level in the differential. Add SAE 90 EP oil up to lower edge of plug hole if necessary.

(*vi*) Check oil level in the steering box through filler plug seat. Add SAE 90 EP oil up to 1 cm from lower edge of plug seat.

Every 12000 km

(*i*) Wash well in kerosene oil rear springs and inject some graphitized oil between the leaves.

(*ii*) Remove front wheel hub cups and pack with wheel bearing grease.

Every 24000 km

(*i*) Lubricate drive end ball bearing of generator with high melting point grease. Pack the pocket between bush and commutator head inner end with the same grease. Pull out lubricator wick, soak with thick oil and refit in place after filling high melting point grease in the cap.

(*ii*) Replace oil in the gearbox.

(*iii*) Replace oil in the differential.

(*iv*) Lubricate free-wheel components of self-starter with special grease and electromagnet plunger with some thin oil.

Methods employed in vehicle lubrication :

Keeping in view the convenience in maintenance, *automatic lubrication system* was employed in automobiles. Under this system, lubricant is fed to all lubrication points in the chassis by pressing a pedal provided in the driver's compartment. Owing to recent advancement in motor-vehicle technology, tendency to provide minimun lubrication points is receiving wide acclaim. There are only two lubrication points in Fiat 600 car.

Before starting lubrication work, clean the grease nipples and filler plugs thoroughly. Clean all containers, funnels etc. before using them. There are different methods employed in lubrication. Follow the lubrication instructions strictly.

The most common tool in lubrication is *grease gun* by means of which grease is fed to the lubrication points provided with grease nipples. The *grease nipple* acts as a non-return valve. It prevents the injected grease to come out of the bearing surfaces. The grease gun acts as pump to inject grease into the bearing surfaces through grease nipples.

At service stations, *power lubrication method* is used to feed lubrication at the grease nipples under air pressure. This makes the job easy and quick. There are bucket type hand operated pumps also to inject grease. Such pumps are also used to fill oil in the gearbox, differential, steering gear-box etc. *Direct pouring method* is employed in filling oil into the engine by means of a funnel. Various cables, pins, pivots etc. are oiled through *oil-can*. Upper cylinder lubrication is fed direct through the carburettor air horn. Grease is packed in the wheel hub bearings and the method is known as *pack-method*. The grease is also packed in lubricators mostly provided at distributor shank. To sum up, the following methods are employed in automobile lubrication.

(*i*) **Pack method** through which lubricant is packed over the bearing surfaces.

(*ii*) **Direct pouring method** through which oil is directly filled into the machine housing.

(*iii*) **Hand operated pump method** through which grease or oil is pumped out.

(*iv*) **Simple grease gun method** employed to feed grease through grease nipples.

(*v*) **Power lubrication method** to feed lubrication under air pressure.

(*vi*) **Automatic lubrication method** for feeding lubrication automatically to all the points.

(*b*) **Cleaning, inspection and adjustment.**

Again, for this part of maintenance, the manufacturers provide instruction manuals and charts to carry out the job systematically. Maintenance instructions should be strictly followed in doing the job. A sample of instructions as applicable to Premier cars is given below.

Every 500 km

Check water level in radiator. Add water if level is low. If water level is very low and engine rather warm, do not pour in cold water. When temperature drops close to 0°C, change to anti-freeze mixture.

In case of refilling radiator, pull up cock control knob of heater radiator, fill engine radiator to maximum level, run engine at idle speed for few minutes and then top up once more.

Every 1000 km

Check electrolyte level with battery at rest and cold. Add distilled water upto 5 mm. above separators if necessary.

Every 6000 km

(*i*) Remove filter gauge from oil delivery filter and wash it in kerosene. When the engine is new, the filter must be cleaned after 5500—6500 km.

(*ii*) Replace cartridge of bypass oil filter after every two oil renewals. When engine is new, the cartridge must be replaced after first 5500—6500 km.

(*iii*) Check up fan belt tension. Adjust if necessary. Belt sag is 1 to 1·5 cm when under a pressure of 10 kg.

(*iv*) Check up C.B. point gap which should be 0·42 to 0·48 mm. Adjust, if necessary, by sliding the stationary C. B. carrier plate after slackening its screw. Relock screw after adjustment.

(*v*) Check the condition of spark plugs. Clean and adjust gap at 7 mm.

(*vi*) Check level of brake fluid in master cylinder and add fluid if required.

(*vii*) Check battery terminals and clamps for tightness and cleanliness. Coat them with pure ropy vaseline.

(*viii*) Remove air cleaner rear cover and clean cartridge with air under pressure. If restriction is excessive, replace.

(*ix*) Exchange tyres in criss-cross manner to ensure long life and equalize wear.

Every 12000 km

(*i*) Replace air cleaner cartridge. In dusty conditions, clean and replace cartridge more often.

(*ii*) Check up clutch pedal free play which should be 15 to 20 mm. Adjust if necessary.

(*iii*) Check up and adjust brake shoes.

(*iv*) Inspect shock-absorbers. Add shock-absorber fluid if necessary.

(*v*) Inspect, clean and lubricate ball articulations of steering rods.

(*vi*) Check up front wheel bearing play. Adjust if necessary.

Every 24000 km

(*i*) Clean generator commutator. Check brushes for wear and contact conditions and replace if necessary.

(*ii*) Inspect the entire valve gear. Reface valve seats and decarbonise combustion chambers.

(*iii*) Check up water pump seal and replace if necessary.

(*iv*) Check up the operation of thermostat valve. It should open only when the coolant temperature reaches 80°—87°C.

(*v*) Check the differential bearings and final drive gears and adjust if necessary.

(*vi*) Clean self-starter commutator. Check wear and contact conditions of brushes and replace if necessary.

Occasional inspections

(*i*) Check and if necessary, correct idle running adjustment. Inspect and clean all carburrettor parts if required.

(*ii*) Check back-lash in steering gear. Adjust if necessary.

(*iii*) Check worm-screw roller bearings and adjust by removing shims between lower cover and box.

Maintenance schedule differs from vehicle to vehicle. Hence the particular instructions should be followed for the vehicle for which they are meant. Maintenance is done periodwise or on the basis of mileage. Periodwise maintenance is split up into different periods such as weekly, monthly. quarterly, yearly, half-yearly, etc. Daily checks are usually as under :—

(*i*) Check engine oil level.

(*ii*) Check water level in the radiator.

(*iii*) Check fan belt tension.

(*iv*) Check air cleaner.

(*v*) Check tyre pressures.

(*vi*) Check brake pedal toe-board clearance.

(*vii*) Check clutch pedal toe-board clearance.

(*viii*) Check tightness of wheel nuts.

(*ix*) Check lights and horn.

(*x*) Check tools and equipment.

If daily distance to be covered by the vehicle is not fixed which is generally not possible in case of private vehicles, periodwise maintenance is not proper. Maintenance on mileage basis is quite in order in such cases and the vehicle could be properly looked after. It is owing to this reason that the manufacturers usually prescribe maintenance on the basis of distance travelled by the vehicle. Those vehicles which are used for short distance traffic or which have to work under unfavourable operating conditions (dumping operation, vehicles using engine as a stationary power unit with power take off for pumping sets, threshers, cane crushers etc.) should not be maintained on mileage basis. In such cases, oil changing should be done on the basis of operational hours or fuel consumption should be used as a guide. The general rule is to change engine oil after 60 operational hours or after a fuel consumption of about 800 litres. Apart from scheduled routine maintenance, the vehicle should be looked after and saved from the climatic and seasonal effects. The vehicle should be prevented from any damage which could be foreseen.

2. **Preventive maintenance**. It is the care to prevent troubles which often lead to a major overhaul. For the trouble free working of the vehicle, keep in view the following points relating to preventive maintenance.

(*i*) Keep moisture out of fuel system. In view of this, keep the fuel tank full. Fill fuel tank up at the end of day's journey to drive out moisture laden air during winter or damp weather. Drain out about one litre fuel from the tank and fuel filter so that water trapped in there, is removed. The drained out fuel could be reused after filtering.

(*ii*) In order to prevent freezing of the cooling water, drain out the cooling system completely. Wash the system with washing soda solution and flush with clean water. Fill the system with anti-freeze solution. Check up all joints and rectify immediately any leakage.

(*iii*) Use oils of lower viscosity during winter as prescribed by the manufacturer keeping in view the temperature.

(*iv*) Keep batteries well charged. Don't bring any naked flame before them lest the emitted gases catch fire. Keep the battery top and terminals clean to ensure minimum resistance and prevent corrosion.

(*v*) Spot paint external parts from where paint has chipped off in order to avoid rust formation.

(*vi*) Don't let water particles stay under the floor mat but keep clean lest rusting creeps in.

(*vii*) Keep head lamp rings tight, lest water traps into the reflector and leads to rusting. Replace the defective gaskets.

(*viii*) Occasionally remove the flints which may have got embedded in the tyre treads.

(*ix*) Keep the engine warm during extreme cold temperatures by blanking the radiator with water-proof curtain rather than cutting out fan.

(*x*) In case of air brakes, condensed water should be drained out of the compressed air tank everyday otherwise the braking efficiency shall be affected by the formation of ice in the pipelines and the brake valves due to low atmospheric temperatures. During extreme cold, anti-freeze material should be filled into the brake system.

(*xi*) Mount anti-skid chains on the rear wheels of a heavy vehicle to prevent skidding over icy, heavily snowed up or slippery roads.

(*xii*) Dilute the diesel fuel with superior kerosene in the following ratio for use in different temperature ranges :—

Outside temperature	*Diesel fuel*	*Kerosene*
(*a*) Between 0°C and −10°C	80–90%	20–10 %
(*b*) Between −10°C and −18°C	70%	30%

(*xiii*) After the day's journey, remove the batteries from the vehicle and store them in a heated or warm room when the atmospheric temperature is below minus 15°C because starting capacity of a warm battery is considerably higher.

3. Laying up vehicle for a long period. Corrosion affects the life of vehicle. If a vehicle is laid up for a long period, it should be properly preserved to avoid corrosion. Even in a closed garage, the vehicle is exposed to harmful effects of changing temperatures, corrosion, dust and air humidity. Hence the following measures are suggested to protect the vehicle :—

(*i*) Wash, clean and lubricate the vehicle thoroughly.

(*ii*) Check all painted parts of the body for damage and touch up, if necessary.

(*iii*) Apply chromium protective paste on the chromium plated parts.

(*iv*) Grease all unpainted parts including springs and spring suspensions with anti-corrosion grease.

(*v*) Drain out oil from the engine while hot and replace by anti-corrosion oil and SAE 10 for engines.

(*vi*) Remove spark plugs or glow plugs in case of C.I. engine and inject about 20 c c. of anti-corrosion engine oil SAE 10 into each cylinder. At the time of oil injection, the piston should be at B.D.C. and after oil injection, screw on spark plugs/glow plugs and crank the engine for few seconds. Before cranking diesel engines, disconnect fuel injection pipes from the F.I. pump so that no fuel is injected into the cylinders.

(*vii*) Spray the unpainted engine parts with anti-corrosion engine oil SAE 10. Cover fan belt and rubber hoses before spraying oil otherwise rubber parts shall be affected by oil.

(*viii*) In case of diesel vehicle, add about 10% anti-corrosion engine oil SAE 10 in the fuel tank and oil in F.I. pump housing. After adding oil operate engine for few minutes. Then, remove pump cover and spray pump elements and springs with engine oil containing 5—10% anti-corrosion oil. Screw on pump cover and spray outside of the pump with this mixture.

In case of petrol vehicle too, add about 10% anti-corrosion engine oil in the fuel tank and operate the engine for few minutes.

If anti-corrosion oil is not available, drain out petrol from the fuel tank, fuel filter, fuel pump and carburettor and preserve it in a sealed container for reuse.

(*ix*) Spray the rocker arms, valves, springs etc. with anti-corrosion engine oil SAE 10, after removing valve chamber cover. Fit the cover back in position after spraying oil.

(*x*) Mix 1% water-soluble anti-corrosion oil in radiator water. Then start and run the engine till it attains a temperature of

60°C so that anti-corrosion oil is distributed throughout the cooling system. After this, drain out the cooling system completely.

(*xi*) If the vehicle is to be laid up for more than six months, drain off transmission and differential oils when hot, after the vehicle has been run for sometime. Fill anti-corrosion oil and drive the vehicle for some time without load.

Anti-corrosion oil should be replaced with fresh transmission oil at the time of vehicle reuse.

(*xii*) Remove the batteries and store them in a cool and dry place. Check battery condition every fortnight and recharge, if necessary.

(*xiii*) Jack up the vehicle to relieve load from the tyres.

(*xiv*) To prevent damage from moths, sprinkle the upholstery with naphthalene, camphor or similar products.

(*xv*) Keep the vehicle in dry and well ventilated place and cover with tarpauline.

4. Hints on vehicle maintenance :—

(*i*) Follow maintenance instructions given in the operation and maintenance manual supplied with the vehicle.

(*ii*) Always use lubricants of correct grade and quality at the correct interval as specified by the manufacturer.

(*iii*) Replace engine oil as under :—

(*a*) drain out engine oil while warm.

(*b*) fill in flushing oil up to "Min" (Minimum) mark on dip stick and run the engine at about 1000 rpm for 10 minutes.

(*c*) drain out flushing oil.

(*d*) refill with engine oil of specified grade.

(*iv*) Use purest available water in the radiator.

(*v*) Be careful in using de-scaling compounds based on inorganic salts such as borates, carbonates, phosphates etc., lest aluminium cylinder heads are affected by them.

(*vi*) In order to avoid cracking of cylinder block, do not top up radiator with cold water after switching off engine when water level is very low and the engine is rather hot.

(*vii*) The radiator cap should be removed in two steps to check up water level in case engine is hot. Initially turn the cap to allow the discharge of pressure from the cooling system and then, only when the pressure is released.

(*viii*) For replacing water in the cooling system, drain water through the drain cocks located under the radiator and on cylinder block. Flush out radiator thoroughly if the drained water is verv dirty and full of sediments.

(*ix*) Flush out radiator as below :—

(*a*) Drain out water from the cooling system.

(*b*) Fill in water containing a 4% solution of sodium bicarbonate and run the engine at about 1000 rpm for ten minutes.

(*c*) Switch off engine and let the water stand in the radiator for half an hour.

(*d*) Start the engine again and drain out water while the engine is running.

(*e*) After draining water, switch off engine and let the engine cool down.

(*f*) Keeping drain cocks open, push in water hose from external water supply line into the radiator neck and circulate running water in the radiator and engine water jackets.

(*g*) Close the cocks and fill radiator with clean water.

(*h*) Run the engine for a while and drain the radiator again.

(*i*) Close the cocks and fill radiator to normal level.

(*x*) Use good quality anti-freeze mixture containing corrosion inhibitors when engine temperature reaches 0°C.

(*xi*) Never use sand paper for refacing CB-points but use oil stone.

(*xii*) While adjusting C.B. points or lubricating C B. arm, be careful lest the points are upset resulting in arcing, rapid wear and ignition irregularities.

(*xiii*) Keep the clutch pedal properly adjusted to avoid clutch slip which may result in overheating, unnecessary wear of facings and release bearing.

(*xiv*) In case of excessive free play in brake pedal and braking unbalance, get the brake system checked and adjusted.

(*xv*) Use non-mineral brake fluid which should not corrode or affect rubber parts.

(*xvi*) In order to avoid excessive tyre wear and steering irregularities, get the wheel alignment checked. In a properly adjusted steering system, the vehicle will continue to run straight ahead when hands are taken off the steering wheel and the steering wheel shall return to straight ahead position after turning.

(*xvii*) Keep the tyres inflated to specified pressures.

(*xviii*) Don't spill water on battery top but keep it clean and dry.

(*xix*) Add distilled water in battery when cold.

(*xx*) During garaging the vehicle for long periods, recharge the battery once a month.

(*xxi*) Check the wear and contact condition of dynamo and self-starter brushes and replace them if required.

(*xxii*) Don't tamper with cut out regulator adjustments.

(*xxiii*) Substitute burnt bulbs with bulbs of same type and capacity (wattage).

(*xxiv*) Wash and clean the vehicle as under :—

(*a*) Wash and clean the lower part of the vehicle first, including wheels, with water jet, using a sponge.

(*b*) Wash the body shell avoiding excessive pressure water spray. Use sponge in washing. Rub it gently to avoid scratching.

(*c*) Dry the vehicle with a clean chamois.

(*d*) Remove grease, oil and tar spots from the painted parts by wetting with petrol and wiping immediately with a dry cloth.

(*e*) For cleaning chrome plated parts, apply some natural vaseline and then rub with a soft and clean rag as well as chamois.

(*f*) Clean windows and windshield with a linen cloth or a very soft chamois. Use windshield washer solvent or water containing some alcohol for very dirty glass panes.

(*g*) Clean rubber mats with soap and water.

(*h*) Clean leather seats or other parts with a wet and soapy sponge or cloth. Rinse carefully with clean water and dry well with chamois.

(*xxv*) Avoid parking of vehicle in the sun lest the body finish is affected.

(*xxvi*) Never wash or polish the vehicle in sun especially in summer.

(*xxvii*) Avoid spilling of petrol, alcohol, sodium compounds, windshield solvent or brake fluid on body finish.

(*xxviii*) Periodically clean water drain-holes at the bottom of doors to avoid rust formation.

(*xxix*) Paint the under-chassis at least once a year to protect it from corrosion.

(*xxx*) Park the vehicle under cover during nights, rainy days and winter months to avoid excessive condensation of water on vehicle body which would affect its finish.

(*xxxi*) Wipe out all drops of water penetrated inside the body as a result of moisture condensation otherwise water drops shall lead to rusting.

(*xxxii*) Touch the chipped off or scratched point to avoid corrosion.

(*xxxiii*) Polish the vehicle with good quality wax polish after washing.

(*xxxiv*) Don't scuff painted surfaces with any rough object which may create scratches.

(*xxxv*) Dust off body before washing or wiping with moist cloth.

QUESTIONS

1. The life of an automobile depends upon maintenance and proper operation. Discuss.

2. What is maintenance ? Explain daily and weekly maintenance tasks for an owner driver car.

3. Explain the qualities of a good lubricant. Which types of lubricants are required for different parts of an automobile ?

4. Which methods are employed in lubricating motor-vehicles ? Explain.

5. Which grade and type of lubricant is required for the following parts ?

(*i*) Wheel bearings.
(*ii*) Distributor shaft.
(*iii*) Brake pedal link joints.
(*iv*) Steering gear.
(*v*) Engine working below 0°C.
(*vi*) Universal joint.
(*vii*) Air cleaner.
(*viii*) Differential.

6. What sort of information you get from a lubrication chart ? Explain the various tasks given in the lubrication chart of a car.

7. Special attention is needed during running-in period. Explain

8. What is preventive maintenance ? Explain the various steps in this regard.

9. Which precautionary measures are required in laying up a diesel truck for a long period ?

10. Mention the hints on vehicle maintenance.

20

Trouble Shooting

Trouble shooting. In order to keep the vehicle rolling, the trouble must be shot (removed) immediately because a stitch in time saves nine. The trouble can't be shot unless discovered. Once the trouble is visible, it is easy to remove. So, first diagnose the disease then cure it. The disease can't be diagnosed until and unless one is fully familiar with the construction and working of the machine. An automobile is composed of frame and body ; springs, shock absorbers, axles and wheels ; engine, clutch, gearbox and propeller shaft ; steering and linkages ; battery, horn, fuel tank and radiator ; accelerator, brake and clutch pedals ; switches, controls and gauges etc. A petrol vehicle contains the following systems :—

(*i*) Fuel system. (*ii*) Ignition system. (*iii*) Lubrication system. (*iv*) Cooling system. (*v*) Transmission system. (*vi*) Suspension system. (*vii*) Steering system. (*viii*) Brake system. (*ix*) Lighting and horn system. (*x*) Starting and charging system.

Although general principle of working of each system is the same yet there is variation in construction from vehicle to vehicle. The manufacturers supply fault-finding charts for their different makes and models. These charts act as a guide in trouble shooting. Here a general chart of petrol vehicles is given to give an idea of usual troubles in automobiles, their probable causes and remedies. Actual diagnosis and removal of defect needs through knowledge, practical experience and scientific approach to the problem.

Trouble No. 1. Engine does not start.

Probable causes	*Remedies*
(*i*) No fuel in tank.	Fill fuel.
(*ii*) Choked fuel supply.	Clean and clear the obstruction in fuel line.
(*iii*) Loose or leaking fuel line connection.	Tighten loose connection or remove leakage.
(*iv*) Sticking needle valve in carburettor.	Release.
(*v*) Flooded carburettor.	(*i*) Release sticking float. (*ii*) Repair or replace punctured float. (*iii*) Repair or replace defective needle valve. (*iv*) Check and adjust float level. (*v*) Open choke valve if closed. (*vi*) Remove liquid petrol from induction manifold.
(*vi*) Ignition off.	Switch on ignition/replace defective switch
(*vii*) Spark plugs dirty or bridged by soot or oil.	Clean, check and adjust gap.
(*viii*) Damaged spark plugs.	Replace.
(*ix*) Loose or dirty battery terminals.	Clean and tighten battery terminals.
(*x*) Discharged battery.	Recharge.
(*xi*) Loose or leaking H.T. lead from ign. coil.	Tighten or replace.
(*xii*) Defective or missing rotor.	Replace/fix on rotor.

Probable causes	*Remedies*
(*xiii*) Dirty, sitted or burned C.B. points.	Clean or replace.
(*xiv*) C.B. points not opening.	(*i*) Adjust gap. (*ii*) Repair or replace defective C.B. arm.
(*xv*) Defective ign. coil or condenser.	Replace.
(*xvi*) Loose or leaking connection in ignition system.	Remove leakage or tighten loose connection.
(*xvii*) Upset ignition-timing.	Reset.
(*xviii*) Distributor cap cracked.	Replace.
(*xix*) Distributor drive gear pin sheared.	Replace pin.
(*xx*) Poor compression.	(*i*) Tighten loose spark-plugs, cylinder head. Replace defective gaskets. (*ii*) Check and adjust tappet clearance. (*iii*) Replace worn out or broken piston rings. (*iv*) Repair or replace sticking or leaking valves. (*v*) Replace worn out pistons.

Trouble No. 2. Engine starts but stops immediately.

(*i*) Faulty fuel pump.	Check up and repair.
(*ii*) Restricted or leaking fuel line.	(*i*) Clean/clear restriction. (*ii*) Tighten loose connection. (*iii*) Repair or replace defective pipe line.

Probable causes	*Remedies*
(*iii*) Choked air-vent in fuel tank cap.	Clear the air-vent.
(*iv*) Choked fuel filter.	Clean or replace filtering element.
(*v*) Carburettor overflow.	Check as in trouble No. 1.
(*vi*) Sticking C.B. points.	Repair or replace.
(*vii*) Loose connection in ignition circuits.	Tighten.
(*viii*) Choked silencer.	Repair or replace.
Trouble No. 3. Starter does not turn engine.	
(*i*) Discharged battery.	Recharge.
(*ii*) Loose connections in the starting system.	Check up and tighten.
(*iii*) Defective push button or solenoid switch.	Repair or replace.
(*iv*) Defective self-starter.	Repair.
(*v*) Sticking starter pinion.	Release or repair.
(*vi*) Defective leads.	Repair or replace.
(*vii*) Frozen water pump.	(*i*) Warm the engine and use anti-freeze mixture in the cooling system.
	(*ii*) Repair defective water pump.
(*viii*) Frozen engine oil.	(*i*) Warm the engine and replace engine oil.
	(*ii*) Use correct grade of oil.
(*ix*) Seized engine.	Disassemble, find out fault and carry out necessary repairs and replacements.
(*x*) Engaged gear.	Disengage.

Trouble No. 4. Engine misfiring.

Probable causes	*Remedies*
(*i*) Faulty carburettor.	Clean and adjust.
(*ii*) Clogged air cleaner.	Clean or replace defective element.
(*iii*) Restricted fuel supply.	Clean and clear restriction.
(*iv*) Choked silencer.	Clean.
(*v*) Misadjusted valves.	Adjust tappet clearance.
(*vi*) Sticking or leaking valves.	Repair.
(*vii*) Leaking cylinder head, induction manifold or carburettor flange.	(*i*) Tighten loose bolts or nuts. (*ii*) Replace defective gaskets.
(*viii*) Defective spark plugs.	(*i*) Clean and adjust gap. (*ii*) Replace defective spark plugs.
(*ix*) Spark plugs of incorrect heat-value.	Use plugs of prescribed heat-value.
(*x*) Dirty or pitted C.B. points.	Clean/reface and adjust gap.
(*xi*) Defective ignition coil, condenser, distributor cap, rotor or H.T. leads.	Replace the defective parts.
(*xii*) Loose connection or defective wiring.	Repair or replace.
(*xiii*) Incorrect timing.	Adjust.
(*xiv*) Defective automatic advance and retard mechanism.	Repair or replace.
(*xv*) Water in fuel.	Clean the fuel system and fill in fresh petrol.
(*xvi*) Unbalanced compression.	Check up the cause and repair accordingly.

Probable causes	*Remedies*
(*xvii*) Tight engine.	(*i*) Check up the cause of partial seizure and repair accordingly. (*ii*) Check up cause for lack of lubrication and make necessary corrections.
(*xviii*) Carbon deposit in engine combustion-chamber.	Decarbonise.
(*xix*) Vacuum pipe connections loose at induction manifold.	Tighten.

Trouble No. 5. Engine continues to run after switching off.

(*i*) Spark plugs of incorrect heat-value.	Use spark-plugs of correct grade.
(*ii*) Carbon deposit in combustion chamber.	Decarbonise.
(*iii*) Defective ignition switch.	Replace.
(*iv*) Badly burnt or over-heating exhaust-valves.	Replace.
(*v*) Engine overheating.	Check up cause of overheating and make necessary repairs.

Trouble No. 6. Engine overheating.

(*i*) Retarded timing.	Check up and adjust.
(*ii*) Distributor governo weights stuck in retarded position.	Check up, clean, release and lubricate sticking fly-weights.
(*iii*) Too lean fuel mixture.	Clean and adjust carburettor.
(*iv*) Less water in radiator.	Add to level.
(*v*) Leakage of water in the cooling system.	Repair the leakage.

Probable causes	*Remedies*
(*vi*) Choked radiator or water jackets.	Flush out and clean.
(*vii*) Loose fan belt.	Adjust.
(*viii*) Defective thermostat valve or water pump.	Repair or replace.
(*ix*) Lack of lubrication.	(*i*) Top up engine oil. (*ii*) Use correct grade and quality of oil. (*iii*) Clean or replace choked filter element. (*iv*) Check up serviceability of oil pump.
(*x*) Tight engine.	If new or overhauled engine, run carefully using correct grade of oil. If otherwise tight, check up mechanical failure and carry out necessary repairs.
(*xi*) Weak compression.	(*i*) Tighten loose cylinder head, spark plugs and manifold. (*ii*) Replace defective gaskets for above. (*iii*) Adjust tappet clearance. (*iv*) Repair or replace defective valves. (*v*) Replace defective pistons and rings.
(*xii*) Choked silencer.	Clean.
(*xiii*) Clutch slipping.	Adjust. If still slipping, dismantle and make out necessary repairs and replacements.

Probable causes	*Remedies*
(*ix*) Brakes binding.	(*i*) Adjust. (*ii*) Lubricate the link and cable joints. (*iii*) Replace defective return/retracting springs. (*iv*) Replace defective rubber cups in master or wheel cylinders. (*v*) Clean air vent and compensating port in master cylinder. (*vi*) Replace choked brake hose pipes.
(*x*) Heat control valve sticking.	Repair or replace.
(*xi*) Excessive carbon deposit in engine.	Decarbonise.
(*xii*) Incorrect use of gears.	Avoid driving for too long period in low gears unless road conditions demand.
(*xiii*) Inferior grade of petrol used, leading to detonation.	Use proper grade of petrol
(*xiv*) Exhaust valves burned.	Replace.
(*xv*) Exhaust valves not opening to full extent.	Adjust tappet clearance.
(*xvi*) Radiator partly blanked off.	Remove curtain from the radiator core.

Trouble No. 7. Engine producing abnormal noises.

(*i*) Worn out gudgeon pins and small end bearings.	Replace
(*ii*) Worn out piston and rings.	Replace
(*iii*) Piston rings broken in grooves.	Make necessary repairs and replacements.

Probable causes	*Remedies*
(*iv*) Gudgeon pin circlips broken.	Repairs and replacements
(*v*) Worn out big-end bearings.	Replace.
(*vi*) Worn out crank-shaft main bearings.	Replace.
(*vii*) Excessive carbon deposit in engine.	Decarbonise.
(*viii*) Ignition too far advanced.	Adjust.
(*ix*) Inferior grade of fuel.	Use proper grade of fuel.
(*x*) Wrong grade of spark-plugs.	Use spark plugs of correct heat value.
(*xi*) Loose flywheel or crank pulley.	Tighten.
(*xii*) Worn out tappets and valve guides.	Replace.
(*xiii*) Worn rockers (O.H.V. engine) and shaft or bent push rods.	Replace.
(*xiv*) Excessive tappet clearance.	Adjust.
(*xv*) Worn out timing gears or chain and sprockets.	Replace.
(*xvi*) Worn out distributor or oil pump drive gears.	Replace.
(*xvii*) Loose timing chain	(*i*) Adjust. (*ii*) Replace defective chain tensioner.
(*xviii*) Dry generator or fan bearing.	Lubricate.
(*xix*) Defective water pump bearing.	Replace.
(*xx*) Lack of lubrication at distributor shaft bearing.	Lubricate.
(*xxi*) Hard or glazed dynamo brushes.	Lap or replace.

Probable causes	Remedies
(*xxii*) Fan brakes striking radiator or hose.	Make necessary adjustments.
(*xxiii*) Slipping fan belt.	(*i*) Adjust. (*ii*) Replace torn out fan belt. (*iii*) Align buckling crank pulley.
(*xxiv*) Cranked or loose engine mountings.	Replace, or tighten.

Trouble No. 8. Excessive fuel consumption.

Probable causes	Remedies
(*i*) Improper adjustments of carburettor.	Make necessary adjustments properly.
(*ii*) Dirty air cleaner.	clean
(*iii*) Leakage in the fuel line.	(*i*) Check up and remove leakage at different connections. (*ii*) Replace leaking pipes. (*iii*) Repair or remove leakage at fuel tank, filter, pump and carburettor.
(*iv*) Sticking controls.	(*i*) Release sticking choke and throttle in carburettor. (*ii*) Lubricate the link joints and shafts. (*iii*) Make necessary adjustments.
(*v*) Excessive idling.	Do not allow the engine to idle unnecessary for long periods when vehicle is stationary.
(*vi*) Engine overheating.	Avoid overheating.
(*vii*) Dragging brakes.	Adjust or make necessary repairs and replacements.
(*viii*) Under inflated tyres.	Inflate to proper pressure.

Probable causes	*Remedies*
(*ix*) Engine not tuned properly.	**Tune up properly.**
(*x*) Vehicle overloaded.	**Avoid overloading.**
(*xi*) Unnecessary use of low gear.	**Use gears intelligently.**
(*xii*) Operating with partially closed choke.	**Check and adjust choke.**
(*xiii*) Oval or worn out jets in the carburettor.	**Replace.**
(*xiv*) Fierce acceleration and sudden braking.	**Drive carefully.**
(*xv*) Tight wheel bearings.	**Adjust.**
(*xvi*) Slipping clutch	**Adjust or repair.**
(*xvii*) Incorrect wheel alignment.	**Correct the alignment.**
(*xviii*) Worn out piston and rings.	**Replace.**
(*xix*) Incorrect spark plugs.	**Use plugs of proper range.**

Trouble No. 9. Low or high oil pressure and excessive oil consumption in the engine.

(*a*) Low oil pressure.	**(*i*) Top up engine oil if level low.** **(*ii*) Ensure that proper grade and quality of oil is used.** **(*iii*) Check up oil pressure gauge or indicator and replace if defective.** **(*iv*) Remove any restriction in pipe leading to oil gauge.** **(*v*) Clean and adjust oil relief valve.** **(*vi*) Replace defective ball seat or spring in relief valve.**

Probable causes	*Remedies*
	(*vii*) Adjust or replace loose or worn big-end and main bearings.
	(*viii*) Clean oil strainer on suction side of oil pump.
	(*ix*) Replace or repair worn out oil pump.
	(*x*) Remove leakage on suction side of oil pump.
	(*xi*) Repair or replace leaking oil pipes.
(*b*) High oil pressure.	(*i*) Replace thick oil in winter,
	(*ii*) Check up oil pressure gauge or indicator and replace if defective.
	(*iii*) Check up adjustment of oil relief valve.
	(*iv*) Remove restriction or replace choked oil pipe.
	(*v*) Check up engine bearing failure and as a result blocking of oil passages.
(*c*) Excessive oil consumption.	(*i*) Replace worn pistons and rings.
	(*ii*) Rebore worn or scored cylinders.
	(*iii*) Replace broken piston rings.
	(*iv*) Scatter away if ring gaps in line.
	(*v*) Replace worn valve guides.
	(*vi*) Replace defective oil seals on valve stems in O.H.V. engines.
	(*vii*) Tighten bolts or replace valve chamber or oil sump gaskets if oil leaks through them.

Probable causes	Remedies
	(*viii*) Replace front and rear main bearing oil seals if oil leaks through them.
	(*ix*) Replace damaged gasket of oil-sum drain plug.
	(*x*) Remove oil leakage at mechanical fuel pump flange.
	(*xi*) Repair or replace defective positive crankcase ventilation (PVC) valve.
	(*xii*) Remove ony other external oil leakage.
	(*xiii*) Drain out excess oil. Maintain oil at the proper level.

Trouble No. 10. Engine does not run slow.

Probable causes	Remedies
(*i*) Throttle valve open.	Adjust.
(*ii*) Choke valve partly closed.	(*i*) Push down choke. (*ii*) Adjust if still closed.
(*iii*) Idle passage in carburettor restricted.	Clear off restriction.
(*iv*) Incorrect idle adjustment.	Adjust correctly.
(*v*) Worn out pilot slow running jet.	Replace.
(*vi*) Damaged or worn idle adjustment screw.	Replace.
(*vii*) Air cleaner dirty.	Clean.
(*viii*) Leakage at carburettor flange or induction manifold.	(*i*) Tighten bolts and puts (*ii*) Replace damaged gaskets.
(*ix*) Restriction in air-bleed holes in carburettor.	Clean and blow out.
(*x*) Damaged idle passage.	Repair.
(*xi*) Incorrect float level adjustment.	Adjust correctly.

Trouble No. 11. Engine does not pick up speed.

Probable causes	*Remedies*
(*i*) Restriction in throttle valve operation.	Lubricate and remove any restriction.
(*ii*) Metering rod in carburettor set too low.	Adjust properly,
(*iii*) Restriction in fuel supply.	Check up and remove.
(*iv*) Float level too low.	Adjust.
(*v*) Carburettor overflowing.	(*i*) Adjust float level. (*ii*) Replace worn out needle valve. (*iii*) Replace punctured float.
(*vi*) Restriction in main or high speed circuit of carburettor.	Clean and blow out restriction.
(*vii*) Water in fuel.	(*i*) Drain out fuel. (*ii*) Clean the fuel system. (*iii*) Fill new petrol.
(*viii*) Retarded ignition.	Adjust timing.
(*ix*) Wrong size main-jet installed.	Replace by correct size.
(*x*) Poor compression.	(*i*) Remove leakage at cylinder head. (*ii*) Replace worn out pistons and rings. (*iii*) Replace leaking valves.

Trouble No. 12. Engine popping and spitting.
(Explosions in carburettor and silencer)

Probable causes	Remedies
(*i*) Incorrect carburettor adjustment.	Adjust properly.
(*ii*) Leaking valves.	Repair or replace.
(*iii*) Incorrect tappet clearance.	Adjust.
(*iv*) Incorrect ignition timing.	Adjust.

Probable causes	*Remedies*
(*v*) Incorrect or defective spark plugs.	Replace with plugs of correct heat value.
(*vi*) Excessive carbon deposit in engine and silencer.	Decarbonise.
(*vii*) Inferior grade of fuel.	Use proper grade fuel.
(*viii*) Lean mixture.	Adjust and correct mixture ratio.
(*ix*) Weak valve springs.	Replace.
(*x*) Defective ignition coil, condenser or C.B. points.	Replace.
(*xi*) Leakage at exhaust joint.	Remove leakage.
(*xii*) Heat control valve defective or seized.	(*i*) Replace. (*ii*) Install spring properly and free up valve.
(*xiii*) Exhaust valve head too thin, leading to overheating.	Replace.

Trouble No. 13. Clutch slipping.

(*i*) No free play in pedal	Adjust pedal free travel (about 1″).
(*ii*) Weak thrust springs.	Replace.
(*iii*) Worn clutch plate.	Replace.
(*iv*) Worn splines on primary shaft (clutch shaft).	Replace shaft.
(*v*) Oil, grease or other such material on clutch facings.	(*i*) Clean clutch plate. (*ii*) Replace oil soaked clutch plate.
(*vi*) Pressure plate binding in cover.	Free up and lubricate.
(*vii*) Excessively worn pressure-plate and flywheel.	Reface or replace.

Probable causes	*Remedies*
iii) Sticking or misadjusted release fingers.	Free up, lubricate and adjust.
x) Siezed clutch linkage or release-mechanism.	Free up and lubricate.

Trouble No. 14. Clutch dragging or spinning.

Oil or grease on clutch plate-linings.	(*i*) Clean. (*ii*) Replace oil soaked plate.
) Clutch shaft out of line with engine.	(*i*) Replace worn out pilot bearing in crankshaft tail. (*ii*) Replace worn out primary shaft bearing. (*iii*) Tighten loose holding down bolts of gearbox.
i) Warped or damaged pressure plate.	Replace.
) Clutch plate hub binding on splines.	Replace.
Clutch plate distorted.	Replace.
Cushion springs damaged.	Replace clutch plate.

Trouble No. 15. Clutch judder.

Linings not making even contact.	(*i*) Replace clutch plate. (*ii*) Reface flywheel and pressure plate.
Pressure plate not parallel with flywheel.	(*i*) Tighten loose bolts holding pressure plate with cover. (*ii*) Replace distorted pressure plate cover.
) Buckled clutch plate.	Replace.
Bent clutch shaft.	Replace.

Trouble No. 16. Clutch rattling.

Probable causes	*Remedies*
(*i*) Worn release bearing.	Replace.
(*ii*) Worn release bearing carrier.	Replace.
(*iii*) Worn release fork.	Replace.
(*iv*) Disconnected pedal return spring.	Connect.
(*v*) Release fork loose on ball stud.	Replace ball stud.

Trauble No. 17. Gear slipping out of mesh.

Probable causes	*Remedies*
(*i*) Damaged ball, plunger or spring locking selector shaft or fork.	(*i*) Replace damaged ball. (*ii*) Replace weak or broken spring.
(*ii*) Worn grooves of selector shaft.	Repair or replace defective shaft.
(*iii*) Worn shifter fork or gears in case of partial engagement.	Replace.
(*iv*) Damaged main-shaft pilot bearing.	Replace.
(*v*) Bent shifter fork.	Replace.

Trouble No. 18. Gear changing difficult.

Probable causes	*Remedies*
(*i*) Defective synchronizing unit.	Repair or replace.
(*ii*) Worn gearshift interlock guide plate.	Replace.
(*iii*) Worn reverse gear latch in case of reverse gear.	Replace.
(*iv*) Gear shift lever lower end worn.	Replace.
(*v*) Damaged spring or positioning pins at the ball joint (fulcrum) of shift lever.	Replace.
(*vi*) Clutch dragging.	Remove trouble.

Probable causes	Remedies
(*vii*) Faulty selector mechanism.	(*i*) Lock the loose selector fork. (*ii*) Lock the loose selector rods.
(*viii*) Gearbox cover misaligned.	Lift cover, fix in position and tighten bolts.
(*ix*) Misaligned linkage in case of remote control gear lever.	Adjust and align properly.
(*x*) Binding in linkage of steering column gear lever.	Free up and lubricate.

Trouble No. 19. Noisy gearbox operation.

Probable causes	Remedies
(*i*) Worn bearings.	Replace.
(*ii*) Worn gears.	Replace.
(*iii*) Foreign matter in gearbox.	Clean the gearbox.
(*iv*) Damaged ball races.	Replace
(*v*) Worn thrust washers.	Replace.
(*vi*) Lack of lubrication.	(*i*) Top up oil. (*ii*) Use oil of specified grade.
(*vii*) Gearbox loose and out of alignment.	Tighten holding down bolts of gearbox and clutch housing.
(*viii*) Gears loose on shafts.	Replace gears and shafts.
(*ix*) Clutch shaft out of alignment.	(*i*) Replace worn bearings. (*ii*) Replace defective shaft.
(*x*) Damaged rollers/needle bearing at the front end of main shaft.	Replace.

Trouble No. 20. Propeller shaft gives noise and vibrates at speed.

Probable causes	*Remedies*
(*i*) Bent propeller shaft.	Replace.
(*ii*) Universal joints misaligned.	(*i*) Check up fitting and correct. (*ii*) Replace worn spider and needle bearings.
(*iii*) Worn hanger bearing or cushion.	Replace.
(*iv*) Worn splines in slip joint.	Replace shaft.
(*v*) Unbalanced propeller shaft.	Replace.
(*vi*) Loose 'U'-joint bolts.	Tighten.

Trouble No. 21. Noisy drive axle.

(*i*) Worn gears and pinions.	Replace.
(*ii*) Improper adjustment of final drive.	Adjust back-lash.
(*iii*) Lack of lubrication.	(*i*) Replace defective oil seal. (*ii*) Top up oil.
(*iv*) Worn bearings.	Replace.
(*v*) Foreign matter in axle casing.	Remove and clean.
(*vi*) Worn splines of axle shafts.	Replace.
(*vii*) Loose axle shafts.	Tighten and hold properly.
(*viii*) Loose or worn wheel bearings.	Adjust or replace.
(*ix*) Worn thrust washers on differential pinions.	Replace.
(*x*) Differential cage bolts loose.	Tighten and lock properly.

Trouble No. 22. Hard steering.

Probable causes	*Remedies*
(*i*) Front axle shifted away from seat.	Put on seat, tighten and lock loose "U" bolt nuts.
(*ii*) Lack of lubrication.	(*i*) Replace leaking oil seal of steering gearbox. (*ii*) Top up oil in steering gear-box. (*iii*) Lubricate steering knuckles and steering linkage.
(*iii*) Improper front-end alignment.	Adjust properly.
(*iv*) Bent frame.	Straighten.
(*v*) Steering gear tight.	Adjust back-lash.
(*vi*) Tyres under inflated.	Inflate to specified pressure
(*vii*) Worn swivel pins and bushes.	Replace.
(*viii*) Pull and push rod joints tight.	Adjust.
(*ix*) Bent or twisted steering arm.	Replace.
(*x*) Distorted steering column.	(*i*) Align steering gearbox with mountings. (*ii*) Tighten loose bolts.

Trouble No. 23. Steering skimmy or wheel wobble.

(*i*) Knuckle bearings loose or worn.	Adjust or replace.
(*ii*) Loose tie rod.	Tighten or replace worn ends.
(*iii*) Front axle shifted.	Re-locate and tighten "U"-bolt nuts.

Probable causes	*Remedies*
(*iv*) Insufficient toe-in.	Adjust.
(*v*) Excessive or insufficient caster.	Adjust.
(*vi*) Improper load distribution.	Equalize load.
(*vii*) Unevenly inflated tyres.	Inflate as specified.
(*viii*) Excessive back-lash in steering gear.	Adjust.
(*ix*) Unbalanced wheels.	Balance and fit correct weights.
(*x*) Wheel rims buckled.	Align.
(*xi*) Worn steering linkage connections.	Adjust or replace.
(*xii*) Uneven wear of tyres.	(*i*) Rotate tyres. (*ii*) Adjust camber.

Trouble No. 24. Vehicle wanders.

Probable causes	*Remedies*
(*i*) Low or unequal tyre pressure at front and rear.	Inflate as specified.
(*ii*) Uneven tyre pressure.	—Do—
(*iii*) Rear tyre pressures lower than front.	—Do—
(*iv*) Vehicle overloaded at rear.	Equalise load.
(*v*) Binding at some point in steering linkage.	Check up and adjust/lubricate.
(*vi*) Insufficient caster angle.	Check up and adjust.
(*vii*) Front and rear axles shifted away.	Relocate and tighten U-bolts.
(*viii*) Tight steering gear.	(*i*) Adjust back lash. (*ii*) Top up oil in steering gear-box.

Probable causes	*Remedies*
(*ix*) Front wheel bearings loose.	Adjust.
(*x*) Twisted axle.	Replace.
(*xi*) Loose or worn spring shackles.	Adjust or replace worn parts.
(*xii*) Worn kingpins and bushes.	Replace.
(*xiii*) Loose steering gear-box.	Tight mounting-bolts.
(*xiv*) Loose wheel nuts.	Tighten.
(*xv*) Pitman or drop arm loose.	Tighten.

Trouble No. 25. Poor (inefficient) brakes.

(*i*) Excessive clearance between brake shoes and drum.	Adjust.
(*ii*) Air in hydraulic brake system.	Bleed out air.
(*iii*) New linings not bedded in.	Reface.
(*iv*) Oil or grease on brake linings.	(*i*) Clean. (*ii*) Replace oil soaked linings.
(*v*) Badly scored brake drums.	Skim or replace.
(*vi*) Worn linings.	Replace.
(*vii*) Low fluid level in master cylinder.	Fill in fluid.
(*viii*) Rubber cups in master- and wheel cylinders damaged.	Replace damaged cups.
(*ix*) Worn pistons in wheel-cylinders and master-cylinder.	Replace worn pistons.
(*x*) Leakage in the brake lines.	Remove leakage.
(*xi*) Binding cables and brake-expanders.	Free up and lubricate.

Probable causes	*Remedies*
(*xii*) Shoes incorrectly centred in brake drums.	Centre properly.
(*xiii*) Brakes not balanced.	Adjust.
(*xiv*) Excessive brake pedal free play.	Adjust.
(*xv*) Excessive play in wheel bearings.	Adjust.

Trouble No. 26. Brakes dragging.

Probable causes	*Remedies*
(*i*) Improper adjustment of brake linkage.	Adjust properly.
(*ii*) Weak pedal return spring.	Replace.
(*iii*) Weak return spring in master cylinder.	Replace.
(*iv*) Mineral oil in the system.	Flush out system and use correct grade of fluid.
(*v*) Weak or broken brake shoe return spring in case of one wheel dragging.	Replace return spring.
(*vi*) Tight brake shoe anchoring pins.	(*i*) Clean and lubricate. (*ii*) Replace defective parts.
(*vii*) Improper adjustment of brake shoes.	Adjust properly.
(*viii*) Improper adjustment of wheel bearings.	Adjust properly.
(*ix*) Swollen or distorted piston cups.	Replace rubber cups.
(*x*) Restricted or damaged brake hoses or brake pipes.	Replace.
(*xi*) Oil or grease soaked linings.	Replace.
(*xii*) Choked air vent in master-cylinder filler cap.	Clean.
(*xiii*) No free play in brake pedal.	Adjust.

Trouble No. 27. Brake judder or squeal.

(**Judder** is caused by low-frequency vibrations and **squeal** is caused by high frequency vibrations.)

Probable causes	*Remedies*
(*i*) Grease or oil on linings.	Clean or replace oil soaked linings.
(*ii*) Leading edge of lining picking up.	Chamfer the edge.
(*iii*) Improper fitting of brake linings.	Fit-in properly. If damaged, replace linings.
(*iv*) Brake shoes out of line.	Align.
(*v*) Rivets on shoes touching the brake drums.	Replace linings.
(*vi*) Worn shoe anchor-pins.	Replace.
(*vii*) Loose back plate.	Tighten.
(*viii*) Weak brake drums due to excessive grinding.	Replace.

Trouble No. 28. Brakes spongy.

(*i*) Air in hydraulic brake system.	Bleed out air.
(*ii*) Improper brake fluid.	Flush out brake system and use proper grade of fluid.
(*iii*) Misadjusted brake shoes	Adjust properly.

Trouble No. 29. Noisy brakes.

(*i*) Bent or distorted back plate.	Repair or replace.
(*ii*) Bent or distorted brake shoes.	Replace.
(*iii*) Dirty linings.	Clean.
(*iv*) Loose lining rivets.	Replace.
(*v*) Improper linings.	Replace with genuine ones.
(*vi*) Worn drums.	Skim or replace.

Trouble No. 30. Vehicle pulls to one side.

Probable causes	*Remedies*
(*i*) Unequal adjustment of brake shoes.	Adjust evenly.
(*ii*) Oil or grease soaked linings of one wheel or one side.	Replace.
(*iii*) Loose back plate on one side.	Tighten.
(*iv*) Wheel bearings improperly adjusted.	Adjust properly.
(*v*) Improper tyre inflation.	Inflate as specified.
(*vi*) Improper linings.	Replace with genuine ones.
(*vii*) Loose leaf spring clips or U-bolts.	Avoid shifting of axles by tightening loose clips or U-bolts.
(*viii*) Dust and grit in drums.	Clean.
(*ix*) Oval or scored drums.	Skim or replace.
(*x*) Broken spring centre-bolt.	Replace.

Trouble No. 31. Vehicle sags to one side.

Probable causes	*Remedies*
(*i*) Weak road springs.	Replace.
(*ii*) Broken spring leaves.	Replace.
(*iii*) Incorrect adjustment of coil springs or torsion bars.	Adjust properly.
(*iv*) Defective suspension units.	Repair or replace.
(*v*) Defective shock absorbers.	Replace.

Trouble No. 32. Harsh suspension or excessive road shocks transmitted to vehicle.

Probable causes	*Remedies*
(*i*) Spring leaves corroded causing excessive friction.	(*i*) Clean and lubricate. (*ii*) Replace.
(*ii*) Wear depressions in spring leaves.	Replace springs.

Probable causes	*Remedies*
(*iii*) Wrong fitting of extra-load leaves.	Fit properly.
(*iv*) Seized shackle pins.	(*i*) Free up, clean and lubricate. (*ii*) Replace damaged parts.
(*v*) Defective shock-absorbers.	Replace.
(*vi*) Suspension arm pivots seized.	(*i*) Free up, clean and lubricate. (*ii*) Replace defective pins and bushes.
(*vii*) Torsion bars incorrectly adjusted.	Adjust properly.
(*viii*) Coil springs stiff.	Replace.

Trouble No. 33. Excessive tyre wear.

(*i*) Tyres improperly inflated.	Inflate as specified.
(*ii*) Buckled wheel.	Repair or replace.
(*iii*) Incorrect toe-in.	Adjust.
(*iv*) Uneven caster.	Adjust.
(*v*) Fierce braking and acceleration.	Cultivate good driving habits.
(*vi*) Loose wheel nuts.	Tighten.
(*vii*) Sheared centre bolts of springs.	Replace.
(*viii*) Loose U-bolts of springs and axle	Tighten
(*ix*) Oil soaked tyres.	Replace
(*x*) Steering knuckle bearings worn	Replace.
(*xi*) Loose steering connections.	Adjust, tighten or replace.
(*xii*) Loose or worn spring shackles.	Tighten or replace.

Prabable causes	*Remedies*
(*xiii*) Bent or twisted front axle.	Replace.
(*xiv*) Unbalanced wheels.	Balance properly.
(*xv*) Dragging brakes.	Repair.
(*xvi*) Worn or loose wheel bearings.	Replace or adjust.
(*xvii*) Bent frame.	Straighten or repair.
(*xviii*) Vehicle overloaded.	Keep load under permissible limits.
(*xix*) Worn kingpins and bushes.	Replace.
(*xx*) Bent steering knuckles.	Replace.

Trouble No. 34. Discharged battery.

(*i*) Ignition switch left on.	Remove switch key for safety.
(*ii*) Loose or dirty battery terminals.	Clean and tighten terminals.
(*iii*) Low electrolyte level.	Fill in distilled water to correct level.
(*iv*) Excessive use of self-starter.	Check up cause of difficult engine starting.
(*v*) Generator not charging.	Check up generator and voltage regulator.
(*vi*) Leakage in wiring.	Check up and repair.
(*vii*) Cut out points stuck.	(*i*) Check up and clean.
	(*ii*) Replace defective cut out regulator.
(*viii*) Damaged battery.	Replace.
(*ix*) Misadjusted or defective voltage regulator.	Adjust or replace.
(*x*) Wrong connections of regulator.	Set right.

Trouble No. 35. **Ammeter shows no charge.**

Probable causes	*Remedies*
(*i*) Break in charging circuit.	(*i*) Check up any loose connection or short circuit and repair.
	(*ii*) Replace leaking or damaged leads.
(*ii*) Generator producing weak or no current.	(*i*) Clean dirty or greasy commutator.
	(*ii*) Replace worn out brushes or weak springs.
	(*iii*) Release sticking brushes in holder.
	(*iv*) Resolder broken connections at commutator.
	(*v*) True up oval or worn commutator and undercut mica.
	(*vi*) Replace damaged commutator.
	(*vii*) Repair break in field coil circuit.
	(*viii*) Replace burnt or damaged field coils.
	(*ix*) Clean out slots in commutator bars.
	(*x*) Replace damaged armature.
(*iii*) Cut out points burnt.	Replace cut-out.
(*iv*) Cut out points remain open.	(*i*) Replace damaged cut-out.
	(*ii*) Clean dirty commutator.
(*v*) Defective ammeter.	Replace.
(*vi*) Loose generator drive belt.	Tighten.

Trouble No. 36. Ammeter shows over-charge.

Probable causes	*Remedies*
(*i*) High setting of voltage regulator.	Check up and adjust.
(*ii*) Generator field-windings shorted or grounded.	Repair.
(*iii*) Short or open circuit in regulator.	Repair or replace regulator.
(*iv*) Defective ammeter.	Replace.

Trouble No. 37. Lighting system troubles.

(*i*) Lamps give insufficient illuminations.	(*i*) Charge, repair or replace run-down or defective battery. (*ii*) Check up alignment of lamps and focussing of bulbs and correct if required. (*iii*) Clean or replace dirty or damaged reflectors. (*iv*) Replace discoloured bulbs. (*v*) Check up earthing of lamps and clean dirty contact surfaces between lamp and mounting ; and mounting and chassis/body.
(*ii*) Lamps gradually fade out after switching on.	Charge, repair or replace run-down or defective battery.
(*iii*) Lights flicker.	Check up loose connection and tighten.
(*iv*) Brilliance varies with vehicle speed.	(*i*) Charge, repair or replace run-down or defective battery. (*ii*) Tighten loose bulbs or connections. (*iii*) Repair or replace defective cables.

Probable causes	*Remedies*
(*v*) No lights.	(*i*) Replace blown out fuses. (*ii*) Tighten loose connection at supply point of lighting switch (*iii*) Replace defective switch.

Trouble No. 38. Horn does not sound or sound improperly.

Probable causes	Remedies
(*i*) Discharged battery.	Recharge.
(*ii*) Loose connection.	Tighten.
(*iii*) Break in circuit.	Repair.
(*iv*) Short circuit.	Trace out and repair.
(*v*) Defective switch.	Repair or replace.
(*vi*) Blown off fuse.	Replace.
(*vii*) Poor earthing.	Clean the base and clamp horn properly.
(*viii*) Damaged C.B. points, resistor or field coil in horn.	Replace damaged parts.
(*ix*) Incorrect tone adjustment.	Adjust properly.
(*x*) Dust or water on horn diaphragm.	Clean.
(*xi*) Cracked diaphragm.	Replace.
(*xiii*) Weak electromagnet.	Replace.

Trouble No. 39. Air-conditioning system not producing cool air

Probable causes	Remedies
(*i*) Compressor drive belt loose or broken.	Replace the broken belt.
(*ii*) Compressor does not operate; belt slips on pulley.	Repair or replace compressor
(*iii*) Compressor valves inopera tive. High and low gauge readings show only slight variation at different engine speeds.	Repair or replace compressor valves. or replace compressor valves.
(*iv*) Expansion valve stuck open. Low gauge reading high, evaporator flooding.	Replace expansion valve.
(*v*) Fuse blown; wire broken or disconnected; switch or blower motor not functioning.	Replace fuse; repair or replace wire, switch or blower motor.
(*vi*) Refrigerant line broken or leakage in the system. High or low gauge reads zero.	Repair the line/system. Replace receiver drier.
(*vii*) Clogged screen in receiver – drier or expansion valve; blocked hose or coil.	Repair.
Frosting occurring at the point of restriction.	Replace receiver drier.

Trouble No. 40. Air-conditioning system not producing sufficeint amount of cool air at the discharge side of blower.	
(*i*) Compressor clutch slipping.	Repair or replace
(*ii*) Insufficeint air from discharge passage.	Clean or replace air filter, remove obstruction in the passage.
(*iii*) Blower motor sluggish.	Replace the motor.
(*iv*) Outside air vents open.	Close the vents.
(*v*) Insufficient air circulation over condenser coils.	Clean engine radiator and condenser.
(*vi*) Evaporator clogged.	Clean evaporator coils and fins.
(*vii*) Evaporator control valves misadjusted or defective.	Adjust or replace.
(*viii*) Insufficient refrigerant. Bubbles appear at sight glass.	Recharge system until bubbles disappear.
(*ix*) Expansion valve not working properly.	Replace.
(*x*) Receiver – drier screen clogged.	Purge the system; replace receiver drier.
(*xi*) Air in the system; sight glass cloudy.	Purge, evacuate and charge the system.

Trouble No. 41 Air-conditioning system cools intermittently.	
(*i*) Compressor clutch slipping.	Repair.
(*ii*) Defective circuit breaker, blower motor or switch.	Repair or replace.
(*iii*) Moisture in the system.	Replace expansion valve and receiver drier.
(*iv*) Thermostatic control defective.	Replace.
(*v*) Loose connection of compressor clutch coil or solenoid.	Repair.
(*vi*) Evaporator control valve stuck.	Purge and evacuate the system. Repair or replace the stuck valves. Charge the system.

Trouble No. 42. Noise in air-conditioning system.	
(*i*) Worn or loose V-belt.	Tighten or replace.
(*ii*) Compressor mounting bracket loose.	Tighten.
(*iii*) Compressor defective.	Repair or replace.
(*iv*) Compressor oil level low.	Fill to correct level.
(*v*) Blower motor loose or worn.	Repair or replace.
(*vii*) Excessive charge in the system causing rumbling and thumbling noise.	Discharge excess refrigerant until pressure gauge readings drop to specified pressure and bubbles disappear from the sight glass.
(*vii*) Low charge in the system causing hissing at expansion valve.	Locate leakage in the system, purge & repair. Evacuate system & replace receiver drier; charge the system.
(*viii*) Moisture in the system causing noise at expansion valve.	Purge and evacuate the system; replace receiver drier; charge the system.

Trouble No. 43. Diesel engine does not start.

(*i*) Weak or discharged battery.	Recharge or replace.
(*ii*) Battery & starter connections loose.	Tighten.
(*iii*) Starter switch defective.	Replace.
(*iv*) Defective self starter.	Repair or replace.
(*v*) Loose connections or defective glow plug switch.	Tighten loose connections; replace defective switch.
(*vi*) Fuel injection pump solenoid fuse blown.	Replace
(*vii*) Insufficient fuel in the tank.	Fill more fuel.
(*viii*) Water, air or dirt in the fuel.	Flush out the fuel system and fill in correct grade of fuel; bleed out air.
(*ix*) Plugged fuel return line.	Clean.
(*x*) No fuel to nozzles or injection pump.	Remove blockage or leakage.
(*xi*) Injection timing upset.	Reset.
(*xii*) Idle incorrectly set.	Reset.
(*xiii*) Dirtry defective injector nozzles.	Clean or adjust or replace.
(*xiv*) Defective fuel transfer or injection pump.	Repair or replace.
(*xv*) Plugged fuel tank vent.	Clean.
(*xvi*) Blocked fuel filter.	Clean and replace element.
(*xvii*) Low compression.	(*i*) Check position of compression release. (*ii*) Remove leakage of cylinder head or valve cage gaskets. (*iii*) Repair or replace leaking engine valves. (*iv*) Replace worn out pistons and rings.

Trouble No. 44. Diesel engine emitting smoky exhaust and misfiring.

(*i*) **Faulty injector/injectors if black smoke**	Clean, reset or replace.
(*ii*) Leaking sticking engine valves.	Repair or replace.
(*iii*) Defective fuel injection pump.	Repair or replace.
(*iv*) Leaking cylinder head induction manifold gaskets.	Replace.
(*v*) Air in the fuel line.	Bleed out air, remove leakage.
(*vi*) Water or dirt in the fuel.	Flush out the fuel system and fill in clean fuel.
(*vii*) Chocked air or fuel filters or strainers	Clean or replace.
(*viii*) Poor quality fuel.	Drain out & fill proper grade fuel.
(*ix*) Excess engine oil if smoke is blue.	Bring to correct level.
(*x*) Uneven distribution of fuel to engine cylinders.	Repair fuel supply system.
(*xi*) Injection timing upset.	Reset.

Appendix I

Primary and Secondary Quantities for S I. Units

The International System of Units is known as S.I. (*Systeme International des* Unites, Primary and secondary quantities for S.I. are as under.

Primary quantities

Quantity	*S I. Unit*	*Abbreviation*
Length	metre	m
Mass	kilogram	kg
Time	second	s
Electric current	ampere	A
Temperature	degree kelvin	K
Luminous intensity	candela	cd

Secondary quantities

Plane angle	radian	rad
Area	square metre	m^2
Volume	cubic metre	m^3
Frequency	hertz	Hz
Density	kilogram/cubic metre	kg/m^3
Speed	metre/second	m/s
Force	newton	N
Pressure	newton/square metre	N/m^2
Work, energy or quantity of heat	joule	J
Power	watt	W

Appendix II

CONVERSION TABLE FOR STANDARD UNITS
(F.P.S. to C.G.S.)

	British Unit (F.P.S.)		*S.I. Unit (C.G.S.)*
Length	1 mile	=	1.609 km
	1 yd.	=	0.914 m
	1 ft.	=	0.304
	1 inch	=	2.54 cm
Area	1 ft^2	=	0.092 m^2
	1 in^2	=	6.451 cm^2
Volume	1 $ft.^3$	=	28.316 dm^2
	1 in^3	=	16.387 cm^2
	1 Imp. gallon	=	4.546 litre
Velocity	1 mph	=	1.609 km/h
	1 ft/s	=	0.304 m/s
Acceleration	1 ft/s^2	=	0.304 m/s^2
Mass	1 ton	=	1016.05 kg
	1 lb.	=	0.453 kg
	1 oz	=	28.349 g
Pressure	1 ton f/in^2	=	15.44 MN/m^2
	1 lb f/in^2	=	6894.76 N/m^2
Density	1 $lb./ft^3$	=	16.018 kg/m^3
	1 lb/in^3	=	27.679 g/cm^3
Force	1 ton f	=	9964.02 N
	1 lb f	=	4.448 N
Torque	1 lb f ft	=	1.355 Nm
	1 lb f in.	=	0.112 Nm
Energy	1 hph	=	2.584 MJ
	1 Btu	=	1.055 KJ
	1 ft lb f	=	1.355 J
Power	1 hp	=	745.7 W
Temperature	F°	=	$C° \times \frac{2}{5} + 32$

Appendix III

CONVERSION TABLE FOR STANDARD UNITS
(C.G.S. to F.P.S.)

	S.I. Unit (C.G.S.)		*British Unit (F.P.S.)*
Length			
	1 km	=	0·621 mile
	1 m	=	1·093 yard
	1 dm	=	0·328 ft.
	1 cm	=	0·393 in.
	1 mm	=	0·039 in.
Area			
	1 m^2	=	1·195 $yd.^2$
	1 cm^2	=	0·155 $in.^2$
Volume			
	1 m^3	=	1·307 $yd.^3$
	1 dm^3	=	0·035 $ft.^3$
	1 cm^3	=	0·061 $in.^3$
	1 litre	=	0·22 Imp. gallon
Velocity			
	1 km/h	=	0·621 mph
	1 m/s	=	3·28 ft/s
Acceleration			
	1 m/s^2	=	3·28 ft/s^2
Mass			
	1 kg.	=	2·204 lb.
	1 g	=	0·035 oz
Density			
	1 kg/m^3	=	0·062 lb/ft^3.
	1 g/cm^3	=	0·036 lb/in^3.
Force			
	1N	=	0·737 lb f

S.I. Unit *British Unit*
(C.G.S.) *(F.P.S.)*

Torque

1 Nm = 0·737 lbf ft.

Pressure

1 N/m^2 = ·000145 lbf/in^2.

1 kN/m^2 = 20·885 lbf/ft^2.

Energy

1 J = 0·737 ft lb f

1 k J = 0·277 Wh

Temperature

$C^c = (F^\circ - 32) \times \frac{5}{9}$

Torque

1 Nm = 0·102 kg f m = 0·737 lb f ft.

1 kg f m = 9·807 Nm = 7·233 lb. f ft.

1 lb f ft = 1·356 Nm = 0·138 kg f m.

Appendix IV

(A) Technical data of Indian Cars (Petrol)

	Contessa Classic 1.8 GL	*Premier 118NE*	*Maruti 800*	*Maruti' 1000 (MF 410)*
Engine	4 Cyl. SOHO (Single Over-head Camshaft)	4 Cyl. In line	3 Cyl SOHC, Transverse	4 Cyle SOHC, Transverse
Bore & Stroke	84 × 82 mm	73 × 70 mm	68.5 × 72 mm	65.5 × 72 mm
Piston Displacement	1817 cc	1171 cc	796 cc	970 cc
Max. output	55.2 kW/ 5000 rpm (DIN)	52 HP / 5600 rpm (DIN)	39.5 HP / 5500 rpm	46 HP/ 5500 rpm (SAE, Net)
Max. Torque	130 Nm/ 3000 rpm -(DIN)	8.1 kgm/4000 (DIN)	6 kgm/3000 rpm	7.2 kgm/35 rpm (SAE, Net)
Compression ratio	8.5 : 1	9 : 1	8.7 : 1	8 .8 : 1
Tappet clearance				
Intake	0.20 mm (Hot)	0.35 mm (Hot)	0.23-0.28 mm (Hot)	0.23-0.28 mm (Hot)
Exhaust	0.30 mm (Hot)	0.35 mm (Hot)	0.23-0.28 mm (Hot)	0.23-0.28 mm (Hot)
Ignition Timing	6° BTDC at 800 rpm	5° BTDC /800 rpm	7° BTDC at 900 rpm	8° BTDC/800 rpm
Firing order	1-3-4-2	1-3-4-2	1-3-2	1-3-4-2
C.B. Point gap	0.2 to 0.4 mm	0.45 to 0.55 mm	0.4 to 0.5 mm	0.4 to 0.5 mm
Spark Plug gap	1.0 to 1.1 mm	0.8 to 0.9 mm	0.7 to 0.8 mm	0.7 to 0.8 mm
Clutch	Dry single plate	Dry single plate	Dry,single plate	Dry single plate
Gearbox	Synchromesh 5-speed	Synchromesh 4-speed	Synchromesh 4-speed	Synchromesh 5-speed
Gear Ratio :				
1st gear	3.736 : 1	3.757 : 1	3.585 : 1	3.416 : 1
2nd gear	1.963 : 1	2.169 : 1	2.166 : 1	1.894 : 1
3rd gear	1.364 : 1	1.404 : 1	1.333 : 1	1.357 : 1
4th gear	1 : 1	1:1	0.900 : 1	1.030 : 1
5th dear	0.775 : 1	—	—	0.870 : 1
Reverse	3.402 : 1	3.640 : 1	3.363 : 1	3.272 : 1
Final Drive gear ratio	3.42 : 1	3.9 : 1	4.351 : 1	4.388 : 1
Steering	Rack and pinion type	Worm and roller type	Rack and pinion type	Rack and pin-ion type
Turning radius	—	5.35 m	4.4 m	4.8 m
Tyre size	6.40 × 13	5.60 ×13	5.65 ×12	155(mm)— 80 R-13 (in)
Tyre pressure :				
Front	24 lb/in²	1.5 kg/cm²	1.8 kg/cm²	1.8 kg/cm²

	Contessa Classic 1.8 GL	*Premier 118NE*	*Maruti 800*	*Maruti' 1000 (MF 410)*
Rear	26 lb/in^2	2 kg/cm^2	1.8 kg/cm^2	1.8 kg/cm^2
Brakes: Front	Hydraulic Disc brake	Hydraulic, Disc brake	Hydraulic; Disc brake	Hydraulic; Disc brake
Rear	Drum brake	Drum brake	Drum brake	Drum brake
Suspension Front	Independent	Independent	Independent Strut (Pillar)	Independent Strut type
Rear	Coil springs Live axle	Coil springs, Live axle	Leaf Spring	Strut type
Electrical system	12 Volt, Battery 60 Ah.	12 Volt Battery v45 Ah.	12 Volt, Battery 30 Ah	12 Volt, Battery 35 Ah
Wheel Base	2667 mm	2420 mm	1215 mm	2365 mm
Kerb(unladen) weight	1138 kg	900 kg	620 kg	790 kg
Gross Vehicle Weight (GVW)	1588 kg	1300 kg	980 kg	1240 kg
Fuel tank capacity	65 Litres	39 Litres	30 Litres	40 Litres
Drive	Rear wheel (4 × 2)	Rear wheel (4 × 2)	Front wheel (4 × 2)	Front wheel (4 × 2)
Manufactured by	Hindustan Motors Ltd. Calcutta	Premier Automobiles Ltd. Bombay	Maruti Udyog Ltd. Gurgaon (Haryana)	Maruti Udyog Ltd. Gurgaon (Haryana)

Appendix IV B

(B) Technical data of Indian Cars (Diesel)

	Ambassador Nova (D)	*Tata Estate*	*Premier 1.37D*
Engine	4 Cyl. Inline	4 Cyl. Indirect Injection Diesel (IDI)	4 Cyl. Inline
Bore × stroke	73 × 88.9 mm	83 × 90 mm	78 × 71.5 mm
Piston Displacement	1489 cc	1948 cc	1366 cc
Max. Output	23.6 kW/4000 rpm	68 PS/4500 rpm	41 HP/4500 prm (DIN)
Max. torque	75 Nm/2000 rpm	12 mkg /2500 rpm	6.9 kgm/2800-3500 rpm (DIN)
Compression ratio	23 : 1	22.5 : 1	21 : 1
Firing Injection order	1-3-4-2	1-3-4-2	1-3-4-2
Fuel Injection	Rotary distributor	Rotary type with	Rotary distributor

	Ambassador Nova (D)	*Tata Estate*	*Premier 1.37D*
Pump	type with stop solenoid	electric stop solenoid	type, with stop solenoid
Clutch	Coil spring friction plate	Single Dry Disc, Diaphragm	Coil spring friction plate
Gearbox	4-speed synchromesh	5-speed synchromesh	4-speed synchromesh
Gear ratio :			
1st gear	3.807 : 1	3.86 : 1	3.86 : 1
2nd gear	2.253 : 1	2.18 : 1	2.38 : 1
3rd gear	1.506 : 1	1.37 : 1	1.57 : 1
4th gear	1:1	1 : 1	1 : 1
Over drive/ 5th speed gear	—	0.80 : 1	—
Reverse	4.875 : 1	3.13 : 1	3.86 : 1
Final drive gear ratio	4.555 : 1	5.44 : 1	4.30 : 1
Steering	Rack and pinion type	Power assisted, Recirculating ball and nut type	Worm and roller type
Turing radius	5.4 m	5.65 m	5.3 m
Suspension :			
Front	Independent	Independent	Independent
Rear	Semi floating live axle with semi elliptic leaf springs	Coil Springs, semi-floating live axle	Live Axle, semi elliptic leaf springs
Brakes (Service)	Drum brakes, Hydraulic	Dual circuit servo, Front disc and rear drum brakes	Drum brakes, hydraulic
Tyre size	5.90 × 15	195(mm) -70 R-15(inch)	5.20 × 14
Tyre pressure:			
Front	24 lb/in²	24 lb/in²	24 lb/in²
Rear	26 lb/in²	30 lb/in²	28 lb/in²
Electrical system	12 Volts , Battery 60 Ah	12 Volt Negative earth, Battery (N 50 Z)	12 Volt , Battery 45 Ah.
Wheel Base	2464 mm	2825 mm	2340 mm
Kerb Weight (Unladen)	1200 kg	1640 kg	965 kg
Gross Vehicle Weight (GVW)	1650 kg	2100 kg	1340 kg
Fuel Tank Capacity	54 Litres	50 Litres	38 Litres
Drive	Rear wheel drive 4 × 2	Rear wheel drive 4×2	Rear wheel drive 4 × 2
Manufactured by	Hindustan Motors Ltd. Calcutta	Tata Engineering and Locomotive Co. Ltd (Telco) Pune	Premier Autom-obiles Ltd. Bombay

Appendix V

Technical data of Indian heavy diesel vehicles

	Tata 1210 SE/ 1210 E	*Ashok Leyland Comet*
1	2	3
Engine	In-line, 6-cylinder Water cooled, Direct injection C.I. engine.	In-line, 6 cylinder Water cooled, Direct injection C.I. engine.
Bore/stroke	92 mm/120 mm	103·38 mm/120·7 mm
Capacity	4788 CC	6079 CC.
Compression ratio	17 : 1	16 : 1
Engine output	112 P. S at 2800 rpm (DIN)*	110 BHP at 2400 rpm.
Firing order	1—5—3—6—2—4	1—5—3—6—2—4
Fuel injection begins.	23° before T.D.C.	29° before T.D.C.
Tappet clearance—		
Intake valve	·20 mm } Warm engine	·508 mm } Cold engine
Exhaust valve	·30 mm } Warm engine	·508 mm } Cold engine
Clutch	single dry disc	single dry disc
Gearbox	5 speed synchro-mesh	5 speed synchro-mesh
Rear axle	Fully floating	Fully floating
Rear axle gear ratio	6·857 : 1	6·016 : 1
Gear ratio	GBS—40 gearbox	
1st gear	7·51 : 1	6·988 : 1
2nd gear	3·99 : 1	4·308 : 1
3rd gear	2·30 : 1	2·655 : 1
4th gear	1·39 : 1	1·605 : 1
5th gear	1 : 1	1 : 1

1	2	3
Reverse gear	6·93 : 1	6·343 : 1
Susoension	Conventional. Semi-elliptic leaf springs.	Conventional. Semi elliptic leaf springs.
Tyre size	9·00×20	9·00×20
Tyre pressure—		
Front	6 kg/cm² (85 p.s.i.)	6 kg/cm²
Rear	6 kg/cm² (85/p.s.i.)	6 kg cm²
Brakes— Service	Hydraulic, assisted by air pressure booster.	Air pressure operated
Parking	Mechanical, acting on rear wheels.	Mechanical, acting on rear wheels.
Steering	Recirculating ball type.	Marles cam and double roller type.
Steering gear ratio	34·2 : 1	24·7 : 1
Battery and electrical system	12 volt	Two 12 volt batteries in 24 volts circuit.
Max. speed in top gear	75·6 km/hr.	
Normal fuel consumption	25 litre/100 km. for gross vehicle weight of 15 tonnes.	
Wheel base	3225 mm (1210 SE/32)	5·334 m (AL—Cop 3/1)
Manufactured by	Tata Engineering and Locomotive Company Ltd., Jamshedpur.	Ashok Leyland Ltd., Madras.

* DIN=Deutsch Industrie Normale (German).

Appendix VI

Technical data of Indian light diesel vehicles

	Mahindra Nissan Allwyn Cabstar-576	*DCM Toyota Dyna 600*	*Tata-407*	*Swaraj Mazda WV-26*	*Matador Max F-307*
Engine : Type	SD-25 / 4 Cyl.	14B-4 Cylinder	Tata 497-Sp/4 Cyl.	4 Cyl. OHV, In line	4 Cyl. IDI(OM616)
Bore and Stroke	89×100 mm	102 × 112 mm	97 × 100 mm	100 × 110 mm	90.9 × 92.4 mm
Piston Displacement	2488 cc	3660 cc	2956 cc	3455 cc	2399 cc
Compression Ratio	21.4 : 1	18 : 1	17 : 1	17 : 1	21 : 1
Max. Output	75/4000 rpm (SAE. Gross)	94HP/3400 rpm (DIN)	65 PS/3200 rpm (DIN)	—	65HP/4000 rpm
Max. Torque	16.2 kgm/2000 rpm	24 kgm/1800 rpm (DIN)	16 kgm/2200 rpm (DIN)	20.6 kgm/1750 (DIN)	12.8 kgm/2400 rpm
Transmission : Clutch	Single dry plate, diaphragm,	Single dry plate diaphragm	Single dry plate diaphragm	Single dry plate, diaphragm	Single dry plate, diaphragm
Gearbox	Synchromesh 5-speed	Synchromesh 5-speed	Synchromesh 5-speed	Synchromesh 5-speed	synchromesh 5-speed
Gear ratio I gear	5.429 : 1	5.657 : 1	6.01 : 1	5.833 : 1	4.07 : 1
Gear ratio II gear	3.048 : 1	2.818 : 1	3.46 : 1	2.855 : 1	2.273 : 1
Gear ratio III gear	1.780 : 1	1.587 : 1	1.97 : 1	1.651 : 1	1.37 : 1
Gear ratio IV gear	1 : 1	1 : 1	1.37 : 1	1 : 1	0.87 : 1
Gear ratio V gear	0.820 : 1	0.835 : 1	1 : 1	0.800 : 1	—
Reverse	6.544 : 1	5.657 : 1	5.69 : 1	5.372 : 1	3.9 : 1
Final drive gear ratio	6.833 : 1	6.167 : 1	4.625 : 1	6.571 : 1	7.37 : 1 (Front wheeldrive)
Steering	Recirculating ball and nut type	Recirculating ball and nut type	Recirculating ball and nut type	Recirculating ball and nut	Worm and roller type

(Contd.)

(Contd.)

	Mahindra Nissan Allwyn Cabstar-576	*DCM Toyota Dyna* 600	*Tata-407*	*Swaraj Mazda WV-26*	*Matador Max F-307*
Suspension	Conventional, semi elliptic leaf springs	Conventional, semi elliptic leaf springs	Conventional, semi-elliptic leaf springs	Conventional semi- elliptic leaf springs	Front- Torsion bar, independent., Rear- semi elliptic leaf springs, solid bar axle
Brakes (service)	Vacuum assisted hydraulic, dual circuit	Vacuum assisted hydraulic, with tandem master cylinder	Vacuum assisted hydraulic with tandem master cylinder	Vacuum assisted hydraulic, dual circuit (tandem master cyl.)	Vacuum assisted hydraulic, dual circuit.
Tyre size	7.00 ×16	7.50 × 16	7.50 × 16	7.50 × 16	7.00 × 15
Turning radius	5.3 m	6.1 m	5.95 m	5.5 m	6 m
Electrical system	12 Volt, Battery 70 AH	12 Volt, Battery 70 Ah	12 Volt, Battery 70 Ah.	12Volt, Battery 70 Ah	12 Volt, Battery 60 Ah
Fuel Tank Capacity	60 litres	90 litres	60 litres	70 litres	35 litres
Wheel Base	2500 mm	3290 mm	2750 mm	2815 mm	2400 mm
Gross Vehicle Weight (GVW)	5350 kg	5990 kg	5300 kg	5480 kg	3500 kg
Manufactured by	Mahindra Nissan Allwyn Ltd. Secundrabad (A.P.)	DCM Toyota Ltd. New Delhi	Tata Engg. and Locomotive Co. Ltd., Pune	Swaraj Mazda Ltd. Asron-Ropar (Punjab)	Bajaj Tempo Ltd. Pune

Appendix VII

Survey Report

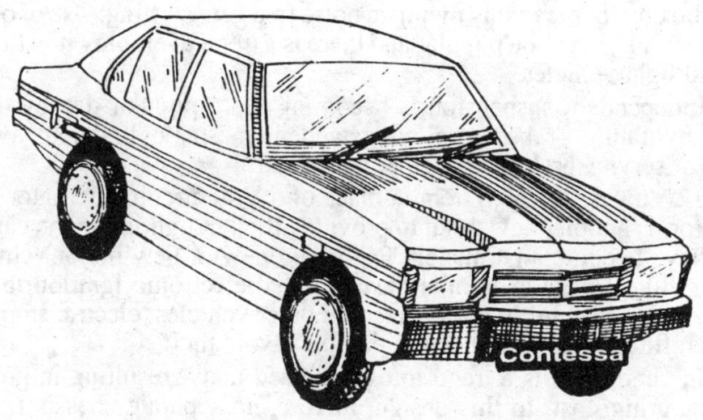

Modern trends in automobile manufacturing. Automobile manufacturing is a business. The object of business is to earn money, provide employment to the people and help in the development of the country. Profit earning depends upon the following factors:

(*i*) Lesser manufacturing cost. (*ii*) Higher sale price.

The manufacturing cost depends upon raw materials used, plant and techniques employed in manufacturing, direct and indirect labour, over-head expenses etc. Apart from this, the design of the product plays an important role to attract customers and fetch maximum profit. The design should appeal to the customers and suit the conditions prevalent in the country.

Owing to high cost of petrol, there is a great demand to manufacture automobiles running on cheaper fuel. That is why there is a trend to employ diesel engines in the cars although heavy and light transport vehicles are mostly running on diesel. Ambassador and Premier cars, running on petrol are now having diesel models also such as Ambassador Nova (D), Premier 1.37 and 1.38 (D). Telco has been producing Tatamobile pick up, Tata Sierra, and Tata Estate models running on diesel.

To combat air resistance, the modern cars are better streamlined, such as in Maruti 800, 1000 and Zen models, Contessa and almost all the other makes of cars.

In order to provide maximum space in the car, Maruti cars are employing transverse engines and front wheel drive power pack system wherein engine power unit is directly mounted over the front axle. Owing to this design, propeller shaft has been eliminated, resulting in lowering of centre of gravity (C.G.) for better stability. This has helped to provide more space inside the car without raising its height

Maruti 800 car comes in the small range which employs three cylinder in-line, water cooled, petrol engine. Its single plate dry friction clutch and 4-speed synchromesh gearbox fulfils the needs of transmission of power to the front wheels through drive shafts. Rack and pinion steering gear box provides ability to the car to take a turn at 4.4 metre radius. Independent suspension system cushions out road shocks. Hydraulic disc brakes at the front and drum brakes at the rear wheels enable the running car to stop efficiently to avoid accident.

Whether it is a petrol or diesel car, the modern trend is to employ diaphragm type of clutch pressure plate instead of coil springs. In most light transport vehicles, the clutch is operated hydraulically. Tata series of cars employ power assisted steering. In almost all the vehicles, synchromesh type of gearbox has come to stay owing to noise free gear shifting. Use of over-drive (5th gear) is going to be popular and there is a trend to employ it in the modern cars and light vehicles.

Independent suspension is becoming more popular day-by-day. Dual circuit hydraulic brakes employing tandem master cylinder to avoid total failure of service brakes, is another new trend.

12 Volt electrical system in place of 6 Volt circuit, alternator in lieu of generator, is another new trend to provide efficient lighting in the cars. Signal lights have become a permanent feature in view of new motor vehicle laws. Scooters like Vespa and Bajaj have adopted electronic ignition in place of breaker point system. In diesel cars and other vehicles, electric stop solenoid has been fitted with fuel injection pump to avoid theft.

In cars, there is a trend to use unitized body, resulting in good cut in manufacturing cost. In this design, there is no separate chassis frame. The suspension and other units such as engine, transmission etc. are attached directly to the car body-cum-frame unit.

As far as material is concerned, aluminium is mostly used in the contruction of casings, housings, blocks etc. in place of cast iron. Engine cylinder heads, clutch and gearbox housings, fuel pump casings, brake drums and number of other parts are made of aluminium, resulting in reduction of weight. Plastic is widely used in making washers, spacers, tiny blocks etc. Steel castings, forgings of higher tensile strength helping in weight reduction and improving strength of the parts, are used. There is no doubt that there is great improvement in ruber, plastic, paint, sheet metal, casting and moulding, welding technology etc but as far as design is concerned, we lag behind the countries like Japan, Germany, America etc. Indian automobile industry started with the collaboration of foreign manufacturers and still we are looking for their help to start new ventures. In spite of some bottle-necks, it is heartening to note that many new plants have been set up with foreign collaboration to manufacture light commercial vehicles such as Swaraj Mazda, DCM Toyota, Bajaj Tempo, Mahindra Nissan Allwyn, Eicher Mitsubishi, etc. and other industrialists too are girding up their loins to jump into the trade. The expansion of old units like Telco, Mahindra and Mahindra is playing an important role in putting new models on Indian roads.

Glossary of Technical Terms

Accelerator : A pedal or lever for controlling engine speed.

Accumulator : Another name for storage battery.

Ackerman Principle : The principle adopted in steering system wherein a line intersecting each kingpin and tie rod end intersects at or near the differential.

Additive : A material added to oil or fuel to improve its quality.

Air bleed : An opening (hole) in the fuel passage of carburettor through which air can bleed or pass into the fuel as it moves through the passage.

Air brake : A braking system to which compressed air is used to supply necessary force to apply brakes.

Air cleaner : A device for filtering and cleaning air.

Air Cooled Engine : An engine containing fins around the cylinder head, cylinder block, exhaust ports etc. and cooled by surrounding air contacting the fins.

Air Fuel Ratio : The ratio of air and fuel (by weight) of mixture prepared by carburettor.

Air Horn : Air inlet of the carburettor to which air cleaner is usually attached.

Air Lock : A bubble of air trapped in a fluid line which obstructs the normal flow of fluid.

Air Spring : A flexible bag filled with compressed air which compresses to absorb shocks in the suspension system.

Alignment : An adjustment to bring into a line.

Alternating Current (A.C.) : An electrical current alternating back and forth in the direction of flow.

Alternator : Alternating current generator used in automobile in which alternating current is changed into direct current by a rectifier.

Aluminium : A metal noted for its lightness and white in colour.

Ammeter : A gauge for measuring the flow of electric current.

Ampere : The unit of measurement for the flow of electric current.

Ampere Hour Capacity : A term used to indicate the capacity of a storage battery.

Anti-Clockwise : Opposite direction of the hands on a clock.

Anti-Freeze Mixture : A mixture alcohol of or glycerine or denatured spirit etc. and water used in the cooling system for avoiding freezing of water.

Armature : The core as in generator, which rotates within the pole shoes surrounded by the field coils.

Asbestos : A natural fibrous mineral having great heat resisting ability.

Aspect ratio : The ratio of type section height to width section.

Atmospheric Pressure : Pressure exerted by the weight of air in all directions upon all objects on the earth. Average atmospheric pressure at sea level is 14.7 lbs. per sq. inch.

Atomize : To split up to fine particles.

Automatic transmission : A transmission (gear box) in which gears are shifted automatically.

Axle : The shaft or beam over which wheels are fitted and which forms axis for them.

Axle ratio : The ratio between drive pinion and crown wheel of differential known as Final-Drive gear ratio.

Back Fire : Ignition of mixture in the induction manifold by flame from the cylinder due to leaking intake valve.

Backlash : The clearance or play between two parts such as gears meshing with each other.

Back Pressure : Resistance to free flow as in exhaust manifold and silencer.

Baffle Plate : An obstruction plate for deflecting the flow of liquid, gas or sound.

Ball Bearing : An anti-friction bearing consisting of a hardened inner and outer race with hardened steel balls interposed between the two races.

Battery : Electrical cells assembled in one case.

B.D.C. : Bottom dead centre.

Bead : That part of tyre which is shaped to fit into the wheel-rim.

Bearing : A part in which a journal, pivot or a similar thing turns or moves.

Bendix Drive : A gear mounted on a screwed shaft attached to the armature shaft of starting motor which automatically engages and disengages the self-starter.

B.H.P. (Brakes Horse Power) : A measurement of the power developed by an engine crank shaft to drive the vehicle.

Bias-ply tyre : A tyre in which the plies are laid diagonally, criss-crossing one another at an angle of about 35 degrees.

Bleeding : A process by means of which air is removed from hydraulic system of brake, power steering or fuel.

Bonding Material : Material to act as a cement for uniting particles or parts.

Booster : A device to boost the power applied by the operator in operating clutch, brake, steering etc.

Bore : Diameter of engine cylinder or any hole in which a bushing fits.

Boss : Projections within the piston which support the gudgeon pin.

Brake Band : A band surrounding a brake drum, to which lining is attached.

Brake Drum : A metal drum enclosing brake shoes over which the road wheel is mounted.

Brake fluid : Fluid used in the hydraulic brake system which has no effect on rubber and other parts.

Brake Lights : Rear lights of the vehicle which indicate the application of brakes.

B.T.D.C. : Abbreviation for "Before top dead centre".

Brake Lining : Strip of frictional material attached to the brake shoe which comes in contact with the rotating brake drum for retarding its speed.

Brake Shoe : The carrier to which the brake lining is attached and which is used to force the lining in contact with the brake drum.

Breaker Points : Two separable points usually faced with silver, tungsten or platinum, which interrupt the primary circuit.

By-pass : An alternate passage for the flow of liquid or gas.

Cab : Separate driver's cabin provided on trucks and tractors.

Calibrate : To determine or adjust the graduation of any instrument giving quantitative measurements.

Calliper : A housing for pistons and brake-shoes with hydraulic disc brake system which holds the brake-pads (shoes) to straddle the disc.

Cam : An eccentric projection on a revolving shaft designed to give some requisite linear motion to a follower.

Camber : Curvature of road or spring or angle or inclination of road wheel.

Camshaft : The shaft containing lobes or cams to operate engine valves.

Carbon : A common non-metallic element.

Carburettor : A device for automatically mixing fuel with air in correct proportions to produce a combustible mixture.

Casing : Another name for housing.

Caster : Backward tile of front axle.

Catalytic converter : A device like a muffler used in the exhaust system to convert harmful exhaust gases into harmless gases.

Cell : Unit of battery containing a group of positive and negative plates, separators and electrolyte.

Centrifugal Advance : A rotating weight mechanism in the distributor which advances and retards ignition timing through centrifugal force.

Centrifugal Force : A force which tends to move a body away from its centre of rotation.

Cetene number : A measure of ignition quality of diesel fuel.

Charging rate : The rate of current in amperes, flowing from the generator/alternator into the battery.

Chassis : Machine portion or framework of automobile excluding cab and body.

Check Valve : Non-return valve which gives passage to a liquid or gas in one direction only.

Choke Valve : A butterfly valve placed in the carburettor air intake passage to control the entrance of air.

Circuit : The path of electrical current, liquids or gases.

Circuit Breaker : A device for interrupting an electrical circuit.

Clearance : The space allowed between two parts.

Clockwise : In the same direction as the hands of a clock.

Clutch : A mechanism for connecting and disconnecting driving member from driven member.

Coefficient of Friction : A measurement of the amount of friction developed between two surfaces pressed together and moved one on the other.

Combustion Chamber : Chamber or space above the cylinder of an engine in which fuel-air mixture is burned.

Commutator : A ring of adjacent copper bars insulated from one another to which the windings are attached.

Compensating Port : A small hole in the master cylinder to allow fluid to return to the reservoir.

Compound : A mixture of two or more ingredients.

Compression : The reduction in volume of a gas.

Compression Ignition (C.I.) Engine : An engine working on diesel cycle in which injected fuel ignites, owing to heat of compressed air in the combustion chamber.

Compression Ratio : Relationship between the volume of cylinder and combustion chamber when the piston is at B.D.C. and the clearance volume when piston is at T.D.C.

Concentric : Having the same centre.

Condenser : An electric condenser is a device for temporarily collecting and storing a surge of electrical current for later discharge.

Conductor : A material through which electricity can flow with slight resistance.

Connecting Rod : Rod which connects piston with the crankshaft.

Constant Mesh Transmission : An arrangement of gearing wherein the gears remain constantly mesh instead of being slid into and out of engagement with one another.

Constant Velocity Joint : Closely coupled type of universal joint in which acceleration-deceleration effects are cancelled out.

Contact Breaker : Circuit breaker for interrupting an electrical circuit.

Contact Points : Two separable points usually faced with platinum, silver or tungsten, which interrupt the primary circuit in ignition system.

Contraction : Squeeze in size.

Convection : Transfer of heat by circulating heated air or liquid.

Cooling system : The system which removes heat from the engine to keep it at the working temperature.

Core : The centre portion of an electromagnet or armature around which the wire is coiled.

Corrode : To eat away gradually by chemical action.

Counter Clockwise : Anti-clockwise.

Coupling : A connecting means for transferring movement from one part to another.

Crankcase : The housing within which crankshaft and allied parts of the engine operate.

Crankcase Dilution : Constituents of combustion escaping from combustion chamber into crankcase and oil sump where they dilute or thin the lubricating oil.

Crank Case Ventilation : Circulation of air through the crankcase of running engine, to remove blow by water and other vapors to prevent oil-dilution, contamination, sludge formation and pressure build-up.

Crankshaft : The main shaft of the engine which in conjunction with the connecting rods, converts reciprocating motion of the pistons into rotary motion.

Current : The flow of electricity.

Cut-out : Automatic switch for opening and closing the electrical circuit.

Cycle : A series of events or operations taking place according to established sequence.

Cylinder : The bore in the engine in which piston operates.

Cylinder Block : The main engine block which contains cylinders.

Cylinder Head : Usually a detachable portion of an engine which covers the cylinder and provides combustion chambers in the engine.

Dash Board : Instrument panel fitted before the driver's seat.

Dead Centre : The extreme upper or lower position of the crankshaft throw at which the piston is not moving in either direction.

Dead Axle : An axle which does not turn.

Degree : 1/360 part of a circle and indicated by a small "°" placed alongside a figure.

Demagnetize : To remove magnetization of a pole which has previously been magnetized.

Denatured Alcohol : Ethyl alcohol to which a denaturant has been added.

Detergent : A compound added in lubricating oil to avoid deposits.

Detonation : Hammer like too rapid blowing fuel in the engine due to its inferior grade.

Diagonal brake system : A dual brake system having separate hydraulic circuits connecting diagonal wheels together (right front to left-rear and right rear to left-front).

Diaphragm : A flexible partition or wall separating two cavities.

Diesel Cycle : Cycle of events in an engine in which air alone is compressed and fuel is injected at the end of compression stroke. Ignition takes place due to the heat produced by compressing air.

Diesel Engine : Named after its developer, Dr. Rudolph Diesel, operating on Diesel Cycle, having no spark plug and carburettor.

Differential : The mechanism housed in the drive axle, which drives outer wheel faster than the inner one in order to take a turn.

D.I.N. : Deutsch Industriell Norm (German Industrial Standard).

Dip-stick : A graduated stick for measuring oil level in engine or gearbox.

Direct Current : Electric current which flows continuously in one direction.

Directional Signals : Lighting system on motor vehicles which flashes lights to indicate the direction to which the driver intends to turn.

Disc brake : A type of brake in which brake-shoes grip the revolving disc like a calliper in order to stop it.

Distributor : An electrical distributor which distributes high tension current to the engine spark plugs according to firing order.

Dog-Clutch : A type of clutch in which driving and driven members are engaged and disengaged through dogs or jaws provided on them.

Double Reduction Axle : A drive axle construction in which two sets of reduction gears are used for extreme reduction of gear ratio.

Down-Draft-Carburettor : A type of carburettor wherein the mixture flows downward to the engine.

Drag-Link : Pull and push rod connecting steering gear drop arm and steering knuckle arm.

Drive Line : Propeller shafts and other parts connecting transmission with the driving axles.

Drive Axle : Live axle which transmits power to the road wheels.

Drum brake : A type of brake in which revolving drum is stopped by internal expanding curved shoes or external contracting brake-band.

Dwell : Number of degrees of distributor-cam-rotation when conract-points remain closed before they open again.

Dynamo : A generator of electricity.

Dynamic-balance : The balance of an object in motion.

Earth : Another term for ground.

Eccentric : Off the centre.

Electrode : In automobiles, the electrode usually refers to the insulated centre rod of spark plug.

Electrolyte : A mixture of sulphuric acid and distilled water used in storage batteries.

Electromagnet : A magnet formed by passing an electrical current through wire wound around a core.

Elliot steering knuckle : Type of axle wherein the ends of the axle beam straddle the spindle.

Emission control : A system to reduce air polluting emissions of exhaust gases.

En-Block : Block cast in one section.

Energy : The capacity of doing work.

Engine : Power unit which is prime source of power generation.

Engine Displacement : The sum of displacement of all cylinders of the engine.

Exhaust Gas Analyzer : An instrument for analyzing exhaust gas.

Exhaust Manifold : Engine part which provides a series of passages through which exhaust gases flow out.

Exhaust Pipe . The pipe connecting exhaust manifold to the muffler for carrying exhaust gases.

Expansion : An increase in size due to heat effect.

F-Head Engine : An engine which contains one valve in the cylinder head and the other by the side of cylinder.

Field : In a generator or electric motor, the area in which magnetic flow occurs.

Field Coil : Coil of insulated wire which magnetizes a field.

Filter : A unit containing filtering element to eliminate foreign particles from the fluid being filtered.

Final drive : The final speed reduction gears (drive pinion and crown wheel) in the power transmission line of an automobile.

Firing Order : The order in which firing takes place in the various cylinders of the engine.

Flange : A projecting collar on an object for keeping it in place.

Float : A bulk for bulk lighter unit used to control automatically a valve in the passage of fluid.

Float Chamber : A small fuel reservoir contained with the carburettor which contains a float operated valve at its entrance.

Float Level : The predetermined height of the fuel in the float chamber.

Floating Piston or Gudgeon Pin : A gudgeon or piston pin which is not locked in the connecting rod or the piston but is free to turn in both the connecting rod and the piston.

Fluid Coupling : A coupling which couples the driving member with driven member through fluid media.

Flux, Electric or Magnetic : Lines of magnetic force flowing in a magnetic field.

Flywheel : A heavy wheel in which energy is absorbed and stored by means of momentum.

Free Wheeling Device : A mechanical device which engages driving member with the driven member in one direction only.

Front end geometry : Angular relationship between the front wheels, wheel attaching parts and vehicle frame (caster, camber, king-pin inclination, toe-in, toe-out etc.)

Fuel Gauge : The instrument which indicates the amount of fuel in the tank.

Fuel Pump : A pump for delivering fuel to the carburettor or injection pump.

Fulcrum : The support on which a lever turns in moving a body.

Full Floating Axle : A drive axle construction wherein axle shafts do not carry any weight of the vehicle.

Fuse : A piece of wire placed in an electrical circuit as a safety measure to avoid damage in case of excessive current flow when it melts away and disconnects the circuit.

Gasket : A packing placed between two metal surfaces to act as a seal.

Gasoline : Liquid blend of hydrocarbons obtained from crude oil, known as petrol.

Glow Plug : A small electric heater fitted at the pre-combustion chamber of diesel engine to pre-heat the chamber for easy starting in cold weather.

Gear Ratio : Ratio between the driving and driven gears.

Generator : A device consisting of an armature, field coils and other parts which when rotated produces electricity.

Governor : A device to control and regulate speed.

Ground : A term for connecting one end of an electrical circuit to the frame work to complete the circuit through one wire.

Gudgeon Pin : Also known as wrist pin or piston pin. It is the pin which connect piston with the connecting rod.

Harmonic Balancer : A device to reduce torsional vibrations in the crankshaft. It is also known as vibration damper.

Heat Control Valve : A valve in the exhaust manifold for varying heat to induction manifold.

Helical Gear : A gear design wherein the gear teeth are cut at an angle to the shaft.

Herring-bone Gear : A pair of helical gears designed to operate together in which the angle of the pair of gears forms a V

High Tension : The secondary or induced high voltage electrical current.

Horse Power : A unit for measuring power. It is the rate for doing work at 33,000 ft. lb. per minute.

Hotchkiss Drive : A design wherein torque reaction and driving thrust are taken through road springs.

Hot Spot : A spot in the engine manifold which is surrounded by exhaust gases to heat the incoming mixture.

Hydrocarbon : Any compound composed entirely of carbon and hydrogen product.

Hydrometer : An instrument to find specific gravity of electrolyte.

Hypoid Gears : A design of drive pinion and ring gear in final drive wherein the centre line of the pinion is offset from the centre line of ring gear.

Idle Port : Outlet of the idle passage in the carburettor.

Idler Gear : A gear placed between a driving and driven gears to make them rotate in the same direction.

Idling Speed : The lowest speed for the engine without load.

Ignition Coil : An electrical coil which transforms low tension current into high voltage surge to produce a spark at the spark plug gap.

Ignition Lag : Delay in time between injection of fuel and commencing of combustion.

Ignition System : The system of providing igniting in the engine.

Ignition timing : Delivery of spark in the engine combustion chamber at the proper time, relative to piston-position.

Inclucded angle : Camber angle plus steering axis ares inclination angle.

I.H.P. : Indicated horsepower developed in an engine.

Impeller : The rotating disc having wings to set up circulation of liquid or gas.

Independent Suspension : Type of suspension in which the wheel on one side of the vehicle rises or falls independently of the wheel on the other side.

Induction : The influence of magnetic fields of different strength not electrically connected to one another.

Induction Coil : Ignition coil which through the action of induction, creates a high tension current.

Induction Manifold : Engine part which provides a series of passages for the flow of fuel mixture from carburettor to the engine.

Inertia : A physical law which tends to keep a motionless body at rest or also tends to keep a moving body in motion; effort is thus needed to start a mass moving or to stop it once it is set in motion.

Injector : A device by means of which fuel is injected into the engine combustion chamber.

Inlet Valve : Also known as intake valve. It permits the entrance of fuel air mixture or air into the engine cylinder.

Insulator : An electrical conductor covered with a non-conducting material.

Internal Combustion Engine : An engine in which combustion of fuel takes place inside the engine.

Jackets : Pockets around the cylinder and combustion chamber through which water circulates.

Jack Shaft : Half shaft or axle shaft through which power flows from differential to wheels.

Jet : A calibrated passage to control the flow of liquid or gas.

Journal : The finished part of a shaft which rotates in a bearing.

King Pin : The pin around which the steering spindle or arm moves.

King-Pin Inclination : The angle at which the kingpin is inclined.

Knock : A general term for describing various noises in the engine.

L-Head Engine : Type of engine in which both intake and exhaust valves are located on one side of the cylinder.

Limousine : A type of automobile body wherein the driver is separated from the passengers by a partition.

Liner : Usually a thin section placed between two parts.

Linkage : A series of rods, yokes and levers etc. used to transmit motion from one unit to another.

Liquid : Any substance which assumes the shape of the vessel in which it is placed without changing volume.

Live Axle : Drive axle through which power flows towards the wheels.

Low Tension : Relatively low voltage.

L.P.G :(Liquefied Petroleum Gas): Made usable as fuel for I.C. engines by compressing volatile petroleum gases to liquid form.

Magnet : Permanent magnet is a piece of hard steel so as to have opposite poles.

Magnetic Field : The flow of magnetic force or magnetism between the opposite poles of a magnet.

Magneto : An electrical device which generates electric current when rotated.

Make and Break Mechanism : The mechanism by means of which electrical circuit is opened and closed.

Manifold : A combination of multichannels used to connect various cylinders to one inlet or outlet.

Master Cylinder : The main cylinder in the hydraulic brake system which is used to force hydraulic fluid to the wheel cylinders in applying the brakes.

Mechanical Efficiency (Engine) : The ratio between the indicated horsepower and the brake horse power of an engine.

Metering rod : A stepped rod in the carburettor which measures quantity of fuel going out of the float chamber into the main nozzle.

Mixing Chamber : The part of the carburettor between the venturi and the throttle valve in which fuel and air are mixed up.

Mono-Block : Enblock in which all the cylinders of an engine are contained in one casting.

Motor : An electric motor for converting electrical energy into mechanical power.

Muffler : A chamber in the exhaust system which allows exhaust gases to expand and cool in order to silence them.

Needle Valve : Valve at the entrance of carburettor float chamber which consists of a tapered needle and a conical seat and which is operated by the float.

Nozzle : Fuel nozzle or jet through which fuel passes.

Octane rating : A measure of anti-knock properties of petrol.

Oil Pan : Oil sump fitted to the engine which contains lubricating oil.

Oil Pressure Gauge : The gauge which indicates the pressure of oil in the engine lubrication system.

Oil Seal : A seal placed around a rotating shaft in order to prevent the leakage of oil.

Orifice : A small hole into a cavity.

Otto Cycle : Cycle of engine operation named after its inventor Dr. Nikolas Otto, in which four events – suction, compression, power and exhaust, occur one after the other in four strokes of the piston.

Output Shaft : The shaft which delivers the power.

Overdrive : Arrangement of gearing by means of which the output shaft is driven at higher speed than the driving shaft.

Over-Head Valve Engine : Type of engine in which the valves are located above the cylinders in the cylinder head.

Oxidize : To combine an element with oxygen to convert into its oxide.

Parking Brake : Hand brake which is used during parking so that the brake may remain applied indefinitely.

Pinion : The smaller of two meshing gears.

Pinion Carrier : A bracket for carrying the pinions.

Piston : A cylindrical part closed at one end which moves inside a cylinder.

Piston Displacement : The volume of air moved or displaced by moving the piston from T.D.C. to B.D.C.

Piston Head : That part of the piston which is above the rings. Also known as piston crown.

Piston Lands : Parts of piston between the piston rings.

Piston Pin : The pin which holds the connecting rod.

Piston Ring : An expanding ring placed in the grooves of the piston to provide a seal for preventing leakage of gases past the piston.

Piston Ring Expander : A spring placed below the piston ring in the groove to increase the pressure of ring against the cylinder wall.

Piston Ring Gap : The clearance between the ends of piston ring and measured by putting ring inside the cylinder.

Piston Ring Grooves : The grooves or slots in the piston in which piston rings are placed.

Piston Skirt : That part of the piston which is below the piston rings.

Pitman Arm : Also known as drop arm. Placed at steering gear shaft, this lever gives movement to the drag link.

Pivot : A pin or short shaft upon which another part rests, or turns or upon and about which another part moves.

Planetary Gears : A system of gearing which resembles the solar system. A sun gear is surrounded by an internal ring gear and star pinions carried by the planet carrier. All the gears are in constant mesh with one another between the ring gear.

Planet Carrier : In planetary gearing, the carrier or bracket which contains shafts upon which the pinions or planet gears revolve.

Planet Gears : The pinions fitted on the planet carrier and meshing with sun gear and ring gear.

Plies : Layers of cord in the tyre casing. Every layer is a ply.

Polarity : Positive or negative terminal of battery or an electrical circuit and also the north or south poles of a magnet.

Popper Valve : A mushroom shaped valve having a circular head attached to a stem which is designed to open and close a circular hole or port.

Port : The openings in the cylinder block for valves, inlet and exhaust pipes, or water connections etc.

Positive Terminal : The terminal from which the current enters the circuit.

Power : The capacity to do work or the rate at which it is done.

Power Steering : The application of hydraulic, vacuum or air power in addition to manual power in automobile steering.

Pre-Combustion Chamber : A separate small chamber in some diesel engines where combustion starts.

Pre-Ignition : Ignition taking place earlier than intended.

Pressurised Cap : The type of a radiator cap which contains a pressure relief valve and a vacuum valve.

Pressure Feed : A type of engine lubricating system in which oil is forced to the working parts under pressure developed by an oil pump.

Pressure Relief Valve : A valve in the engine lubrication system which opens to relieve excessive oil pressure developed in the system.

Primary Brake Shoe : The brake shoe in a set which generates self-energizing action.

Primary Winding : Windings in an ignition coil or magnet carrying low tension current which is to be stepped up by induction into high tension current in the secondary winding.

Progressive Transmission : A type of transmission in which there is neutral position between every two gear positions and it is necessary to go through all the intermediate gears when shifting from the lowest to the highest gear speeds and *vice versa.*

Propeller Shaft : The shafts which connects transmission with the drive axle.

P.S.I. : A measurement of pressure in pounds per square inch. Also indicates as lbs. "□".

Push and Push Rod : Another name for drag link. It pulls or pushes steering spindle to turn the front wheels in an automobile.

Push Rod : A connecting link in an operating mechanism such as rod placed between the tappet and rocker arm of over head valves.

R.A.C. : Royal Automobile Club of England.

Radial Engine: An engine in which the cylinders are arranged in a circle around the crankcase.

Radiator : A device for cooling down hot water passing through it.

Radiation : The transfer of heat by rays.

Radial tyre : A tyre in which plies are placed radially perpendicularly to the rim.

Radius Rods : Rods attached to the axle and chassis frame to maintain correct horizontal position of the axle and yet permit vertical motion.

Ratio : The relation or proportion between two numbers.

Reactor : In torque converter, a fixed wheel having curved blades interposed between the pump and turbine.

Rear end torque : Torque reaction acting on rear axle housing when torque is applied to the wheels.

Raciprocating : Back and forth movement such as the action of piston in cylinder.

Rectifier : An electrical device for transforming alternating current into direct current.

Refrigerant : A substance which carries away heat.

Regulator : An automatic regulating valve or switch.

Relay : A cut out relay for interrupting an electrical circuit.

Retard : Opposite of advance.

Reverse Elliot Front Axle : Type of front axle in which the steering spindle straddles the ends of axle beam.

Rheostat : A variable resistance controlled by a sliding contact.

Ring Gear : Outer gear in planetary gearing or final drive.

Rocker Arm : An arm linked to the diaphragm in a fuel pump or in valve operating mechanism which rocks the diaphragm or valve.

Roller Bearing : Bearing consisting of an inner and outer race upon which hardened steel rollers operate.

Rotary Engine : An engine of Wankel design in which there is a rotor which moves inside an oval chamber and which contains no pistons.

Rotary Valve : A type of valve in which ported holes coincide to provide entrance and exit of fluids or gases.

Rotor : Rotating valve or conductor for carrying fluid or electrical current from a central source to the individual outlets as required.

R.M.P. : Revolutions per minute.

S.A.E. : Society of Automotive Engineers.

Sealed Beam Lamp : Type of electric lamp in which reflector, lens and filament are sealed into one unit.

Secondary Brake Shoe : The brake shoe in a set which is energized by the primary shoe and increase self-energization of the brake.

Secondary Winding : In an ignition coil or magneto, a wire in which a secondary or high tension current is induced by the interruption in the primary circuit.

Selective Transmission : Type of transmission in which it is possible to select any gear speed directly from neutral position.

Self Energization : In automobile brakes, placing of brake shoes in such position so that the brake drum tends to drag the shoe along with it, resulting in wedging action between the anchor and the drum and hence the self-energization.

Semi - Floating Axle : A type of drive axle in which the axle shafts support weight of the vehicle.

Separators : Sheets of rubber, wood or plastic inserted between the positive and negative plates of battery cell to keep them out of contact with each other.

Servo Action : In brake construction wherein a primary shoe push a secondary shoe to generate self-energization.

Service Brake : Foot brake in the vehicle.

Shackle : A swinging connection for a road-spring which permits it to vary in length as it deflects.

Shock Absorber : A device which controls excessive deflection of road springs by providing mechanical or hydraulic friction.

Shroud : A hood placed around an engine fan to improve air flow.

Shunt Winding : An electric winding which forms a bypass or alternate path for electric current.

S.I. : International System of Units (Systeme International des Unites).

Sleeve Valve : A reciprocating sleeve with ported openings to act as a valve.

Solenoid : An iron core surrounded by a coil of wire which moves due to magnetic attraction when current flows into the coil.

Solid Injection : The fuel injection system in which fuel in the fluid state is injected into the cylinder.

Spark Ignition Engine : Engine operating on Otto cycle in which ignition takes place when spark is produced by the spark plugs.

Spark Plug : A device containing an insulated central electrode for carrying high tension current from distributor and to provide spark in the engine combustion chamber.

Spark Plug Heat Range : The distance heat must travel from the central electrode to reach the outer shell of spark plug and enter the engine cylinder head.

Specific Gravity : The ratio between the weight of a substance to the weight of an equal volume of water at 4°C.

Speedometer : A device for measuring and indicating the speed of a vehicle.

Spiral Bevel Gear : A ring gear and pinion wherein the mating teeth are curved and placed at an angle with the pinion shaft.

Spline : A long keyway.

Spur Gear : A gear in which the teeth are cut parallel to the shaft.

Static Balance : The balance of an object while stationary.

Steering : A device for controlling directional change of a vehicle.

Steering Axis : Centre-line of ball-joints through which front wheels are turned.

Steering Geometry : The related angles assumed by the front wheels of a vehicle when turning.

Steering Kunckle : Steering spindle by means of which wheels are turned.

Stream lining : Shaping of vehicle body to minimize air resistance, so that it may move through air with less power.

Stroke : The distance from B.D.C. to T.D.C. in an engine cylinder as travelled by the piston.

Strut : A bar in the suspension system which connects the lower control over to the vehicle frame.

Sun Gear : The central gear around which the other gear revolves in planetary gear system.

Supercharger : A blower which forces air into the cylinder at higher than atmospheric pressure.

Synchromesh Gear Box : Transmission in which gear speeds are obtained through synchronizing unit.

Synchronizing unit : A device which facilitates the meshing of two gears by causing the speed of both gears to coincide at the same time.

Tappet : A sort of valve lifter which rides against the cam and lifts the valve or push rod.

T.D.C. : Top dead centre.

T-Head Engine : Type of engine in which intake valves are located on one side of the cylinders and the exhaust valve on the other side.

Thermal Efficiency : It is the ratio of work accomplished compared to the total quantity of heat combined in the fuel.

Thermostat : A heat controlled valve used in the engine cooling system.

Thermo-Syphon System : A method of engine cooling which utilize the difference in specific gravity of hot and cold water.

Third Brush Generator : Generator in which an auxiliary brush is placed on the commutator in such relation to the other two main brushes so that it may control the current output of the generator.

Throw : The distance from the centre of crankshafts main bearing to the centre of crankpin.

Throttle Valve : A valve in the carburettor, operated by the accelerator, which regulates the flow of fuel air mixture from the carburettor.

Tie Rod : Track rod which connects the steering spindle arms on opposite sides of the vehicle.

Timing Chain : Chain used to drive camshaft.

Timing Gears : Group of gears through which movement is affected to the camshaft for operating valves, ignition distributor or magneto, and other parts of the engine at the scheduled time during the engine cycle.

Toe-In : The difference in measurement between the front of the wheels and the rear of the wheels.

Torque : A twisting or turning efforts.

Torque converter : A device by means of which torque is multiplied while it is transmitted from the driving to driven number by hydraulic action.

Transmission : A device for changing the ratio between the engine revolutions and the driving wheel revolutions.

Turbine : A series of blades located on a wheel at an angle with the shaft against which fluid or gases are impelled to impart rotary motion to the shaft.

Turbulence : A disturbed motion of fluid and gases.

Turning Radius : The diameter of a circle within which a vehicle can be turned around.

Two-Stroke Engine : A type of engine in which the cycle is completed in two strokes of the piston and in which there is one power stroke in every revolution of the crankshaft.

Universal Joint : A type of joint which is capable to transmit power through varied angles.

Updraft Carburrettor : A type of carburettor in which fuel air mixture flows upward to the engine.

Upper Cylinder Lubrication : A method of introducing lubricant into the fuel for providing lubrication to the upper cylinder, valve guides etc.

Valve : A door like device for opening and closing an aperture.

Valve-In-Head Engine : Overhead valve engine.

Valve Overlap : The period expressed in degree, during which both intake and exhaust valves of the engine are open.

Valve Spring : A spring attached to a valve to return it to the seat after it has been released by the lifting means.

Valve Stem : The portion of the valve which moves within a guide.

Valve Timing : Timing of opening and closing of engine valves in relation to piston-positions.

Vapor Lock : A condition wherein the fuel boils in the fuel system forming bubbles which hold the flow of fuel.

Vent : An opening from an enclosed chamber through which air can flow.

Venturi : A tube with a narrowing throat for increasing the velocity of gas or fluid through it.

Vibration Damper : A device to absorb tortional vibrations which occurs along the length of crankshaft.

Viscosimeter : An instrument for determining the viscosity of an oil.

Viscosity : The resistance to the flow of an oil.

Volatility : The tendency of a fluid to evaporate rapidly.

Volt : A unit of electrical force which will cause a current of one ampere to flow through a resistance of one ohm.

Voltage Regulator : An electrical device for controlling or regulating voltage.

Volumetric Efficiency : It is the ratio between the volume of fuel air mixture that could be spaced in the engine and the actual volume of mixture drawn in.

V-Shape Engine : Type of engine in which there are two rows of cylinders arranged to form the shape of 'V'.

Wander : A condition wherein the steering tends to turn slowly first in one direction and then to other, interfering with directional stability.

Wankel engine : A rotary combustion engine in which a three-lobe rotor turns accentrically in an oval chamber to produce power.

Water Jackets : Spaces between inner and outer shells of cylinder block or head through which water circulates.

Weight Transfer : As the centre of gravity of the vehicle is located above the wheel centres, so sudden braking tends to cause the centre of gravity to move forward thus throwing more weight on the front wheels and less on the rear.

Wheel Base : Distance between centre lines of front and rear axle wheels.

Wheel Cylinders : Cylinders in the hydraulic brake system which are operated by the hydraulic pressure to force the brake shoes in contact with the brake drum.

Worm Gear : A shaft having an extremely course thread which is designed to operate in engagement with a toothed wheel.

Wrist Pin : Piston pin or gudgeon pin which holds piston with the connecting rod.

IMPORTANT BOOKS ON ENGINEERING

HEAT AND MASS TRANSFER

R.K. Rajput

The book is meant for the students preparing for engineering undergraduate, AMIE, U.P.S.C. and other competitive examinations.

CONTENTS: Basic Concepts • **Part I: Heat Transfer by conduction** • Conduction-Steady State one dimension • Conduction-Steady State Two Dimensions and three dimensions Conduction-Unsteady State (Transient) **Part-II: Heat Transfer by Convection** • Thermal Radiation-Basic Relations • **Part III: Radiation Exchange** • Between Surfaces **Part-IV: Mass Transfer** • **Part-V: Objective Type Questions Bank** • Index

10 202 **ISBN:81-219-1777-8** **pp. 580**

FLUID MECHANICS & HYDRAULIC MACHINES

R.K. Rajput

The book contains comprehensive treatment of the subject-matter in simple & lucid language, supported by self-explaining figures and a large number of worked out examples. At the end of each chapter, there are (i) Highlights (ii) Objective Type Questions (iii) Theoretical Questions and (iv) Unsolved Problems. The book is specially suited to engineering undergraduate, AMIE (section B) and other competitive examinations.

CONTENTS: Part I: Fluid Mechanics— Properties of Fluids • Pressure Measurememt • Hydrostatic Forces on Surfaces • Buoyancy and Floatation • Fluid Kinematics • Fluid Dynamics • Dimensional & Model Analysis • Flow through Orifices andMouthpieces• Flow Over Notches & Weirs • Laminar Flow • Turbulent Flow in Pipes • Flow Through Pipes • Boundary Layer Theory • Flow Around Submerged Bodies Drag and Lift • Compressible Flow • Flow in Open Channels • **Part II: Hydraulic Machines** -Impact of Free Jets • Hydraulic Turbines • Centrifugal Pumps • Reciprocating Pumps • Misc. Hydraulic Machines • Water Power Development

10 185 **ISBN:81-219-1666-6** **pp. 1088**

A TEXTBOOK OF FLUID MECHANICS

R.K. Rajput

The book covers comprehensively the subject of "Fluid Mechanics" in 16 chapters. Each chapter starts with needed text, supported by simple and self-explanatory figures, and a large number of worked out examples, including typical ones, suited for Competitive Examinations. At the end of each chapter, there are (i) Highlights (ii) Objective Type Questions (iii) Theoretical Questions and (iv) Unsolved Problems. The book is specially suited to students appearing for Engineering Undergraduate Examination, Section B of AMIE Examination and Diploma Examinations.

CONTENTS: Properties of Fluids • Pressure Measurement • Hydrostatic Forces on

Surfaces • Buoyancy and Floatation • Fluid Kinematics • Fluid Dynamics • Dimensional and Model Analysis • Flow through Orifices and Mouthpieces • Flow over Notches and Weirs • Laminar Flow • Turbulent Flow • Flow Through Pipes • Boundary Layer Theory • Flow Around Submerged Bodies -Drag and Lift • Compressible Flow • Flow in Open Channels • Index

10 192 **ISBN:81-219-1667-4** **pp. 784**

A TEXTBOOK OF HYDRAULIC MACHINES

(Fluid Power Engineering)

R.K. Rajput

The book covers comprehensively the subject of "Hydraulic Machines" (Fluid Power Engineering). All the chapters start with the much needed text, supported by simple and self-explanatory figures, and a large number of worked out examples, including typical one, suited for competitive examinations. At the end of each chapter there are (i) Highlights (ii) Objective Type Questions (iii) Theoretical Questions and (iv) Unsolved Problems. Thus the book is a comprehensive treatise on the subject and is specially suited to students appearing for Engineering Undergraduate Examination, Section B of AMIE (India) Examination and other Competitive Examinations.

CONTENTS: Impact of Free Jets • Hydraulic Turbines • Centrifugal Pumps • Reciprocating Pumps • Miscellaneous Hydraulic Machines • Water Power Development • Index

10 194 **ISBN:81-219-1668-2** **pp. 320**

A TEXTBOOK OF HYDRAULICS

R.K. Rajput

This treatise contains comprehensive treatment of the subject-matter in simple and lucid language and envelops a large number of solved problems properly graded including typical examples from examination point of view.

CONTENTS: Properties of Fluids • Pressure Measurement • Hydrostatic Forces on Surfaces • Buoyancy and Floatation • Fluid Kinematics • Fluid Dynamics • Dimensional and Model Analysis • Flow Through Orifices and Mouthpieces • Flow over Notches and Weirs • Laminar Flow • Flow Through Pipes • Flow Around Submerged Bodies-Drag and Lift • Flow in Open Channels • Impact of Free Jets • Hydraulic Turbines • Centrifugal Pumps • Reciprocating Pumps • Miscellaneous Hydraulic Machines • Experiments • Index

10 198 **ISBN:81-219-1731-X** **pp. 546**

A TEXTBOOK OF ENGINEERING MECHANICS

(Applied Mechanics)

R. S. Khurmi

A large number of worked examples have been given in a systematic manner and logical sequence to assist the student to understand the text of the subject. At the end of each chapter highlights and exercises have been added.

CONTENTS: Introduction • Composition and Resolution of Forces • Moments and their Applications • Parallel Forces and Couples • Equilibrium of Forces • Centre of Gravity • Moment of Inertia • Friction • Principles of Lifting Machines • Simple Lifting Machines • Support Reactions • Analysis of Perfect Frames (Analytical Method & Graphical Method) • Equilibrium of Strings • Virtual Work • Plane Motion • Motion Under Variable Acceleration • Relative Velocity • Projectiles • Motion of Rotation • Combined Motion of Rotation and Translation • Simple Harmonic Motion • Laws of Motion • Motion of Connected Bodies • Helical Springs and Pendulums • Collision of Elastic Bodies • Motion Along a Circular Path • Balancing of Rotating Masses • Work, Power and Energy • Kinetics of Motion of Rotation • Motion of Vehicles • Transmission of Power by Belts and Ropes • Gear Trains • Hydrostatics • Equilibrium of Floating Bodies • Index

10 023 **ISBN:81-219-0651-2** **pp. 848**

APPLIED MECHANICS AND STRENGTH OF MATERIALS

R.S. Khurmi

This book is suitable for B.Sc. Engg., A.M.I.E. and Diploma courses. A large number of worked out examples, highlights and illustrations have been given.

CONTENTS: Introduction • Composition and Resolution of Forces • Moments and their Application • Parallel Forces and Couples • Equilibrium of Forces • Centre of Gravity • Moment of Inertia • Principles of Friction • Applications of Friction • Principles of Lifting Machines • Simple Lifting Machines • Linear Motion • Circular Motion • Projectiles • Laws of Motion • Work, Power and Energy • Simple Stresses and Strains • Thermal Stresses and Strains • Elastic Constants • Strain Energy and Impact Loading • Bending Moment and Shear Force • Bending Stresses in Beams • Shearing Stresses in Beams • Deflection of Beams • Deflections of Cantilevers • Torsion of Circular Shafts • Riveted Joints • Thin Cylindrical and Spherical Shells • Analysis of Perfect Frames (Analytical Method & Graphical Method)

10 025 **ISBN:81-219-1077-3** **pp. 580**

A TEXTBOOK OF STRUCTURAL MECHANICS

R.S. Khurmi

10 187 **ISBN:81-219-1642-9** **pp. 352**

A TEXTBOOK OF APPLIED MECHANICS

R.S. Khurmi

10 191 **IBN:81-219-1643-7** **pp. 352**

STRENGTH OF MATERIALS

R.S. Khurmi

This student-oriented book has been widely adopted by undergraduates in Engineering throughout India. Subject-matter has been amply supported with illustrations and solved, unsolved and well-graded examples.

CONTENTS: Introduction • Principles of Simple Stresses and Strains • Thermal Stresses and Strains • Elastic Constants • Principal Stresses and Strains • Strain Energy and Impact

Loading • Centre of Gravity • Moment of Intertia • Analysis of Perfect Frames (Analytical Method & Graphical Method) • Shear Force and Bending Moment of Beams • Bending Stresses in Beams •Shearing Stresses in Beams • Deflection of Beams • Deflection of Cantilevers • Deflection by Conjugate Beam Method • Propped Cantilevers and Beams • Fixed Beams • Theorem of Three Moments • Moment Distribution Method • Torsion of Circular Shafts • Springs • Riveted Joints • Welded Joints • Thin and Thick Cylindrical and Spherical Shells • Direct and Bending Stresses • Dams and Retaining Walls • Columns and Struts • Introduction to Reinforced Concrete • Mechanical Properties of Materials • Index

10 024 **ISBN:81-219-0533-8** **pp. 1000**

10 156 **ISBN:81-219-0898-1** **Hardbound** **pp. 1000**

STRENGTH OF MATERIALS

R.K. Rajput

The book deals the subject of Strength of Materials exhaustively in a lucid, direct and easily understandable style. It contains a large number of worked out simple, problems arranged in a scientific, and a graded manner to enable the students to grasp the subject effectively from the examination point of view.

Another salient feature of the book is "Experiments" at the end of the chapters to enable the students to have an access to the practical aspects of the subject.

CONTENTS: Simple Stresses and Strains • Principal Stresses and Strains • Centroid and Moment of Inertia • Bending Stresses • Combined Direct and BendingStresses • Shear Stresses in Beam • Thin Shells • Thick Shells • Riveted and Welded Joints • Torsion of Circular and Non-Circular Shafts • Springs • Strain Energy and Deflection due to Shear & Bending • Columns and Struts • Analysis of Framed Structures • Theories of Failure • Rotating Discs and Cylinders • Bending of Curved Bars • Unsymmetrical Bending • Material Testing Experiments • Index

10 174 **ISBN:81-219-1381-0** **pp.1056**

A TEXTBOOK OF THERMAL ENGINEERING

R.S. Khurmi & J.K.Gupta,

The entire text of the book has been presented in SI units and arranged in a systematic manner. The text has been profusely illustrated by incorporating a number of solved, unsolved and well-graded examples.

CONTENTS: Introduction • Properties of Perfect Gases • Thermodynamic Processes of Perfect Gases • Entropy of Perfect Gases • Thermodynamic Air Cycles • Formation and Properties of Steam • Entropy of Steam • Thermodynamic Processes of Vapour • Thermodynamic Vapour Cycles • Fuels • Combustion of Fuels • Steam Boilers • Boiler Mountings and Accessories • Performance of Steam Boilers • Boiler Draught • Simple Steam Engines • Compound Steam Engines • Performance of Steam Engines • Steam Condensers • Steam Nozzles • Impulse Turbines • Reaction Turbines • Performance of Steam Turbines • Modern Steam Turbines • Internal Combustion Engines • Testing of Internal Combustion Engines • Reciprocating Air Compressors • Air Motors • Gas Turbines • Performance of Gas Turbines • Introduction to Heat Transfer • Air Refrigeration Systems • Vapour Compression Refrigeration • Psychrometry Air Conditioning Systems • Index

10 172 **ISBN:81-219-1381-0** **pp. 846**

A TEXTBOOK OF HYDRAULICS

R.S. Khurmi

This revised edition continues to cater to the needs of the undergraduate students of Engineering. Latest research works conducted in various countries have been incorporated.

CONTENTS: Introduction • Fluid Pressure and its Measurement • Hydrostatics • Applications of Hydrostatics • Equilibrium of Floating Bodies • Hydrokinematics • Bernoulli's Equation and its Applications • Flow Through Orifices (Measurement of Discharge) • Flow Through Orifices (Measurement of Time) • Flow Through Mouthpieces • Flow Over Notches •Flow Over Weirs • FlowThrough Simple Pipes • Flow Through Compound Pipes • Flow Through Nozzles • Uniform Flow Through Open Channels • Non-Uniform Flow Through Open Channels • Viscous Flow • Viscous Resistance • Impact of Sets • Hydraulic Turbines • Hydraulic Pumps • Pumping Devices • Hydraulic System • Index

10 027 **ISBN:81-219-0135-9** **pp.424**

A TEXTBOOK OF HYDRAULICS, FLUID MECHANICS AND HYDRAULIC MACHINES

R.S. Khurmi

About 50 new examples have been included in this revised edition. The subject matter has been amply illustrated by incorporating a large number of solved and well-graded examples.

CONTENTS: Introduction • Fluid Pressure and its Measurement • Hydrostatics • Applications of Hydrostatics • Equilibrium of Floating Bodies • Hydro-kinematics • Bernoulli's Equation and its Applications • Flow through Orifices (Measurement of Discharge & time • Flow through Mouthpieces • Flow over Notches • Flow over Weirs • Flow through Simple Pipes • Flow through Compound Pipes • Flow through Nozzles • Uniform Flow through Open Channels • Non-uniform flow through Open Channels • Viscous Flow • Viscous Resistance Fluid Masses Subjected to Acceleration • Vortex Flow • Mechanics of Compressible Flow • Compressible Flow of Fluids • Flow Around Immersed Bodies • Dimensional Analysis • Model Analysis (Undistorted Models and Distorted Models) • Non-Dimensional Constants • Impact of Jets • Jet Propulsion • Water Wheels • Impulse Turbines • Reaction Turbines • Performance of Turbines • Centrifugal Pumps • Reciprocating Pumps • Performance of Pumps • Pumping Devices • Hydraulic Systems • Index

10 026 **ISBN:81-219-0162-6** **pp. 656**

THEORY OF MACHINES

R.S. Khurmi & J.K. Gupta

This book is written for the students of Degree, Diploma and A.M.I.E. courses. The subject matter has been amply illustrated by incorporating a large number of solved, unsolved well-graded examples.

CONTENTS: Introduction • Kinematics of Motion • Kinetics of Motion • Simple Harmonic Motion • Simple Mechanisms • Velocity in Mechanisms (Instantaneous Centre Method) • Velocity in Mechanisms (Relative Velocity Method) •Acceleration in Mechanisms • Mechanisms with Lower Pairs • Friction • Belt, Rope and Chain Drives • Toothed Gearing • Gear Trains • Gyroscopic Couple and Precessional Motion • Inertia Forces in Reciprocating Parts • Turning Moment Diagrams and Flywheel • Steam Engine

Valves and Reversing Gears • Governors • Brakes and Dynamometers • Cams • Balancing of Rotating Masses • Balancing of reciprocating Masses • Longitudinal and Transverse Vibrations • Torsional Vibrations • Computer Aided Analysis and Synthesis of Mechanisms • Index

10 013 ISBN:81-219-0132-4 pp. 1010

A TEXTBOOK OF HYDRAULICS & FLUID MECHANICS

R.S. Khurmi

This book is suited to students of Degree, Diploma and AMIE Courses. It has been written in a lucid and crystal clear language and the subject has been made comprehensive to the students.

CONTENTS: Introduction • Fluid Pressure and Its Measurement • Hydrostatics • Applications of Hydrostatics • Equilibrium of Floating Bodies • Hydrokinematics • Bernoulli's Equation and its Applications • Flow Through Orifices (Measurement of Discharge) • Flow Through Orifices (Measurement of Time) • Flow Through Mouthpieces • Flow Over Notches •Flow Over Weirs • Flow Through Simple Pipes • Flow Through Compound Pipes • Flow Through Nozzles • Uniform Flow Through Open Channels • Non-Uniform Flow Through Open Channels • Viscous Flow • Viscous Resistance • Fluid Masses Subjected to Acceleration • Vortex Flow • Mechanics of Compressible Flow • Compressible Flow of Fluids • Flow Around Immersed Bodies •Dimensional Analysis • Model Analysis (Undistorted Models) • Model Analysis (Distorted Models) • Non Dimensional Constants • Index

10 195 ISBN:81-219-1676-3 pp. 496

STEAM TABLES

(With Mollier Diagram in S.I. Units)

R.S. Khurmi

The book presents the various properties of Water and Steam, in a very concise, compact and in a lucid manner. All data is in S.I. Units, now in use in all countries.

CONTENTS: Rules for S.I. Units • Introduction to Steam Tabels and Mollier Diagrams • Saturated Water & Steam (Temperature) Tables • Saturated Water & Steam Tables (Pressure) • Specific Volume of Super-heated Steam • Specific Enthalpy of Super-heated Steam • Entropy of Super-heated Steam • Specific Volume of Super-critical Steam • Specific Enthalpy of Superecritical Steam • Specific Entropy of Super Critical Steam

10 044 ISBN:81-219-0654-7 pp. 32

A TEXTBOOK OF HYDRAULIC MACHINES

(In M.K.S. and S.I. Units)

R.S. Khurmi

The subject matter has been amply illustrated by incorporating a large number of solved and well-graded examples, the rigorous mathematical steps have been replaced by simple treatments.

CONTENTS: Impact of Jets • Jet Propulsion • Water Wheels • Impulse Turbines • Reaction Turbines • Performance of Turbines • Centrifugal Pumps • Reciprocating Pumps • Per formance of Pumps • Pumping Devices • Hydraulic Systems • Index

10 028 ISBN:81-219-0075-1 pp. 272

STEAM TABLES AND MOLLIER DIAGRAM
(IN S I UNITS)

K.L. Kumar

This Standard booklet "Steam Tables and Mollier Diagram in S.I. Units" presents the various Properties of water and Steam in a most Concise, Compact to the point and lucid manner.

CONTENTS: Saturated Water and Steam Tables (Temperature-Based) • Saturated Water and Steam Tables (Pressure-Based) • Superheated Steam Tables • Supercritical Tables • Appendix (Mollier Diagram for Water and Steam)

10 165 ISBN:81-219-1258-X pp. 16

MATERIAL SCIENCE

R.S. Khurmi & R.S. Sedha

The book presents a systematic study of the fundamental concepts of Material Science in a very lucid and comprehensive language. The book is meant for AMIE, Diploma students in particular and others in general.

CONTENTS: Part I: Science of Metals: Introduction • Structure of Atom • Crystal Structure • Bonds in Solids • Electron Theory of Metals • **Part II: Mechanical Behaviour of Metals:** Mechanical Properties of Metals • Mechanical Tests of Metals • Deformation of Metals Fracture of Metals • **Part III: Engineering Metallurgy:** Iron-Carbon Alloy System • Heat Treatment • Corrosion of Metals • **Part IV : Engineering Materials** •Ferrous and Non-FerrousAlloys : Organic Materials • Composite Materials • Semiconductors • Insulating Materials • Magnetic Materials • Appendices • Index

10 109 ISBN:81-219-0146-4 pp. 384

A TEXTBOOK OF MATERIAL SCIENCE

K.G. Aswani

The book is aimed to explain the principles and applications of Material Science, specially to suit the requirements of B.Sc., B.E. and AMIE students, besides the practical aspects. So it can also be used as a reference book by practising engineers.

CONTENTS: Selection of Materials • Structure of Atoms, Crystal Structure and Bonds in Solids • Mechanical Properties of Materials • Deformation of Metals • Alloy Systems and Phase Diagrams • Cast Iron • Carbon and Alloy Steels • Heat Treatment of Carbon and Alloy Steels • Non-Ferrous Metals and Alloys • Polymers and Polymeric Products • Corrosion and its Prevention • Composites • Electrical and Magnetic Properties of Materials • Manufacturing Processes • Refractories • Cementitious Materials • Tables • Index

10 157 ISBN:81-219-1199-0 pp. 390

A TEXTBOOK OF PRODUCTION ENGINEERING

P.C. Sharma

It is written for Mechanical, Production Engineering and A.M.I.E. students. In order to test the student's understanding and concept of the subject, a number of unsolved problems are provided.

CONTENTS: Jigs and Fixtures • Press Tool Design • Forging Die Design • Cost Estimation • Economics of Tooling • Process Planning • Tool Layout for Capstans and Turrets • Tool Layout for Automatics • Limits, Tolerances and Fits • Gauges and Gauge

Design • Surface Finish • Measurement • Analysis of Metal Forming Process • Theory of Metal Cutting • Hardbound Cutting Tool Materials and Design of CuttingTools • Gear Manufacturing • Thread Manufacturing •Numerically Controlled Machines Tools • Tracer Controlled Machine Tools • Design of Product for Economical Production • Statistical Quality Control • Kinematics of Machine Tools • Production Planning and Control • Design of Machine Tool Elements • Machine Tool Testing • Machine Tool installation and Maintenance •Appendices • Index

10 038 **ISBN:81-219-0421-8** **pp. 1000**

ENGINEERING FLUID MECHANICS

K.L. Kumar

This book is suited to Mechanical, Civil, Chemical, and Aeronautical Engineering students. Clear and neat diagrams have been used to explain the text. A large number of solved examples, and problems for solution have been given for the benefit of the students.

CONTENTS: Introduction • Fluid Statics • Fluid Kinematics • Fluid Dynamics • Flow around Immersed Bodies • Laminar Flow • Flow through Pipes • Flow through Open Channels • Boundary Layer Flow • Dimensional Analysis and Similitude • Laboratory Experiments in Fluid Mechanics

10 020 7 **ISBN:81-219-0100-6** **pp. 542**

A TEXTBOOK OF MACHINE DESIGN

R.S. Khurmi & J.K. Gupta

Completely covers the course of Mechanical Engineering Design for engineering students. Fully illustrated with examples and diagrams.

CONTENTS: Introduction • Engineering Materials and their Properties • Manufacturing Considerations in Machine Design • Simple Stresses in Machine Parts • Torsional and Bending Stresses in Machine Parts • Variable Stresses in Machine Parts • Pressure Vessels • Pipes and Pipe Joints • Riveted Joints • Welded joints • Screwed Joints • Cotter and Knuckle Joints • Keys and Coupling • Shafts • Levers • Columns and Struts • Power Screws • Flat belt Drives • Flat Belt Pulleys • V-belt and Rope Drives • Flywheel • Springs • Clutches • Brakes • Bearings • Spur Gears • Helical and Bevel Gears • Worm Gears • Index

10 012 **ISBN:81-219-0501-X** **pp. 928**

A TEXTBOOK OF REFRIGERATION AND AIR-CONDITIONING

R.S. Khurmi & J.K. Gupta

This book is suitable for students preparing for U.P.S.C. (Engg. Services) and A.M.I.E. (I) examinations. The complete solutions of recent examination papers have been included. There are worked out examples and objective-type questions and a few exercises at the end of each chapter.

CONTENTS: Introduction • Air Refrigeration Cycles • Air Refrigeration Systems • Simple Vapour Compression Systems • Compound Vapour Compression Systems • Multi-evaporator and Compressor Systems • Vapour Absorption Refrigeration Systems • Refrigerants • Refrigerant Compressors • Condensers • Evaporators • Expansion Devices • Food Preservation • Low Temperature Refrigeration(Cryogenics) • Steam-Jet Refrigeration system • Psychrometry • Comfort Conditions • Air Conditioning Systems • Cooling Load Estimation • Ducts • Fans • Refrigeration Tables and Charts • Index

10 097 **ISBN:81-219-0268-1** **pp. 656**

MECHANICAL ENGINEERING
(Objective Type)

R.S. Khurmi & J.K. Gupta

The object of this book is to present the subject with multiple choice questions and answers. No effort has been spared to enrich the book with objective-type questions of different types.

CONTENTS: Applied Mechanics • Strength of Materials • Hydraulics and Fluid Mechanics • Hydraulic Machines • Thermodynamics • Steam Boilers, Engines, Nozzles and Turbines • I.C. Engines and Nuclear Power Plants • Compressor, Gas Dynamics and Gas Turbines • Heat Transfer, Refrigeration and Air Conditioning • Theory of Machines • Machine Design • Engineering Materials • Workshop Technology • Production Engineering • Industrial Engineering and Production Management

10 042 **ISBN:81-219-0628-8** **pp. 536**

INDUSTRIAL MAINTENANCE

H.P. Garg

This book is very useful for engineers, supervisors, executives and managers of the industrial units. Repair technology of the machines, planned maintenance, etc. have been elaborately explained in this book.

CONTENTS: Part I: Restoration and Manufacture of Machine Parts • Fits, Tolerances and Surface Finish • Materials for Machine Parts and Heat Treatment of Steels • Guide Surface • Gear Transmission • Bush Bearings, their Shank and Housing • Ball and Roller Bearings their Shank and Housing• Key Fitting • Spline Fitting • Coupling • Clutches • Lead Screw and Nut • Machine Spindle • Vee Belt Drive • Chains and Sproket Wheels • Machine Hydraulics • Tailstock Repairs • Repair of Three Jaw Chucks • Repair of Cracks in C.I. Body • Restoration of Parts by Welding, Metalization, Chromium Plating • Threads and Threaded Joints • Seals and Packings • Special Features of the Repair of Cranes, Hammers, Power Presses/Shears and Furnaces • Fitter's Common Tools, Appliances and Devices, Handling Facilities and Measuring Instruments • Typical Manufacture and Machining Norms • Part II: Planned Maintenance: Preventive Maintenance Planning • Repair Cycle • Repair Complexity • Maintenance Stages • Machine Kinematics and Schedule of Complete Overhaul • Pert in Maintenance • Lubrication and Lubricants • Materials and Standard Spares Planning • Spare Parts Stock Pate • Accuracy and Technological Test Charts • Organisation of Maintenance Department • Depreciation and Machine Life • Appendices • Index

10 021 **ISBN:81-219-0168-5** **pp. 624**

INSTALLATION, SERVICING AND MAINTENANCE

S.N. Bhattacharya

Prepared after consulting lots of authentic books both of Indian and Foreign authors, this book has been made invaluable and indispensable to the students of Mechanical Engineering. The contents have been designed in such a way so that not only the Diploma students, but the Degree students will also be benefited.

CONTENTS: Maintenance and Maintenance Planning • General Hand Tools • Fits, Tolerance and Surface Topography • Seals, Packing and Gaskets • Generalized Procedure of Installation • Power Transmission Devices • Maintenance and Repair of Guide Surfaces • Machine Parts: Repair, Restoring and Rectification • Lubrication •

Corrosion, its Control and Chemical Cleaning • Pumps and Air Compressors • Boiler Maintenance • Heat Engines • Miscellaneous Maintenance • Engineering Materials

10 141 **ISBN:81-219-0831-0** **pp. 300**

ELEMENTS OF STRUCTURAL ANALYSIS

S.A. Bari

The book will benefit the students preparing for Diploma, B.Sc. Engineering and A.M.I.E. Examinations. As a special features the book contains a large number of well graded solved questions in view of the need of average students.

CONTENTS: Simple Stresses and Strains • Analysis of Complex Stresses • Strain Energy • Thin Walled Pressure Vessels • Shearing Force and Bending Moment • Moment of Inertia • Stresses in Beams • Elastic Deflection of Beams • Statically Indeterminate Beams • Combined Direct and Bending Stresses • Torsion of Shafts • Springs • Columns and Struts • Analysis of Simple Trusses • Appendices • Index

10 188 **ISBN:81-219-1662-3** **pp. 504**

INDUSTRIAL MAINTENANCE MANAGEMENT

S.K. Srivastava

The book is intended for all streams of Engineering i.e., Civil, Mechanical, Electrical, Automobile and Mining at Degree and Diploma level. The Contents of the book embodies the inter-disliplinory concepts of various aspects of maintenance and connected production funtions.

CONTENTS: Introduction and Objectives • Quality, Reliability and Maintainability (QRM) • Maintenance Jobs and Technologies • Defect List Generation & Defect/Failure Analysis • Maintenance Types/Systems • Condition Monitoring (Equipment Health Monitoring) • Maintenance Planning and Scheduling • Systematic Maintenance • Computer Managed Maintenance System (CMMS) • Total Productive Maintenance (TPM) • Other Concepts of Maintenance Types/Systems • Maintenance Organisation • Maintenance Effectiveness & Performance Evaluation/Audit • Maintenance Budgeting, Costing and Cost-Control • Training (HRD) of Maintenance Personnel • Bibliography • Index

07 300 **ISBN:81-219-1663-1** **pp. 248**

AEROSPACE MATERIALS

Balram Gupta et al.

A set of 3 volumes, provides details of metallurgy etc of aero space metallic materials, developed and produced, for on-going manufacturing, as well as "development aerospace projects", undertaken by DRDO/aerospace establishments in India. The book will be very useful to practising and design engineers.

CONTENTS: **Vol I:** Certification of Aerospace Materials— A Global Concept • Super Alloys • Titanium Alloys • **Vol II:** Carbon Steels • Alloy Steels • Stainless & Heat Resistant Steels • Indigenised Steels • Stainless Steel Seamless Tubes • Maraging Steels • Cast Irons • **Vol III:** Aluminium Alloys • Magnesium Alloys • Copper Alloys • Futuristic Materials • AR & DB (M & PP) Projects

10 178 **ISBN:81-219-1360-8** **pp. 1100**

A TEXTBOOK OF WORKSHOP TECHNOLOGY

(Manufacturing Process)

R.S. Khurmi and J.K. Gupta

The objective of this book is to present the subject matter in most concise, compact, to the point and lucid manner to the students of Degree and Diploma in Engineering.

This Fifth revised edition has five new chapters and about 500 objective type questions in three groups: Multiple choice questions, fill in the blanks and True or False.

CONTENTS: Introduction • Ferrou Metals and Alloys • Structure of Iron and Steel • Non-Ferrous Metals and Alloys • Mechanical Working Metals • Carpentry• Pattern Making • Foundary Tools and Equipment • Moulding and Core-Making • Casting Methods • Smithy and Forging • Welding • Bench Work and Fitting • Rivets and Screws • Limit System and Surface Finish • Measuring Instruments and Gauges • Powder Metallurgy • Plastics • Metallic and Non-Metallic Coatings • Pipes and Pipe Joints • Machine Tools (Introduction) • Safety in Workshops • Objective Type Questions • Index

10 155 **ISBN:81-219-00868-X** **pp. 560**

INDUSTRIAL ENGINEERING AND PRODUCTION MANAGEMENT

Martand Telsang

The book is intended to introduce various tools and techniques of industrial engineering and production management. Changing Technoeconomic scenario because of liberalisation policy of our government is putting pressure on Indian indudustries to be competitive for their survival and growth. Quality, cost timeliness and productivity have cecome the watch words of India industries. This book comprehensively discribes all kinds of solutions to the above problems.

The book covers the syllbus offered at the undergraduate & Postgraduate levels in B.E. & Diploma Engineering. It is also sueful guide for examination of IIE and other competitive examinations and management courses.

CONTENTS: Part I: Industrial Engineering • Introduction to Industrial Engineering • Productivity • Work-Study • Method Study • Work Measurement • Value Engineering • Plant Location • Material Handing • Job Evaluation and Merit Rating • Wages and Incentives • Ergonomics • **Part II: Production Management** • Production System • New Production Design • Demand Forecasting • Production Planning and Control • Capacity Planning • Material Requirement Planning (MRP) • Process Planning • Project Scheduling with CPM and Pert • Production Control • Inventory Control • Production Cost Concepts and Break-even Analysis • Maintenance Management • Make or Buy Decisions • Planning and Control of Batch Production • Part III: Advanced Topics In Production Management • Application of Linear Programming Technique In Production Management • Management Information System • Business Process Engineering • Group Technology • Just in Time (JIT) Manufacturing • Appendices

10 197 **ISBN:81-219-1773-5** **pp. 430**

A TEXTBOOK OF ENGINEERING DRAWING

(Geometrical Drawing)

R. K. Dhawan

This book covers the syllabi of the Engineering and Technical institutions. The objective is to train an engineering student to give him the basic knowledge of the treatment of simple to complicated problems in design and drawing by geometrical methods.

CONTENTS: Section I: Introduction & Drawing Instruments • Layout of Drawing Sheet • Conventions • Lettering • Dimensioning Scales • Geometrical Constructions • **Section II:** Loci of Points • Conic Sections • Plane and Space Curves • **Section III:** Theory of Projection & Orthographic Projection • Orthographic Reading or Interpretation of Views • Identification of Surfaces • Missing Lines & Views • Sectional Views • Isometric Projections •Auxiliary Views • Freehand Sketching • **Section IV:** Projections of Points • Projections & Traces of Straight Lines Projections of Planes • Projections of Solids • Sections of Solids • Intersection of Surfaces • Development of Surfaces • Perspective Projections

10 147 **ISBN:81-219-1431-0** **pp.560**

A TEXTBOOK OF MACHINE DRAWING

(In first Angle Projections)

R.K. Dhawan

The book is specially designed to meet the requirements of B.Sc. (Engg.), B.E., B.Tech., AMIE (India), production engineeing; Automobile Engg; Textile Engg. I.I.T. (Draftsman Course – Mech. Engg.) C.T.I. and other Engg. examinations. It is in 'first angle projections', as desired by the Bureau of Indian Standards: SP-46-1988. However, the third angle projections have also been included in it. Each chapter contains questions for self-test (viva-voce).

CONTENTS: Section I:Introduction & Drawing Instruments • Layout of Drawing Sheet•Conventions•Lettering • Dimensioning • Scales • Section II: Theory of Projection & Orthographic projection • Orthographic Reading or Interpretation of views • Indentification of Surfaces • Missing Lines & Views • Sectional Views • Isometric Projections • Auxiliary Views • Freehand Sketching • Section III: Detail & Assembly Drawings • Limits, Fits & Machining Symbols • Rivets and Riveted Joints • Welding • Screw Threads • Fastenings • Keys, Cotters and Joints • Shaft Couplings • Bearings • Brackets • Pulleys • Pipe Joints • Steam Engine Parts • I.C. Engine Parts • Valves • Gears • Cams • Jigs & Fixtures • Miscellaneous Drawings

10 148 **ISBN:81-219-0824-8** **pp. 366**

10 166 **Ist Edn. 1996** **ISBN:81-219-1168-0** **Hardbound** **pp. 366**